A DICTIONARY AND BIBLIOGRAPHY OF DISCRETE DISTRIBUTIONS

A DICTIONARY AND BIBLIOGRAPHY OF DISCRETE DISTRIBUTIONS

GANAPATI P. PATIL
M.Sc., M.S., Ph.D.(Mich.)., A.I.S.I.
Professor of Mathematical Statistics
The Pennsylvania State University
University Park, Pennsylvania

SHARADCHANDRA W. JOSHI
M.Sc. (Poona)
Graduate Research Assistant
The Pennsylvania State University
University Park, Pennsylvania

with a foreword by
C. RADHAKRISHNA RAO
Sc.D.(Camb.)., F.N.I., F.R.S.
Director, Research and Training School
Indian Statistical Institute
Calcutta

Published for
The International Statistical Institute
by
HAFNER PUBLISHING COMPANY
NEW YORK

With Fondness and Gratitude

To our Parents who introduced us to the art of how to identify, assemble and classify

and

To the Research Scholars who by their contributions inspired us to practise the art on discrete distributions

OLIVER AND BOYD LTD
Tweeddale Court Edinburgh 1

First published 1968

05 001730 6

Published in
U.S.A. by Hafner Publishing Company

Printed in Great Britain by T. and A. Constable Ltd
Edinburgh, Scotland

FOREWORD

The theory of Probability and Statistical Methodology originated in the study of random counts or discrete random variables. Historically speaking, the famous normal distribution was introduced as a continuous distribution providing an approximation to the discrete binomial distribution. Endowed with a wide and rapidly expanding literature, the subject of discrete distributions offers a fascinating area of study and research. However, the lack of suitable textbooks and the widely scattered literature have been serious obstacles to a systematic study of the subject. In this context, the service done by Professor G. P. Patil and his associates by providing us with the much needed dictionary and classified bibliography is highly commendable.

The project on the bibliography was conceived by Professor Patil at the time of the first International Symposium on classical and contagious discrete distributions organized by him in 1963. The symposium and its published proceedings were the first successful attempts in consolidating the numerous theoretical developments and the wide variety of practical applications, and in establishing the subject of discrete distributions as an important discipline within the realm of Probability and Statistics. Now the *Dictionary and Bibliography* is another laudable attempt in this direction bringing as it does the scattered subject matter area of discrete distributions to a cohesive focus.

I have always insisted on the inclusion of a self-contained and full-length course on discrete distributions to advanced students at the Indian Statistical Institute as an essential requisite of their statistical education mainly for two reasons. First, it provides a natural and a simple way of introducing the students to the concept of a stochastic model in the study of probability distributions. Second, it offers an excellent scope for discussing live examples to demonstrate applications of mathematical results and limitations, if any, of the theoretical developments. Both these aspects are important for a proper training of research workers. I believe Professor Patil's own interest in discrete distributions developed when he was a research scholar at the Indian Statistical Institute. I hope the *Dictionary and Bibliography* will be useful in giving a good graduate level course on discrete distributions.

The International Statistical Institute also deserves credit for realizing the importance of the subject of discrete distributions and for deciding to bring out this volume under its auspices. With introductory notes and comments on over 100 distributions and a classified bibliography of over 3000 relevant publications, the volume should be of great value and service, in many ways, to students, consultants and research workers not only in statistics but in diverse fields of applied science.

CALCUTTA, INDIA
1968

C. R. RAO

CONTENTS

PREFACE

The present volume is a multistage outgrowth of its origin in the idea of organizing a display of available literature on discrete distributions during the international symposium on classical and contagious distributions held at McGill University in 1963. At that time, a selected bibliography of about 1200 entries (Patil, 1965e) was prepared for the purpose of its inclusion in the symposium proceedings (Patil, 1965f). The second stage provided a technical report for a bibliography of about 2000 entries (Patil and Joshi, 1966). In that report, the majority of the entries appeared with their users listings and also with their classifications according to major discrete distributions, inference topics and applications. The third stage picked up about 1000 additional entries from several miscellaneous sources and demonstrated a definite need for the preparation of an extensive dictionary of discrete distributions and their inter-relations. In response to suggestions towards improving the usefulness of the book further, an additional chapter grouped the publications discussing common distributions and inference topics. Thus, on the expanding horizon of statistical literature, has emerged the present volume, which, to our knowledge, is the first of its kind devoted to the field of discrete distributions. While it does cover the items which ordinarily come under the scope of a dictionary and a bibliography, it provides also a sizable amount of related material which can be of value to the user of such a publication.

The first chapter constitutes a dictionary of discrete distributions and their inter-relations. It discusses 115 models. Preference has been given, in general, to the distributions that arise in ' statistical ' problems rather than in ' probability ' problems. The material consists of sections on introduction, notation and terminology, alphabetical index of distributions, and individual distributions and their inter-relations. The section on notation and terminology provides a listing of the symbols and abbreviations used. The section on alphabetical index of distributions lists the distributions appearing in the last section along with the page numbers on which they appear. The last section provides, for each distribution covered, its brief description, including, at times, its genesis, explicit probability function, probability generating function, mean, variance and covariance as may be applicable. Inter-relations between distributions are indicated. Many of the listed inter-relations are known in literature at one place or another. A few others appear to be new. A couple of major references are also provided for each distribution. Most of the distributions listed are established in literature generally under the names in which they appear. A few distributions have been added to the list to make it as complete as possible.

The second chapter constitutes a bibliography of discrete distributions of about 3000 entries. It consists of sections on introduction, list of classes, list of scanned periodicals, list of other periodicals appearing in the bibliography and individual entries. An effort has been made to make the bibliography current. A systematic search was made for most of the entries published in English during the period ending 1966. Later publications, noticed and found relevant, are also listed. The original objective of achieving complete coverage of non-English publications was partially abandoned due to lack of adequate resources. We are reasonably sure, however, that it has been possible to recover most of the relevant non-English publications from one source or another.

Papers in statistics and probability discussing discrete distributions in general or those providing explicit treatment of one or more of them on individual basis were considered relevant. Publications in other fields discussing applications of these distributions have been also included. A few relevant topics such as matching, occupancy, run length and contingency tables are omitted since these have been covered in the *Bibliography of non-parametric statistics*, by I. R. Savage (1962), Harvard University Press. Publications on the general theory of queues and other stochastic processes have been excluded to a large extent. These are available in the *Bibliography on time series and stochastic processes*, by H. Wold (1966), Oliver and Boyd Ltd. The scope of the present bibliography, including topics covered, may be more apparent from the discussion on matters related to the classification schemes as discussed in the text.

A prominent feature of the bibliography lies in the use of a citation scheme as discussed and used by I. R. Savage in his *Bibliography of non-parametric statistics*. This technique gives a list of all entries that cite a particular item (entry). The citation lists (users listings) can serve as a detailed indexing and can also help bring one up to date in a sense. Another prominent feature of the present bibliography is in the extensive classification of the individual entries according to 44 distributions, 22 inference topics and 8 fields of application.

The third chapter groups the entries of the bibliography of the second chapter according to the 44 classes of distributions and 22 classes of statistical inference as listed in the text. The classification here is complete to the extent of the individual entries for which classification is available in the second chapter. We are hopeful that one would come to know of most of the entries not classified here through the users listings available for the individual entries in the second chapter.

Care has been taken to see that the volume would be useful to the scientists as well as to the statisticians with both theoretical and applied interests ranging over a wide spectrum of disciplines of knowledge where one is faced with the problems of statistical counts and what you do with them. This book should help one to pick up the vocabulary of the field of discrete distributions and also to lead him to the relevant literature that may be available.

Our interest in the subject of discrete distributions continues. We would much appreciate it if any omissions or errors found in the present volume are brought to our attention. We would also be very thankful if reprints and/or preprints of publications relevant to discrete distributions are sent to us as and when they are ready. Copies of publications in the past are also welcome. This would greatly help us in the effort of keeping the library and the bibliography on discrete distributions complete and current.

UNIVERSITY PARK
PENNSYLVANIA

G. P. PATIL
S. W. JOSHI

ACKNOWLEDGEMENTS

We wish to take this opportunity to express our deep appreciation to the following colleagues for their advice and assistance in the preparation of the bibliography.

Australia:	Douglas, J. B.; Lancaster, H. O.
Belgium:	Martin, L. J.
Canada:	Kotz, Samuel; Pielou, E. C.; Seshadri, V.; Shanfield, Florence; Sprott, D. A.
Denmark:	Hald, A.
England:	Forrester, Colin; Healy, M. J. R.; Herdan, G.; Kendall, M. G.; Pearson, E. S.
France:	Dugué, D.
Germany:	Fels, E. M.; Heinisch, O.
India:	Bildikar, Sheela; Mitra, S. K.; Rao, C. R.
Ireland:	Kemp, C. D.
Italy:	Brambilla, F.; Castellano, V.
Japan:	Kitagawa, T.; Kobayashi, S.; Matusita, K.
Netherlands:	Hemelrijk, J.; Molenaar, W.
Norway:	Rasch, G.
South Africa:	Kerrich, J. E.
Sweden:	Sandelius, D. Martin; Matern, B.
Switzerland:	Linder, A.
U.S.A.:	Bartko, John J.; Batschelet, E.; Blischke, W. R.; Bliss, C. I.; Cohen, A. C. Jr.; Crow, E. L.; David, H. A.; Greville, T. N. E.; Haight, Frank A.; Hartley, H. O.; Katti, S. K.; Krishnaiah, P. R.; Lord, F. M.; Martin, Paul S.; Mosimann, J. E.; Mosteller, Frederick; Neyman, J.; Rider, P. R.; Savage, I. Richard; Schultz, Vincent; Shantaram, R.; Shenton, L. R.; Waters, William E.; Watson, G. S.
U.S.S.R.:	Linnik, Yu. V.; Prohorov, Yu. V.

The initial work on the bibliography was carried out with the partial support of the Aerospace Research Laboratories, Office of Aerospace Research, United States Air Force, which is gratefully acknowledged here.

The work on the dictionary began around a course on discrete statistical models and methods given by the senior author at Penn State during the summer term of 1966. The enthusiastic participation of E. Atzinger and A. Godambe in the project is greatly appreciated.

The present version of both the dictionary and the bibliography has been prepared with the partial support of the United States Department of Agriculture, Forest Service, under a co-operative project. We are indebted to Dr William E. Waters for his support and encouragement. We record our thanks also to A. Godambe, K. G. Janardan and S. Mehta for their assistance with the final version.

We also wish to take this opportunity to thank Professor J. B. Bartoo and his office for providing congenial working facilities at the Department of Mathematics at Penn State.

Our acknowledgements would not be complete without an explicit expression of our gratitude and appreciation for Professor C. R. Rao who has so kindly written a fitting foreword to this volume. Our thanks also go to the International Statistical Institute and to Oliver and Boyd Ltd. for their active and efficient interest in the present publication. And, at the same time' we wish to record our profound appreciation and indebtedness to Professor Paul S. Dwyer of the University of Michigan for his valuable personal advice in 1959 to the senior author for going 'slow' on the projected plans.

Finally, we record our appreciation to Shirley Bradford, Leona Ollinger, and Susan Yanez for their assistance in the preparation of the manuscript.

UNIVERSITY PARK
PENNSYLVANIA

G. P. PATIL
S. W. JOSHI

Chapter 1
A DICTIONARY OF DISCRETE DISTRIBUTIONS AND THEIR INTER-RELATIONS

1.1 Introduction

In this chapter, we provide a dictionary of classical and contagious discrete distributions. Preference has been given, in general, to the distributions that arise in ' statistical ' problems rather than in ' probability ' problems. Particularly since the terms like ' classical ' and ' contagious ' remain to be clearly defined, the coverage of distributions here may appear arbitrary.

This chapter consists of sections on introduction, notation and terminology, alphabetical index of distributions and individual distributions and their inter-relations. The section on notation and terminology provides a listing of the symbols and abbreviations used. The section on alphabetical index of distributions lists the distributions appearing in the last section along with the page numbers on which they appear. The last section provides, for each distribution covered, its brief description including, at times, its genesis, explicit probability function, probability generating function, mean, variance and covariance as may be applicable. Inter-relations between distributions are indicated. Many of the listed inter-relations are known in literature at one place or another. A few others appear to be new. A couple of major references are also provided for each distribution. Most of the distributions listed are established in literature generally under the names in which they appear. A few distributions have been added to the list to make it as complete as possible.

The following publications, among several others as cited in the dictionary, may be of special value to the user of the dictionary:

1. *Classical and Contagious Discrete Distributions*, edited by G. P. Patil, Pergamon Press, Oxford, New York, Toronto, Paris, Frankfurt; and Statistical Publishing Society, Calcutta-35 (1965).
2. *Handbook of the Poisson Distribution*, by Frank A. Haight, John Wiley and Sons (1966).
3. *Formulae and Tables for Statistical Work*, edited by C. R. Rao, S. K. Mitra and A. Mathai, Statistical Publishing Society, Calcutta-35 (1966).

1.2 Notation and Terminology

1. All logarithms are to the base e.

2. Abbreviations:

pf: probability function (of a discrete distribution).
pdf: probability density function (of a continuous distribution).
pgf: probability generating function.
gf: generating function.

fmgf: factorial moment generating function.
fcgf: factorial cumulant generating function.
PSD: power series distribution.

3. $$\Gamma(a) = \int_0^\infty x^{a-1}e^{-x}dx, \quad a>0.$$

$$a! = \Gamma(a+1), \qquad a>-1.$$

In particular,

$n! = 1.2.3.....(n-1)n$, if n is a positive integer.

$0! = 1.$

4. $$\binom{n}{x} = \frac{n!}{x!(n-x)!}, \ n \text{ and } x \text{ integers.}$$

$$\frac{(-u)!}{(-u-v)!} = \frac{(-1)^v(u+v-1)!}{(u-1)!}, \ v \text{ an integer.}$$

5. $$\left.\begin{aligned} a^{(r)} &= a(a-1)(a-2)...(a-r+1) \\ a_{(r)} &= a(a+1)(a+2)...(a+r-1) \end{aligned}\right\}, \ a \text{ real and } r \text{ a positive integer.}$$

$a^{(0)} = 1.$

$a_{(0)} = 1.$

5a. S_n^r = Stirling number of the first kind with arguments r and $n = \frac{1}{r!}\left[\frac{d^r}{dx^r}\{x^{(n)}\}\right]_{x=0}$ = the coefficient of x^r in the expansion of $x^{(n)}$.

6. If f is a function of the real variable x, then the difference operator Δ is defined by

$$\Delta f(x) = f(x+1) - f(x).$$

$$\Delta^r f(x) = \Delta(\Delta^{r-1}f(x)) = \Delta^{r-1}f(x+1) - \Delta^{r-1}f(x), \ r = 2, 3, \ldots$$

$$\Delta^r 0^n = \Delta^r(x^n)\big|_{x=0}.$$

$$(1+z\Delta)^k f(x) = \sum_{r=0}^{k} \binom{k}{r} z^r \Delta^r f(x).$$

7. $\mathfrak{S}_n^r$ = Stirling number of the second kind with arguments r and n, $n \geqq r$,

$$= \frac{\Delta^r 0^n}{r!} = \frac{(-1)^r}{r!} \sum_{k=0}^{\infty} \binom{r}{k} (-1)^k k^n.$$

$\mathfrak{S}_n^r = 0$ for $n<r$.

8. If X is a random variable having one-dimensional discrete distribution with the pf $P(x)$, then

$$\mu = \text{the expected value or the mean of } X = E(X) = \sum_x xP(x).$$

$$\sigma^2 = \text{variance of } X = E(X-\mu)^2 = \sum_x (x-\mu)^2 P(x).$$

$G(z)$ = The pgf (i.e. the same as gf) of X [or, of the distribution of X]

$$= E(z^X) = \sum_x z^x P(x).$$

The fmfg of X [or, of the distribution of X] $= G(z+1)$.
The fcgf of X [or, of the distribution of X] $= \log G(z+1)$.

9. If the random vector $(X_1, X_2, \ldots, X_s)$ has the s-dimensional discrete distribution with the pf $P(x_1, x_2, \ldots, x_s)$, then

μ_i = the expected value or the mean of $X_i = E(X_i)$

$$= \Sigma_{(x_1, x_2, \ldots, x_s)} x_i P(x_1, x_2, \ldots, x_s), \qquad i = 1, 2, \ldots, s.$$

$\sigma_{ii} = \sigma_i^2$ = variance of $X_i = E(X_i - \mu_i)^2$

$$= \Sigma_{(x_1, x_2, \ldots, x_s)} (x_i-\mu_i)^2 P(x_1, x_2, \ldots, x_s), \quad i = 1, 2, \ldots, s.$$

$\sigma_{ij}(i \neq j) = \sigma_{ji}$ = covariance between X_i and X_j

$$= E(X_i-\mu_i)(X_j-\mu_j)$$

$$= \Sigma_{(x_1, x_2, \ldots, x_s)} (x_i-\mu_i)(x_j-\mu_j) P(x_1, x_2, \ldots, x_s), \quad i = 1, 2, \ldots, s.$$

$$j = 1, 2, \ldots, s.$$

$G(z_1, z_2, \ldots, z_s)$ = the pgf (i.e. the same as gf) of $(X_1, X_2, \ldots, X_s)$ [or, of the distribution of $(X_1, X_2, \ldots, X_s)$]

$$= E(z_1^{X_1} z_2^{X_2} \ldots z_s^{X_s})$$

$$= \Sigma_{(x_1, x_2, \ldots, x_s)} z_1^{x_1} z_2^{x_2} \ldots z_s^{x_s} P(x_1, x_2, \ldots, x_s).$$

The fmgf of $(X_1, X_2, \ldots, X_s)$ [or, of the distribution of $(X_1, X_2, \ldots, X_s)$]

$$= G(z_1+1, z_2+1, \ldots, z_s+1).$$

The fcgf of $(X_1, X_2, \ldots, X_s)$ [or, of the distribution of $(X_1, X_2, \ldots, X_s)$]

$$= \log G(z_1+1, z_2+1, \ldots, z_s+1).$$

10. $${}_1F_1(a;\,b;\,t) = \sum_{r=0}^{\infty} \frac{a_{(r)}}{b_{(r)}} \cdot \frac{t^r}{r!}.$$

$${}_2F_1(a,\,b;\,c;\,t) = \sum_{r=0}^{\infty} \frac{a_{(r)}b_{(r)}}{c_{(r)}} \cdot \frac{t^r}{r!}.$$

$${}_{(s+1)}F_1(a,\,b_1,\,b_2,\,\ldots,\,b_s;\,c;\,z_1,\,z_2,\,\ldots,\,z_s)$$

$$= \sum_{r_1, r_2, \ldots, r_s = 0}^{\infty} \frac{a_{(r)}b_{1(r_1)}b_{2(r_2)}\ldots b_{s(r_s)}}{c_{(r)}} \cdot \frac{z_1^{r_1}}{r_1!} \cdot \frac{z_2^{r_2}}{r_2!} \cdot \frac{z_s^{r_s}}{r_s!},$$

where $r = \sum_{i=1}^{s} r_i$ and $s = 2, 3, \ldots$. The F-functions are called hypergeometric functions.

11. $\zeta(s)$ = Riemann zeta function with the argument s

$$= \sum_{n=1}^{\infty} \frac{1}{n^s} \text{ where } s > 1.$$

12. The beta distribution with parameters α and β is defined by the pdf

$$f(x) = \frac{1}{B(\alpha,\,\beta)}\, x^{\alpha-1}(1-x)^{\beta-1},$$

where $B(\alpha,\,\beta) = \dfrac{\Gamma(\alpha)\Gamma(\beta)}{\Gamma(\alpha+\beta)}$ with $0<x<1$, $0<\alpha<\infty$ and $0<\beta<\infty$.

13. The gamma distribution [or (Pearson's) type III distribution] with parameters α and β is defined by the pdf

$$f(x) = \frac{\alpha^\beta}{\Gamma(\beta)}\, x^{\beta-1}e^{-\alpha x},$$

where $0<x<\infty$, $0<\alpha<\infty$ and $0<\beta<\infty$.

14. The rectangular [or, the uniform] distribution over the interval $(0,\,\theta)$ is defined by the pdf

$$f(x) = \frac{1}{\theta},$$

where $0<x<\theta$ and $0<\theta<\infty$.

15. The exponential distribution with parameter θ is defined by the pdf

$$f(x) = \theta e^{-\theta x},$$

where $0<x<\infty$ and $0<\theta<\infty$.

16. The normal distribution with parameters μ and σ is defined by the pdf

$$f(x) = \frac{1}{\sqrt{(2\pi)}\sigma} \exp\left\{-\frac{(x-\mu)^2}{2\sigma^2}\right\},$$

where $-\infty<x<\infty$, $-\infty<\mu<\infty$ and $0<\sigma<\infty$.

17. The normal distribution with parameters μ and σ truncated at left at zero has the pdf

$$f(x) = \frac{1}{\sqrt{(2\pi)}\sigma\Phi\left(\frac{\mu}{\sigma}\right)} \exp\left\{-\frac{(x-\mu)^2}{2\sigma^2}\right\},$$

where $\Phi(u) = \frac{1}{\sqrt{(2\pi)}}\int_{-\infty}^{u} e^{-x^2/2}\,dx$ with $0<x<\infty$, $-\infty<\mu<\infty$ and $0<\sigma<\infty$.

18. The random variable X is said to have the lognormal distribution with parameters μ and σ if $Y = \log X$ has the normal distribution with parameters μ and σ.

19. The s-variate Dirichlet distribution with parameters $\alpha_1, \alpha_2, \ldots, \alpha_s$ and β is defined by the pdf

$$f(x_1, x_2, \ldots, x_s) = \frac{\Gamma\left(\sum_{i=1}^{s} \alpha_i + \beta\right)}{\Gamma(\alpha_1)\Gamma(\alpha_2)\ldots\Gamma(\alpha_s)\Gamma(\beta)} x_1^{\alpha_1-1}x_2^{\alpha_2-1}\ldots x_s^{\alpha_s-1}(1-x_1-x_2-\ldots-x_s)^{\beta-1}$$

where, $0<x_i<1$, $i = 1, 2, \ldots, s$, $\sum_{i=1}^{s} x_i < 1$, $0<\alpha_i<\infty$; $i = 1, 2, \ldots, s$ and $0<\beta<\infty$.

20. If X is a random variable with the pf $P(x)$, $x = 0, 1, 2, \ldots$, then the corresponding zero-truncated distribution is defined by the pf $\frac{P(x)}{1-P(0)}$, $x = 1, 2, \ldots$.
If $(X_1, X_2, \ldots, X_s)$ is an s-dimensional random vector with the pf $P(x_1, x_2, \ldots, x_s)$, $x_i = 0, 1, 2, \ldots$, then the corresponding origin-truncated distribution is defined by the pf

$$\frac{P(x_1, x_2, \ldots, x_s)}{1-P(0, 0, \ldots, 0)}, \; x_i = 0, 1, 2, \ldots; \; i = 1, 2, \ldots, s, \text{ but } \sum_{j=1}^{s} x_j \neq 0.$$

21. If X_1 and X_2 are two independent random variables with the gf's $g_1(z)$ and $g_2(z)$ respectively, then the random variable X having the gf $g_1(g_2(z))$ is called X_1 variable generalized by [or, with respect to] X_2. The distribution of X is the generalized distribution [Gurland (1957)].

22. Suppose X is a random variable with the pf $P_\theta(x)$, where, θ is a parameter and θ itself considered a random variable has the pdf $f(\theta)$ [or, the pf $f(\theta)$]. Then the distribution of a random variable Y is said to be the mixture on θ of the distribution of X if the pf $P_1(y)$ of Y is given by $\int P_\theta(y)f(\theta)d\theta$ [or, by $\sum_\theta p_\theta(y)f(\theta)$, if θ is discrete]. Alternatively, the distribution of Y is said to be the compound distribution of X with respect to the compounder θ. Clearly, the distribution of Y is the marginal distribution of X.

23. Let $\phi = (\phi_1, \phi_2, \ldots, \phi_m)$, $\zeta = (\zeta_1, \zeta_2, \ldots, \zeta_m)$ and $\psi = (\psi_1, \psi_2, \ldots, \psi_k)$. The discrete probability distribution A with pf $P_A(x, \phi)$ is said to tend to the discrete probability distribution B with pf $P_B(x, \psi)$ as $\phi\to\zeta$ if it is true that for every x, limit $P_A(x, \phi) = P_B(x, \psi)$ as $\phi\to\zeta$, where ψ is a function of ζ. It is quite possible that there may be one or more side conditions on the limiting process of ϕ.

1.3 Alphabetical Index of Distributions

DICTIONARY

1.4 Individual Distributions and their Inter-relations

1. THE BINOMIAL DISTRIBUTION

(Bernoulli distribution, Bernoullian distribution, Point Binomial distribution)
The number of successes in n independent trials with the constant probability p of success at each trial has the binomial distribution with parameters n and p and has the pf

$$P(x) = \binom{n}{x} p^x(1-p)^{n-x}, \qquad x = 0, 1, 2, \ldots, n,$$

$$n = 1, 2, \ldots;\ n<\infty,$$

$$0<p<1.$$

$$G(z) = (q+pz)^n = [1+p(z-1)]^n, \quad \text{where } q = 1-p.$$

$$\mu = np,\ \sigma^2 = npq = \mu q.$$

Some inter-relations

(i) Occasionally, the binomial distribution with parameters 1 and p has been called the Bernoulli distribution with parameter p.

(ii) The Polya distribution with parameters a, b, 0 and n is the same as the binomial distribution with parameters n and $\frac{a}{(a+b)}$.

(iii) The mixture on p of the binomial distribution with parameters n and p is the negative hypergeometric distribution with parameters $a+b$, a and n if p has the beta distribution with parameters a and b.

(iv) As $N\to\infty$, $M\to\infty$ such that $\frac{M}{N}\to p$, $0<p<1$, the hypergeometric distribution with parameters N, M and n tends to the binomial distribution with parameters n and p.

(v) As $n\to\infty$, $p\to 0$ such that $np\to\lambda$, $0<\lambda<\infty$, the binomial distribution with parameters n and p tends to the Poisson distribution with parameter λ.

(vi) The binomial distribution with parameters n and p is a special case of the generalized binomial distribution of Poisson with parameters $n, p_1, p_2, \ldots, p_n$ when $p_1 = p_2 = \ldots = p_n = p$.

(vii) The binomial distribution with parameters n and p is a special case of the modified binomial distribution (i) with parameters n, m and p when $m = 1$.

(viii) As $N \to \infty$, $M \to \infty$, such that $\frac{M}{N} \to p$, the negative hypergeometric distribution with parameters N, M and n tends to the binomial distribution with parameters n and p.

(ix) The binomial distribution with parameters n and p is the PSD with parameter $\theta = \frac{p}{(1-p)}$ and the series function $f(\theta) = (1+\theta)^n$.

(x) As $a \to \infty$, $b \to \infty$ such that $\frac{a}{(a+b)} \to p$, the generalized hypergeometric distribution type IA(i) with parameters a, b, and n tends to the binomial distribution with parameters n and p.

(xi) The mixture on p of the binomial distribution with parameters n and p is the generalized hypergeometric distribution type IIA with parameters $-a$, $-b$ and n if p has a beta distribution with parameters a and b.

(xii) As $a \to -\infty$ and $b \to -\infty$ such that $\frac{a}{(a+b)} \to p$, the generalized hypergeometric distribution type IIA with parameters a, b and n tends to the binomial distribution with parameters n and p.

(xiii) Let n_1, $n_1+n_2 \to \infty$ such that $\frac{n_1}{(n_1+n_2)} \to p$, $0<p<1$, and let $\phi = p\theta/(q+p\theta)$, $q = 1-p$. Then the extended hypergeometric distribution with parameters n_1, n_2, r and θ tends to the binomial distribution with parameters r and ϕ.

Some references Feller (1957), p. 136; Wilks (1962), p. 137.

2. THE GENERALIZED BINOMIAL DISTRIBUTION OF POISSON

(Poisson binomial distribution)

The number of successes in n independent trials with probability p_j of success in the jth trial has the generalized binomial distribution of Poisson with parameters n, $p_1, p_2, \ldots, p_n$. It has the pf

$$P(x) = \Sigma\, p_{j_1}p_{j_2}\cdots p_{j_x}q_{i_1}q_{i_2}\cdots q_{i_{n-x}}, \quad x = 0, 1, 2, \ldots, n,$$
$$n = 1, 2, \ldots;\ n<\infty,$$
$$0<p_j<1,\ q_j = 1-p_j,$$
$$j = 1, 2, \ldots, n.$$

where the summation extends over all combinations $\{j_1, j_2, \ldots, j_x\}$ of exactly x members of $\{1, 2, \ldots, n\}$, $\{i_1, i_2, \ldots, i_{n-x}\}$ being the complementary combinations.

$$G(z) = \prod_{j=1}^{n} (q_j+p_jz) = \prod_{j=1}^{n} [1+p_j(z-1)].$$

$$\mu = \sum_{j=1}^{n} p_j, \quad \sigma^2 = \sum_{j=1}^{n} p_jq_j.$$

Some inter-relations

(i) The binomial distribution with parameters n and p is a special case of the generalized binomial distribution of Poisson with parameters $n, p_1, p_2, \ldots, p_n$ when $p_1 = p_2 = \ldots = p_n = p$.

(ii) The generalized binomial distribution of Poisson with parameters $n, p_1, p_2, \ldots, p_n$ tends to the Poisson distribution with parameter λ if the parameters p_j depend on n in such a way that the largest $p_j \to 0$, but $p_1+p_2+\ldots+p_n \to \lambda$ as $n \to \infty$.

Some references Cramér (1946), p. 206; Feller (1957), p. 263.

3. THE POISSON LEXIS SERIES

The number of successes in n independent trials with the probability of success in the jth trial being p_j (where for $j = 1, 2, \ldots, n$, $p_j = \phi(p, c_j)$ is a function of a random variable p with distribution function $F(p)$ and a constant c_j) has the Poisson Lexis series distribution with parameters $n, p_1, p_2, \ldots, p_n$. It has the pf

$$P(x) = \int \{\Sigma p_{j_1} p_{j_2} \ldots p_{j_x} q_{i_1} q_{i_2} \ldots q_{i_{n-x}}\} dF(p), \quad x = 0, 1, 2, \ldots, n, \quad n = 1, 2, \ldots; \ n<\infty, \quad 0<p_j<1,\ q_j = 1-p_j.$$

where the integration is taken over the range of p and the summation is taken over all combinations $\{j_1, j_2, \ldots, j_x\}$ of exactly x members of $\{1, 2, \ldots, n\}$, $\{i_1, i_2, \ldots, i_{n-x}\}$ being the complementary combinations.

$$G(z) = \int \prod_{j=1}^{n} (q_j + p_j z) dF(p),$$

$$\mu = \int \sum_{j=1}^{n} p_j dF(p), \quad \sigma^2 = \int \sum_{j=1}^{n} p_j q_j dF(p) + \left[V\left(\sum_{j=1}^{n} p_j\right)\right],$$

where the integration is taken over the range of p and V stands for variance.

A reference Ottestad (1943).

4. THE MODIFIED BINOMIAL DISTRIBUTION (i)

Consider a sequence of n trials with constant probability p of success in an individual trial with the restriction that as soon as a success is experienced, the subsequent $m-1$ trials result in failure. Then the number of successes in the above sequence has the modified binomial distribution (i) with parameters n, m and p and with the pf

$$P(x) = F(x) - F(x-1), \quad x = 0, 1, 2, \ldots \text{ such that } n-xm>0,$$

where the distribution function $F(x)$ is defined by

$$F(-1) = 0,$$

$$F(x) = q^{n-xm}\left[1+(n-xm)p+\frac{(n-xm)(n-xm+1)}{2!}p^2+\ldots \frac{(n-xm)(n-xm+1)\ldots(n-xm+x-1)}{x!}p^x\right],$$

and $F(x) = 1$ for $n-xm<0$,

$$n = 1, 2, \ldots, \quad m = 1, 2, \ldots, \quad 0<p<1.$$

Simple forms for the pgf, mean and variance are not known.

Some inter-relations

(i) The binomial distribution with parameters n and p is a special case of the modified binomial distribution (i) with parameters n, m and p when $m = 1$.

(ii) As $n, m \to \infty$ and $p \to 0$ such that $np \to \lambda$ and $\frac{m}{n} \to k$, λ and k being constants, the modified binomial distribution (i) with parameters n, m and p tends to the modified Poisson distribution (i) with parameters k and λ.

Some references Basu (1955); Dandekar (1955).

5. THE MODIFIED BINOMIAL DISTRIBUTION (ii)

Consider an experiment as in the modified binomial distribution (i). However, instead of assuming that the probability of a success in the first trial is p, we regard our first observation in the experiment as a random start in an infinite sequence of similar trials. Thus we allow the possbility that there might be a success within m trials before our observations start. Then, the number of successes in n trials has the modified binomial distribution (ii) with parameters n, m and p. It has the pf

$$P(x) = F(x) - F(x-1),$$

where the distribution function $F(x)$ is given by

$$F(-1) = 0,$$

$$F(x) = \frac{1}{1+(m-1)p} \sum_{j=0}^{x} p'[(x-j),\ n-(j+1)(m-1)],$$

with $p'[(x-j),\ n-(j+1)(m-1)] \equiv p'(x-j)$, p' being the pf of the modified binomial distribution (i) with parameters $n-(j+1)(m-1)$, m and p. Simple forms of the pgf, mean and variance are not known.

Some inter-relations

(i) The binomial distribution with parameters n and p is a special case of the modified binomial distribution (ii) with parameters n, m and p when $m = 1$.

(ii) As $n, m \to \infty$ and $p \to 0$ such that $np \to \lambda$ and $\frac{m}{n} \to k$, $0 < \lambda < \infty$, $0 < k < \infty$, the modified binomial distribution (ii) with parameters n, m and p tends to the modified Poisson distribution (ii) with parameters k and λ.

Some references Basu (1955); Dandekar (1955).

6. THE POISSON DISTRIBUTION

Suppose that points are distributed in space (or, in time) in accordance with the conditions: (i) The probability of k points being in a region R depends only on the volume v of the region R

and not on its shape nor its position in space. Let this probability be $p_k(v)$. (ii) The numbers of points falling in non-overlapping regions are independent random variables. (iii)

$$p_o(\Delta v)+p_1(\Delta v) = 1-0(\Delta v).$$

Then the number of points over a specified region has the Poisson distribution. The pf of the Poisson distribution with parameter λ is given by

$$P(x) = \frac{e^{-\lambda}\lambda^x}{x!} \qquad x = 0, 1, 2, \ldots \quad 0<\lambda<\infty.$$

$$G(z) = e^{\lambda(z-1)}.$$

$$\mu = \lambda, \; \sigma^2 = \lambda = \mu$$

Some inter-relations

(i) If a random variable Y has a distribution of Poisson with parameter λ, then the random variable $X = aY+b$, where $a \neq 0$ and b are constants, is said to have the Poisson type distribution with parameters λ, a and b.

(ii) Suppose that $X_o, X_1, X_2, \ldots, X_s$ have $s+1$ independent Poisson distributions with parameters $\lambda_o, \lambda_1, \lambda_2, \ldots, \lambda_s$ respectively. Then, given $\sum_{i=0}^{s} X_i = n$, the conditional distribution of $(X_1, X_2, \ldots, X_s)$ is s-variate multinomial with parameters $n, p_1, p_2, \ldots, p_s$ where

$$p_i = \frac{\lambda_i}{\lambda_0+\lambda_1+\lambda_2+\ldots+\lambda_s}, \qquad i = 1, 2, \ldots, s.$$

(iii) The negative binomial distribution with parameters k and p results if the Poisson distribution with parameter $-k \log p$ is generalized by the logarithmic distribution with parameter $1-p$.

(iv) The mixture on λ of the Poisson distribution with parameter λ is the negative binomial distribution with parameters k and p if λ has the Pearson type III distribution with parameters $p/(1-p)$ and k.

(v) The generalized binomial distribution of Poisson with parameters $n, p_1, p_2, \ldots, p_n$ tends to the Poisson distribution with parameter λ if the parameters p_j depend on n in such a way that the largest $p_j \to 0$, but $p_1+p_2+\ldots+p_n \to \lambda$ as $n \to \infty$.

(vi) As $n \to \infty$, $p \to 0$ such that $np \to \lambda$, $0<\lambda<\infty$, the binomial distribution with parameters n and p tends to the Poisson distribution with parameter λ.

(vii) The hypergeometric distribution with parameters N, M and n tends to the Poisson distribution with parameter λ as $M \to \infty$, $N \to \infty$, $n \to \infty$ such that $\frac{M}{N} \to 0$ and $n\frac{M}{N} \to \lambda$.

(viii) The distribution function $F(x)$ of the modified Poisson distribution (i) with parameters k and λ is equal to the sum of the first $x+1$ terms of the pf of the Poisson distribution with parameter $(1-xk)\lambda$.

(ix) The hyper-Poisson distribution with parameters 1 and θ is the Poisson distribution with parameter θ.

(x) The Poisson distribution with parameter λ is a special case of the displaced Poisson distribution with parameters λ and r, when $r = 0$.

(xi) The Polya distribution with parameters p, γ and n tends to the Poisson distribution with parameter λ as $n \to \infty$, $p \to 0$ and $\gamma \to 0$ such that $np \to \lambda$ and $n\gamma \to 0$.

(xii) As $\theta \to 0$, the Polya Aeppli distribution with parameters λ and θ tends to the Poisson distribution with parameter λ.

(xiii) The grouped Poisson probability function is the sum of (xk)th, $(xk+1)$th, ..., $(xk+k-1)$th terms of the Poisson series with parameter λ.

(xiv) The Poisson distribution with parameter λ is the PSD with parameter $\theta = \lambda$ and the series function $f(\theta) = e^{\theta}$.

(xv) As $a \to \infty$, $b \to \infty$ and $n \to \infty$ such that $\dfrac{na}{(a+b)} \to \lambda$, the generalized hypergeometric distribution type IA (i) with parameters a, b and n tends to the Poisson distribution with parameter λ.

(xvi) As $a \to \infty$, $b \to \infty$ and $n \to \infty$ such that $\dfrac{na}{(a+b)} \to \lambda$, the generalized hypergeometric distribution type IA(ii) tends to the Poisson distribution with parameter λ.

(xvii) As $a \to \infty$, $b \to \infty$ and $n \to \infty$ such that $\dfrac{na}{(a+b)} \to \lambda$, the generalized hypergeometric distribution type IB tends to the Poisson distribution with parameter λ.

(xviii) As $a \to -\infty$, $b \to -\infty$, $n \to \infty$ such that $\dfrac{na}{(a+b)} \to \lambda$, the generalized hypergeometric distribution type IIA with parameters a, b and n tends to the Poisson distribution with parameter λ.

(xix) As $a \to \infty$, $b \to -\infty$ and $n \to -\infty$ such that $\left|\dfrac{na}{(a+b)}\right| \to \lambda$, the generalized hypergeometric distribution type IIIA with parameters a, b and n tends to the Poisson distribution with parameter λ.

(xx) As $a \to -\infty$, $b \to \infty$ and $n \to -\infty$ such that $\left|\dfrac{na}{(a+b)}\right| \to \lambda > 0$, the generalized hypergeometric distribution type IV with parameters a, b and n tends to the Poisson distribution with parameter λ.

(xxi) Let n_1, n_2, $r \to \infty$ such that $\dfrac{n_1 r}{(n_1+n_2)} \to \lambda$, $0 < \lambda < \infty$. Then the extended hypergeometric distribution with parameters n_1, n_2, r and θ tends to the Poisson distribution with parameter $\lambda\theta$.

(xxii) The Hermite distribution with parameters λ_1 and λ_2 results if the Poisson distribution with parameter $\lambda > \lambda_1 + \lambda_2$ is generalized by the distribution having the pgf

$$\left(1 - \frac{\lambda_1+\lambda_2}{\lambda}\right) + \frac{\lambda_1}{\lambda} z + \frac{\lambda_2}{\lambda} x^2.$$

(xxiii) If X_1 and X_2 have two independent Poisson distributions with parameters λ_1 and λ_2, then the random variable $X_1 + 2X_2$ has the Hermite distribution with parameters λ_1 and λ_2.

Some references Cramér (1946); Feller (1957), p. 146.

7. THE POISSON TYPE DISTRIBUTION

If a random variable Y has the Poisson distribution with parameter λ, and $a \neq 0$, b are constants, then the random variable $X = aY+b$ is said to have the Poisson type distribution with parameters λ, a, and b. It has the pf

$$P(x) = \mathrm{e}^{-\lambda}\frac{\lambda^y}{y!}, \qquad y = \frac{x-b}{a},$$

$$x = b,\ a+b,\ 2a+b,\ \ldots,$$

$$a \neq 0,$$

$$-\infty < b < \infty,$$

$$0 < \lambda < \infty.$$

$$G(z) = z^b \mathrm{e}^{\lambda(z^a-1)} = z^b H(z^a),$$

where $H(z)$ is the generating function of the Poisson random variable Y with parameter λ.

$$\mu = a\lambda+b,\ \sigma^2 = a^2\lambda.$$

Some inter-relations

The Poisson distribution with parameter λ is a special case of the Poisson type distribution with parameters λ, 1 and 0.

A reference Lukacs, E. (1960*a*), pp. 83, 84.

8. THE DISPLACED POISSON DISTRIBUTION

The number of events in a Poisson process in excess of a threshold value r, when it is assumed that at least r events do occur, has the displaced Poisson distribution with parameters λ and r. It has the pf

$$P(x) = \frac{\mathrm{e}^{-\lambda}\lambda^{x+r}}{I(r, \lambda)(x+r)!} \qquad x = 0, 1, 2, \ldots,$$

$$0 < \lambda < \infty,$$

$$r = 0, 1, 2, \ldots,$$

where $I(r, \lambda) = \sum_{y=r}^{\infty} \frac{e^{-\lambda}\lambda^y}{y!}$.

$$\mu = \lambda - r[1-P(0)],\ \sigma^2 = \lambda - \mu r P(0).$$

Some inter-relations

The Poisson distribution with parameter λ is a special case of the displaced Poisson distribution with parameters λ and r when $r = 0$.

A reference Staff (1964). *Note:* Allowing $I(r, \lambda) = \frac{1}{\Gamma(r)}\int_0^{\lambda} \mathrm{e}^{-u}u^{r-1}du$, Staff enlarges the parameter space for r to $r > -1$.

9. THE MODIFIED POISSON DISTRIBUTION (i)

A modified Poisson process is an ordinary Poisson process with the additional restriction that as soon as an event takes place, a period $k(0<k<\infty)$ of immunity follows during which no further event can take place. The number of events occurring during a unit of time is a modified Poisson process having the Poisson distribution (i) with parameters k and λ. It has the pf

$$P(x) = F(x) - F(x-1), \qquad x = 0, 1, 2, \ldots, xk \leqq 1,$$

where

$$F(x) = e^{-(1-xk)\lambda}\left[1+(1-xk)\lambda + \frac{(1-xk)^2}{2!}\lambda^2 + \ldots + \frac{(1-xk)^x}{x!}\lambda^x\right], \quad 0<\lambda<\infty$$
$$0<k<\infty.$$

Simple forms for the pgf, mean and variance are not known.

Some inter-relations

(i) The distribution function $F(x)$ of the modified Poisson distribution (i) with parameters k and λ is equal to the sum of the first $x+1$ terms of the pf of the Poisson distribution with parameter $(1-xk)\lambda$.

(ii) As $n, m\to\infty$ and $p\to 0$ such that $np\to\lambda$ and $\frac{m}{n}\to k$, where $0<\lambda<\infty$, $0<k<\infty$, the modified binomial distribution (i) with parameters n, m and p tends to the modified Poisson distribution (i) with parameters k and λ.

Some references Basu (1955); Dandekar (1955).

10. THE MODIFIED POISSON DISTRIBUTION (ii)

Consider an experiment as in the modified Poisson distribution (i). In this case, assume that the starting time is a random point of the infinite axis of time for which the process has been going on. Then the number of events occurring during a unit of time in a modified Poisson process has the modified Poisson distribution (ii) with parameters k and λ and has the pf

$$P(x) = F(x) - F(x-1), \qquad x = 0, 1, 2, \ldots \text{ such that } xk \leqq 1,$$

where

$$F(-1) = 0,$$
$$F(x) = \Phi(x) - \Phi(x-1),$$
$$\Phi(-1) = 0,$$
$$\Phi(x) = \frac{\exp[-(1-[x+1]k)\lambda]}{1+k\lambda}\sum_{j=0}^{x}\sum_{i=0}^{x-j}\frac{(1-[x+1]k)^i}{i!}\lambda^i, \quad 0<\lambda<\infty$$
$$0<k<\infty.$$

Simple forms for the pgf, mean and variance are not known.

Some inter-relations

As n, $m \to \infty$ and $p \to 0$ such that $np \to \lambda$ and $\frac{m}{n} \to k$, where $0<\lambda<\infty$, $0<k<\infty$, the modified binomial distribution (ii) with parameters n, m and p tends to the modified Poisson distribution (ii) with parameters k and λ.

Some references Basu (1955); Dandekar (1955).

11. THE HYPER-POISSON DISTRIBUTION

The hyper-Poisson distribution with parameters θ and λ has the pf

$$P(x) = \frac{\Gamma(\lambda)\theta^x}{{}_1F_1[1;\ \lambda;\ \theta]\Gamma(\lambda+x)}, \qquad x = 0, 1, 2, \ldots,$$

$$0<\theta<\infty,$$

$$G(z) = \left(\frac{{}_1F_1[1;\ \lambda;\ \theta z]}{{}_1F_1[1;\ \lambda;\ \theta]}\right). \qquad 0<\lambda<\infty.$$

$$\mu = 1-\lambda+\theta-\frac{1-\lambda}{{}_1F_1[1;\ \lambda;\ \theta]}, \qquad \sigma^2 = \theta+\frac{\mu(1-\lambda)}{{}_1F_1[1;\ \lambda;\ \theta]},$$

Some inter-relations

(i) The hyper-Poisson distribution with parameters θ and 1 is the Poisson distribution with parameter θ.

(ii) The hyper-Poisson distribution with parameters θ and λ is the generalized power series distribution with two parameters θ and λ and the series function $f(\theta, \lambda) = {}_1F_1[1;\ \lambda;\ \theta]$.

Note: The hyper-Poisson distribution with parameters θ and λ is called sub-Poisson or super-Poisson according as $\lambda<1$ *or* $\lambda>1$.

Some references Bardwell (1961); Bardwell and Crow (1964).

12. GROUPED POISSON DISTRIBUTION

In the pure birth process if we assume that the time between two events has the gamma distribution then the population size at a given time has the grouped Poisson distribution with parameters k and λ. It has the pf

$$P(x) = \sum_{j=1}^{k} e^{-\lambda} \frac{\lambda^{xk+j-1}}{(xk+j-1)!}, \qquad x = 0, 1, 2, \ldots,$$

$$k = 1, 2, \ldots;\ k<\infty,$$

$$0<\lambda<\infty.$$

$$G(z) = 1+(z-1)\sum_{j=1}^{\infty} z^{j-1}\frac{I(kj, \lambda)}{\Gamma(kj)},$$

where
$$I(n, x) = \int_0^x e^{-t}t^{n-1}dt.$$

$$\mu = \sum_{j=1}^{\infty} \frac{I(kj, \lambda)}{\Gamma(kj)}.$$

Some inter-relations

The grouped Poisson probability function is the sum of (xk)th, $(xk+1)$th, ..., $(xk+k-1)$th terms of the Poisson series with parameter λ.

Some references Goodman (1952*a*); Haight (1959*a*).

13. THE GEOMETRIC DISTRIBUTION

Consider a sequence of independent trials with the probability of success being p at each trial. Then the number of failures encountered in order to obtain the first success has the geometric distribution, form (i), with parameter p; whereas, the number of trials required to get the first success has the geometric distribution, form (ii), with parameter p. The pf is given by

$$\text{Form (i):}\quad P(x) = pq^x, \qquad x = 0, 1, 2, \ldots, \quad 0<p<1,\ q = 1-p,$$

or

$$\text{Form (ii):}\quad P(x) = pq^{x-1}, \qquad x = 1, 2, \ldots, \quad 0<p<1,\ q = 1-p.$$

Unless otherwise stated, geometric distribution will mean form (i).

Form (i)	Form (ii)
$G(z) = \dfrac{p}{1-qz}.$	$G(z) = \dfrac{pz}{1-qz}.$
$\mu = \dfrac{q}{p}.$	$\mu = \dfrac{1}{p}.$
$\sigma^2 = \dfrac{q}{p^2} = \dfrac{\mu}{p}.$	$\sigma^2 = \dfrac{q}{p^2} = \mu\dfrac{q}{p}.$

Some inter-relations

(i) The geometric distribution with parameter p is a special case of form (ii) of the Pascal distribution with parameters k and p when $k = 1$.

(ii) The Pascal distribution with parameters k and p, form (ii), is the distribution of the sum of k independent random variables each having the geometric distribution with parameter p.

(iii) The Polya Aeppli distribution with parameters λ and q results if the Poisson distribution with parameter λ is generalized by the geometric distribution with parameter p, form (ii).

A reference Feller (1957), p. 156.

14. THE PASCAL DISTRIBUTION

(Binomial waiting time distribution)

Consider a sequence of independent trials with the probability of success being p at each trial. Then the number of trials required to get k successes has the Pascal distribution, form (i), with parameters k and p; where as, the number of failures encountered in order to obtain k successes has the Pascal distribution, form (ii), with parameters k and p. The pf is given by

$$\text{Form (i):}\quad P(x) = \binom{x-1}{k-1} p^k q^{x-k}, \qquad x = k,\ k+1,\ k+2,\ \ldots,\quad k = 1, 2, \ldots,\quad 0<p<1,\ q = 1-p.$$

$$\text{Form (ii):}\quad P(x) = \binom{x+k-1}{k-1} p^k q^x, \qquad x = 0, 1, 2, \ldots,\quad k = 1, 2, \ldots,\quad 0<p<1,\ q = 1-p.$$

Unless otherwise stated, Pascal distribution will mean form (i).

Form (i)	Form (ii)
$G(z) = \left(\frac{pz}{1-qz}\right)^k.$	$G(z) = \left(\frac{p}{1-qz}\right)^k.$
$\mu = \frac{k}{p}.$	$\mu = \frac{kq}{p}.$
$\sigma^2 = \frac{kq}{p} = \mu q.$	$\sigma^2 = \frac{kq}{p^2} = \frac{\mu}{p}.$

Some inter-relations

(i) The Pascal distribution with parameters k and p, form (ii), is a special case of the negative binomial distribution with parameters k and p where k is an integer.

(ii) The geometric distribution with parameter p is a special case of form (ii) of the Pascal distribution with parameters k and p when $k = 1$.

(iii) The Pascal distribution with parameters k and p, form (ii), is the distribution of the sum of k independent random variables each having the geometric distribution with parameter p.

(iv) The mixture on k of the Pascal distribution with parameters k and $1-\theta$ is the Polya Aeppl distribution with parameters λ and θ if k has the Poisson distribution with parameter λ.

Note A few more inter-relations may be found under negative binomial distribution.

Some references Anscombe (1950*a*); Feller (1957), p. 156; Bartko (1961*b*); Haight (1961*b*); Patil (1963*c*).

15. THE NEGATIVE BINOMIAL DISTRIBUTION

The negative binomial distribution with parameters k and p arises as the mixture on λ of the Poisson distribution with the parameters λ if λ has the gamma distribution with parameters $p/(1-p$ and k.) It has the pf

$$P(x) = \frac{\Gamma(x+k)}{x!\Gamma(k)} p^k q^x$$

$$= \binom{-k}{x} p^k(-q)^x, \qquad x = 0, 1, 2, \ldots,$$

$$0<k<\infty,$$

$$G(z) = \frac{p^k}{(1-qz)^k} \qquad 0<p<1,\ q = 1-p.$$

$$\mu = \frac{kq}{p}, \quad \sigma^2 = \frac{kq}{p^2} = \frac{\mu}{p}.$$

Some inter-relations

(i) When k is a positive integer, the negative binomial distribution with parameters k and p is the same as form (ii) of the Pascal distribution with parameters k and p.

(ii) The negative binomial distribution with parameters $\frac{h}{\theta}$ and $\frac{1}{(1+\theta)}$ is called the Polya Eggenberger distribution with parameters h and θ.

(iii) The mixture on p of the negative binomial distribution with parameters k and p is the generalized hypergeometric distribution type IV with parameters $-a$, b and $-k$ if p has the beta distribution with parameters $-a+b+1$ and a.

(iv) The negative binomial distribution with parameters k and p results if the Poisson distribution with parameter $-k \log p$ is generalized by the logarithmic distribution with parameter $1-p$.

(v) When k is a positive integer, the negative binomial distribution with parameters k and p is a special case of the univariate negative multinomial distribution with parameters $s, k, p_1, p_2, \ldots, p_s$, where $s=1$ and $p_1=1-p$.

(vi) As $n\to\infty$, $p\to0$, $\gamma\to0$ such that $np\to h$ and $n\gamma\to\theta$, the Polya distribution with parameters p, γ and n tends to the negative binomial distribution with parameters h/θ and $1/(1+\theta)$.

(vii) As $k\to0$, the zero-truncated negative binomial distribution with parameters k and p tends to the logarithmic distribution with parameter $1-p$.

(viii) As $a\to\infty$, $b\to-\infty$ such that $\frac{(a+b)}{b}\to p$, $0<p<1$, the generalized hypergeometric distribution type IIIA with parameters a, b and n tends to the negative binomial distribution with parameters $|n|$ and p.

(ix) As $a\to-\infty$, $b\to\infty$ such that $a+b>0$ and $\frac{(a+b)}{b}\to p$, $0<p<1$, the generalized hypergeometric distribution type IV with parameters a, b and n tends to the negative binomial distribution with parameters $|n|$ and p.

(x) As $N\to\infty$, $M\to\infty$ such that $\frac{M}{N}\to p$, the inverse hypergeometric distribution, form (ii), with parameters N, M and k tends to the negative binomial distribution with parameters k and p.

(xi) If $N = n$ and $n, N\to\infty$ with M remaining finite, then the distribution of exceedances with parameters N, n and M tends to the negative binomial distribution with parameters M and $\frac{1}{2}$.

(xii) The factorial distribution with parameters n and λ is the mixture on p of the negative binomial distribution with parameters $k = \lambda-n+1$ and p where p has the beta distribution with parameters $n-1$ and 1. (Negative binomial beta distribution.)

(xiii) The negative binomial distribution with parameters k and p is the PSD with parameter $\theta = 1-p$ and the series function $f(\theta) = (1-\theta)^{-k}$.

(xiv) As $N\to\infty$, $n\to\infty$ such that $\dfrac{N}{(N+n)}\to p$, $0<p<1$, the negative hypergeometric distribution with parameters N, M and n tends to the negative binomial distribution with parameters M and p.
Note Inter-relations listed under Pascal distribution also apply to the negative binomial distribution with integral k.

Some references Anscombe (1950); Bartko (1961*b*); Feller (1957), p. 155; Haight (1961*b*); Patil (1960*c*).

16. THE POLYA EGGENBERGER DISTRIBUTION

The Polya Eggenberger distribution with parameters h and θ is a reparametrized form of the negative binomial distribution with parameters k and p when $k = \dfrac{h}{\theta}$ and $p = \dfrac{1}{(1+\theta)}$ and its pf is given by

$$P(x) = \frac{\Gamma\left(x+\dfrac{h}{\theta}\right)}{x!\,\Gamma\left(\dfrac{h}{\theta}\right)}\left(\frac{1}{1+\theta}\right)^{h/\theta}\left(\frac{\theta}{1+\theta}\right)^{x},$$

$$x = 0, 1, 2, \ldots,$$

$$0<h<\infty$$

$$0<\theta<\infty.$$

$$G(z) = (1+\theta-\theta z)^{-h/\theta}.$$

$$\mu = h,\ \sigma^2 = h(1+\theta) = \mu(1+\theta).$$

In view of the defining relation between the two distributions, further details about this distribution will be found under the negative binomial distribution with parameters k and p. Originally, the Polya Eggenberger distribution was obtained as a limit of the Polya distribution with parameters p, γ and n as $p\to 0$, $\gamma\to 0$ and $n\to\infty$ such that $np\to h$ and $n\gamma\to\theta$.

A reference Anscombe (1950*a*).

17. THE LOGARITHMIC DISTRIBUTION

(Fisher's logarithmic distribution, Logarithmic series distribution)
The limit of the zero-truncated negative binomial distribution with parameters k and $1-\theta$ as $k \to 0$ is the logarithmic distribution with parameter θ and has the pf

$$P(x) = \frac{\alpha\theta^x}{x}, \qquad x = 1, 2, \ldots$$

$$0<\theta<1,$$

$$\text{where } \alpha = \frac{-1}{\log(1-\theta)}.$$

$$G(z) = \frac{\log(1-\theta z)}{\log(1-\theta)}$$

$$\mu = \frac{\alpha\theta}{1-\theta}, \quad \sigma^2 = \frac{\alpha\theta(1-\alpha\theta)}{(1-\theta)^2} = \mu\left(\frac{1}{1-\theta} - \mu\right).$$

Some inter-relations

(i) The negative binomial distribution with parameters k and p results if the Poisson distribution with parameter $-k \log p$ is generalized by the logarithmic distribution with parameter $1-p$.

(ii) The logarithmic distribution with parameter θ is the PSD with parameter $\theta = \theta$ and the series function $f(\theta) = -\log(1-\theta)$.

Some references Fisher, Corbet and Williams (1943); Patil, Kamat and Wani (1964); Patil and Shorrock (1966).

18. THE MODIFIED LOGARITHMIC DISTRIBUTION

(Modified logarithmic series distribution)
The modified logarithmic distribution with parameters δ and θ results if the binomial distribution with parameters $n = 1$ and $p = 1-\delta$ is generalized by the logarithmic distribution with parameter θ. It has the pf

$$P(x) = \begin{cases} \delta, & x = 0, \\ (1-\delta)\dfrac{\alpha\theta^x}{x}, & x = 1, 2, \ldots, \end{cases}$$

$$0 \leqq \delta < 1,$$

$$0<\theta<1,$$

$$\text{where } \alpha = \frac{-1}{\log(1-\theta)}.$$

$$G(z) = \delta + (1-\delta)\frac{\log(1-\theta z)}{\log(1-\theta)}.$$

$$\mu = (1-\delta)\left(\frac{\alpha\theta}{1-\theta}\right), \quad \sigma^2 = \frac{(1-\delta)\alpha\theta}{(1-\theta)^2}[1-\alpha\theta(1-\delta)]$$

$$= \frac{\mu}{(1-\theta)}[1-\mu(1-\theta)].$$

Some inter-relations

(i) The mixture on n of the binomial distribution with parameters n and p, where n has the logarithmic distribution with parameter θ, is the modified logarithmic distribution with parameters δ and ϕ, where $\delta = -\alpha \log[1-\theta(1-p)]$ and $\phi = \theta p/[1-\theta(1-p)]$.

(ii) If $X_1, X_2, \ldots, X_s$ have the s-dimensional logarithmic distribution with parameters $\theta_1, \theta_2, \ldots, \theta_s$, then X_1 has the modified logarithmic distribution with parameters

$$\delta = [-\log(1-\theta_2-\theta_3-\ldots-\theta_s)]/L \text{ and } \theta,$$

where $L = -\log(1-\theta_1-\theta_2-\ldots-\theta_s)$ and $\theta = \theta_1/(1-\theta_2-\theta_3-\ldots-\theta_s)$.

A reference Patil (1964).

19. THE POWER SERIES DISTRIBUTION

Let $f(z) = \sum_{x=0}^{\infty} a_x z^x$ for $|z|<r$, $0<r<\infty$, where a_x are real ($x = 0, 1, 2, \ldots$). Suppose either (i) a_x are all non-negative, or (ii) $(-1)^x a_x$ are all non-negative, where in the first case, we will take $0<z<r$, $\theta = z$ and in the second case, we will take $-r<z<0$, $\theta = -z$ and absorb the negative sign of z in a_x. Then the power series distribution (PSD) with parameter θ and the series function $f(\theta)$ is defined by the pf

$$P(x) = \frac{a_x\theta^x}{f(\theta)}, \qquad x = 0, 1, 2, \ldots, \quad 0<\theta<r$$

$$r>0, \text{ depending on } f(\theta).$$

$$G(z) = \frac{f(\theta z)}{f(\theta)}.$$

$$\mu = \theta\frac{d}{d\theta}[\log f(\theta)], \qquad \sigma^2 = \mu+\theta^2\frac{d^2}{d\theta^2}[\log f(\theta)].$$

Some inter-relations

(i) The binomial distribution with parameters n and p is a PSD with parameter $\theta = \frac{p}{(1-p)}$ and the series function $f(\theta) = (1+\theta)^n$.

(ii) The Poisson distribution with parameter λ is a PSD with parameter $\theta = \lambda$ and the series function $f(\theta) = e^\theta$.

(iii) The negative binomial distribution with parameters k and p is a PSD with parameter $\theta = 1-p$ and the series function $f(\theta) = (1-\theta)^{-k}$.

(iv) The logarithmic distribution with parameter θ is a PSD with parameter $\theta = \theta$ and the series function $f(\theta) = -\log(1-\theta)$.

(v) When $\mu = \sigma^2$, the Poisson truncated normal distribution with parameters μ and σ^2 reduces to

$$P(x) = \frac{e^{-\mu/2}}{\mu\Phi(\mu^{1/2})}\cdot\frac{\left(\frac{\mu}{2}\right)^{\frac{x}{2}+1}}{\Gamma\left(\frac{x}{2}+1\right)},$$

which is a power series distribution.

(vi) The power series distribution with parameter λ is a special case of the general Dirichlet series distribution, where $\lambda_x = x$ and $e^{-\theta} = \lambda$.

Some references Kosambi (1949); Noack (1950); Patil (1959*a*); Patil (1962*d*).

20. THE GENERALIZED POWER SERIES DISTRIBUTION

Let T be a countable subset of the set of real numbers without any limit point. Define $f(\theta) = \Sigma a_x \theta^x$ where the summation extends over T and $a_x > 0$, $\theta \geqq 0$ with $\theta \in \Theta$, the parameter space, such that $f(\theta)$ is finite and differentiable. Then the generalized power series distribution with range T and the series function $f(\theta)$ is defined by the pf

$$P(x) = \frac{a_x \theta^x}{f(\theta)}, \qquad x \in T,\ \theta \in \Theta.$$

$$G(z) = \frac{f(\theta z)}{f(\theta)}.$$

$$\mu = \theta \frac{d}{d\theta}[\log f(\theta)], \quad \sigma^2 = \mu + \theta^2 \frac{d^2}{d\theta^2}[\log f(\theta)].$$

Note that when T is restricted to non-negative integers, $\Theta = \{\theta\colon\ 0 \leq \theta < r\}$ where r is the radius of convergence of the power series of $f(\theta)$. The inter-relations listed under the power series distribution also apply to the generalized power series distribution.

Some references Patil (1959*a*); Patil (1962*d*).

21. THE GENERALIZED POWER SERIES DISTRIBUTION WITH TWO PARAMETERS

The generalized power series distribution with two parameters θ and λ has the pf

$$P(x) = \frac{a_x(\lambda)\theta^x}{f(\theta, \lambda)}, \qquad x \in T,$$

where the series function $f(\theta, \lambda) = \Sigma a_x(\lambda)\theta^x$, the summation extending over T, the range of the generalized power series distribution, such that $f(\theta, \lambda)$ is positive, finite and differentiable for all admissible values of the two parameters θ and λ and the positive coefficients $a_x(\lambda)$ depend on x and λ.

$$G(z) = \frac{f(\theta z, \lambda)}{f(\theta, \lambda)}$$

$$\mu = \theta \frac{\partial}{\partial\theta} \log [f(\theta, \lambda)], \quad \sigma^2 = \theta \frac{\partial\mu}{\partial\theta}.$$

Some inter-relations

(i) The binomial distribution with parameters n and p is the generalized power series distribution with parameters θ and λ and the series function $f(\theta, \lambda) = (1+\theta)^\lambda$ when $\theta = \dfrac{p}{(1-p)}$ and $\lambda = n$.

(ii) The negative binomial distribution with parameters k and p is the generalized power series distribution with parameters θ and λ and the series function $f(\theta, \lambda) = (1-\theta)^{-\lambda}$ when $\theta = 1-p$ and $\lambda = k$.

(iii) The hyper-Poisson distribution with parameters θ and λ is the generalized power series distribution with parameters θ and λ and the series function $f(\theta, \lambda) = {}_1F_1[1;\ \lambda;\ \theta]$.

(iv) The Stirling distribution of the first kind with parameters θ and n is the generalized power series distribution with two parameters θ and λ and the series function $f(\theta, \lambda) = [-\log(1-\theta)]^{\lambda}$, when $\lambda = n$.

(v) The Stirling distribution of the second kind with parameters θ and n is the generalized power series distribution with two parameters θ and λ and the series function $f(\theta, \lambda) = (e^{\theta}-1)^{\lambda}$, when $\lambda = n$.

Some references Patil (1957); Patil (1964*d*).

22. THE GENERAL DIRICHLET SERIES DISTRIBUTION

Consider the general Dirichlet series defined as

$$f(\theta) = \sum_{x=1}^{\infty} a_x \exp(-\lambda_x \theta)$$

where $a_x \geqq 0$, $\theta > 0$ and $\{\lambda_x\}$ is a sequence of real positive increasing numbers whose limit is infinity. Then $f(\theta)$ is positive, finite and differentiable for all $\theta > r$, the abscissa of convergence, that is,

$$r = \limsup_{n\to\infty} \left\{ \frac{\log \sum_{x=k}^{n} a_x}{\lambda_n} \right\}.$$

Then the general Dirichlet series distribution with parameter θ and the series function $f(\theta)$ is defined by the pf

$$P(x) = \frac{a_x \exp(-\lambda_x \theta)}{f(\theta)} \qquad x = 1, 2, \ldots.$$

The pgf, mean and variance are not known.

Some inter-relations

(i) The Dirichlet series distribution is a special case of the general Dirichlet series distribution, where

$$\lambda_x = \log x, \text{ such that } f(\theta) = \sum_{x=1}^{\infty} a_x x^{-\theta} \text{ and } a_x \geqq 0,\ \theta > 0 \text{ and } \Sigma a_x \text{ is bounded.}$$

(ii) The discrete Pareto distribution is a special case of the Dirichlet series distribution, where $a_x = a$ and $\theta > 1$.

(iii) The power series distribution with parameter λ is a special case of the general Dirichlet series distribution, where $\lambda_x = x$ and $e^{-\theta} = \lambda$.

A reference Siromoney (1964).

23. THE STIRLING DISTRIBUTION OF THE FIRST KIND

The Stirling distribution of the first kind with paramters θ and n has the pf

$$P(x) = \frac{\frac{n!}{x!} S_x^n \theta^x}{[-\log(1-\theta)]^n}, \qquad x = n, n+1, \ldots, \quad 0<\theta<1, \quad n = 1, 2, \ldots,$$

where S_x^n is the Stirling number of the first kind with arguments n and x.

$$G(z) = \left[\frac{\log(1-\theta z)}{\log(1-\theta)}\right]^n$$

$$\mu = n\left(\frac{\alpha\theta}{1-\theta}\right), \qquad \sigma^2 = n\left(\frac{\alpha\theta}{1-\theta}\right)\left(\frac{1-\alpha\theta}{1-\theta}\right) \text{ where } \alpha = \frac{-1}{\log(1-\theta)}.$$

Some inter-relations

(i) If $X_1, X_2, \ldots, X_n$ is a random sample of size n from the logarithmic distribution with parameter θ, then the sample total $X_1 + X_2 + \ldots + X_n$ has the Stirling distribution of the first kind with parameters θ and n.

(ii) The Stirling distribution of the first kind with parameters θ and n is a special case of the power series distribution with parameter θ, having the series function $f(\theta) = [-\log(1-\theta)]^n$. It it also a special case of the generalized power series distribution with two parameters.

A reference Patil and Wani (1965).

24. THE STIRLING DISTRIBUTION OF THE SECOND KIND

The Stirling distribution of the second kind with parameters θ and n has the pf

$$P(x) = \frac{\frac{n!}{x!} \mathfrak{S}_x^n \theta^x}{(e^\theta - 1)^n}, \qquad x = n, n+1, \ldots, \quad 0<\theta<\infty, \quad n = 1, 2, \ldots,$$

where $\mathfrak{S}_x^n$ is the Stirling number of the second kind with arguments n and x.

$$G(z) = \left(\frac{e^{\theta z} - 1}{e^\theta - 1}\right)^n$$

$$\mu = \frac{\theta e^\theta}{e^\theta - 1}, \qquad \sigma^2 = \mu(1+\theta-\mu)$$

Some inter-relations

(i) If $X_1, X_2, \ldots, X_n$ is a random sample of size n from the zero-truncated Poisson distribution with θ, then the sample total $X_1 + X_2 + \ldots + X_n$ has the Stirling distribution of the second kind with parameters θ and n.

(ii) The Stirling distribution of the second kind with parameters θ and n is a special case of the power series distribution with parameter θ, having the series function $f(\theta) = (e^{\theta}-1)^n$. It is also a special case of the generalized power series distribution with two parameters.

Some references Roy and Mitra (1957); Tate and Goen (1958).

25. THE INVERSE FACTORIAL SERIES DISTRIBUTION

Suppose $\phi(t)$ has the power series expansion in $(1-t)$, i.e.

$$\phi(t) = a_0 + a_1(1-t) + \frac{a_2(1-t)^2}{2!} + \dots.$$

Then

$$\int_0^1 t^{\theta-1}\phi(t)dt = \frac{a_0}{\theta} + \frac{a_1}{\theta(\theta+1)} + \frac{a_2}{\theta(\theta+1)(\theta+2)} + \dots.$$

Let

$$\Omega(\theta) = \int_0^1 t^{\theta-1}\phi(t)dt.$$

If

$$\frac{a_i}{\theta(\theta+1)(\theta+2)\dots(\theta+i)\Omega(\theta)} \geqq 0, \quad i = 0, 1, 2, \dots,$$

then the inverse factorial series distribution with the series function ϕ and parameter θ is defined by the pf

$$P(x) = \frac{a_x}{\theta(\theta+1)(\theta+2)\dots(\theta+x)\Omega(\theta)} \qquad x = 0, 1, 2, \dots, \quad 0<\theta<\infty.$$

Some inter-relations

(i) The Waring distribution with parameters k and ρ is a special case of the inverse factorial series distribution with series function ϕ and parameter θ, when $\phi(t) = t^{-k}$ and $\theta = \rho+k$.

(ii) The Yule distribution with parameter ρ is a special case of the inverse factorial series distribution with series function ϕ and parameter θ, when $\phi(t) = t^{-1}$ and $\theta = \rho+1$.

Some references: Irwin (1963); Irwin (1965).

26. THE HYPERGEOMETRIC DISTRIBUTION

The number of objects of type 1 in a random sample of size n taken without replacement has the hypergeometric distribution with parameters N, M and n if the sample is taken from a population of N objects of which M are of type 1. It has the pf

$$P(x) = \frac{\binom{M}{x}\binom{N-M}{n-x}}{\binom{N}{n}}, \qquad x = 0, 1, 2, \dots, n \text{ such that } x \leqq M \text{ and } (n-x) \leqq (N-M),$$

$$N = 1, 2, \ldots,$$

$$M = 1, 2, \ldots;\ M < N,$$

$$n = 1, 2, \ldots;\ n < N.$$

$$G(z) = \frac{(N-M)^{(n)}}{(N)^{(n)}}\ {}_2F_1(-n, -M;\ N-M-n+1;\ z), \text{ if } N-M-n+1>0.$$

$$\mu = n\left(\frac{M}{N}\right), \qquad \sigma^2 = n\left(\frac{M}{N}\right)\left(1-\frac{M}{N}\right)\left(\frac{N-n}{N-1}\right)$$

$$= \mu\left(1-\frac{M}{N}\right)\left(\frac{N-n}{N-1}\right).$$

Some inter-relations

(i) The hypergeometric distribution with parameters N, M and n is a special case of the Polya distribution with parameters M, $N-M$, c and n when $c = -1$.

(ii) The hypergeometric distribution with parameters N, M and n tends to the binomial distribution with parameters n and p as $N \to \infty$, $M \to \infty$ such that $\frac{M}{N} \to p$, $0<p<1$.

(iii) The hypergeometric distribution with parameters N, M and n tends to the Poisson distribution with parameter λ as $M \to \infty$, $N \to \infty$, $n \to \infty$ such that $\frac{M}{N} \to 0$ and $n\frac{M}{N} \to \lambda$.

(iv) The hypergeometric distribution with parameters N, M, n is a special case of the extended hypergeometric distribution with parameters M, $N-M$, n, θ when $\theta = 1$.

(v) The generalized hypergeometric distribution type IA (ii) with parameters a, b and n is the same as the hypergeometric distribution with parameters $a+b$, a and n when a, b and n are positive integers.

Some references Feller (1957), p. 41; Kendall and Stuart (1958), p. 133; Tsao (1965).

27. THE INVERSE HYPERGEOMETRIC DISTRIBUTION

(The hypergeometric waiting time distribution)

Consider an urn containing N balls, M white and $N-M$ black. Then the number of random draws required without replacement to obtain k white balls has the inverse hypergeometric distribution, form (i) with parameters N, M and k. In form (ii), the inverse hypergeometric distribution with parameters N, M and k is the distribution of the number of black balls encountered in order to obtain k white balls. The two pfs are given by

Form (i): $$P(x) = \frac{\binom{x-1}{k-1}\binom{N-x}{M-k}}{\binom{N}{M}}, \qquad \begin{aligned} &x = k, k+1, k+2, \ldots, N-M+k, \\ &N = 1, 2, \ldots, \\ &M = 1, 2, \ldots; M<N, \\ &k = 1, 2, \ldots, M. \end{aligned}$$

Form (ii): $$P(x) = \frac{\binom{M}{k-1}\binom{N-M}{x}(M-k+1)}{\binom{N}{k+x-1}(N-k+1-x)}, \qquad \begin{aligned} &x = 0, 1, 2, \ldots, N-M, \\ &N = 1, 2, \ldots, \\ &M = 1, 2, \ldots; M<N, \\ &k = 1, 2, \ldots, M. \end{aligned}$$

Unless otherwise stated the inverse hypergeometric distribution will mean form (i).

Form (i)	Form (ii)
$\mu = \dfrac{k(N+1)}{M+1}.$	$\mu = \dfrac{k(N-M)}{M+1}.$
$\sigma^2 = k\left(\dfrac{N-M}{M+1}\right)\left(\dfrac{(M+1-k}{M+1}\right)\left(\dfrac{N+1}{M+2}\right)$	$\sigma^2 = k\left(\dfrac{N-M}{M+1}\right)\left(\dfrac{M+1-k}{M+1}\right)\left(\dfrac{N+1}{M+2}\right)$

Some inter-relations

(i) The inverse hypergeometric distribution, form (ii), with parameters N, M and k is the same as the generalized hypergeometric distribution type IIIA with integral valued parameters $N-M$, $-N-1$ and $-k$.

(ii) As $N\to\infty$, $M\to\infty$ such that $\dfrac{M}{N}\to p$, the inverse hypergeometric distribution, form (ii), with parameters N, M and k tends to the negative binomial distribution with parameters k and p.

(iii) The inverse hypergeometric distribution, form (ii), with parameters N, M and k is the same as the negative hypergeometric distribution with parameters $M+1$, k and $N-M$.

(iv) The inverse hypergeometric distribution with parameters N, M and k is a special case of the univariate negative factorial distribution with parameters s, N, M_1, M_2, ..., M_s, k when $s=1$ and $M_1=N-M$.

Some references Kemp and Kemp (1956*b*); Wilks (1962), p. 141.

28. THE NEGATIVE HYPERGEOMETRIC DISTRIBUTION

The mixture on p of the binomial distribution with parameters n and p is the negative hypergeometric distribution with parameters N, M and n if p has the beta distribution with parameters M and $N-M$. It has the pf

$$P(x) = \frac{\binom{-M}{x}\binom{-N+M}{n-x}}{\binom{-N}{n}}, \qquad x = 0, 1, 2, \ldots, n,$$

$$0 < N < \infty,$$

$$0 < M < N,$$

$$n = 1, 2, \ldots.$$

A simple form of the pgf is not known.

$$\mu = n\left(\frac{M}{N}\right), \qquad \sigma^2 = n\left(\frac{M}{N}\right)\left(\frac{N-M}{N}\right)\left(\frac{N+n}{N+1}\right).$$

Some inter-relations

(i) The Polya distribution with parameters a, b, 1 and n is the same as the negative hypergeometric distribution with parameters $a+b$, a and n.

(ii) The generalized hypergeometric distribution type IIA with parameters a, b and n is the same as the negative hypergeometric distribution with parameters $-a-b$, $-a$ and n, where a and b are integers.

(iii) Let (X_1, X_2) have the double hypergeometric distribution with parameters N, M, n_1 and n_2 and let M be a random variable having a uniform distribution over the range $(0, N)$. Then the conditional distribution of X_2 given $X_1 = x_1$ is negative hypergeometric with parameters n_1+2, x_1+1 and n_2.

(iv) As $N \to \infty$, $M \to \infty$ such that $\dfrac{M}{N} \to p$, the negative hypergeometric distribution with parameters N, M and n tends to the binomial distribution with parameters n and p.

(v) The inverse hypergeometric distribution, form (ii), with parameters N, M and k is the same as the negative hypergeometric distribution with parameters $M+1$, k and $N-M$.

(vi) As $N \to \infty$, $n \to \infty$ such that $\dfrac{N}{N+n} \to p$, $0<p<1$, the negative hypergeometric distribution with parameters N, M and n tends to the negative binomial distribution with parameters M and p.

Some references Kemp and Kemp (1956*b*); Särndal (1965); Skellam (1948).

29. THE POLYA DISTRIBUTION

Consider an urn containing a white balls and b black balls. If n balls ared drawn at random, from the urn, where each ball drawn is replaced together with additional c balls of the same colour before the next ball is drawn, then the number of white balls drawn in the sample of size n has the Polya distribution with parameters a, b, c and n with the pf

$$P(x) = \binom{n}{x} \frac{a[a+c]\ldots[a+(x-1)c]b[b+c]\ldots[b+(n-x-1)c]}{[a+b][a+b+c]\ldots[a+b+(n-1)c]}, \quad x = 0, 1, 2, \ldots, n,$$

$$a = 1, 2, \ldots,$$

$$b = 1, 2, \ldots,$$

$$c = 1, 2, \ldots,$$

$$n = 1, 2, \ldots.$$

Letting $p = \dfrac{a}{(a+b)}$, $q = \dfrac{b}{(a+b)}$, $\gamma = \dfrac{c}{(a+b)}$, we obtain a more convenient form with parameters p, γ and n, namely,

$$P(x) = \frac{\binom{-p/\gamma}{x}\binom{-q/\gamma}{n-x}}{\binom{-1/\gamma}{n}}, \quad x = 0, 1, 2, \ldots, n,$$

$$0<\gamma<\infty,$$

$$0<p<1, \; q = 1-p,$$

$$n = 1, 2, \ldots.$$

$$G(z) = \frac{(-q/\gamma)^{(n)}}{(-1/\gamma)^{(n)}} {}_2F_1(-n, p/\gamma; \; -q/\gamma-n+1; \; z).$$

$$\mu = n\left(\frac{a}{a+b}\right) = np, \quad \sigma^2 = \frac{nab}{(a+b)^2}\frac{(a+b+nc)}{(a+b+c)} = \frac{npq(1+n\gamma)}{1+\gamma} = \frac{\mu q(1+n\gamma)}{1+\gamma}$$

Note Bosch (1963) considers a more general case by allowing $c \leqq 0$.

Some inter-relations

(i) The Polya distribution with parameters a, b, 0 and n is the same as the binomial distribution with parameters n and $\dfrac{a}{(a+b)}$.

(ii) The Polya distribution with parameters a, b, -1 and n is the same as the hypergeometric distribution with parameters $a+b$, a and n.

(iii) The Polya distribution with parameters a, b, c, n is the same as the uniform distribution with range $\{0, 1, 2, \ldots, n\}$ when $a = b = c$.

(iv) The Polya distribution with parameters a, b, 1 and n is the same as the negative hypergeometric distribution with parameters $a+b$, a and n.

(v) The Polya distribution with parameters p, γ and n tends to the negative binomial distribution with parameters $\frac{h}{\theta}$ and $\frac{1}{(1+\theta)}$ as $n\to\infty$, $p\to 0$ and $\gamma\to 0$ such that $np\to h$ and $n\gamma\to\theta$.

(vi) The Polya distribution with parameter p, γ and n tends to the Poisson distribution with parameter λ as $n\to\infty$, $p\to 0$ and $\gamma\to 0$ such that $np\to\lambda$ and $n\gamma\to 0$.

(vii) The Polya distribution with parameters a, b, 1, n has the same pf as the distribution of exceedances with parameters $a+b-1$, n and a.

(viii) The Polya distribution with parameters a, b, 1 and n is the same as the generalized hypergeometric distribution type IIA with parameters $-a$, $-b$ and n.

Some references Bosch (1963); Sarkadi (1957*b*).

30. THE INVERSE POLYA DISTRIBUTION

The number of black balls drawn from an urn containing a white balls and b black balls, where the balls are drawn at random replacing each ball along with c additional balls of the same colour before drawing the next ball, until k white balls are drawn has the inverse Polya distribution with parameters a, b, c and k. It has the pf

$$P(x) = \binom{k+x-1}{x}\frac{a[a+c]\ldots[a+(k-1)c]b[b+c]\ldots[b+(x-1)c]}{[a+b][a+b+c]\ldots[a+b+(k+x-1)c]}, \quad x = 0, 1, 2, \ldots,$$

$$a = 1, 2, \ldots,$$

$$b = 1, 2, \ldots,$$

$$c = 1, 2, \ldots,$$

$$k = 1, 2, \ldots.$$

Letting $p = \frac{a}{(a+b)}$, $q = \frac{b}{(a+b)}$, and $\gamma = \frac{c}{(a+b)}$, we obtain

$$P(x) = \left(\frac{k}{k+x}\right)\frac{\binom{-p/\gamma}{k}\binom{-q/\gamma}{x}}{\binom{-1/\gamma}{k+x}}$$

$$= \frac{\binom{-q/\gamma}{x}\binom{1/\gamma-1}{-k-x}}{\binom{p/\gamma-1}{-k}}$$

which is the inverse Polya distribution with parameters p, γ and k. A more general case of $c \neq$ can be considered.

$$G(z) = \frac{\left(\frac{-p}{\gamma}\right)^{(k)}}{\left(-\frac{1}{\gamma}\right)^{(k)}} \cdot {}_2F_1\left(k, \frac{q}{\gamma}; \frac{1}{\gamma}+k; z\right)$$

$$\mu = \begin{cases} \dfrac{kb}{a-c} = \dfrac{kq}{p-\gamma} \text{ if } a>c, \text{ i.e. } p>\gamma. \\ \text{does not exist otherwise.} \end{cases}$$

$$\sigma^2 = \begin{cases} \left(\dfrac{kb}{a-c}\right)\left(\dfrac{a+b-c}{a-c}\right)\left[\dfrac{a+(k-1)c}{a-2c}\right], \text{ if } a>2c, \text{ i.e. if } p>2\gamma. \\ \text{does not exist otherwise.} \end{cases}$$

Some inter-relations

The inverse Polya distribution with parameters a, b, 1 and k is the same as the generalized hypergeometric distribution type IV with integer-valued parameters $-b$, $a+b-1$ and $-k$.

Some references Sarkadi (1957*b*); von Schelling (1949*a*).

31. THE GENERALIZED HYPERGEOMETRIC DISTRIBUTION TYPE IA (i)

The pf of the generalized hypergeometric distributian type IA (i) with parameters a, b and n is given by

$$P(x) = \frac{\binom{a}{x}\binom{b}{n-x}}{\binom{a+b}{n}}, \qquad x = 0, 1, 2, \ldots, n,$$
$$a>n-1,$$
$$b>n-1,$$
$$n = 1, 2, \ldots.$$

$$G(z) = \frac{(a+b-n)!\,b!}{(a+b)!\,(b-n)!} \cdot {}_2F_1(-a, -n; b-n+1; z).$$

$$\mu = \frac{an}{a+b}, \quad \sigma^2 = \frac{abn}{(a+b)^2}\frac{(a+b-n)}{(a+b-1)} = \frac{\mu b}{(a+b)}\frac{(a+b-n)}{(a+b-1)}.$$

Some inter-relations

(i) The generalized hypergeometric distribution type IA (i) with parameters a, b and n is the same as the hypergeometric distribution with parameters $a+b$, a and n when a, b and n are positive integers.

(ii) As $a \to \infty$, $b \to \infty$ such that $\dfrac{a}{(a+b)} \to p$, the generalized hypergeometric distribution type IA (i) with parameters a, b and n tends to the binomial distribution with parameters n and p.

(iii) As $a \to \infty$, $b \to \infty$ and $n \to \infty$, such that $\dfrac{na}{(a+b)} \to \lambda$, the generalized hypergeometric distribution type IA (i) with parameters a, b and n tends to the Poisson distribution with parameter λ.

(iv) The generalized hypergeometric distribution type IA (i) with parameters a, b and n is the same as the generalized hypergeometric distribution type IA (ii) with parameters n, a and $a+b-n$.

Some references Kemp and Kemp (1956*b*); Sarkadi (1957*b*).

32. THE GENERALIZED HYPERGEOMETRIC DISTRIBUTION TYPE IA (ii)

The pf of the generalized hypergeometric distribution type IA (ii) with parameters a, b and n is given by

$$P(x) = \frac{\binom{a}{x}\binom{b}{n-x}}{\binom{a+b}{n}} \qquad \begin{array}{l} x = 0, 1, 2, \ldots, a, \\ a = 1, 2, \ldots; \; a<n-1, \\ b>n-1. \end{array}$$

$$G(z) = \frac{(a+b-n)!\,b!}{(a+b)!\,(b-n)!}\cdot {}_2F_1(-a,-n;b-n+1;z).$$

$$\mu = \frac{an}{a+b}, \quad \sigma^2 = \frac{abn}{(a+b)^2}\,\frac{(a+b-n)}{(a+b-1)} = \frac{\mu b}{(a+b)}\,\frac{(a+b-n)}{(a+b-1)}.$$

Some inter-relations

(i) The generalized hypergeometric distribution type IA (ii) with parameters a, b and n is the same as the hypergeometric distribution with parameters $a+b$, a and n when a, b and n are positive integers.

(ii) As $a\to\infty$, $b\to\infty$ and $n\to\infty$ such that $\dfrac{na}{(a+b)}\to\lambda$, the generalized hypergeometric distribution type IA (ii) tends to the Poisson distribution with parameter λ.

(iii) The generalized hypergeometric distribution type IA (ii) with parameters a, b and n is the same as the generalized hypergeometric distribution type IA (i) with parameters b, $n+a-b$ and a.

Some references Kemp and Kemp (1956*b*); Sarkadi (1957*b*).

33. THE GENERALIZED HYPERGEOMETRIC DISTRIBUTION TYPE IB

The pf of the generalized hypergeometric distribution type IB with parameters a, b and n is given by

$$P(x) = \frac{\binom{a}{x}\binom{b}{n-x}}{\binom{a+b}{n}} \qquad \begin{array}{l} x = 0, 1, 2, \ldots, \\ n-b-1<0, \\ [a] = [n]. \end{array}$$

$$G(z) = \frac{(a+b-n)!\,b!}{(a+b)!\,(b-n)!}\cdot {}_2F_1(-a,-n;b-n+1;z).$$

$$\mu = \begin{cases} \dfrac{an}{a+b} & \text{if } a+b>0 \\ \text{does not exist otherwise.} \end{cases} \qquad \sigma^2 = \begin{cases} \dfrac{abn}{(a+b)^2}\,\dfrac{(a+b-n)}{(a+b-1)} & \text{if } a+b>1 \\ \text{does not exist otherwise.} \end{cases}$$

Some inter-relations

As $a \to \infty$, $b \to \infty$ and $n \to \infty$ such that $\dfrac{na}{(a+b)} \to \lambda$, the generalized hypergeometric distribution type IB tends to the Poisson distribution with parameter λ.

Some references Kemp and Kemp (1956*b*); Sarkadi (1957*b*).

34. THE GENERALIZED HYPERGEOMETRIC DISTRIBUTION TYPE IIA

The pf of the generalized hypergeometric distribution type IIA with parameters a, b and n is given by

$$P(x) = \frac{\binom{a}{x}\binom{b}{n-x}}{\binom{a+b}{n}} \qquad \begin{array}{l} x = 0, 1, 2, \ldots, n, \\ a<0, \\ b<0;\ b \neq -1, \\ n = 1, 2, \ldots. \end{array}$$

$$G(z) = \frac{(a+b-n)!\,b!}{(a+b)!\,(b-n)!} \cdot {}_2F_1(-a, -n; b-n+1; z).$$

$$\mu = \frac{an}{a+b}, \qquad \sigma^2 = \frac{abn}{(a+b)^2}\,\frac{(a+b-n)}{(a+b-1)}$$

Some inter-relations

(i) The generalized hypergeometric distribution type IIA with parameters a, b and n is the same as the negative hypergeometric distribution with parameters $-a-b$, $-a$ and n.

(ii) The Polya distribution with parameters a, b, 1 and n is the same as the generalized hypergeometric distribution type IIA with parameters $-a$, $-b$, $-a$ and n.

(iii) The mixture on p of the binomial distribution with parameters n and p is the generalized hypergeometric distribution type IIA with parameters a, $-b$ and n if p has a beta distribution with parameters a and b.

(iv) As $a \to -\infty$ and $b \to -\infty$ such that $\dfrac{a}{(a+b)} \to p$, the generalized hypergeometric distribution type IIA with parameters a, b and n tends to the binomial distribution with parameters n and p.

(v) As $a \to -\infty$, $b \to -\infty$, $n \to \infty$ such that $\dfrac{na}{(a+b)} \to \lambda$, the generalized hypergeometric distribution type IIA with parameters a, b and n tends to the Poisson distribution with parameter λ.

(vi) The generalized hypergeometric distribution type IIA with parameters a, b and n is the same as the generalized hypergeometric distribution type IIIA with parameters n, a and $a+b-n$.

Some references Kemp and Kemp (1956*b*); Sarkadi (1957*b*).

35. THE GENERALIZED HYPERGEOMETRIC DISTRIBUTION TYPE IIB

The pf of the generalized hypergeometric distribution type IIB with parameters a, b and n is given by

$$P(x) = \frac{\binom{a}{x}\binom{b}{n-x}}{\binom{a+b}{n}} \qquad \begin{aligned} &x = 0, 1, 2, \ldots, \\ &a<0, \\ &0<a+b+1, \\ &[n-b-1] = [n]. \end{aligned}$$

$$G(z) = \frac{(a+b-n)!\,b!}{(a+b)!\,(b-n)!} \cdot {}_2F_1(-a, -n; b-n+1; z).$$

The mean and the variance do not exist.

Some inter-relations

The generalized hypergeometric distribution type IIB with parameters a, b and n is the same as the generalized hypergeometric distribution type IIIB with parameters n, a and $a+b-n$.

Some references Kemp and Kemp (1956*b*); Sarkadi (1957*b*).

36. THE GENERALIZED HYPERGEOMETRIC DISTRIBUTION TYPE IIIA

The pf of the generalized hypergeometric distribution type IIIA with parameters a, b and n is given by

$$P(x) = \frac{\binom{a}{x}\binom{b}{n-x}}{\binom{a+b}{n}} \qquad \begin{aligned} &x = 0, 1, 2, \ldots, a, \\ &a = 1, 2, \ldots, \\ &b<n-a;\ b \neq n-a-1, \\ &n<0. \end{aligned}$$

$$G(z) = \frac{(a+b-n)!\,b!}{(a+b)!\,(b-n)!} \cdot {}_2F_1(-a, -n; b-n+1; z).$$

$$\mu = \frac{an}{a+b}, \qquad \sigma^2 = \frac{abn}{(a+b)^2}\,\frac{(a+b-n)}{(a+b-1)}.$$

Some inter-relations

(i) The generalized hypergeometric distribution type IIIA with parameters a, b and n is the same as the generalized hypergeometric distribution type IIA with parameters n, a and $a+b-n$.

(ii) As $a \to \infty$, $b \to -\infty$ such that $\frac{(a+b)}{b} \to p$, $0<p<1$ the generalized hypergeometric distribution type IIIA with parameters a, b and n tends to the negative binomial distribution with parameters $|n|$ and p.

(iii) As $a \to \infty$, $b \to -\infty$ and $n \to -\infty$ such that $\left|\frac{na}{a+b}\right| \to \lambda$, the generalized hypergeometric distribution type IIIA with parameters a, b and n tends to the Poisson distribution with parameter λ.

(iv) The inverse hypergeometric distribution form (ii) with parameters N, M and k is the same as the generalized hypergeometric distribution type IIIA with integral valued parameters $N-M$, $-N-1$ and $-k$.

Some references Kemp and Kemp (1956*b*); Sarkadi (1957*b*).

37. THE GENERALIZED HYPERGEOMETRIC DISTRIBUTION TYPE IIIB

The pf of the generalized hypergeometric distribution type IIIB with parameters a, b and n is given by

$$P(x) = \frac{\binom{a}{x}\binom{b}{n-x}}{\binom{a+b}{n}} \qquad \begin{aligned} &x = 0, 1, 2, \ldots, \\ &0<a+b+1, \\ &n<0, \\ &[n-b-1] = [a]. \end{aligned}$$

$$G(z) = \frac{(a+b-n)!\,b!}{(a+b)!\,(b-n)!} \cdot {}_2F_1(-a,-n;b-n+1;z).$$

The mean and variance do not exist.

Some inter-relations

The generalized hypergeometric distribution type IIIB with parameters a, b and n is the same as the generalized hypergeometric distribution type IIB with parameters n, a and $a+b-n$.

Some references Kemp and Kemp (1956*b*); Sarkadi (1957*b*).

38. THE GENERALIZED HYPERGEOMETRIC DISTRIBUTION TYPE IV

The pf of the generalized hypergeometric distribution type IV with parameters a, b and n is given by

$$P(x) = \frac{\binom{a}{x}\binom{b}{n-x}}{\binom{a+b}{n}} \qquad \begin{aligned} &x = 0, 1, 2, \ldots, \\ &a<0, \\ &0<a+b+1, \\ &n<0. \end{aligned}$$

$$G(z) = \frac{(a+b-n)!\,b!}{(a+b)!\,(b-n)!} \cdot {}_2F_1(-a,-n;b-n+1;z)$$

$$\mu = \begin{cases} \dfrac{an}{a+b} \text{ if } a+b>0 \\ \text{does not exist otherwise.} \end{cases} \qquad \sigma^2 = \begin{cases} \dfrac{abn}{(a+b)^2}\dfrac{(a+b-n)}{(a+b-1)} \text{ if } a+b>1 \\ \text{does not exist otherwise.} \end{cases}$$

Some inter-relations

(i) The generalized hypergeometric distribution type IV with parameters a, b and 1 where a and b are integers is known as the Miller distribution.

(ii) The inverse Polya distribution with parameters a, b, 1 and k is the same as the generalized hypergeometric distribution type IV with integer valued parameters $-b$, $a+b-1$ and $-k$.

(iii) The Waring distribution with parameters k and ρ is a special case of the generalized hypergeometric distribution type IV with parameters a, b and n when $a = -1$, $b = \rho$ and $n = -k$.

(iv) The mixture on p of the negative binomial distribution with parameters k and p is the generalized hypergeometric distribution type IV with parameters $-a$, b and $-k$ if p has the beta distribution with parameters $-a+b+1$ and a. (Negative binomial beta distribution; Pascal beta distribution.)

(v) As $a \to -\infty$, $b \to \infty$ such that $a+b>0$ and $\dfrac{(a+b)}{b} \to p$, $0<p<1$, the generalized hypergeometric distribution type IV with parameters a, b and n tends to the negative binomial distribution with parameters $|n|$ and p.

(vi) As $a \to -\infty$, $b \to \infty$ and $n \to -\infty$ such that $\left|\dfrac{na}{a+b}\right| \to \lambda > 0$, the generalized hypergeometric distribution type IV with parameters a, b and n tends to the Poisson distribution with parameter λ.

(vii) The factorial distribution with parameters n and λ is a special case of the generalized hypergeometric distribution type IV with parameters -1, $n-1$, $-\lambda+n-1$.

Some references Kemp and Kemp (1956*b*); Sarkadi (1957*b*); Dubey (1966).

39. THE UNIVARIATE MULTINOMIAL DISTRIBUTION

Consider a sequence of n independent trials, each trial having $s+1$ mutually exclusive outcomes $A_0, A_1, A_2, \ldots, A_s$ with probabilities $p_0, p_1, p_2, \ldots, p_s$ respectively. Suppose occurrence of A_i in a single trial is interpreted as i successes, $i = 1, 2, \ldots, s$ and that of A_0 in a trial as a failure. If X is the number of successes in the n trials, then X has the univariate multinomial distribution with parameters $s, n, p_1, p_2, \ldots, p_s$.

$$P(x) = \Sigma \frac{n!}{r_0!r_1!r_2!\ldots r_s!} p_0^{r_0} p_1^{r_1} p_2^{r_2} \ldots p_s^{r_s}, \qquad x = 0, 1, 2, \ldots, ns,$$

$$s = 1, 2, \ldots,$$

$$n = 1, 2, \ldots,$$

$$0<p_i<1; \; i = 1, 2, \ldots, s,$$

$$\sum_{i=1}^{s} p_i<1,$$

where $p_0 = 1-p_1-p_2-\ldots p_s$ and the summation extends over all non-negative integers r_0, r_1, r_2, $\ldots$, r_s such that $\sum_{i=0}^{s} r_i = n$ and $\sum_{i=1}^{s} ir_i = x$.

$$G(z) = (p_0+p_1z+p_2z^2+\ldots+p_sz^s)^n.$$

$$\mu = n \sum_{i=1}^{s} ip_i.$$

$$\sigma^2 = n\left[\sum_{i=1}^{s} i^2p_i - \left(\sum_{i=1}^{s} ip_i\right)^2\right].$$

Some inter-relations:

(i) The binomial distribution with parameters n and p is a special case of the univariate multinomial distribution with parameters $s, n, p_1, p_2, \ldots, p_s$ when $s = 1$ and $p_1 = p$.

(ii) The univariate factorial multinomial distribution with parameters $s, N, M_1, M_2, \ldots, M_s, n$ tends to the univariate multinomial distribution with parameters $s, n, p_1, p_2, \ldots, p_s$ as $N \to \infty$, $M_i \to \infty$ such that $\frac{M_i}{N} \to p_i$, $0 < p_i < 1$, $i = 1, 2, \ldots, s$.

A reference Steyn (1956).

40. THE UNIVARIATE NEGATIVE MULTINOMIAL DISTRIBUTION

Suppose a trial has $s+1$ mutually exclusive outcomes $A_o, A_1, A_2, \ldots, A_s$ with probabilities p_0 $p_1, p_2, \ldots, p_s$ respectively, where A_0 is interpreted as a failure and A_i as i successes, $i = 1, 2, \ldots, s$. Consider a sequence of such independent trials until exactly k failures are obtained. Then the number X of successes obtained in the sequence has the univariate negative multinomial distribution with parameters $s, k, p_1, p_2, \ldots, p_s$. It has the pf

$$P(x) = \Sigma \frac{(k+r-1)!}{r_1! r_2! \ldots r_s! (k-1)!} p_0^k p_1^{r_1} p_2^{r_2} \ldots p_s^{r_s}, \quad x = 0, 1, 2, \ldots,$$

$$s = 1, 2, \ldots,$$
$$k = 1, 2, \ldots,$$
$$0 < p_i < 1; \; i = 1, 2, \ldots, s,$$
$$\sum_{i=1}^{s} p_i < 1,$$

where $p_0 = 1 - p_1 - p_2 - \ldots - p_s$, $r = r_1 + r_2 + \ldots + r_s$ and the summation extends over all non-negative integers $r_1, r_2, \ldots, r_s$ such that $\sum_{i=1}^{s} i r_i = x$.

$$G(z) = p_0^k (1 - p_1 z - p_2 z^2 - \ldots - p_s z^s)^{-k}.$$

$$\mu = \frac{k}{p_0} \sum_{i=1}^{s} i p_i.$$

$$\sigma^2 = \frac{k}{p_0} \left[\left(\sum_{i=1}^{s} i p_i \right)^2 + p_0 \sum_{i=1}^{s} i^2 p_i \right].$$

Some inter-relations

(i) The Pascal distribution with parameters k and p is a special case of the univariate negative multinomial distribution with parameters $s, k, p_1, p_2, \ldots, p_s$, when $s = 1$ and $p_1 = 1 - p$.

(ii) The univariate negative factorial multinomial distribution with parameters $s, N, M_1, \ldots, M_s, k$ tends to the univariate negative multinomial distribution with parameters $s, k, p_1, p_2, \ldots, p_s$ as $N \to \infty$, $M_i \to \infty$ such that $\frac{M_i}{N} \to p_i$, $0 < p_i < 1$, $i = 1, 2, \ldots, s$; $\sum_{i=1}^{s} p_i < 1$.

A reference Steyn (1956).

41. THE UNIVARIATE FACTORIAL MULTINOMIAL DISTRIBUTION

Consider a finite population of N individuals divided into $s+1$ classes $A_0, A_1, A_2, \ldots, A_s$ containing $M_0, M_1, M_2, \ldots, M_s$ individuals respectively. Suppose selection of an individual from A_i is interpreted as i successes, $i = 1, 2, \ldots, s$ and that from A_0 is interpreted as a failure. Suppose n individuals are selected at random from the population without replacement. Then, the number of successes in the n observations has the univariate factorial multinomial distribution with parameters s, N, $M_1, M_2, \ldots, M_s, n$. It has the pf

$$P(x) = \sum \frac{\binom{M_0}{r_0}\binom{M_1}{r_1}\binom{M_2}{r_2}\cdots\binom{M_s}{r_s}}{\binom{N}{n}}, \qquad \begin{aligned} &x = 0, 1, 2, \ldots, \\ &s = 1, 2, \ldots, \\ &N = 1, 2, \ldots, \\ &M_i = 1, 2, \ldots;\ i = 1, 2, \ldots, s, \\ &\sum_{i=1}^{s} M_i < N, \\ &n = 1, 2, \ldots, \end{aligned}$$

where $M_0 = N - M_1 - M_2 - \ldots - M_s$ and the summation extends over all non-negative integers $r_0, r_1, r_2, \ldots, r_s$ such that $\sum_{i=0}^{s} r_i = n$ and $\sum_{i=0}^{s} ir_i = x$.

$$G(z) = \frac{M_0^{(n)}}{N^{(n)}}\left[{}_{(s+1)}F_1(-n, -M_1, -M_2, \ldots, -M_s;\ M_0 - n + 1;\ z, z^2, \ldots, z^s)\right].$$

$$\mu = n \sum_{i=1}^{s} \frac{iM_i}{N}.$$

$$\sigma^2 = \left(\frac{N-n}{N-1}\right) n \left[\sum_{i=1}^{s} \frac{i^2 M_i}{N} - \left(\sum_{i=1}^{s} \frac{iM_i}{N}\right)^2\right].$$

Some inter-relations

(i) The hypergeometric distribution with parameters N, M, n is a special case of the univariate factorial multinomial distribution with parameters $s, N, M_1, M_2, \ldots, M_s, n$ when $s = 1$ and $M_1 = M$.

(ii) The univariate factorial multionomial distribution with parameters $s, N, M_1, M_2, \ldots, M_s, n$ tends to the univariate multinomial distribution with parameters $s, n, p_1, p_2, \ldots, p_s$ as $N \to \infty$, $M_i \to \infty$ such that $\frac{M_i}{N_i} \to p_i$, $0 < p < 1$, $i = 1, 2, \ldots, s$.

A reference Steyn (1956).

42. THE UNIVARIATE NEGATIVE FACTORIAL MULTINOMIAL DISTRIBUTION

Consider a finite population of N individuals divided into $s+1$ classes $A_0, A_1, A_2, \ldots, A_s$ containing $M_o, M_1, M_2, \ldots, M_s$ individuals respectively. Suppose selection of an individual from A_i is interpreted as i successes, $i = 1, 2, \ldots, s$ and that from A_0 is interpreted as a failure. Suppose now that sampling without replacement from the population is continued until exactly k failures are obtained. Then the number of successes obtained in this sampling process has the univariate negative factorial multinomial distribution with parameters $s, N, M_1, M_2, \ldots, M_s, k$. Its pf is given by

$$P(x) = \sum \frac{\binom{M_0}{k-1}\binom{M_1}{r_1}\binom{M_2}{r_2}\cdots\binom{M_s}{r_s}}{\binom{N}{r+k-1}}\left(\frac{M_0-k+1}{N-r-k+1}\right), \qquad x = 0, 1, 2, \ldots, \sum_{i=1}^{s} iM_i,$$

$$s = 1, 2, \ldots,$$

$$N = 1, 2, \ldots,$$

$$M_i = 1, 2, \ldots; \; i = 1, 2, \ldots, s,$$

$$\sum_{i=1}^{s} M_i < N,$$

$$k = 1, 2, \ldots,$$

where $M_0 = N-M_1-M_2-\ldots-M_s$, $r = r_1+r_2+\ldots+r_s$ and the summation extends over all non-negative integers $r_1, r_2, \ldots, r_s$ such that $\sum_{i=1}^{s} ir_i = x$.

$$G(z) = \frac{M_0^{(k)}}{N^{(k)}}\left[{}_{(s+1)}F_1(k, -M_1, -M_2, \ldots, -M_s; -N+k; z, z^2, \ldots, z^s)\right].$$

$$\mu = \frac{k}{M_0+1}\sum_{i=1}^{s} iM_i.$$

$$\sigma^2 = \frac{k(M_0-k+1)}{(M_0+2)(M_0+1)^2}\left[\left(\sum_{i=1}^{s} iM_i\right)^2 + (M_0+1)\sum_{i=1}^{s} i^2M_i\right].$$

Some inter-relations

(i) The inverse hypergeometric distribution with parameters N, M and k is a special case of the univariate negative factorial multinomial distribution with parameters $s, N, M_1, M_2, \ldots, M_s, k$ when $s = 1$ and $M_1 = N-M$.

(ii) The univariate negative factorial multinomial distribution with parameters $s, N, M_1, M_2, \ldots, M_s, k$ tends to the univariate negative multinomial distribution with parameters $s, k, p_1, p_2, \ldots, p_s$ as $N\to\infty$, $M_i\to\infty$ such that $\frac{M_i}{N}\to p_i$, $0<p_i<1$, $i = 1, 2, \ldots, s$.

A reference Steyn (1956).

43. THE POISSON RECTANGULAR DISTRIBUTION

The mixture on λ of the Poisson distribution with parameters $\alpha+(\beta-\alpha)\lambda$, $0<\alpha<\beta<\infty$ where λ has the rectangular distribution over the interval (0, 1) is the Poisson rectangular distribution with parameters α and β. Its pf is given by

$$P(x) = \frac{1}{(\beta-\alpha)}\left\{e^{-\alpha}\sum_{i=0}^{x}\frac{\alpha^i}{i!} - e^{-\beta}\sum_{i=0}^{x}\frac{\beta^i}{i!}\right\}$$

$$G(z) = \frac{1}{(\beta-\alpha)(z-1)}\{e^{\beta(z-1)}-e^{\alpha(z-1)}\}$$

$$\mu = \frac{\alpha+B}{2}$$

$$\sigma^2 = \frac{\alpha+\beta}{2}+\frac{(\beta-\alpha)^2}{12}$$

A reference Bhattacharya and Holla (1965).

44. THE HERMITE DISTRIBUTION

(Poisson normal distribution)

The pgf of a generalized Poisson distribution can be expressed as

$$G(z) = \exp\{\lambda_1(z-1)+\lambda_2(z^2-1)+\ldots\}$$

where $\sum_{i=1}^{\infty}\lambda_i = \lambda$, the Poisson parameter. This generalized distribution tends to the Poisson distribution with parameter λ_1 if $\lambda_2, \lambda_3, \ldots$ become negligible compared to λ_1 in some limiting process. If λ_2 does not become negligible compared to λ_1, then the limiting distribution is the Hermite distribution with parameters λ_1 and λ_2. It has the pf

$$P(x) = \exp\{-(\lambda_1+\lambda_2)\}\sum_{j=0}^{[x/2]}\frac{\lambda_1^{x-2j}\lambda_2^j}{(x-2j)!j!}, \quad x = 0, 1, 2, \ldots, \quad 0<\lambda_1<\infty, \quad 0<\lambda_2<\infty.$$

$$= e^{-(a_1+a_2)}H_x^*(\beta)\frac{\alpha^x}{x!},$$

where $\alpha = \sqrt{(2a_2)}$, $\beta = \frac{a_1}{\alpha}$ and $H_x^*(\beta)$ is the modified Hermite Polynomial defined in Kemp and Kemp (1965).

$$G(z) = \exp\{\lambda_1(z-1)+\lambda_2(z^2-1)\}.$$

$$\mu = \lambda_1+2\lambda_2, \quad \sigma^2 = \lambda_1+4\lambda_2.$$

Some inter-relations

(i) The Hermite distribution with parameters λ_1 and λ_2 results if the Poisson distribution with parameters $\lambda \geqq \lambda_1+\lambda_2$ is generalized by the distribution having the pgf

$$\left(1-\frac{(\lambda_1+\lambda_2)}{\lambda}\right)+\frac{\lambda_1}{\lambda}z+\frac{\lambda_2}{\lambda}z^2.$$

(ii) If X_1 and X_2 have two independent Poisson distributions with parameters λ_1 and λ_2, then the random variable X_1+2X_2 has the Hermite distribution with parameters λ_1 and λ_2.

(iii) If X_1, X_2 have the bivariate Poisson distribution with parameters λ_1, λ_2 and λ_{12}, then X_1+X_2 has the Hermite distribution with parameters $\lambda_1+\lambda_2$ and λ_{12}.

(iv) The Hermite distribution with parameters $2\lambda p(1-p)$ and λp^2 results as a special case of the Poisson binomial distribution with parameters λ, n and p, when $n = 2$.

(v) The mixture on λ of the Poisson distribution with parameter λ is the Hermite distribution with parameters $\mu-\sigma^2$ and $\frac{1}{2}\sigma^2$ if λ has the normal distribution with parameters μ and σ^2 such that $\mu>\sigma^2$.

(vi) As $\dfrac{\mu}{\sigma^2}\to\infty$, the Poisson truncated normal distribution with parameters μ and σ^2 tends to the Hermite distribution with parameters $\mu-\sigma^2$ and $\frac{1}{2}\sigma^2$.

Some references Kemp and Kemp (1965); Kemp and Kemp (1966).

45. THE POISSON LOGNORMAL DISTRIBUTION

(Discrete lognormal distribution)

The Poisson lognormal distribution with parameters m and θ is the mixture on λ of the Poisson distribution with parameter λ where λ has the lognormal distribution. It has the pf

$$P(x) = \frac{m^x\theta^{\frac{1}{2}x(x-1)}}{x!\sqrt{(2\pi\log\theta)}}\int_{-\infty}^{\infty}\exp\left\{-\frac{u^2}{2\log\theta}-m\theta^{x-\frac{1}{2}}e^u\right\}du; \qquad x = 0, 1, 2, \ldots, \quad 0<m<\infty, \quad \theta>1.$$

An explicit expression for the generating function is not known.

$$\mu = m,\ \sigma^2 = m+m^2(\theta-1) = \mu+\mu^2(\theta-1).$$

Note If the logarithm of a positive random variable X has (approximately) normal distribution then, in certain quarters, the random variable X is said to have the discrete lognormal distribution.

Some references Anscombe (1950); Cassie (1962); Preston (1948*b*).

46. THE POISSON TRUNCATED NORMAL DISTRIBUTION

The Poisson truncated normal distribution with parameters μ and σ^2 is the mixture on λ of the Poisson distribution with parameter λ where λ has the normal distribution with parameters μ and σ^2 truncated at left at 0. Its pf is given by

$$P(x) = \frac{\exp\left\{-\left(\mu-\frac{\sigma^2}{2}\right)\right\}}{x!2\sqrt{(2\pi)}\phi\left(\frac{\mu}{\sigma}\right)}s(x)(\mu-\sigma^2+\sqrt{(2\sigma^2)})^x, \qquad x = 0, 1, 2, \ldots, \quad -\infty<\mu<\infty, \quad \sigma^2>0,$$

where

$$s(x) = \sum_{r=0}^{x} \left\{ \binom{x}{r} p^r (1-p)^{x-r} \Gamma\left(\frac{r+1}{2}\right) \left[1 - I_a\left(\frac{r+1}{2}\right)\right] \right\},$$

$$\phi(u) = \frac{1}{\sqrt{(2\pi)}} \int_{-\infty}^{u} e^{-t^2/2}\, dt, \quad I_a(m) = \int_0^a \frac{e^{-x} x^{m-1}}{\Gamma(m)}\, dx,$$

$$a = \frac{(\mu-\sigma^2)^2}{2\sigma^2}, \quad p = \frac{\sqrt{(2\sigma^2)}}{\mu-\sigma^2+\sqrt{(2\sigma^2)}}.$$

Note that μ and σ^2 appearing above are the mean and variance respectively of the complete normal distribution.

Some inter-relations

(i) When $\mu = \sigma^2$, the Poisson truncated normal distribution reduced to

$$P(x) = \frac{e^{-\mu/2}}{\mu\phi(\mu^{\frac{1}{2}})} \cdot \frac{\left(\frac{\mu}{2}\right)^{x/2+1}}{\Gamma\left(\frac{x}{2}+1\right)}$$

which is a power series distribution.

(ii) As $\frac{\mu}{\sigma^2} \to \infty$, the Poisson truncated normal distribution with parameters μ and σ^2 tends to the Hermite distribution with parameters $\mu-\sigma^2$ and $\frac{1}{2}\sigma^2$.

Some references Berljand, Nazarov and Pressman (1962); Patil (1964*b*); Kemp and Kemp (1966).

47. THE NEYMAN'S TYPE A DISTRIBUTION WITH TWO PARAMETERS

The Neyman's type A distribution with two parameters λ_1 and λ_2 results if the Poisson distribution with parameter λ_1 is generalized by the Poisson distribution with parameter λ_2. Its pf is given by

$$P(x) = e^{-\lambda_1} \frac{\lambda_2^x}{x!} \sum_{j=0}^{\infty} \frac{j^x}{j!} (\lambda_1 e^{-\lambda_2})^j, \quad x = 0, 1, 2, \ldots,$$

$$0<\lambda_1<\infty,$$

$$0<\lambda_2<\infty.$$

$$G(z) = \exp\{\lambda_1[e^{\lambda_2(z-1)}-1]\}.$$

$$\mu = \lambda_1\lambda_2, \quad \sigma^2 = \lambda_1\lambda_2(1+\lambda_2) = \mu(1+\lambda_2).$$

Some inter-relations

(i) If X_1 has the Neyman's type A distribution with two parameters $\frac{\lambda_1}{2}$ and λ_2 and X_2 has the Neyman's type A distribution with two parameters $\frac{\lambda_1}{2}$ and λ_3 and further, X_1 and X_2 are independent, then X_1+X_2 has the Neyman's type A distribution with three parameters λ_1, λ_2 and λ_3.

(ii) The mixture on X_1 of the Poisson distribution with parameters $\lambda_2 X_1$ is the Neyman's type A distribution with two parameters λ_1 and λ_2 if X_1 has the Poisson distribution with parameter λ_1.

(iii) The Neyman's type B distribution with two parameters λ_1 and λ_2 results if the Poisson distribution with parameter λ_1 is generalized by the distribution which in itself is the mixture on t of the Poisson distribution with parameter $\lambda_2 t$ where t has the rectangular distribution over (0, 1).

(iv) The Neyman's type C distribution with two parameters λ_1 and λ_2 results if the Poisson distribution with parameter λ_1 is generalized by the distribution which in itself is the mixture on t of the Poisson distribution with parameter $\lambda_2 t$ where t has the beta distribution with parameters 1 and 2.

(v) For large values of λ_2, the Thomas distribution with parameters λ_1 and λ_2 can be reasonably approximated by the Neyman's type A distribution with two parameters λ_1 and λ_2.

Some references Anscombe (1950); Feller (1943); Neyman (1939).

48. THE NEYMAN'S TYPE A DISTRIBUTION WITH THREE PARAMETERS

Consider two independent random variables X_1 and X_2 having the Neyman's type A distribution with two parameters $\frac{\lambda_1}{2}$, λ_2 and $\frac{\lambda_1}{2}$, λ_3 respectively. Then the random variable $X = X_1 + X_2$ has the Neyman's type A distribution with three parameters λ_1, λ_2 and λ_3. It has the pf

$$P(x) = e^{-\lambda_1} \sum_{j=0}^{\infty} \sum_{i=0}^{\infty} \frac{\left(\frac{\lambda_1}{2} e^{-\lambda_2}\right)^i}{i!} \frac{\left(\frac{\lambda_1}{2} e^{-\lambda_3}\right)^j}{j!} \frac{(\lambda_2 i + \lambda_3 j)^x}{x!}, \quad x = 0, 1, 2, \ldots,$$

$$0<\lambda_1<\infty,$$
$$0<\lambda_2<\infty,$$
$$0<\lambda_3<\infty.$$

$$G(z) = \exp\left\{-\lambda_1 + \frac{\lambda_1}{2}\left[e^{\lambda_2(z-1)} + e^{\lambda_3(z-1)}\right]\right\}.$$

$$\mu = \frac{\lambda_1}{2}(\lambda_2+\lambda_3), \quad \sigma^2 = \frac{\lambda_1}{2}\{\lambda_2(1+\lambda_2)+\lambda_3(1+\lambda_3)\}.$$

Some references Anscombe (1950); Neyman (1939).

49. THE NEYMAN'S TYPE B DISTRIBUTION WITH TWO PARAMETERS

The Neyman's type B distribution with two parameters λ_1 and λ_2 results if the Poisson distribution with parameter λ_1 is generalized by the distribution which in itself is the mixture on t of the Poisson distribution with parameter $\lambda_2 t$ where t has the rectangular distribution over (0, 1). Its pgf is given by

$$G(z) = \exp\left\{s\lambda_1\left[\frac{e^{\lambda_2(z-1)}-1}{\lambda_2(z-1)} - 1\right]\right\}, \quad 0<\lambda_1<\infty$$
$$0<\lambda_2<\infty.$$

$$\mu = \frac{\lambda_1\lambda_2}{2}, \quad \sigma^2 = \frac{\lambda_1\lambda_2}{2}(1+\tfrac{2}{3}\lambda_2) = \mu(1+\tfrac{2}{3}\lambda_2).$$

Some references Anscombe (1950); Feller (1943); Neyman (1939).

50. THE NEYMAN'S TYPE C DISTRIBUTION WITH TWO PARAMETERS

The Neyman's type C distribution with two parameters λ_1 and λ_2 results if the Poisson distribution with parameter λ_1 is generalized by the distribution which in itself is the mixture on t of the Poisson distribution with parameter $\lambda_2 t$ where t has the beta distribution with parameters 1 and 2. Its pgf is given by

$$G(z) = \exp\left\{\lambda_1\left[\frac{e^{\lambda_2(z-1)}-1-\lambda_2(z-1)}{\frac{1}{2}[\lambda_2(z-1)]^2}-1\right]\right\}, \quad 0<\lambda_1<\infty$$
$$0<\lambda_2<\infty.$$

$$\mu = \frac{\lambda_1\lambda_2}{3}, \quad \sigma^2 = \frac{\lambda_1\lambda_2}{3}\left(1+\frac{\lambda_2}{2}\right) = \mu\left(1+\frac{\lambda_2}{2}\right).$$

Some references Anscombe (1950); Neyman (1939).

51. THE BEALL-RESCIAS' GENERALIZATION OF NEYMAN'S DISTRIBUTION

The pgfs of Neyman's type A, type B and type C distributions with two parameters λ_1, λ_2 are

$$G_A(z) = \exp\{\lambda_1[e^{\lambda_2(z-1)}-1]\},$$

$$G_B(z) = \exp\{\lambda_1[e^{\lambda_2(z-1)}-1-\lambda_2(z-1)]/[\lambda_2(z-1)]\},$$

$$G_C(z) = \exp\left\{\lambda_1[e^{\lambda_2(z-1)}-1-\lambda_2(-1)-\lambda_2^2(z-1)^2/2!]/\left[\frac{\lambda_2^2(z-1)^2}{2!}\right]\right\}.$$

These suggest a general form

$$G(z) = \exp\left\{\lambda_1\Gamma(\beta+1)\sum_{i=0}^{\infty}\frac{\lambda_2^{i+1}(z-1)^{i+1}}{\Gamma(\beta+i+2)}\right\}.$$

This is actually a pgf and defines Beall-Rescias' generalization of Neyman's distribution with parameters β, λ_1, λ_2, $0<\beta<\infty$, $0<\lambda_1<\infty$, $0<\lambda_2<\infty$ over the range $\{0, 1, 2, \ldots\}$.

The pf satisfies the recurrence relation

$$P(x+1) = \frac{\lambda_1}{x+1}\sum_{k=0}^{x} F_k P(x-k),$$

where

$$F_k = \frac{1}{k!}\lambda_2^{k+1}\sum_{j=0}^{\infty}\frac{(j+k+1)!(-\lambda_2)^j}{j!\Gamma(\beta+j+k+2)}.$$

$$G(z) = \exp\left\{\lambda_1\Gamma(\beta+1)\sum_{i=0}^{\infty}\frac{\lambda_2^{i+1}(z-1)^{i+1}}{\Gamma(\beta+i+2)}\right\}.$$

$$\mu = \frac{\lambda_1\lambda_2}{\beta+1}.$$

$$\sigma^2 = \left(\frac{\lambda_1\lambda_2}{\beta+1}\right)\left(1+\frac{2\lambda_2}{\beta+2}\right) = \mu\left(1+\frac{2\lambda_2}{\beta+2}\right).$$

Some inter-relations

(i) Beall-Rescias' generalization of Neyman's distribution with parameters β, λ_1, λ_2 is a special case of Gurland's generalization of Neyman's distributions with parameters $\lambda_1, \lambda_2, \alpha, \beta$ when $\alpha = 1$.

(ii) Neyman's type A distribution with two parameters λ_1, λ_2 is a special case of Beall-Rescias' generalization of Neyman's distributions with parameters β, λ_1, λ_2 when $\beta = 0$.

(iii) Neyman's type B distribution with two parameters λ_1, λ_2 is a special case of Beall-Rescias' generalization of Neyman's distributions with parameters β, λ_1, λ_2 when $\beta = 1$.

(iv) Neyman's type C distribution with two parameters λ_1, λ_2 is a special case of Beall-Rescias' generalization of Neyman's distributions with parameters β, λ_1, λ_2 when $\beta = 2$.

Some references Beall and Rescia (1953); Gurland (1958).

52. THE GURLAND'S GENERALIZATION OF NEYMAN'S DISTRIBUTION

The mixture on p of Neyman's type A distribution with two parameters λ_1 and $\lambda_2 p$ where p has the beta distribution with parameters α and β has the Gurland's generalization of Neyman's distribution with parameters λ_1, λ_2, α and β where all four parameters are positive. The pf is not available in a simple explicit form.

$$G(z) = \exp\{\lambda_1[q(z)-1]\},$$

where $$q(z) = {}_1F_1[\alpha, \alpha+\beta, \lambda_2(z-1)].$$

$$\mu = \lambda_1\lambda_2\left(\frac{\alpha}{\alpha+\beta}\right), \quad \sigma^2 = \lambda_1\lambda_2\left(\frac{\alpha}{\alpha+\beta}\right)\left[1+\frac{\lambda_2(\alpha+1)}{\alpha+\beta+1}\right] = \mu\left[1+\lambda_2\frac{(\alpha+1)}{(\alpha+\beta+1)}\right].$$

Note: Gurland's generalization of Neyman's distribution with parameters λ_1, λ_2, α and β is obtained by generalizing the Poisson distribution with parameter λ_1 by the distribution having gf $q(z)$ given above. The mixture on p of the Poisson distribution with parameter $\lambda_2 p$ where p has the beta distribution with parameters α and β has the gf $q(z)$.

Some inter-relations

(i) The Gurland's generalization of Neyman's distribution with parameters λ_1, λ_2, α and β when $\alpha = 1$ is the same as Beall-Rescia's generalization of Neyman's distribution with parameters λ_1, λ_2 and β.

(ii) With $\alpha = 1$ and $\beta = 0$, 1 and 2 respectively, the Gurland's generalization of Neyman's distribution with parameters λ_1, λ_2, α and β reduces to Neyman's type A, B and C distributions, respectively, with parameters λ_1 and λ_2.

(iii) The Gurland's generalization of Neyman's distribution with parameters λ_1, λ_2, α and β tends to the Neyman's type A distribution with two parameters under each of the following conditions

(*a*) $\alpha\to\infty$, $\beta\to\infty$ such that $\frac{\beta}{\alpha}\to\gamma$,

(*b*) $\alpha\to\infty$, β fixed,

(*c*) $\alpha\to\infty$, $\beta\to\infty$ but μ and σ^2 are fixed,

(*d*) $\alpha\to\infty$, μ, σ^2 and β fixed.

The parameters of the limiting distribution in the above cases are:

(*a*) λ_1 and $\lambda_2/(1+\gamma)$,

(*b*) λ_1 and λ_2,

(*c*) $\dfrac{\mu^2}{\sigma^2-\mu}$ and $\dfrac{\sigma^2-\mu}{\mu}$

(*d*) $\dfrac{\mu^2}{\sigma^2-\mu}$ and $\dfrac{\sigma^2-\mu}{\mu}$.

(iv) As $\beta\to\infty$ with α being fixed, the Gurland's generalization of Neyman's distribution tends to the degenerate distribution having the unit mass at zero.

(v) For large β, the Gurland's generalization of Neyman's distribution with parameters λ_1, λ_2, α and β can be approximated by the Poisson negative binomial distribution with parameters α and $(\alpha+\beta)/(\alpha+\beta+\lambda)$.

(vi) As $\beta\to\infty$ with μ, σ^2 and α remaining fixed, the Gurland's generalization of Neyman's distribution with parameters λ_1, λ_2, α and β tends to the Poisson negative binomial distribution with parameters $\left(\dfrac{\alpha+1}{\alpha}\right)\left(\dfrac{\mu}{c-1}\right)$, α and $(\alpha+1)/(\alpha+c)$, $c=(\sigma^2-\mu)/\mu$.

A reference Gurland (1958).

53. THE THOMAS DISTRIBUTION

(The double Poisson distribution)

Consider two independent random variables X_1 and X_2 having Poisson distributions with parameters λ_1 and λ_2 respectively. Then the Thomas distribution with parameters λ_1 and λ_2 results if the random variable X_1 is generalized by the random variable $1+X_2$. Its pf is given by

$$P(x)=\begin{cases} e^{-\lambda_1} & x=0,\\ e^{-\lambda_1}\displaystyle\sum_{j=1}^{x}\frac{\lambda_1^j}{j!}\frac{(j\lambda_2)^{x-j}}{(x-j)!}e^{-j\lambda_2}, & x=1,2,\ldots,\end{cases}$$

$$0<\lambda_1<\infty,$$

$$0<\lambda_2<\infty.$$

$$G(z)=\exp\{[\lambda_1 z e^{\lambda_2(z-1)}-1]\}.$$

$$\mu=\lambda_1(\lambda_2+1),\quad \sigma^2=\mu(\lambda_2+1)+\lambda_1\lambda_2.$$

Some inter-relations

For large values of λ_2, the Thomas distribution with parameters λ_1 and λ_2 can be reasonably approximated by the Neyman's type A distribution with two parameters λ_1 and λ_2.

Some references Anscombe (1950); Thomas (1949).

54. THE POISSON BINOMIAL DISTRIBUTION

The Poisson binomial distribution with parameters λ, n and p is obtained by generalizing the Poisson distribution with parameter λ by the binomial distribution with parameters n and p. It also results as the mixture on y of the binomial distribution with parameters ny and p, where y has the Poisson distribution with parameter λ. Its pf is given by

$$P(x) = e^{-\lambda} \sum_{j=0}^{\infty} \binom{nj}{x} p^x q^{nj-x} \frac{\lambda^j}{j!}, \qquad x = 0, 1, 2, \ldots,$$

$$0<\lambda<\infty,$$

$$n = 1, 2, \ldots,$$

$$0<p<1, \; q = 1-p.$$

$$G(z) = \exp\{\lambda[(q+pz)^n - 1]\}$$

$$\mu = n\lambda p, \; \sigma^2 = n\lambda p(np+q) = \mu(np+q).$$

Some inter-relations

(i) With $n = 1$, the Poisson binomial distribution with parameters λ, n and p is the Poisson distribution with parameter λp.

(ii) With $n = 2$, the Poisson binomial distribution with parameters λ, n and p is the Hermite distribution with parameters $2\lambda pq$ and λp^2.

(iii) As $n \to \infty$ and $p \to 0$ such that $np \to \lambda_2$, $0<\lambda_2<\infty$, the Poisson binomial distribution with parameters λ_1, n and p tends to the Neyman's type A distribution with two parameters λ_1 and λ_2.

Some references Shumway and Gurland (1960*a*); Katti and Gurland (1962*a*).

55. THE POLYA AEPPLI DISTRIBUTION

(Poisson geometric distribution)

The Polya Aeppli distribution with parameters λ and θ results if the Poisson distribution with parameter λ is generalized by the geometric distribution, form (ii), with parameter $1-\theta$. Its pf is given by

$$P(x) = \begin{cases} e^{-\lambda} & x = 0, \\ e^{-\lambda}\theta^x \sum_{j=1}^{x} \binom{x-1}{j-1} \frac{1}{j!} \left(\frac{\lambda(1-\theta)}{\theta}\right)^j, & x = 1, 2, \ldots, \end{cases}$$

$$0<\lambda<\infty,$$

$$0<\theta<1.$$

$$G(z) = \exp \lambda \left\{\frac{(1-\theta)z}{1-\theta z} - 1\right\}.$$

$$\mu = \frac{\lambda}{1-\theta}, \quad \sigma^2 = \frac{\lambda(1+\theta)}{(1-\theta)^2} = \mu\left(\frac{1+\theta}{1-\theta}\right).$$

Some inter-relations

(i) The mixture on k of the Pascal distribution with parameters k and $1-\theta$ is the Polya Aeppli distribution with parameters λ and θ if k has the Poisson distribution with parameter λ.

(ii) As $\theta \to 0$, the Polya Aeppli distribution with parameters λ and θ tends to the Poisson distribution with parameter λ.

Some references Anscombe (1950a); Evans (1953).

56. THE POISSON NEGATIVE BINOMIAL DISTRIBUTION

(The Poisson Pascal distribution)

The Poisson negative binomial distribution with parameters λ, k and p results if the Poisson distribution with parameter λ is generalized by the negative binomial distribution with parameters k and p. The Poisson negative binomial distribution with parameters λ, k and p arises also as the mixture on y of the negative binomial distribution with parameters ky and p if y has the Poisson distribution with parameter λ. Its pf is given by

$$P(x) = \mathrm{e}^{-\lambda} \sum_{i=0}^{\infty} \frac{\Gamma(ki+x)}{\Gamma(ki)x!} p^{ki} q^{x} \frac{\lambda^{i}}{i!} \qquad x = 0, 1, 2, \ldots,$$

$$0<\lambda<\infty,$$

$$0<k<\infty,$$

$$0<p<1,$$

$$q = 1-p.$$

$$G(z) = \exp\left\{\lambda\left[\left(\frac{p}{1-qz}\right)^{k} - 1\right]\right\}$$

$$\mu = \frac{\lambda kq}{p}, \quad \sigma^2 = \frac{\lambda kq}{p^2}(1+kq) = \frac{\mu(1+kq)}{p}.$$

Some inter-relations

(i) The Polya Aeppli distribution with parameters λ and θ results as the generalization of the Poisson distribution with parameter λ by the geometric distribution, form (ii), with parameter, $(1-\theta)$. If instead, the Poisson distribution with parameter λ is generalized by the geometric distribution, form (i), with parameter $(1-\theta)$, the special case of the Poisson negative binomial distribution with parameters λ, k and $(1-\theta)$ is obtained where $k = 1$.

(ii) As $k\to\infty$ and $p\to 1$ such that $kq\to\lambda_1$, $0<\lambda_1<\infty$, the Poisson negative binomial distribution with parameters λ, k and p tends to the Neyman's type A distribution with two parameters λ and λ_1.

(iii) As $\lambda\to\infty$ and $k\to 0$ such that $\lambda k\to k_1$, $0<k_1<\infty$, the Poisson negative binomial distribution with parameters λ, k and p tends to the negative binomial distribution with parameters k_1 and p.

(iv) As $\lambda\to\infty$ and $p\to 1$ such that $\lambda q\to\lambda_1$, $0<\lambda_1<\infty$, the Poisson negative binomial distribution with parameters λ, k and p tends to the Poisson distribution with parameter $\lambda_1 k$.

(v) For large β, the Gurland's generalization of Neyman's distribution with parameters λ_1, λ_2, α and β can be approximated by the Poisson negative binomial distribution with parameters λ_1, α and $(\alpha+\beta)/(\alpha+\beta+\lambda_2)$.

(vi) As $\beta\to\infty$ with μ, σ^2 and α remaining fixed, the Gurland's generalization of Neyman's distribution with parameters λ_1, λ_2, α and β tends to the Poisson negative binomial distribution with parameters

$$\left(\frac{\alpha+1}{\alpha}\right)\left(\frac{\mu}{c-1}\right), \alpha \text{ and } \frac{\alpha+1}{\alpha+c} \text{ where } c = \frac{\sigma^2-\mu}{\mu}.$$

Some references Skellam (1952); Katti and Gurland (1961).

57. THE WARING DISTRIBUTION

The Waring distribution with parameters k and ρ is the mixture on u of the negative binomial distribution with parameters k and e^{-u}, where u has the exponential distribution with parameter ρ The pf is

$$P(x) = \frac{\rho}{k+\rho}\,\frac{k_{(x)}}{(k+\rho+1)_{(x)}}, \qquad x = 0, 1, 2, \ldots,$$

$$0<k<\infty,$$

$$0<\rho<\infty.$$

$$G(z) = \frac{\rho}{k+\rho}\,{}_2F_1(1,\, k;\; k+\rho+1;\; z).$$

$$\mu = \begin{cases} \dfrac{k}{\rho-1}, & \rho>1 \\ \text{does not exist otherwise.} \end{cases}$$

$$\sigma^2 = \begin{cases} \dfrac{k\rho(k+\rho-1)}{(\rho-1)^2(\rho-2)} = \mu(1+\mu)\dfrac{\rho}{\rho-2}, & \rho>2 \\ \text{does not exist otherwise.} \end{cases}$$

Some inter-relations

(i) The Waring distribution with parameters k and ρ is a special case of the generalized hypergeometric distribution type IV with parameters a, b, n when $a = -1$, $b = \rho$, $n = -k$.

(ii) The factorial distribution with parameters n and λ is the same as the Waring distribution with parameters $\lambda - n + 1$, $n-1$.

(iii) The Yule distribution with parameter ρ is a special case of the Waring distribution with parameters k and ρ when $k = 1$.

(iv) The Waring distribution with parameters k and ρ tends to the geometric distribution with parameter p as $k \to \infty$, $\rho \to \infty$ such that $\dfrac{\rho}{k+\rho} \to p$, $0<p<1$.

(v) The Waring distribution with parameters k and ρ is a special case of the Pascal gamma distribution with parameters k, α and λ when $\alpha = \rho$ and $\lambda = 1$.

(vi) The Waring distribution with parameters k and ρ is a special case of the inverse factorial series distribution with series function ϕ and parameter θ, when $\phi(t) = t^{-k}$ and $\theta = \rho + k$.

A reference Irwin (1965).

58. THE YULE DISTRIBUTION

The mixture on u of the geometric distribution with parameter e^{-u}, where u has the exponential distribution with parameter ρ, is the Yule distribution with parameter ρ. Its pf is

$$P(x) = \frac{\rho x!}{(\rho+1)_{(x+1)}}, \qquad x = 0, 1, 2, \ldots, \quad 0<\rho<\infty.$$

$$G(z) = \frac{\rho}{\rho+1}\, {}_2F_1(1, 1;\ \rho+2;\ z).$$

$$\mu = \begin{cases} \dfrac{1}{\rho-1}, & \rho>1, \\ \text{does not exist otherwise.} \end{cases}$$

$$\sigma^2 = \begin{cases} \dfrac{\rho^2}{(\rho-1)^2(\rho-2)} = \dfrac{\mu^2\rho^2}{\rho-2}, & \rho>2 \\ \text{does not exist otherwise.} \end{cases}$$

Some inter-relations

(i) The Yule distribution with parameter ρ is a special case of the Waring distribution with parameters k and ρ when $k = 1$.

(ii) The Yule distribution with parameter ρ is a special case of the generalized hypergeometric distribution type IV with parameters a, b, n when $a = -1$, $b = \rho$, $n = -1$.

(iii) The factorial distribution with parameters n, λ reduces to the Yule distribution with parameter $n-1$, when $n = \lambda$.

(iv) The Yule distribution with parameter ρ is a special case of the Pascal gamma distribution with parameters k, α, and λ when $k=\lambda=1$ and $\alpha=\rho$.

(v) The Yule distribution with parameter ρ is a special case of the inverse factorial series distribution with series function ϕ and parameter θ, when $\phi(t)=t^{-1}$ and $\theta=\rho+1$.

Some references Simon (1955); Irwin (1965).

59. THE PASCAL GAMMA DISTRIBUTION

The Pascal Gamma distribution with parameters k, α and λ is the mixture on u of the Pascal distribution with parameters k and e^{-u}, where u has the gamma distribution with parameters α and λ. Its pf is given by

$$P(x) = \sum_{j=0}^{x-k} \binom{x-1}{k-1}\binom{x-k}{j}(-1)^j\left(1+\frac{j+k}{\alpha}\right)^{-\lambda}, \qquad x = k, k+1, \ldots, \quad 0<\alpha<\infty, \quad 0<\lambda<\infty.$$

$$\mu = k\left(1-\frac{1}{\alpha}\right)^{-\lambda}, \text{ when } \alpha>1.$$

$$\sigma^2 = k(k+1)\left(1-\frac{2}{\alpha}\right)^{-\lambda} - k\left(1-\frac{1}{\alpha}\right)^{-\lambda} - k^2\left(1-\frac{1}{\alpha}\right)^{-2\lambda}, \text{ when } \alpha>2.$$

Some inter-relations

(i) The Waring distribution with parameters k and ρ is a special case of the Pascal gamma distribution with parameters k, α and λ when $\alpha=\rho$ and $\lambda=1$.

(ii) The Yule distribution with parameter ρ is a special case of of the Pascal gamma distribution with parameters k, α and λ when $k=\lambda=1$ and $\alpha=\rho$.

A reference Dubey (1966).

60. THE LOG-ZERO-POISSON DISTRIBUTION

(Modified logarithmic Poisson distribution)

Writing $p_1 = -\dfrac{1-\delta}{\log(1-\theta)}$, $p_2 = \dfrac{\theta}{1-\theta}$ and $q_2 = 1+p_2$, the pgf of the modified logarithmic distribution with parameters δ and θ reduces to $1-p_1\log(q_2-p_2z)$, which Katti (1965) calls the pgf of the log-zero distribution with parameters p_1 and p_2. The log-zero-Poisson distribution with parameters p_1, p_2 and λ results if the log-zero distribution with parameters p_1 and p_2 is generalized by the Poisson distribution with parameter λ. Its pgf is given by

$$G(z) = 1-p_1\log\{q_2-p_2\exp[\lambda(z-1)]\},\quad 0<p_1<\infty,\ 0<p_2<\infty,\ 0<\lambda<\infty.$$

$$\mu = p_1p_2\lambda,\quad \sigma^2 = \mu[\lambda(1+p_2)+1-\mu].$$

Some references Katti (1965): Katti and Rao (typescript).

61. THE DETERMINISTIC DISTRIBUTION

(The degenerate distribution, the causal distribution)

The deterministic distribution with the mass at k, a real number, is defined by the pf

$$P(x) = \begin{cases}1 & x=k\\ 0 & x\neq k.\end{cases}$$

$$G(z) = z^k.$$

$$\mu = k.$$

$$\sigma^2 = 0.$$

A reference Haight (1961*b*).

62. THE UNIFORM DISTRIBUTION

(The discrete rectangular distribution).

A random observation from a finite population $\{x_1, x_2, \ldots, x_n\}$ has the uniform distribution with range $\{x_1, x_2, \ldots, x_n\}$ and has the pf

$$P(x) = \frac{1}{n},\quad x = x_1, x_2, \ldots, x_n.$$

The pgf, mean and variance of the special case when $x_i = i$, $i = 1, 2, \ldots, n$, are

$$G(z) = \frac{z(1-z^n)}{n(1-z)},$$

$$\mu = \frac{n+1}{2}, \qquad \sigma^2 = \frac{n^2-1}{12}.$$

Some inter-relations

The Polya distribution with parameters a, b, c and n reduced to the uniform distribution with range $\{0, 1, 2, \ldots, n\}$ when $a = b = c$.

A reference Feller (1957), p. 223.

63. THE DISCRETE PARETO DISTRIBUTION

(The zeta distribution)

This is a discrete analogue of the continuous Pareto law. The pf of the discrete Pareto distribution with parameter β is given by

$$P(x) = \frac{1}{\zeta(\beta)x^\beta}, \qquad x = 1, 2, \ldots,$$

$$1<\beta<\infty,$$

where ζ is the Riemann zeta function.

$$\mu = \begin{cases} \dfrac{\zeta(\beta-1)}{\zeta(\beta)}, & \beta>2 \\ \text{does not exist otherwise.} \end{cases}$$

$$\sigma^2 = \begin{cases} \dfrac{\zeta(\beta-2)}{\zeta(\beta)} - \dfrac{\zeta(\beta-1)^2}{\zeta(\beta)}, & \beta>3 \\ \text{does not exist otherwise.} \end{cases}$$

Some references Rider (1965); Seal (1953).

64. THE DISCRETE NORMAL DISTRIBUTION

A discrete analogue of the (continuous) normal distribution, the discrete normal distribution with parameters m and v is defined by the pf

$$P(x) = A \exp\left\{-\frac{(x-m)^2}{2v}\right\}, \qquad x = 0, \pm 1, \pm 2, \ldots,$$

$$-\infty<m<\infty,$$

$$0<v<\infty,$$

where, $A \equiv A(m, v)$ is given by

$$\frac{1}{A} = \Sigma \exp\left\{-\frac{(x-m)^2}{2v}\right\}.$$

When v and m^2/v are large, say both >9, $m \approx \mu$ and $v \approx \sigma^2$.

A reference Haight (1957).

65. THE DISCRETE TYPE III DISTRIBUTION

A discrete analogue of the (continuous) type III distribution, the discrete type III distribution with parameters a, α, k, is defined by the pf

$$P(x) = A(x+a)^k e^{-\alpha x}, \quad x = 0, 1, 2, \ldots, \quad 0<a<\infty, \quad 0<\alpha<\infty, \quad 0<k<\infty,$$

where, $A \equiv A(a, \alpha, k)$ is given by

$$\frac{1}{A} = \sum_{x=0}^{\infty} (x+a)^k e^{-\alpha x}.$$

A reference Haight (1957).

66. THE HARMONIC DISTRIBUTION

The pf of the harmonic distribution with parameter λ is given by

$$P(x) = \frac{1}{2\lambda}\left\{\left[\frac{2\lambda}{2x-1}\right] - \left[\frac{2\lambda}{2x+1}\right]\right\}, \quad x = 1, 2, \ldots, \quad 0<\lambda<\infty.,$$

where $[a]$ denotes the integral part of a.

$$\mu = \frac{1}{2\lambda} \sum_{i=1}^{\left[\frac{2\lambda+1}{2}\right]} \left[\frac{2\lambda}{2i-1}\right].$$

$$\sigma^2 = \sum_{i=1}^{\left[\frac{2\lambda+1}{2}\right]} \left(\frac{2i-1}{2\lambda}\right)\left[\frac{2\lambda}{2i-1}\right] - \mu^2.$$

Some inter-relations

The harmonic distribution with parameter λ tends to the Haight's zeta distribution with parameter 1 as $\lambda \to \infty$.

A reference Haight (typescript).

67. THE HAIGHT'S ZETA DISTRIBUTION

Consider, for fixed $\beta>0$, $z>1$, a table of decimal values of $zN^{-\beta}$, $N = 1, 2, \ldots$. The number $N(x)$ of values nearest the integer x (counting half integers as belonging to the upper category) is the number of N satisfying $\frac{1}{2}(2x-1) \leqq zN^{-\beta} < \frac{1}{2}(2x+1)$. Writing the inequality in the form

$$\left(\frac{2z}{2x-1}\right)^{1/\beta} \geqq N > \left(\frac{2z}{2x+1}\right)^{1/\beta}, \quad N(x) = \left[\left(\frac{2z}{2x-1}\right)^{1/\beta}\right] - \left[\left(\frac{2z}{2x+1}\right)^{1/\beta}\right],$$

where $[a]$ means the integral part of a. Since the total number of non-zero values of $N(x)$ is $(2z)^{1/\beta}$, the quantities $P(x, z) = (2z)^{-1/\beta} N(x)$, $x = 1, 2, 3, \ldots$, have the formal properties of a probability distribution. Letting $z \to \infty$, and writing $\rho = 1/\beta$, the Haight's zeta distribution with parameter ρ is given by the pf

$$P(x) = \frac{1}{(2x-1)^{\rho}} - \frac{1}{(2x+1)^{\rho}}, \qquad x = 1, 2, \ldots$$
$$0 < \rho < \infty.$$

$$\mu = \begin{cases} \left(1 - \dfrac{1}{2^{\rho}}\right) \displaystyle\sum_{n=1}^{\infty} \frac{1}{n^{\rho}} = \left(1 - \frac{1}{2^{\rho}}\right) \zeta(\rho), \ \rho < 1 \\ \text{does not exist otherwise.} \end{cases}$$

$$\sigma^2 = \begin{cases} \left(1 - \dfrac{1}{2^{\rho-1}}\right) \zeta(\rho-1) - \mu^2, \ \rho > 2 \\ \text{does not exist otherwise.} \end{cases}$$

Some inter-relations

The harmonic distribution with parameter λ tends to the Haight's zeta distribution with parameter 1 as $\lambda \to \infty$.

Some references Haight (typescript). *Note* Zipf (1949) argued that the size of cities (as well as an enormous number of other things) would be proportional to $N^{-\beta}$, where N is the rank of the city. If this were true, then the relative frequency of cities with populations near x would be $P(x, z)$, where z depends on how many digits are rounded off in the grouping. If a substantial number were dropped, the zeta distribution would furnish a good approximation to the relative abundance of cities with populations in a fixed range.

68. THE DISTRIBUTION OF EXCEEDANCES

Consider the random samples S_1 and S_2 of sizes n and N respectively from a continuous distribution. Then the number of exceedances X, that is, the number of observations in S_2 which surpass at least $N-M+1$ observations in S_1, has the distribution of exceedances with parameters N, n, and M with the pf.

$$P(x) = \frac{\dbinom{-M}{x}\dbinom{-N-1+M}{n-x}}{\dbinom{-N-1}{n}}, \qquad x = 0, 1, 2, \ldots, n,$$
$$n = 1, 2, \ldots;\ n < \infty,$$
$$N = 1, 2, \ldots;\ N < \infty,$$
$$M = 1, 2, \ldots, N.$$

$$\mu = \frac{Mn}{N+1}, \quad \sigma^2 = \frac{nM(N-M+1)(N+n-1)}{(N+1)^2(N+2)}.$$

Some inter-relations

(i) The Polya distribution with parameters a, b, 1, n has the same pf as the distribution of exceedances with parameters $a+b-1$, n and a.

(ii) If $N = n$ and n, $N\to\infty$ with M remaining finite then the distribution of exceedances with parameters N, n and M tends to the negative binomial distribution with parameters M and $\frac{1}{2}$.

Some references Gumbel (1963); Gumbel and von Schelling (1950); Sarkadi (1957*a*).

69. THE FACTORIAL DISTRIBUTION

The factorial distribution with parameters n and λ is defined by the pf

$$P(x) = (n-1)\,\frac{(\lambda-1)^{(n-1)}}{(\lambda+x)^{(n)}}, \qquad x = 0, 1, 2, \ldots,$$
$$n = 2, 3, 4, \ldots,$$
$$n-1<\lambda<\infty.$$

$$\mu = \frac{\lambda-n+1}{n-2} \text{ for } n>2 \text{ and does not exist otherwise.}$$

$$\sigma^2 = \frac{n-1}{n-3}(\mu^2+\mu) \text{ for } n>3 \text{ and does not exist otherwise.}$$

Some inter-relations

(i) The factorial distribution with parameters n and λ is a special case of the generalized hypergeometric distribution, type IV, with parameters $-1, n-1, -\lambda+n-1$.

(ii) The factorial distribution with parameters n and λ is the same as the Waring distribution with parameters $\lambda-n+1, n-1$.

(iii) The factorial distribution with parameters n and λ is the mixture on p of the negative binomial distribution with parameters $k = \lambda-m+1$ and p where p has the beta distribution with parameters $n-1$ and 1 (Negative binomial beta distribution).

A reference Marlow (1965).

70. THE STEVENS CRAIG DISTRIBUTION

The number of colours represented in a random sample of size n drawn with replacement from an urn containing ka balls, a being the number of each of k different colours, has the Stevens Craig distribution with parameters k and n. It has the pf

$$P(x) = \frac{1}{k^n}\binom{k}{x}\Delta^x 0^n,$$
$$= \frac{1}{k^n}k^{(x)}\mathfrak{S}_n^x, \qquad x = 1, 2, \ldots, k,$$
$$k = 1, 2, \ldots,$$
$$n = 1, 2, \ldots,$$

where $\mathfrak{S}_n^x$ is the Stirling number of the second kind and $\Delta^x 0^n$ is the xth order difference of zero of the nth order.

$$G(z) = \frac{1}{k^n}(1+z\Delta)^k 0^n.$$

$$\mu = \left[1-\left(1-\frac{1}{k}\right)^n\right], \quad \sigma^2 = k\left(1-\frac{1}{k}\right)^n + k(k-1)\left(1-\frac{2}{k}\right)^n - k^2\left(1-\frac{1}{k}\right)^{2n}.$$

Some inter-relations

Let a random vector $(x_1, \ldots, x_k)$ have a k-variate singular multinomial distribution with parameters n, $p_1 . p_2, \ldots, p_k$ where $p_1 = p_2 = \ldots = p_k = \frac{1}{k}$, then the number of components of $(x_1, x_2, \ldots, x_k)$ which are not zero has the Stevens Craig distribution with parameters k and n.

Some references Craig (1953*a*); Stevens (1937).

71. THE MATCHING DISTRIBUTION

Suppose two sets having n objects each are numbered 1, 2, ..., n and each is arranged in a random order so as to form n pairs. Then the number of pairs on which the two numbers are the same has the matching distribution with pf

$$P(x) = \frac{1}{x!}\left(1 - \frac{1}{1!} + \frac{1}{2!} - \ldots + (-1)^{n-x}\frac{1}{(n-x)!}\right), \quad x = 0, 1, 2, \ldots, n,$$

$$n = 1, 2, \ldots.$$

$$\mu = 1, \quad \sigma^2 = 1.$$

Some inter-relations

For relatively small $n (n \geqq 10)$, the probabilities defined by $P(x)$ are very closely approximated by the Poisson pf with parameter 1.

Some references Feller (1957), p. 97; Irwin (1955); Iyer (1954).

72. THE ISING STEVENS DISTRIBUTION

Let $n = n_1 + n_2$ objects of two kinds C_1 and C_2, C_i being n_i in number, $i = 1, 2$, be arranged at random in n positions along a circle. A run is the sequence of objects of the same kind terminated by objects of the other kind. Let X = number of runs of C_1 along the circle
= number of runs of C_2 along the circle.

Then X has the Ising-Stevens distribution with parameters n_1 and n_2. Its pf is

$$P(x) = \frac{\binom{n_1-1}{x-1}\binom{n_2}{x}}{\binom{n_1+n_2-1}{n_1}} = \frac{\binom{n_1}{x}\binom{n_2-1}{x-1}}{\binom{n_1+n_2-1}{n_2}}$$

$$\mu = \frac{n_1 n_2}{(n_1+n_2-1)} \qquad \sigma^2 = \frac{n_1(n_1-1)n_2(n_2-1)}{(n_1+n_2-1)^2(n_1+n_2-2)}$$

A reference Stevens (1939).

73. THE EXTENDED HYPERGEOMETRIC DISTRIBUTION

The extended hypergeometric distribution with parameters n_1, n_2, r and θ arises in testing the hypothesis of the equality $p_1 = p_2$ where p_1 and p_2 are parameters of two independent binomial distributions with parameters n_i and p_i, $i = 1, 2, \ldots$. Its pf is given by

$$P(x) = \frac{g(x)\theta^x}{p(\theta)}, \qquad \begin{aligned} &x = a, a+1, a+2, \ldots, b, \\ &n_1 = 1, 2, \ldots, \\ &n_2 = 1, 2, \ldots, \\ &r = 0, 1, 2, \ldots, n_1+n_2 = n, \\ &0<\theta<\infty, \end{aligned}$$

where $a = \max(0, r-n_2)$, $b = \min(n_1, r)$,

$$g(x) = \frac{\binom{n_1}{x}\binom{n_2}{r-x}}{\binom{n_1+n_2}{r}}, \quad \theta = \frac{p_1(1-p_2)}{p_2(1-p_1)} \text{ and } p(\theta) = \sum_{y=a}^{b} g(y)\theta^y.$$

$$G(z) = \frac{p(\theta z)}{p(\theta)}.$$

$$\mu = \frac{n_1 r\theta}{(n_2-r+1)}\frac{F(1)}{F(0)}, \quad \sigma^2 = \frac{n_1 r\theta - c_n\mu - (1-\theta)\mu^2}{1-\theta}$$

where

$$F(k) = {}_2F_1(-n_1+k, -r+k; n_2-r+1+k; \theta)$$
$$c_n = n-(n_1+r)(1-\theta).$$

Some inter-relations

(i) The extended hypergeometric distribution with parameters n_1, n_2, r and θ is the same as the hypergeometric distribution with parameters n_1+n_2, n_1 and r, when $\theta = 1$.

(ii) Let X, Y have independent binomial distributions with parameters n_1, p_1 and n_2, p_2 respectively. Then, given $X+Y = r$, the conditional distribution of X is the extended hypergeometric distribution with parameters n_1, n_2, r and $\theta = \dfrac{p_1(1-p_2)}{p_2(1-p_1)}$.

(iii) Let $n_1, n_2 \to \infty$ such that $\dfrac{n_1}{(n_1+n_2)} \to p$, $0<p<1$, and let $\phi = p\theta/(q+p\theta)$, $q = 1-p$, then the extended hypergeometric distribution with parameters n_1, n_2, r and θ tends to the binomial distribution with parameters r and ϕ.

(iv) Let n_1, r and $n_2 \to \infty$ such that $\dfrac{n_1 r}{(n_1+n_2)} \to \lambda$, then the extended hypergeometric distribution with parameters n_1, n_2, r and θ tends to the Poisson distribution with parameter $\lambda\theta$.

A reference Harkness (1965).

74. THE BOREL TANNER DISTRIBUTION

In a problem of queues, the number of customers served before the queue vanishes for the first time, given that initially at time $t = 0$ there were r customers in the queue has the Borel Tanner distribution with parameters r and $\alpha(= \lambda\beta)$ if the probability of a customer's arrival during the time interval

$(t, t+\Delta t)$ is $\lambda\Delta t+0(\Delta t)$ and that the time required to serve each customer is constant, say β. The pf is given by

$$P(x) = \frac{r!}{(x-r)!} e^{-\alpha x}\alpha^{x-r}x^{x-r-1}, \quad x = r, r+1, r+2, \ldots,$$
$$r = 1, 2, \ldots,$$
$$0<\alpha<1.$$

$$G(z) = \left[\frac{\alpha(\theta z)}{\alpha(\theta)}\right]^r = z^r \exp(r)\left\{\sum_{j=1}^{\infty}\left(\frac{j^{j-1}\alpha j}{j!e^{\alpha j}}\right)(z^j-1)\right\},$$

where $\alpha(\theta)$ denotes functional dependence defined by $\theta = \alpha e^{-\alpha}$

$$\mu = \frac{r}{1-\alpha}, \quad \sigma^2 = \frac{\alpha r}{(1-\alpha)^3}.$$

A reference Haight and Breuer (1960).

75. THE STER DISTRIBUTION

Consider a non-negative integral valued random variable Y having the pf $P_Y(y)$. Then, the random variable whose pf is defined by sums which are successively truncated from what starts out to be the expectation of the reciprocal of the (zero-truncated) random variable Y has the STER distribution with pf

$$P(x) = \frac{1}{1-P_Y(0)}\sum_{y=x+1}^{\infty}\frac{P_Y(y)}{y}, \quad x = 0, 1, 2, \ldots.$$

$$G(z) = \frac{\theta}{1-z}\int_z^1 \frac{G_Y(t)-P_Y(0)}{t}\,dt,$$

where $\theta = \dfrac{1}{1-P_Y(0)}$ and $G_Y(t)$ is the generating function of the random variable Y.

$$\mu = \mu_x = \frac{\theta\mu_Y-1}{2}, \quad \sigma^2 = \frac{\theta(\sigma_Y^2+\mu_x^2)-1}{3} - \mu(1+\mu),$$

where μ_Y and σ_Y^2 are the mean and the variance of Y respectively.

Some inter-relations

The STER distribution is a special case of the generalization of the STER distribution where $k = 1$.

A reference Bissinger (1965).

76. THE GENERALIZATION OF THE STER DISTRIBUTION

A generalization of the STER distribution corresponding to a non-negative integral valued random variable Y with pf $P_Y(y)$ has the pf

$$P(x) = \frac{\sum_{y=x+k}^{\infty}\dfrac{1}{y-k+1}P_Y(y)}{\sum_{y=k}^{\infty}P_Y(y)}, \quad x = 0, 1, 2, \ldots, \quad k = 1, 2, \ldots.$$

Some inter-relations

The STER distribution is a special case of the generalization of the STER distribution where $k = 1$.

A reference Haight (1964).

77. THE BIVARIATE BINOMIAL DISTRIBUTION

Consider a population of individuals each having one or both or none of the two characters A_1 and A_2. Suppose the probabilities that an individual possesses none of A_1 and A_2, only A_1, only A_2, and both A_1 and A_2 are p_{00}, p_{10}, p_{01} and p_{11} respectively. Suppose n individuals are selected at random from the population. Let X_1 and X_2 denote the numbers of individuals in the random sample possessing characters A_1 and A_2 respectively. Then (X_1, X_2) has the bivariate binomial distribution with parameters $n, p_{10}, p_{01}, p_{11}$. Its pf is given by

$$P(x_1, x_2) = \sum_{i=0}^{\min(x_1, x_2)} \frac{n!}{i!(x_1-i)!(x_2-i)!(n-x_1-x_2+i)!} p_{11}^i p_{10}^{x_1-i} p_{01}^{x_2-i} p_{00}^{n-x_1-x_2+i},$$

$$x_i = 0, 1, 2, \ldots, n;\ i = 1, 2$$
$$n = 1, 2, \ldots,$$
$$0<p_{10}<1,$$
$$0<p_{01}<1,$$
$$0<p_{11}<1,$$
$$0<p_{10}+p_{01}+p_{11}<1,$$

where $p_{00} = 1-p_{10}-p_{01}-p_{11}$.

$$G(z_1, z_2) = (p_{00}+p_{10}z_1+p_{01}z_2+p_{11}z_1z_2)^n.$$

Writing $p_1 = p_{10}+p_{11}, p_2 = p_{01}+p_{11}, q_1 = 1-p_1$ and $q_2 = 1-p_2$,

$$\mu_i = np_i,\ \sigma_{ij} = \begin{cases} np_iq_i, & i = j, \\ n(p_{11}-p_1p_2), & i \neq j. \end{cases}$$

Some inter-relations

(i) If (X_1, X_2) has the bivariate binomial distribution with parameters $n, p_{10}, p_{01}, p_{11}$ then the distribution of X_i is binomial with parameters n and p_i for $i = 1, 2$, where $p_1 = p_{10}+p_{11}$, $p_2 = p_{01}+p_{11}$.

(ii) If (X_1, X_2, X_3) has the 3-dimensional multinomial distribution with parameters n, p_1, p_2 p_3 and if $Y_1 = X_1+X_3$ and $Y_2 = X_2+X_3$, then (Y_1, Y_2) has the bivariate binomial distribution with parameters n, p_1, p_2, p_3.

(iii) The bivariate binomial distribution with parameters $n, p_{10}, p_{01}, p_{11}$ tends to the bivariate Poisson distribution with parameters $\lambda_1, \lambda_2, \lambda_{12}$ as $n\to\infty$, $p_{10}\to 0$, $p_{01}\to 0$, $p_{11}\to 0$ such that $np_{10}\to\lambda_1$, $np_{01}\to\lambda_2$, $np_{11}\to\lambda_{12}$, $0<\lambda_i<\infty$, $i = 1, 2$, $0<\lambda_{12}<\infty$.

Some references Krishnamoorthy (1951); Kendall and Stuart (1958), p. 141.

78. THE BIVARIATE BINOMIAL DISTRIBUTION WITH OVERLAPPING TRIALS

Consider two sequences of n_1 and n_2 independent trials such that these two sequences have exactly n_{12} trials in common. Let the probability of a success and that of a failure in any particular trial be p and $q(=1-p)$ respectively, $0<p<1$. Suppose X_1 and X_2 are the numbers of successes in the two sequences. Then (X_1, X_2) has the bivariate binomial distribution with overlapping trials with parameters n_1, n_2, n_{12}, p and has the pf

$$P(x_1, x_2) = \sum_{i=0}^{n_{12}} \binom{n_1-n_{12}}{x_1-i}\binom{n_{12}}{i}\binom{n_2-n_{12}}{x_2-i} p^{x_1+x_2-i} q^{n_1+n_2-n_{12}+x_1+x_2-i},$$

$$x_i = 0, 1, 2, \ldots, n_i;\ i = 1, 2,$$

$$n_i = 1, 2, \ldots;\ i = 1, 2,$$

$$n_{12} = 1, 2, \ldots;\ n_{12} < n_1, n_2,$$

$$0<p<1,$$

where $q = 1-p$.

$$G(z_1, z_2) = (q+pz_1)^{n_1-n_{12}}(q+pz_2)^{n_2-n_{12}}(q+pz_1z_2)^{n_{12}}.$$

$$\mu_i = n_i p,\ \sigma_{ij} = \begin{cases} n_i pq, & i=j, \\ n_{12}pq, & i \neq j. \end{cases}$$

Some inter-relations

(i) If (X_1, X_2) has the bivariate binomial distribution with overlapping trials with parameters n_1, n_2, n_{12}, p then X_i has the binomial distribution with parameters n_i and p for $i = 1, 2$.

(ii) The bivariate binomial distribution with overlapping trials with parameters n_1, n_2, n_{12}, p tends to the bivariate Poisson distribution with parameters $\lambda_1, \lambda_2, \lambda_{12}$ as $n_1-n_{12}\to\infty$, $n_2-n_{12}\to\infty$, $n_{12}\to\infty$ such that $(n_1-n_{12})p\to\lambda_1$, $(n_2-n_{12})p\to\lambda_2$, $n_{12}p\to\lambda_{12}$, $0<\lambda_1<\infty$, $0<\lambda_2<\infty$, $0<\lambda_{12}<\infty$.

A reference Fischer (1942).

79. THE BIVARIATE POISSON DISTRIBUTION

In the Poisson process with birthrate λ, let X_1 be the number of events in the time interval $(0, t_1)$ and let X_2 be the number of events in the time interval (τ, t_2) where $0<\tau<t_1<t_2$. Then (X_1, X_2) has the bivariate Poisson distribution with parameters $\lambda_1 = \lambda\tau$, $\lambda_2 = \lambda(t_2-t_1)$, $\lambda_{12} = \lambda(t_1-\tau)$. The pf is given by

$$P(x_1, x_2) = \exp\{-(\lambda_1+\lambda_2+\lambda_{12})\} \sum_{i=0}^{\min(x_1, x_2)} \frac{\lambda_1^{x_1-i}\lambda_2^{x_2-i}\lambda_{12}^{i}}{(x_1-i)!(x_2-i)!i!}, \quad x_i = 0, 1, 2, \ldots;\ i = 1, 2,$$

$$0<\lambda_1<\infty,$$

$$0<\lambda_2<\infty,$$

$$0<\lambda_{12}<\infty.$$

$$G(z_1, z_2) = \exp\{\lambda_1(z_1-1)+\lambda_2(z_2-1)+\lambda_{12}(z_1z_2-1)\}$$

$$\mu_i = \lambda_i+\lambda_{12},\quad \sigma_{ij} = \begin{cases} \lambda_i+\lambda_{12}, & i=j, \\ \lambda_{12}, & i\neq j. \end{cases}$$

Some inter-relations

(i) If X_1, X_2, X_3 are three independent Poisson variables with parameters λ_1, λ_2 and λ_{12} respectively and $Y_1 = X_1+X_3$, $Y_2 = X_2+X_3$ then (Y_1, Y_2) has the bivariate Poisson distribution with parameters $\lambda_1, \lambda_2, \lambda_{12}$.

(ii) The bivariate binomial distribution with parameters n, p_{10}, p_{01}, p_{11} tends to the bivariate Poisson distribution with parameters λ_1, λ_2, λ_{12} as $n\to\infty$, $p_{10}\to 0$, $p_{01}\to 0$, $p_{11}\to 0$ such that $np_{10}\to\lambda_1$, $np_{01}\to\lambda_2$ and $np_{11}\to\lambda_{12}$, $0<\lambda_1<\infty$, $0<\lambda_2<\infty$, $0<\lambda_{12}<\infty$.

(iii) If X_1, X_2 have the bivariate Poisson distribution with parameters λ_1, λ_2 and λ_{12}, then X_1+X_2 has the Hermite distribution with parameters $\lambda_1+\lambda_2$ and λ_{12}.

Some references Campbell (1934); Edwards (1962).

80. THE BIVARIATE NEGATIVE BINOMIAL DISTRIBUTION

Suppose given $Y = y$, the distribution of (X_1, X_2) is bivariate Poisson with parameters $\lambda_1 y$, $\lambda_2 y$, $\lambda_{12} y$. Suppose Y has the gamma distribution with parameters α and k. Then the compound bivariate distribution of (X_1, X_2) with respect to the compounder Y is bivariate negative binomial with parameters k, p_{10}, p_{01}, p_{11} where

$$p_{10} = \frac{\lambda_2}{\lambda_1+\lambda_2+\lambda_{12}+\alpha}, \quad p_{01} = \frac{\lambda_1}{\lambda_1+\lambda_2+\lambda_{12}+\alpha} \text{ and } p_{11} = \frac{\alpha}{\lambda_1+\lambda_2+\lambda_{12}+\alpha}.$$

The pf of the bivariate negative binomial distribution with parameters k, p_{10}, p_{01}, p_{11} is given by

$$P(x_1, x_2) = p_{11}^k \sum_{i=0}^{\min(x_1, x_2)} \frac{\Gamma(k+x_1+x_2-i)}{\Gamma(k)(x_1-i)!(x_2-i)!i!} p_{00}^i p_{01}^{x_1-i} p_{10}^{x_2-i}, \qquad \begin{aligned} &x_i = 0, 1, 2, \ldots, \\ &0<k<\infty, \\ &0<p_{10}<1, \\ &0<p_{01}<1, \\ &0<p_{11}<1, \\ &p_{10}+p_{01}+p_{11}<1, \end{aligned}$$

where $\quad p_{00} = 1-p_{10}-p_{01}-p_{11}$.

$$G(z_1, z_2) = p_{11}^k(1-p_{01}z_1-p_{10}z_2-p_{00}z_1z_2)^{-k}.$$

Writing $p_1 = p_{11}+p_{10}$, $p_2 = p_{11}+p_{01}$, $q_1 = 1-p_1$ and $q_2 = 1-p_2$.

$$\mu_i = \frac{kq_i}{p_{11}}, \quad \sigma_{ij} = \begin{cases} \dfrac{kq_i}{p_{11}}\left(1+\dfrac{q_i}{p_{11}}\right) = \mu_i\left(1+\dfrac{q_i}{p_{11}}\right), & i=j, \\[2ex] \dfrac{k}{p_{11}}\left(p_{00}+\dfrac{q_1q_2}{p_{11}}\right), & i\neq j. \end{cases}$$

Some inter-relations

(i) The bivariate Pascal distribution with parameters k, p_{10}, p_{01}, p_{11} is a special case of the bivariate negative binomial distribution with parameters k, p_{10}, p_{01}, p_{11} when k is an integer.

(ii) If (X_1, X_2) has the bivariate negative binomial distribution with parameters k, p_{10}, p_{01}, p_{11} then the distributions of X_1 and X_2 are negative binomial with parameters $k, \dfrac{p_{11}}{(1-p_{10})}$ and $k, \dfrac{p_{11}}{(1-p_{01})}$ respectively.

(iii) Suppose (X_1, X_2, X_3) has the 3-variate negative multinomial distribution with parameters k, p_1, p_2, p_3 and $Y_i = X_i + X_3$, $i = 1, 2$. Then (Y_1, Y_2) has the bivariate negative binomial distribution with parameters k, p_2, p_1, p_0.

(iv) The bivariate negative binomial distribution with parameters $k, p_{10}, p_{01}, p_{11}$ results if the Poisson distribution with parameter $-k \log p_{11}$ is generalized by the bivariate logarithmic distribution with parameters p_{01}, p_{10}, p_{00}.

(v) The bivariate negative binomial distribution with parameters $k, p_{10}, p_{01}, p_{11}$ tends to the bivariate Poisson distribution with parameters $\lambda_1, \lambda_2, \lambda_{12}$ as $k \to \infty$, $p_{10} \to 0$, $p_{01} \to 0$, $p_{00} \to 0$ such that $kp_{01} \to \lambda_1$, $kp_{10} \to \lambda_2$ and $kp_{00} \to \lambda_{12}$, $0 < \lambda_i < \infty$, $i = 1, 2$, $0 < \lambda_{12} < \infty$.

(vi) The origin-truncated bivariate negative binomial distribution with parameters $k, p_{10}, p_{01}, p_{11}$ tends to the bivariate logarithmic distribution with parameters p_{01}, p_{10}, p_{00} as $k \to 0$.

Some references Wishart (1949); Wiid (1957-58).

81. THE BIVARIATE PASCAL DISTRIBUTION

Consider an experiment of observing individuals one after another for two characters A_1 and A_2 until exactly k individuals possessing both the characters are observed. Suppose the probability that any particular individual possesses, independently of other individuals, the character A_1 but not A_2 is p_{10}, the character A_2 but not A_1 is p_{01}, both the characters is p_{11} and none of the two characters is p_{00}. Let X_i denote the number of individuals possessing character A_i observed in the experiment, $i = 1, 2$. Then (X_1, X_2) has the bivariate Pascal distribution with parameters $k, p_{10}, p_{01}, p_{11}$. Its pf is

$$P(x_1, x_2) = p_{11}^k \sum_{i=0}^{\min(x_1, x_2)} \frac{(k+x_1+x_2-i-1)!}{k!(x_1-i)!(x_2-i)!i!} p_{00}^i p_{01}^{x_1-i} p_{10}^{x_2-i} \quad x_i = 0, 1, 2, \ldots;\ i = 1, 2,$$

$$k = 1, 2, \ldots,$$

$$0 < p_{10} < 1,$$

$$0 < p_{01} < 1,$$

$$0 < p_{11} < 1,$$

$$p_{10} + p_{01} + p_{11} < 1,$$

where $\quad p_{00} = 1 - p_{10} - p_{01} - p_{11}$.

$$G(z_1, z_2) = p_{11}^k (1 - p_{01} z_1 - p_{10} z_2 - p_{00} z_1 z_2)^{-k}.$$

Writing $p_1 = p_{11} + p_{10}$, $\quad p_2 = p_{11} + p_{01}$

$q_1 = 1 - p_1$, $\quad q_2 = 1 - p_2$,

$$\mu_i = \frac{kq_i}{p_{11}}, \quad \sigma_{ij} = \begin{cases} \dfrac{kq_i}{p_{11}}\left(1 + \dfrac{q_i}{p_{11}}\right) = \mu_i\left(1 + \dfrac{q_i}{p_{11}}\right), & i = j \\ \dfrac{k}{p_{11}}\left(p_{00} + \dfrac{q_1 q_2}{p_{11}}\right), & i \neq j. \end{cases}$$

Some inter-relations

The bivariate Pascal distribution with parameters $k, p_{10}, p_{01}, p_{11}$ is a special case of the bivariate negative binomial distribution with parameters $k, p_{10}, p_{01}, p_{11}$ when k is an integer.

For other inter-relations see the bivariate negative binomial distribution.

Some references Wishart (1949); Wiid (1957-58).

82. THE BIVARIATE LOGARITHMIC DISTRIBUTION

(Bivariate logarithmic series distribution)

The bivariate logarithmic distribution arises as the limit of the origin-truncated bivariate negative binomial distribution with parameters k, p_{10}, p_{01}, p_{11} as $k \to 0$. The pf of the bivariate logarithmic distribution with parameters θ_1, θ_2, θ_{12} is given by

$$P(x_1, x_2) = \sum_{i=0}^{\min(x_1, x_2)} \frac{(x_1+x_2-i-1)!}{(x_1-i)!(x_2-i)!i!} \theta_1^{x_1-i}\theta_2^{x_2-i}\theta_{12}^i, \quad x_i = 0, 1, 2, \ldots;\ i = 1, 2,$$

$$x_1+x_2 \neq 0,$$

$$0<\theta_i<1;\ i = 1, 2,$$

$$0<\theta_{12}<1,\ \theta_1+\theta_2+\theta_{12}<1.$$

$$G(z_1, z_2) = \frac{\log(1-\theta_1 z_1-\theta_2 z_2-\theta_{12}z_1 z_2)}{\log(1-\theta_1-\theta_2-\theta_{12})}.$$

$$\mu_i = \frac{-\theta_i-\theta_{12}}{(1-\theta_1-\theta_2-\theta_{12})\log(1-\theta_1-\theta_2-\theta_{12})} = \frac{\theta_i+\theta_{12}}{\theta_0 L},$$

where $\theta_0 = 1-\theta_1-\theta_2-\theta_{12}$, $L = -\log\theta_0$.

$$\sigma_{ij} = \begin{cases} \dfrac{(\theta_i+\theta_{12})}{\theta_0^2 L}\left\{\theta_0+\theta_i+\theta_{12}-\dfrac{(\theta_i+\theta_{12})}{L}\right\} = \mu_i\left[\dfrac{\theta_i+\theta_{12}}{\theta_0}-\mu_i+1\right], & i = j, \\ \dfrac{1}{\theta_0^2 L}\left\{\theta_1\theta_2+\theta_{12}-\dfrac{(\theta_1+\theta_{12})(\theta_2+\theta_{12})}{L}\right\} = \mu_i\left[\dfrac{\theta_j+\theta_{12}}{\theta_0}-\mu_j+\dfrac{\theta_{12}}{\theta_i+\theta_{12}}\right], & i \neq j. \end{cases}$$

Some inter-relations

(i) Let (X_1, X_2, X_3) have the 3-variate logarithmic distribution with parameters θ_1, θ_2, θ_3 and let $Y_1 = X_1+X_3$, $Y_2 = X_2+X_3$. Then (Y_1, Y_2) has the bivariate logarithmic distribution with parameters θ_1, θ_2, θ_3.

(ii) The bivariate negative binomial distribution with parameters k, p_{10}, p_{01}, p_{11} results if the Poisson distribution with parameters $-k\log p_{11}$ is generalized by the bivariate logarithmic distribution with parameters p_{01}, p_{10}, p_{00}.

83. THE BIVARIATE HYPERGEOMETRIC DISTRIBUTION

(Fourfold factorial binomial distribution, Fourfold hypergeometric distribution)

Consider a population of N individuals. Suppose M_i of these have the character A_i, $i = 1, 2$, and M_{11} have both the characters A_1 and A_2. Suppose n individuals are selected without replacement from the population, and X_i denotes the number of individuals with the character A_i in the sample. Let $M_{10} = M_1 - M_{11}$, $M_{01} = M_2 - M_{11}$. Then (X_1, X_2) has the bivariate hypergeometric distribution with parameters N, M_{10}, M_{01}, M_{12}, n. Its pf is given by

$$P(x_1, x_2) = \sum_{i=0}^{\min(x_1, x_2)} \frac{\binom{M_{10}}{x_1 - i}\binom{M_{01}}{x_2 - i}\binom{M_{11}}{i}\binom{M_{00}}{n - x_1 - x_2 + i}}{\binom{N}{n}} \quad \begin{array}{l} x_i = 0, 1, 2, \ldots, n;\ i = 1, 2, \\ N = 1, 2, \ldots, \\ M_{10} = 1, 2, \ldots;\ M_{10} < N, \\ M_{01} = 1, 2, \ldots;\ M_{01} < N, \\ M_{11} = 1, 2, \ldots, \\ M_{10} + M_{01} + M_{11} < N, \\ n = 1, 2, \ldots,\ n < N, \end{array}$$

where $M_{00} = N - M_{10} - M_{01} - M_{11}$.

$$G(z_1, z_2) = \frac{M_{00}^{(n)}}{N^{(n)}} [{}_4F_1(-n, -M_{10}, -M_{01} - M_{11};\ M_{00} - n + 1;\ z_1, z_2, z_1 z_2)].$$

$$\mu_i = n\left(\frac{M_i}{N}\right).$$

$$\sigma_{ij} = \begin{cases} n\left(\frac{M_i}{N}\right)\left(1 - \frac{M_i}{N}\right)\left(\frac{N-n}{N-1}\right) = \mu_i\left(1 - \frac{M_i}{N}\right)\left(\frac{N-n}{N-1}\right), & i = j, \\ n\left(\frac{M_{11}}{N} - \frac{M_{10}}{N} \cdot \frac{M_{01}}{N}\right)\left(\frac{N-n}{N-1}\right), & i \neq j. \end{cases}$$

Some inter-relations

(i) The bivariate hypergeometric distribution with parameters N, M_{10}, M_{01}, M_{11}, n tends to the bivariate binomial distribution with parameters n, p_{10}, p_{01}, p_{11} as $N \to \infty$, $M_{10} \to \infty$, $M_{01} \to \infty$, $M_{11} \to \infty$ such that

$$\frac{M_{10}}{N} \to p_{10},\ \frac{M_{01}}{N} \to p_{01},\ \frac{M_{11}}{N} \to p_{11},\ 0 < p_{10} < 1,\ 0 < p_{01} < 1,\ 0 < p_{11} < 1,\ 0 < p_{10} + p_{01} + p_{11} < 1.$$

(ii) The bivariate hypergeometric distribution with parameters N, M_{10}, M_{01}, M_{11}, n tends to the bivariate Poisson distribution with parameters λ_1, λ_2, λ_{12} as $N \to \infty$, $M_{10} \to \infty$, $M_{01} \to \infty$, $M_{11} \to \infty$, $n \to \infty$ such that $\frac{nM_{10}}{N} \to \lambda_1$, $\frac{nM_{01}}{N} \to \lambda_2$, $\frac{nM_{11}}{N} \to \lambda_{12}$, $0 < \lambda_1 < \infty$, $0 < \lambda_2 < \infty$, $0 < \lambda_{12} < \infty$.

(iii) Let (X_1, X_2, X_3) have a 3-variate hypergeometric distribution with parameters N, M_1, M_2, M_3, n and let $Y_1 = X_1 + X_3$, $Y_2 = X_2 + X_3$. Then (Y_1, Y_2) has the bivariate hypergeometric distribution with parameters N, M_1, M_2, M_3, n.

Some references Steyn (1957); Wiid (1957-58).

84. THE DOUBLE HYPERGEOMETRIC DISTRIBUTION

(Double hypergeometric series)

Suppose an urn contains N balls, M of which are white and $N-M$ black. Let n_1 balls be drawn from the urn at random without replacement. Let another sample of n_2 balls be drawn from the urn without replacement and without replacing the n_1 balls previously drawn. If X_1, X_2 denote the number of white balls in the first and in the second sample respectively, then (X_1, X_2) has the double hypergeometric distribution with parameters N, M, n_1, n_2. It has the pf

$$P(x_1, x_2) = \frac{\binom{M}{x_1}\binom{N-M}{n_1-x_1}}{\binom{N}{n_1}} \cdot \frac{\binom{M-x_1}{x_2}\binom{N-n_1-M+x_1}{n_2-x_2}}{\binom{N-n_1}{n_2}}$$

$x_i = a_i, a_i+1, a_i+2, \ldots, b_i;\ i = 1, 2,$

$N = 1, 2, \ldots,$

$M = 1, 2, \ldots, M<N,$

$n_1 = 1, 2, \ldots, n_1<N,$

$n_2 = 1, 2, \ldots, n_2<N-n_2,$

where $a_1 = \max(0, n_1-N+M)$, $b_1 = \min(n_1, M)$,

$a_2 = \max(0, n_2-N+n_1+M-x_1)$, $b_2 = \min(n_2, M-x_1)$.

$$\mu_i = n_i\left(\frac{M}{N}\right).$$

$$\sigma_{ij} = \begin{cases} n_i\left(\frac{M}{N}\right)\left(1-\frac{M}{N}\right)\left(1-\frac{n_i-1}{N-1}\right) = \mu_i\left(1-\frac{M}{N}\right)\left(1-\frac{n_i-1}{N-1}\right), & i=j, \\ \left(\frac{-n_1n_2}{N-1}\right)\left(\frac{M}{N}\right)\left(1-\frac{M}{N}\right), & i \neq j. \end{cases}$$

Some inter-relations

(i) If (X_1, X_2) has the double hypergeometric distribution with parameters N, M, n_1, n_2 then $X = X_1+X_2$ has the hypergeometric distribution with parameters N, M, n_1+n_2.

(ii) Suppose (X_1, X_2) has the double hypergeometric distribution with parameters N, M, n_1, n_2. Then the mixture on M of the conditional distribution of X_2, given $X_1 = x_1$, where M has the uniform distribution with the range $\{0, 1, 2, \ldots, N\}$, is the negative hypergeometric distribution with parameters n_1+2, x_1+1, n_2.

(iii) The double hypergeometric distribution with parameters N, M, n_1, n_2 tends to the joint distribution of two independent binomial variables with parameters n_1 and p and n_2 and p respectively as $N\to\infty$, $M\to\infty$ such that $\frac{M}{N}\to p . 0<p<1$.

Some references Pearson (1924*b*); Särndal (1965).

85. THE BIVARIATE NEYMAN'S TYPE A DISTRIBUTION (i)

The bivariate Neyman's type A distribution (i) with parameters λ, λ_1, λ_2, λ_{12} results if the Poisson distribution with parameter λ is generalized by the bivariate Poisson distribution with parameters λ_1, λ_2, λ_{12}.

$$P(x_1, x_2) = e^{-\lambda} \sum_{n=0}^{\infty} \left\{ \frac{\lambda^n}{n!} e^{-n(\lambda_1+\lambda_2+\lambda_{12})} \sum_{i=0}^{\min(x_1, x_2)} \frac{(n\lambda_1)^{x_1-i}(n\lambda_2)^{x_2-i}(n\lambda_{12}^i)}{(x_1-i)!(x_2-1)!i!} \right\} \quad x_i = 0, 1, 2, \ldots; i = 1, 2,$$

$0<\lambda<\infty$,

$0<\lambda_i<\infty$; $i = 1, 2$,

$0<\lambda_{12}<\infty$.

$$\mu_i = \lambda(\lambda_i+\lambda_{12}).$$

$$\sigma_{ij} = \begin{cases} \lambda(\lambda_i+\lambda_{12})(\lambda_i+\lambda_{12}+1), & i = j, \\ \lambda\{(\lambda_1+\lambda_{12})(\lambda_2+\lambda_{12})+\lambda_{12}\}, & i \neq j. \end{cases}$$

$$G(z_1, z_2) = \exp\left[\lambda\{e^{\lambda_1(z_1-1)+\lambda_2(z_2-1)+\lambda_{12}(z_1z_2-1)} - 1\}\right].$$

A reference Holgate (1966*b*).

86. THE BIVARIATE NEYMAN'S TYPE A DISTRIBUTION (ii)

Suppose $H(z_1, z_2)$ is the pgf of the bivariate Poisson distribution with parameters λ_1, λ_2, λ_{12} and $g(z_1)$ and $g(z_2)$ are pgfs of the two Poisson distributions with parameters θ_1 and θ_2, respectively all the three distributions being mutually independent. Then the bivariate Neyman's type A (ii) distribution with parameters λ_1, λ_2, λ_{12}, θ_1, θ_2 is defined by the pgf $H(g(z_1), g(z_2))$. The components X_1, X_2 of the corresponding random vector can take the values 0, 1, 2, ..., and $0<\lambda_1<\infty$, $0<\lambda_2<\infty$, $0<\lambda_{12}<\infty$, $0<\theta_1<\infty$, $0<\theta_2<\infty$.

$$\mu_i = (\lambda_i+\lambda_{12})\theta_i$$

$$\sigma_{ij} = \begin{cases} (\lambda_i+\lambda_{12})(\theta_i)(\theta_i+1), & i = j, \\ \lambda_{12}\theta_1\theta_2, & i \neq j. \end{cases}$$

$$G(z_1, z_2) = \exp\left[\lambda_1\{e^{\theta_1(z_1-1)} - 1\} + \lambda_2\{e^{\theta_2(z_2-1)} - 1\} + \lambda_{12}\{e^{\theta_1(z_1-1)+\theta_2(z_2-1)} - 1\}\right].$$

A reference Holgate (1966*b*).

87. THE BIVARIATE NEYMAN'S TYPE A DISTRIBUTION (iii)

Suppose Y_1, Y_2, Y_3 are independent random variables, Y_i having the Neyman's type A distribution with two parameters λ_i and θ, respectively, $i = 1, 2, 3$. Let $X_i = Y_i + Y_3$, $i = 1, 2$. Then (X_1, X_2) has the bivariate Neyman's type A (iii) distribution with parameters λ_1, λ_2, λ_3, θ. $x_i = 0, 1, 2, \ldots$; $i = 1, 2$; $0<\lambda_i<\infty$; $i = 1, 2, 3$; $0<\theta<\infty$.

$$\mu_i = (\lambda_i+\lambda_3)\theta$$

$$\sigma_{ij} = \begin{cases} (\lambda_i+\lambda_3)(\theta)(\theta+1), & i = j, \\ \lambda_3\theta(\theta+1), & i \neq j. \end{cases}$$

$$G(z_1, z_2) = \exp\left[\lambda_1\{e^{\theta(z_1-1)} - 1\} + \lambda_2\{e^{\theta(z_2-1)} - 1\} + \theta_3[e^{\theta(z_1 z_2-1} - 1\}\right].$$

A reference Holgate (1966*b*).

88. THE MULTINOMIAL DISTRIBUTION

(The bernoullian frequency function of s-variables)

Consider a sequence of n independent trials each of which has $s+1$ mutually exclusive outcomes $A_0, A_1, A_2, \ldots, A_s$, such that the probability of A_i in a single trial is p_i, $0<p_i<1$, $i = 0, 1, 2, \ldots, s$, $\sum_{i=0}^{s} p_i = 1$. If X_i denotes the number of occurrences of A_i in the sequence of n trials, $i = 0, 1, 2, \ldots, s$, then the random vector $(X_1, X_2, \ldots, X_s)$ has the s-dimensional (or s-variate) multinomial distribution with parameters $n, p_1, p_2, \ldots, p_s$. Its pf is given by

$$P(x_1, x_2, \ldots, x_s) = \frac{n!}{x_0!x_1!x_2!\ldots x_s!} p_0^{x_0}p_1^{x_1}p_2^{x_2}\ldots p_s^{x_s} \quad x_i = 0, 1, 2, \ldots, n;\ i = 1, 2, \ldots, s,$$

$$0 \leqq \sum_{i=1}^{s} x_i \leqq n,$$

$$n = 1, 2, \ldots,$$

$$0<p_i<1;\ i = 1, 2, \ldots, s,$$

$$0< \sum_{i=1}^{s} p_i<1,$$

where $\quad x_0 = n-x_1-x_2-\ldots-x_s$ and $p_0 = 1-p_1-p_2-\ldots-p_s$.

$G(z_1, z_2, \ldots, z_s) = (p_0+p_1z_1+p_2z_2+\ldots+p_sz_s)^n$.

$$\mu_i = np_i,\ \sigma_{ij} = \begin{cases} np_i(1-p_i) = \mu_i(1-p_i), & i=j, \\ -np_ip_j = \dfrac{-\mu_i\mu_j}{n}, & i \neq j. \end{cases}$$

Some inter-relations

(i) The binomial distribution with parameters n and p is a special case of the s-dimensiona multinomial distribution with parameters $n, p_1, p_2, \ldots, p_s$, when $s = 1$ and $p_1 = p$.

(ii) Suppose that $X_0, X_1, X_2, \ldots, X_s$ have $s+1$ independent Poisson distributions with parameters $\lambda_0, \lambda_1, \lambda_2, \ldots, \lambda_s$ respectively. Then, given $\sum_{i=0}^{s} X_i = n$, the conditional distribution of $(X_1, X_2, \ldots, X_s)$ is s-variate multinomial with parameters $n, p_1, p_2, \ldots, p_s$ where $p_i = \dfrac{\lambda_i}{\lambda_0+\lambda_1+\lambda_2+\ldots+\lambda_s}$, $i = 1, 2, \ldots, s$.

(iii) The mixture on n of the s-variate multinomial distribution with parameters $n, p_1, p_2, \ldots, p_s$, where n has the negative binomial distribution with parameters k and p, is the s-dimensional negative multinomial distribution with parameters $k, \dfrac{qp_1}{1-qp_0}, \dfrac{qp_2}{1-qp_0}, \ldots, \dfrac{qp_s}{1-qp_0}$, where $q = 1-p$ and $p_0 = 1-p_1-p_2-\ldots-p_s$.

(iv) The mixture on n of the s-variate multinomial distribution with parameters $n, p_1, p_2, \ldots, p_s$, where n has the Poisson distribution with parameter λ, is the s-variate distribution whose s components are independent Poisson variables with parameters $\lambda p_1, \lambda p_2, \ldots, \lambda p_s$ respectively.

(v) The mixture on n of the s-variate multinomial distribution with parameters $n, p_1, p_2, \ldots, p_s$ where n has the logarithmic distribution with parameter θ is the s-dimensional modified logarithmic distribution with parameters $\frac{\log(1-\theta p_0)}{\log(1-\theta)}, \frac{\theta p_1}{1-\theta p_0}, \frac{\theta p_2}{1-\theta p_0}, \ldots, \frac{\theta p_s}{1-\theta p_0}$,

(vi) The mixture on $(p_1, p_2, \ldots, p_s)$ of the s-variate multinomial distribution with parameters $n, p_1, p_2, \ldots, p_s$, where $(p_1, p_2, \ldots, p_s)$ has the s-variate Dirichlet distribution with parameters $M_1, M_2, \ldots, M_s, N - \sum_{i=1}^{s} M_i$, is the s-variate negative hypergeometric distribution with parameters $N, M_1, M_2, \ldots, M_s, n$.

(vii) The s-dimensional multinomial distribution with parameters $n, p_1, p_2, \ldots, p_s$ tends to the joint distribution of s independent Poisson variables with parameters $\lambda_1, \lambda_2, \ldots, \lambda_s$ respectively as $n \to \infty$, $p_i \to 0$ such that $np_i \to \lambda_i$, $0 < \lambda_i < \infty$, $i = 1, 2, \ldots, s$.

(viii) The s-dimensional hypergeometric distribution with parameters $N, M_1, M_2, \ldots, M_s, n$ tends to the s-variate multinomial distribution with parameters $n, p_1, p_2, \ldots, p_s$ as $N \to \infty$, $M_i \to \infty$ such that $\frac{M_i}{N} \to p_i$, $0 < p_i < 1$, $i = 1, 2, \ldots, s$, $\sum_{i=1}^{s} p_i < 1$.

(ix) The s-dimensional negative hypergeometric distribution with parameters $N, M_1, M_2, \ldots, M_s, n$ tends to the s-variate multinomial distribution with parameters $n, p_1, p_2, \ldots, p_s$ as $N \to \infty$, $M_i \to \infty$ such that $\frac{M_i}{N} \to p_i$, $i = 1, 2, \ldots, s$, $\sum_{i=1}^{s} p_i < 1$.

Some references Kendall and Stuart (1958), p. 141; Patil and Bildikar (1966*a*); Wilks (1962), p. 138.

89. THE SINGULAR MULTINOMIAL DISTRIBUTION

The pf of the s-variate singular multinomial distribution with parameters $n, p_1, p_2, \ldots, p_s$ is given by

$$P(x_1, x_2, \ldots, x_s) = \frac{n!}{x_1! x_2! \ldots x_s!} p_1^{x_1} p_2^{x_2} \ldots p_s^{x_s}, \quad x_i = 0, 1, 2, \ldots, n;\ i = 1, 2, \ldots, s,$$

$$\sum_{i=1}^{s} x_i = n,$$

$$0 < p_i < 1;\ i = 1, 2, \ldots, s,$$

$$\sum_{i=1}^{s} p_i = 1.$$

$$G(z_1, z_2, \ldots, z_s) = (p_1 z_1 + p_2 z_2 + \ldots + p_s z_s)^n.$$

$$\mu_i = np_i, \quad \sigma_{ij} = \begin{cases} np_i(1-p_i), & i = j \\ -np_i p_j, & i \neq j. \end{cases}$$

Some inter-relations

(i) If $(X_1, X_2, \ldots, X_s)$ has the s-variate singular multinomial distribution with parameters n, $p_1, p_2, \ldots, p_s$ then $(X_1, X_2, \ldots, X_{s-1})$ has the $(s-1)$-variate multinomial distribution with parameters $n, p_1, p_2, \ldots, p_{s-1}$. If $(Y_1, Y_2, \ldots, Y_s)$ has the s-variate multinomial distribution with parameters $n, p_1, p_2, \ldots, p_s$ then $(Y_0, Y_1, Y_2, \ldots, Y_s)$ where $Y_0 = Y_1 - Y_2 - \ldots - Y_s$ has the $(s+1)$-variate singular multinomial distribution with parameters $n, p_0, p_1, p_2, \ldots, p_s$.

(ii) If $(X_1, X_2, \ldots, X_s)$ has the s-variate logarithmic distribution with parameters $\theta_1, \theta_2, \ldots, \theta_s$ then the conditional distribution of $(X_1, X_2, \ldots, X_s)$, given $\sum_{i=1}^{s} X_i = n$, is the s-variate singular multinomial with parameters $n, p_1, p_2, \ldots, p_s$, where $p_i = \dfrac{\theta_i}{\theta_1+\theta_2+\ldots+\theta_s}$, $i = 1, 2, \ldots, s$.

Some references Lukacs and Laha (1964), p. 31; Patil and Bildikar (1966*a*).

90. THE NEGATIVE MULTINOMIAL DISTRIBUTION

(Multivariate negative binomial distribution)

Consider a sequence of independent trials in which each trial has $s+1$ mutually exclusive outcomes $A_0, A_1, A_2, \ldots, A_s$ with probabilities $p_0, p_1, p_2, \ldots, p_s$ respectively, $0<p_i<1$, $i = 0, 1, 2, \ldots, s$ $\sum_{i=0}^{s} p_i = 1$. Let $X_1, X_2, \ldots, X_s$ be the numbers of occurrences of $A_1, A_2, \ldots, A_s$ respectively before A_0 occurs exactly k times. Then $(X_1, X_2, \ldots, X_s)$ has the s-variate negative multinomial distribution with parameters $k, p_1, p_2, \ldots, p_s$.

Alternatively the s-variate negative multinomial distribution with parameters $k, p_1, p_2, \ldots, p_s$ arises as the mixture on u of the distribution of $(X_1, X_2, \ldots, X_s)$ where $X_1, X_2, \ldots, X_s$ are independent Poisson variables with parameters $\lambda_1 u, \lambda_2 u, \ldots, \lambda_s u$ and u has the gamma distribution with parameters α and k. Here $p_i = \dfrac{\lambda_i}{\alpha + \sum_{j=1}^{s} \lambda_j}$, $i = 1, 2, \ldots, s$. The pf of the s-variate negative multinomial distribution with parameters $k, p_1, p_2, \ldots, p_s$ is given by

$$P(x_1, x_2, \ldots, x_s) = \frac{\Gamma(k+x_1+x_2+\ldots+x_s)}{\Gamma(k)x_1!x_2!\ldots x_s!} p_0^k p_1^{x_1} p_2^{x_2} \ldots p_s^{x_s}, \quad x_i = 0, 1, 2, \ldots;\ i = 1, 2, \ldots, s,$$

$$0<k<\infty,$$

$$0<p_i<1;\ i = 1;\ 2, \ldots, s,$$

$$\sum_{i=1}^{s} p_i<1,$$

where $\quad p_0 = 1-p_1-p_2-\ldots-p_s$.

$$G(z_1, z_2, \ldots, z_s) = p_0^k(1-p_1z_1-p_2z_2-\ldots-p_sz_s)^{-k}.$$

$$\mu_i = \frac{kp_i}{p_0}, \quad \sigma_{ij} = \begin{cases} \dfrac{kp_i}{p_0}\left(1+\dfrac{p_i}{p_0}\right) = \mu_i\left(1+\dfrac{p_i}{p_0}\right), & i = j, \\ \dfrac{kp_ip_j}{p_0^2} = \dfrac{\mu_i\mu_j}{k}, & i \neq . \end{cases}$$

Some inter-relations

(i) The negative binomial distribution with parameters k and p is a special case of the s-variate negative multinomial distribution with parameters $k, p_1, p_2, \ldots, p_s$, when $s = 1$ and $p_1 = 1-p$.

(ii) If (X_1, X_2, X_3) has the 3-variate negative multinomial distribution with parameters k, p_1, p_2, p_3 and $Y_i = X_i+X_3$; $i = 1, 2$, then (Y_1, Y_2) has the bivariate negative binomial distribution with parameters k, p_2, p_1, p_0.

(iii) The s-variate negative multinomial distribution with parameters $k, p_1, p_2, \ldots, p_s$ results if the Poisson distribution with parameter $-k \log p_0$ is generalized by the s-variate logarithmic distribution with parameters $p_1, p_2, \ldots, p_s$.

(iv) The mixture on n of the s-variate multinomial distribution with parameters $n, p_1, p_2, \ldots, p_s$, where n has the negative binomial distribution with parameters k and p, is the s-variate negative multinomial with parameters $k, \dfrac{qp_1}{1-qp_0}, \dfrac{qp_2}{1-qp_0}, \ldots, \dfrac{qp_s}{1-qp_0}$, where $q = 1-p$.

(v) The origin-truncated s-variate negative multinomial distribution with parameters $k, p_1, p_2, \ldots, p_s$ tends to the s-variate logarithmic distribution with parameters $p_1, p_2, \ldots, p_s$ as $k \to 0$.

(vi) The s-variate negative multinomial distribution with parameters $k, p_1, p_2, \ldots, p_s$ tends to the joint distribution of s independent Poisson variables with parameters $\lambda_1, \lambda_2, \ldots, \lambda_s$ respectively as $k \to \infty$, $p_i \to 0$ such that $kp_i \to \lambda_i$, $0 < \lambda_i < \infty$, $i = 1, 2, \ldots, s$.

(vii) The s-variate inverse hypergeometric distribution with parameters $N, M_1, M_2, \ldots, M_s, k$ tends to the s-variate negative multinomial distribution with parameters $k, p_1, p_2, \ldots, p_s$ as $N \to \infty$, $M_i \to \infty$ such that $\dfrac{M_i}{N} \to p_i$, $0 < p_i < 1$, $i = 1, 2, \ldots, s$, $\sum_{i=1}^{s} p_i < 1$.

Some references Bates and Neyman (1952*a*); Patil (1965*c*); Sibuya, Yoshimura and Shimizu (1964).

91. THE MULTIVARIATE LOGARITHMIC DISTRIBUTION

(The multivariate logarithmic series distribution)

Suppose $(X_1, X_2, \ldots, X_s)$ has the s-variate negative multinomial distribution with parameters $k, p_1, p_2, \ldots, p_s$. Then the conditional distribution of $(X_1, X_2, \ldots, X_s)$ given that $(X_1, X_2, \ldots, X_s) \neq (0, 0, \ldots, 0)$ tends to the s-variate logarithmic distribution with parameters $\theta_1, \theta_2, \ldots, \theta_s$ as $k \to 0$, where $\theta_i = 1-p_i$, $i = 1, 2, \ldots, s$.

The pf of the s-variate logarithmic distribution with parameters $\theta_1, \theta_2, \ldots, \theta_s$ is given by

$$P(x_1, x_2, \ldots, x_s) = \frac{(x_1+x_2+\ldots+x_s-1)!}{x_1!x_2!\ldots x_s!} \cdot \frac{\theta_1^{x_1}\theta_2^{x_2}\ldots\theta_s^{x_s}}{[-\log(1-\theta_1-\theta_2\ldots-\theta_s)]}, \quad x_i = 0, 1, 2, \ldots; \; i = 1, 2, \ldots, s,$$

$$\sum_{i=1}^{s} x_i \neq 0,$$

$$0<\theta_i<1; \; i = 1, 2, \ldots, s,$$

$$\sum_{i=1}^{s} \theta_i<1.$$

$$G(z_1, z_2, \ldots, z_s) = \frac{\log(1-\theta_1 z_1-\theta_2 z_2-\ldots-\theta_s z_s)}{\log(1-\theta_1-\theta_2-\ldots-\theta_s)}.$$

Writing $\theta_0 = 1-\theta_1-\theta_2-\ldots-\theta_s$ and $L = -\log\theta_0$,

$$\mu_i = \frac{\theta_i}{\theta_0 L}, \quad \sigma_{ij} = \begin{cases} \mu_i\left[\dfrac{\theta_i}{\theta_0} - \mu_i + 1\right], & i = j, \\ \mu_i\left[\dfrac{\theta_j}{\theta_0} - \mu_j\right], & i \neq j. \end{cases}$$

Some inter-relations

(i) The s-variate negative multinomial distribution with parameters $k, p_1, p_2, \ldots, p_s$ results if the Poisson distribution with parameter $-k\log p_0$ is generalized by the s-variate logarithmic distribution with parameters $p_1, p_2, \ldots, p_s$.

(ii) If $(X_1, X_2, \ldots, X_s)$ has the s-variate logarithmic distribution with parameters $\theta_1, \theta_2, \ldots, \theta_s$, then for $r<s$, $(X_1, X_2, \ldots, X_r)$ has the r-variate modified logarithmic distribution with parameters $\delta, \theta_1', \theta_2', \ldots, \theta_r'$ where

$$\delta = \frac{\log(1-\theta_{r+1}-\theta_{r+2}-\ldots-\theta_s)}{L},$$

$$\theta_i' = \frac{\theta_i}{(1-\theta_{r+1}-\theta_{r+2}-\ldots-\theta_s)}, \quad i = 1, 2, \ldots, r,$$

$$L = -\log(1-\theta_1-\theta_2-\ldots-\theta_s).$$

(iii) If $(X_1, X_2, \ldots, X_s)$ has the s-variate logarithmic distribution with parameters $\theta_1, \theta_2, \ldots, \theta_s$, then, given $\sum_{i=1}^{s} X_i = n$, the conditional distribution of $(X_1, X_2, \ldots, X_s)$ is singular multinomial with parameters $n, p_1, p_2, \ldots, p_s$, where $p_i = \dfrac{\theta_i}{\theta_1+\theta_2+\ldots+\theta_s}$, $i = 1, 2, \ldots, s$.

(iv) The s-variate modified logarithmic distribution with parameters $\delta, \theta_1, \theta_2, \ldots, \theta_s$ tends to the s-variate logarithmic distribution with parameters $\theta_1, \theta_2, \ldots, \theta_s$ as $\delta \to 0$.

Some references Khatri (1959); Patil and Bildikar (1966*a*).

92. THE MULTIVARIATE MODIFIED LOGARITHMIC DISTRIBUTION

(The multivariate modified logarithmic series distribution)

The s-variate modified logarithmic distribution with parameters $\delta, \theta_1, \theta_2, \ldots, \theta_s$ arises as the mixture on n of the s-variate multinomial distribution with parameters $n, p_1, p_2, \ldots, p_s$ where n has the logarithmic distribution with parameter θ. Here $\delta = \dfrac{\log(1-\theta p_0)}{\log(1-\theta)}$, $\theta_i = \dfrac{\theta p_i}{1-\theta p_0}$, $i = 1, 2, \ldots, s$. More generally, its pf is given by

$$P(x_1, x_2, \ldots, x_s) = \begin{cases} \delta & x_1 = x_2 = \ldots = x_s = 0 \\ \dfrac{(1-\delta)(x_1+x_2+\ldots+x_s-1)!}{x_1!x_2!\ldots x_s!} \dfrac{\theta_1^{x_1}\theta_2^{x_2}\ldots\theta_s^{x_s}}{[-\log(1-\theta_1-\theta_2-\ldots-\theta_s)]}, \end{cases}$$

$$x_i = 0, 1, 2, \ldots;\ i = 1, 2, \ldots, s,$$

$$\sum_{i=1}^{s} x_i > 0,$$

$$0 < \delta < 1,$$

$$0 < \theta_i < 1;\ i = 1, 2, \ldots, s,$$

$$\sum_{i=1}^{s} \theta_i < 1.$$

$$G(z_1, z_2, \ldots, z_s) = \delta + (1-\delta)\left[\frac{\log(1-\theta_1 z_1-\theta_2 z_2-\ldots-\theta_s z_s)}{\log(1-\theta_1-\theta_2-\ldots-\theta_s)}\right].$$

Writing $\quad L = -\log(1-\theta_1-\theta_2-\ldots-\theta_s)$, $\theta_0 = 1-\theta_1-\theta_2-\ldots-\theta_s$,

$$\mu_i = (1-\delta)\frac{\theta_i}{\theta_0 L},\quad \sigma_{ij} = \begin{cases} \mu_i\left(\dfrac{\theta_i}{\theta_0}-\mu_i+1\right), & i = j, \\ \mu_i\left(\dfrac{\theta_j}{\theta_0}-\mu_j\right), & i \neq j. \end{cases}$$

Some inter-relations

(i) If $(X_1, X_2, \ldots, X_s)$ has the s-variate logarithmic distribution with parameters $\theta_1, \theta_2, \ldots, \theta_s$, then for $r<s$, $(X_1, X_2, \ldots, X_r)$ has the r-variate modified logarithmic distribution with parameters $\delta, \theta'_1, \theta'_2, \ldots, \theta'_r$, where

$$\delta = \frac{-\log(1-\theta_{r+1}-\theta_{r+2}-\ldots-\theta_s)}{L},$$

$$L = -\log(1-\theta_1-\theta_2-\ldots-\theta_s),$$

$$\theta'_j = \frac{\theta_i}{1-\theta_{r+1}-\theta_{r+2}-\ldots-\theta_s},\quad i = 1, 2, \ldots, r.$$

(ii) The s-variate modified logarithmic distribution with parameters $\delta, \theta_1, \theta_2, \ldots, \theta_s$ tends to the s-variate logarithmic distribution with parameters $\theta_1, \theta_2, \ldots, \theta_s$ as $\delta \to 0$.

Some references Patil and Bildikar (1966*a*).

93. THE MULTIVARIATE POWER SERIES DISTRIBUTION

Let $f(\theta_1, \theta_2, ..., \theta_s) = \sum_{x_1, x_2, ..., x_s} a_{x_1 x_2 ... x_s} \theta_1^{x_1} \theta_2^{x_2} ... \theta_s^{x}$ be a convergent power series in $\theta_1, \theta_2, ..., \theta_s$ such that $a_{x_1 x_2 ... x_s} \theta_1^{x_1} \theta_2^{x_2} ... \theta_s^{x_s} \geqq 0$, $x_i = 0, 1, 2, ...$; $i = 1, 2, ..., s$ for all $(\theta_1, \theta_2, ..., \theta_s) \in \Theta$ the s-dimensional parameter space. Then the s-variate power series distribution with the series function $f(\theta_1, \theta_2, ..., \theta_s)$ is defined by the pf

$$P(x_1, x_2, ..., x_s) = \frac{a_{x_1 x_2 ... x_s} \theta_1^{x_1} \theta_2^{x_2} ... \theta_s^{x_s}}{f(\theta_1, \theta_2, ..., \theta_s)}, \quad x_i = 0, 1, 2, ...; \ i = 1, 2, ..., s$$

$$(\theta_1, \theta_2, ..., \theta_s) \in \Theta.$$

$$G(z_1, z_2, ..., z_s) = \frac{f(\theta_1 z_1, \theta_2 z_2, ..., \theta_s z_s)}{f(\theta_1, \theta_2, ..., \theta_s)}.$$

$$\mu_i = \theta_i \frac{\partial}{\partial \theta_i} [\log f(\theta_1, \theta_2, ..., \theta_s)].$$

$$\sigma_{ij} = \theta_j \frac{\partial \mu_i}{\partial \theta_j} = \theta_i \frac{\partial \mu_j}{\partial \theta_i}.$$

Some inter-relations

(i) The s-variate power series distribution with the series function $f(\theta_1, \theta_2, ..., \theta_s)$ is a special case of the s-variate generalized power series distribution with range T and the series function $f(\theta_1, \theta_2, ..., \theta_s)$ when $T \subset I_s$, the s-fold cartesian product of $I = \{0, 1, 2, ...\}$.

(ii) The s-variate multinomial distribution with parameters $n, p_1, p_2, ..., p_s$ is the s-variate power series distribution with the series function $(1+\theta_1+\theta_2+...+\theta_s)^n$ where

$$\theta_i = \frac{p_i}{1 - \sum_{j=1}^{s} p_j}, \quad i = 1, 2, ..., s.$$

(iii) The s-variate negative multinomial distribution with $k, p_1, p_2, ..., p_s$ is the s-variate power series distribution with the series function $(1-\theta_1-\theta_2-...-\theta_s)^{-k}$ where $\theta_i = p_i$, $i = 1, 2, ..., s$.

(iv) The s-variate logarithmic distribution with parameters $\theta_1, \theta_2, ..., \theta_s$ is the power series distribution with the series function $-\log(1-\theta_1-\theta_2-...-\theta_s)$.

Some references Khatri (1959); Patil (1965*c*).

94. THE MULTIVARIATE GENERALIZED POWER SERIES DISTRIBUTION

Let T be a countable subset without any limit point of R_s, the s-dimensional Euclidean space. Define

$$f(\theta_1, \theta_2, ..., \theta_s) = \Sigma a_{x_1 x_2 ... x_s} \theta_1^{x_1} \theta_2^{x_2} ... \theta_s^{x_s}$$

where the summation extends over T and $a_{x_1 x_2 ... x_s} > 0$, $\theta_i \geqq 0$ with $(\theta_1, \theta_2, ..., \theta_s) \in \Theta$, the s-dimensional parameter space, so that $f(\theta_1, \theta_2, ..., \theta_s)$ is finite and differentiable. Then the s-variate generalized power series distribution with range T and the series function $f(\theta_1, \theta_2, \ldots, \theta_s)$ is defined by the pf

$$P(x_1, x_2, \ldots, x_s) = \frac{a_{x_1 x_2 \ldots x_s} \theta_1^{x_1} \theta_2^{x_2} \ldots \theta_s^{x_s}}{f(\theta_1, \theta_2, \ldots, \theta_s)} \quad (x_1, x_2, \ldots, x_s) \in T$$

$$(\theta_1, \theta_2, \ldots, \theta_s) \in \Theta.$$

$$G(z_1, z_2, \ldots, z_s) = \frac{f(\theta_1 z_1, \theta_2 z_2, \ldots, \theta_s z_s)}{f(\theta_1, \theta_2, \ldots, \theta_s)}.$$

$$\mu_i = \theta_i \frac{\partial}{\partial \theta_i} [\log f(\theta_1, \theta_2, \ldots, \theta_s)].$$

$$\sigma_{ij} = \theta_j \frac{\partial \mu_i}{\partial \theta_j} = \theta_i \frac{\partial \mu_j}{\partial \theta_i}.$$

Some inter-relations See the inter-relations under the multivariate power series distribution.

A reference Patil (1965*c*).

95. THE MULTIVARIATE HYPERGEOMETRIC DISTRIBUTION

(The factorial multinomial distribution)

Suppose n balls are drawn at random without replacement from an urn which contains N balls of $(s+1)$ different colours, M_i being of the ith colour, $i = 0, 1, 2, \ldots, s$. If X_i is the number of balls of the ith colour, $i = 1, 2, \ldots, s$, drawn in the n draws, then $(X_1, X_2, \ldots, X_s)$ has the s-dimensional hypergeometric distribution with parameters $N, M_1, M_2, \ldots, M_s, n$. Its pf is

$$P(x_1, x_2, \ldots, x_s) = \frac{\binom{M_0}{x_0}\binom{M_1}{x_1}\binom{M_2}{x_2} \cdots \binom{M_s}{x_s}}{\binom{N}{n}}, \quad x_i = 0, 1, 2, \ldots, \min(n, M_i);\ i = 1, 2, \ldots, s,$$

$$n - M_0 \leqq \sum_{i=1}^{s} x_i \leqq n,$$

$$N = 1, 2, \ldots$$

$$M_i = 1, 2, \ldots;\ i = 1, 2, \ldots, s,$$

$$\sum_{i=1}^{s} M_i < N,$$

$$n = 1, 2, \ldots;\ n < N,$$

where $M_0 = N - M_1 - M_2 - \ldots - M_s$ and $x_0 = n - x_1 - x_2 - \ldots - x_s$. When $n \leqq M_0$, the pgf is given by

$$G(z_1, z_2, \ldots, z_s) = \frac{M_0^{(n)}}{N^{(n)}} [{}_{(s+1)}F_1(-n, -M_1, -M_2, \ldots, -M_s;\ M_0 - n + 1;\ z_1, z_2, \ldots, z_s)].$$

$$\mu_i = n\left(\frac{M_i}{N}\right).$$

$$\sigma_{ij} = \begin{cases} n\left(\frac{M_i}{N}\right)\left(1 - \frac{M_i}{N}\right)\left(\frac{N-n}{N-1}\right) = \mu_i\left(1 - \frac{M_i}{N}\right)\left(\frac{N-n}{N-1}\right), & i = j, \\ -n\left(\frac{M_i}{N}\right)\left(\frac{M_j}{N}\right)\left(\frac{N-n}{N-1}\right) = -\frac{\mu_i \mu_j}{n}\left(\frac{N-n}{N-1}\right), & i \neq j. \end{cases}$$

Some inter-relations

(i) The s-dimensional hypergeometric distribution with parameters N, M_1, M_2, ..., M_s, n tends to the s-dimensional multinomial distribution with parameters n, p_1, p_2, ..., p_s as $N\to\infty$, $M_i\to\infty$ such that $\frac{M_i}{N}\to p_i$, $0<p_i<1$, $i = 1, 2, ..., s$, $\sum_{i=1}^{s} p_i<1$.

(ii) The s-variate hypergeometric distribution with parameters N, M_1, M_2, ..., M_s, n tends to the joint distribution of s independent Poisson variables with parameters λ_1, λ_2, ..., λ_s respectively as $N\to\infty$, $M_i\to\infty$ and $n\to\infty$ such that $\frac{M_i}{N}\to 0$ and $\frac{nM_i}{N}\to\lambda_i$, $0<\lambda_i<\infty$, $i = 1, 2, ..., s$.

Some references Ram (1955); Steyn (1955).

96. THE MULTIVARIATE INVERSE HYPERGEOMETRIC DISTRIBUTION

(The negative factorial multinomial distribution)

(The multivariate negative hypergeometric distribution)

Suppose balls are drawn at random without replacement from an urn which contains N balls of $(s+1)$ different colours, M_i being of the ith colour, $i = 0, 1, 2, ..., s$, until exactly k balls of the 0th colour are obtained. If X_i is the number of balls of the ith colour drawn in the process, $i = 1, 2, ..., s$, then $(X_1, X_2, ..., X_s)$ has the s-variate inverse hypergeometric distribution with parameters N, M_1, M_2, ..., M_s, k. Its pf is given by

$$P(x_1, x_2, ..., x_s) = \frac{\binom{M_0}{k-1}\binom{M_1}{x_1}\binom{M_2}{x_2}\cdots\binom{M_s}{x_s}}{\binom{N}{k+x-1}}\left(\frac{M_0-k+1}{N-k-x+1}\right) = \frac{\binom{-N-1}{-k-x}\binom{M_1}{x_1}\binom{M_2}{x_2}\cdots\binom{M_s}{x_s}}{\binom{-M_0-1}{-k}},$$

$x_i = 0, 1, 2, ..., M_i;\ 1, 2, ..., s,$

$N = 1, 2, ...,$

$M_1 = 1, 2, ...;\ i = 1, 2, ..., s,$

$\sum_{i=1}^{s} M_i<N,$

$k = 1, 2, ..., M_0,$

where $x = \sum_{i=1}^{s} x_i$ and $M_0 = N - \sum_{i=1}^{s} M_i$.

$$G(z_1, z_2, ..., z_s) = \frac{M_0^{(k)}}{N^{(k)}}\left[{}_{(s+1)}F_1(k, -M_1, -M_2, ..., -M_s; -N+k; z_1, z_2, ..., z_s)\right].$$

$$\mu_i = k\left(\frac{M_i}{M_0+1}\right).$$

$$\sigma_{ij} = \begin{cases} k\left(\frac{M_i}{M_0+1}\right)\left(1+\frac{M_i}{M_0+1}\right)\left(\frac{M_0-k+1}{M_0+2}\right) = \mu_i\left(1+\frac{\mu_i}{k}\right)\left(\frac{M_0-k+1}{M_0+2}\right), & i=j, \\ k\left(\frac{M_i}{M_0+1}\right)\left(\frac{M_j}{M_0+1}\right)\left(\frac{M_0-k+1}{M_0+2}\right) = \mu_i\left(\frac{\mu_j}{k}\right)\left(\frac{M_0k+1}{M_0+2}\right), & i\neq k. \end{cases}$$

Some inter-relations

(i) The s-variate inverse hypergeometric distribution with parameters $N, M_1, M_2, \ldots, M_s, k$ tends to the s-variate negative multinomial distribution with parameters $k, p_1, p_2, \ldots, p_s$ as $N \to \infty$, $M_i \to \infty$ such that $\frac{M_i}{N} \to p_i$, $0<p_i<1$, $i = 1, 2, \ldots, s$, $\sum_{i=1}^{s} p_i < 1$.

(ii) The s-variate inverse hypergeometric distribution with parameters $N, M_1, M_2, \ldots, M_s$, tends to the joint distribution of s independent Poisson variables with parameters $\lambda_1, \lambda_2, \ldots,$ respectively as $N \to \infty$, $M_i \to \infty$, $k \to \infty$ such that

$$\frac{M_i}{N} \to 0, \quad \frac{kM_i}{N} \to \lambda_i, \quad 0<\lambda_i<\infty, \quad i = 1, 2, \ldots, s.$$

Some references Steyn (1951); Steyn (1955); Sibuya, Yoshimura and Shimizu (1964).

97. THE MULTIVARIATE NEGATIVE HYPERGEOMETRIC DISTRIBUTION

The s-variate negative hypergeometric distribution arises as the compound s-variate multinomial distribution when the parameters $p_1, p_2, \ldots, p_s$ of the s-variate multinomial distribution with parameters $n, p_1, p_2, \ldots, p_s$ have the s-variate Dirichlet distribution. (See inter-relation number (i) below).

The s-variate negative hypergeometric distribution also arises in the following situation. Suppose an urn contains N balls of $(s+1)$ different colours, M_i being of the ith colour, $i = 0, 1, 2, \ldots, s$, $\sum_{i=0}^{s} M_i = N$. Suppose m balls are drawn without replacement. Suppose another sample of n balls without replacement is drawn without replacing the first sample. Let Y_i and X_i be the numbers of balls of the ith colour in the first and the second samples respectively, $i = 1, 2, \ldots, s$. Now suppose $(M_1, M_2, \ldots, M_s)$ is a random vector having the distribution defined by the pf

$$P(m_1, m_2, \ldots, m_s) = \frac{1}{\binom{N+s}{s}}, \quad 0 \leqq m_i \leqq N;\ i = 1, 2, \ldots, s, \quad \sum_{i=1}^{s} M_i \leqq N.$$

Then the distribution of $(X_1, X_2, \ldots, X_s)$ given $(Y_1, Y_2, \ldots, Y_s) = (y_1, y_2, \ldots, y_s)$ is the s-variate negative hypergeometric with parameters $m+s+1, y_1+1, y_2+1, \ldots, y_s+1, n$.

The s-variate negative hypergeometric distribution with parameters $N, M_1, M_2, \ldots, M_s, n$ has the pf

$$P(x_1, x_2, \ldots, x_s) = \frac{\binom{-M_0}{x_0}\binom{-M_1}{x_1}\binom{-M_2}{x_2}\cdots\binom{-M_s}{x_s}}{\binom{-N}{n}}, \quad \begin{array}{l} x_i = 0, 1, 2, \ldots, n;\ i = 1, 2, \ldots, s, \\ 0<N<\infty, \\ 0<M_i<\infty;\ i = 1, 2, \ldots, s, \\ \sum_{i=1}^{s} M_i<N, \\ n = 1, 2, \ldots, \end{array}$$

where $x_0 = n-x_1-x_2-\ldots-x_s$ and $M_0 = N-M_1-M_2-\ldots-M_s$.

$$G(z_1, z_2, \ldots, z_s) = \frac{(-M_0)^{(n)}}{(-N)^{(n)}}\left[{}_{(s+1)}F_1(-n, M_1, M_2, \ldots, M_s;\ -M_0-n+1;\ z_1, z_2, \ldots, z_s)\right]$$

$$\mu_i = n\left(\frac{M_i}{N}\right)$$

$$\sigma_{ij} = \begin{cases} n\left(\frac{M_i}{N}\right)\left(1-\frac{M_i}{N}\right)\left(\frac{N+n}{N+1}\right) = \mu_i\left(1-\frac{M_i}{N}\right)\left(\frac{N+n}{N+1}\right), & i=j, \\ -n\left(\frac{M_i}{N}\right)\left(\frac{M_j}{N}\right)\left(\frac{N+n}{N+1}\right) = -\frac{\mu_i\mu_j}{n}\left(\frac{N+n}{N+1}\right), & i \neq j. \end{cases}$$

Some inter-relations

(i) The s-variate negative hypergeometric distribution with parameters $N, M_1, M_2, \ldots, M_s, n$ is the mixture on $(p_1, p_2, \ldots, p_s)$ of the s-variate multinomial distribution with parameters $n, p_1, p_2, \ldots, p_s$ where $(p_1, p_2, \ldots, p_s)$ has the s-variate Dirichlet distribution with parameters $M_1, M_2, \ldots, M_s, N-\sum_{i=1}^{s} M_i$.

(ii) The s-variate negative hypergeometric distribution with parameters $N, M_1, M_2, \ldots, M_s, n$ tends to the s-variate multinomial distribution with parameters $n, p_1, p_2, \ldots, p_s$ as $N \to \infty$, $M_i \to \infty$ such that $\frac{M_i}{N} \to p_i$, $0<p_i<1$; $i = 1, 2, \ldots, s$, $\sum_{i=1}^{s} p_i<1$.

(iii) The s-variate negative hypergeometric distribution with parameters $N, M_1, M_2, \ldots, M_s, n$ tends to the joint distribution of s mutually independent negative binomial variables with parameters M_i, λ; $i = 1, 2, \ldots, s$ respectively as $N\to\infty$, $n\to\infty$, $M_1, M_2, \ldots, M_s$ remaining fixed, such that $\frac{N}{N+n} \to \lambda$, $0<\lambda<1$.

(iv) If $x_1, x_2, \ldots, x_i, \ldots, x_s$, are independent with x_i having the negative binomial distribution with parameters k_i and p, $i=1, 2, \ldots, s$, then, the conditional distribution of $x_1, x_2, \ldots, x_s-1$ given $\sum_{i=1}^{s} x_i=x$ is the $(s-1)$-variate negative hypergeometric distribution with parameters $\sum_{i=1}^{s} k_i, k_1, k_2, \ldots, k_{s-1}$ and x.

Some references Sarndal (1965); Sibuya, Yoshimura and Shimizu (1964).

98. THE MULTIVARIATE POLYA DISTRIBUTION

Suppose an urn contains N balls of $(s+1)$ different colours, a_i being of the ith colour, $i = 1, 2, \ldots, s$ and b of the $(s+1)$th colour. Suppose n balls are drawn one after another, with replacement, such that at each replacement c new balls of the same colour are added to the urn. If X_i denotes the number of balls of the ith colour in the sample, $i = 1, 2, \ldots, s$, then $(X_1, X_2, \ldots, X_s)$ has the s-variate Polya distribution with parameters $a_1, a_2, \ldots, a_s, b, c, n$. Its pf is

$$P(x_1, x_2, \ldots, x_s) = \frac{n!}{x_0!x_1!x_2!\ldots x_s!} \prod_{i=1}^{s} a_i(a_i+c)\ldots\{a_i+(x_i-1)c\} \frac{b(b+c)\ldots\{b+(x_0-1)c\}}{N(N+c)\ldots\{N+(n-1)c\}},$$

$$x_i = 0, 1, 2, \ldots, n; \; i = 1, 2, \ldots, s,$$

$$\sum_{i=1}^{s} x_i \leqq n,$$

$$a_i = 1, 2, \ldots; \; i = 1, 2, \ldots, s,$$

$$b = 1, 2, \ldots,$$

$$c = 1, 2, \ldots,$$

$$n = 1, 2, \ldots,$$

where $N = a_1+a_2+\ldots+a_s+b$ and $x_0 = n-x_1-x_2-\ldots-x_s$.

Writing $\gamma = \dfrac{c}{N}$, $p_i = \dfrac{a_i}{N}$, $i = 1, 2, \ldots, s$ and $p_0 = \dfrac{b}{N}$, the s-variate Polya distribution with parameters $a_1, a_2, \ldots, a_s, b, c, n$ can be written in the form

$$P(x) = \frac{\prod_{i=0}^{s} \binom{-p_i/\gamma}{x_i}}{\binom{-\frac{1}{\gamma}}{n}} \qquad x_i = 1, 2, \ldots, s; \; i = 1, 2, \ldots, s,$$

$$\sum_{i=1}^{s} x_i \leqq n,$$

$$0 < p_i < 1; \; i = 1, 2, \ldots, s,$$

$$0 < \gamma < \infty,$$

$$n = 1, 2, \ldots,$$

where $p_0 = 1-p_1-p_2-\ldots-p_s$ and $x_0 = n-x_1-x_2-\ldots-x_s$. This defines the s-variate Polya distribution with parameters $p_1, p_2, \ldots, p_s, \gamma, n$.

$$G(z_1, z_2, \ldots, z_s) = \frac{\left(\frac{p_0}{\gamma}+n-1\right)^{(n)}}{\left(\frac{1}{\gamma}+n-1\right)^{(n)}} \left[{}_{(s+1)}F_1\left(-n, \frac{p_1}{\gamma}, \frac{p_2}{\gamma}, \ldots, \frac{p_s}{\gamma}; -\frac{p_0}{\gamma}-n+1; z_1, z_2, \ldots, z_s\right)\right]$$

$$\mu_i = n\left(\frac{a_i}{N}\right) = np_i.$$

$$\sigma_{ij} = \begin{cases} n\left(\dfrac{a_i}{N}\right)\left(\dfrac{N-a_i}{N}\right)\left(\dfrac{N+nc}{n+c}\right) = np_i(1-p_i)\left(\dfrac{1+\gamma n}{1+\gamma}\right), & i = j, \\ -n\left(\dfrac{a_i}{N}\right)\left(\dfrac{a_j}{N}\right)\left(\dfrac{N+nc}{n+c}\right) = -np_ip_j\left(\dfrac{1+\gamma n}{1+\gamma}\right), & i \neq j. \end{cases}$$

A reference Steyn (1951).

99. THE NEGATIVE MULTINOMIAL DIRICHLET DISTRIBUTION

The s-variate negative multinomial Dirichlet distribution with parameters $\alpha_1, \alpha_2, \ldots, \alpha_s, \beta$ and k is the mixture on $p_1, p_2, \ldots, p_s$ of the s-variate negative multinomial distribution with parameters $k, p_1, p_2, \ldots, p_s$ when $p_1, p_2, \ldots, p_s$ have the s-variate Dirichlet distribution with parameters $\alpha_1, \alpha_2, \ldots, \alpha_s$ and β. Its pf is given by

$$P(x_1, x_2, \ldots, x_s) = \frac{\Gamma(k+x_1+x_2+\ldots+x_s)}{\Gamma(k)x_1!x_2!\ldots x_s!} \cdot \frac{\Gamma(\beta+\alpha_1+\alpha_2+\ldots+\alpha_s)}{\Gamma(\beta)\Gamma(\alpha_1)\Gamma(\alpha_2)\ldots\Gamma(\alpha_s)} \cdot \frac{\Gamma(k+\beta)\Gamma(\alpha_1+x_1)\Gamma(\alpha_2+x_2)\ldots\Gamma(\alpha_s+x_s)}{\Gamma(k+\beta+\alpha_1+\alpha_2+\ldots+\alpha_s+x_1+x_2+\ldots+x_s)}$$

$x_i = 0, 1, 2, \ldots;\ i = 1, 2, \ldots, s,$

$0<k<\infty,$

$0<\alpha_i<\infty;\ i = 1, 2, \ldots, s,$

$0<\beta<\infty.$

$$\mu_i = \frac{k\alpha_i}{\beta-i}, \text{ if } \beta>1.$$

$$\sigma_{ij} = \begin{cases} c[\mu_i(\mu_i+k)/k], & \text{if } \beta>2,\ i=j, \\ c[\mu_i\mu_j/k], & \text{if } \beta>2,\ i\neq j. \end{cases}$$

where $c = (k+\beta-1)/(\beta-2)$.

Some references Mosimann (1963); Sibuya, Yoshimura and Shimuzu (1964).

100. THE MULTIVARIATE POISSON RECTANGULAR DISTRIBUTION

Let $X_1, X_2, \ldots, X_s$ be s independent random variables each having the Poisson distribution with parameter $\alpha+(\beta-\alpha)\lambda$, $0<\alpha<\beta<\infty$, where λ has the uniform distribution over the interval (0, 1). Then the mixture on λ of the joint distribution of $(X_1, X_2, \ldots, X_s)$ is the s-variate Poisson rectangular distribution with parameters α, β. Its pf is given by

$$P(x_1, x_2, \ldots, x_s) = \frac{1}{(\beta-\alpha)s^{x+1}\prod_{i=1}^{s} x_i!}\left\{e^{-s\alpha}\sum_{i=0}^{x}\frac{(s\alpha)^i}{i!} - e^{-s\beta}\sum_{i=0}^{x}\frac{(s\beta)^i}{i!}\right\}$$

$x_i = 0, 1, 2, \ldots;\ i = 1, 2, \ldots, s,$

$0<\alpha<\beta<\infty.$

$$G(z_1, z_2, \ldots, z_s) = \frac{1}{(\beta-\alpha)\left(\sum_{i=1}^{s} z_i - s\right)}\left[\exp\left\{\beta\left(\sum_{i=1}^{s} z_i - s\right)\right\} - \exp\left\{\alpha\left(\sum_{i=1}^{s} z_i - s\right)\right\}\right]$$

$$\mu_i = \frac{\alpha+\beta}{2}$$

$$\sigma_{ij} = \begin{cases} \dfrac{\alpha+\beta}{2} + \dfrac{(\beta-\alpha)^2}{12}, & i=j, \\ \dfrac{(\beta-\alpha)^2}{12}, & i\neq j. \end{cases}$$

A reference Bhattacharya and Holla (1965).

101. THE MULTIVARIATE BINOMIAL DISTRIBUTION

Consider a population of individuals each having one or more or none of the s characters A_1, A_2, ..., A_s. Suppose n selections are made independently of each other from the population and the number X_i having the ith character is counted, $i = 1, 2, ..., s$, in the sample. Let the probabilities of observing different characters on a single individual be as follows:

$$\begin{aligned} p_i &= \text{probability of } A_i,\ i = 1, 2, ..., s, \\ p_{ij} &= \text{probability of } A_iA_j,\ i \neq j,\ i, j = 1, 2, ..., s, \\ &\vdots \\ p_{12\cdots s} &= \text{probability of } A_1A_2...A_s. \end{aligned}$$

Then $(X_1, X_2, ..., X_s)$ has the s-variate binomial distribution whose factorial moment generating function is

$$(1+\Sigma\, p_iz_i+\sum_{i<j} p_{ij}z_iz_j+...+p_{12...s}z_1z_2...z_s)^n.$$

$$\mu_i = np_i,\ \sigma_{ij} = \begin{cases} np_i(1-p_i), & i=j, \\ np_{ij}(1-p_ip_j), & i \neq j. \end{cases}$$

Some inter-relations

(i) Suppose the random vector $(X_{i_1i_2...i_s};\ i_j = 0, 1;\ j = 1, 2, ..., s)$ has the (2^s-1)-dimensional multinomial distribution with parameters n, $p_{100...00}$; $p_{010...00}$, ..., $p_{000...01}$, $p_{110...00}$, ..., $p_{000...011}$, ..., $p_{111...11}$. Let $X_j = \Sigma X_{i_1i_2...i_s}$ where the summation is over all the components of the random vector for which $i_j = 1$; $j = 1, 2, ..., s$. Then $(X_1, X_2, ..., X_s)$ has the s-variate binomial distribution with the factorial moment generating function

$$(1+\sum_i p_iz_i+\sum_{i<j} p_{ij}z_iz_j+...+p_{12...s}z_1z_2...z_s)^n,$$

where $p_j = \Sigma p_{i_1i_2...i_s}$, the summation extending over all p's for which $i_j = 1$

$p_{jk} = \Sigma p_{i_1i_2...i_s}$, the summation extending over all p's for which $i_j = i_k = 1$

$\vdots$

$p_{12...s} = p_{111...1}$.

(ii) The s-variate binomial distribution with the factorial moment generating function

$$(1+\sum_i p_iz_i+\sum_{i<j} p_{ij}z_iz_j+...+p_{12...s}z_1z_2...z_sn)$$

tends to the s-variate Poisson distribution with the pgf

$$\exp\{\sum_i \lambda_i(z_i-1)+\sum_{i<j} \lambda_{ij}(z_iz_j-1)+...+\lambda_{12...s}(z_1z_2...z_s-1)\}$$

as $n\to\infty$, $p_i\to 0$; $i = 1, 2, ..., s$ such that

$$\begin{aligned} &n\{p_i-\sum_j p_{ij}+\sum_{j<k} p_{ijk}-...+(-1)^{s-1}p_{12...s}\}\to\lambda_i \\ &n\{p_{ij}-\sum_k p_{ijk}+\sum_{k<l} p_{ijkl}-...+(-1)^{s-2}p_{12...s}\}\to\lambda_{ij} \\ &\vdots \\ &np_{12...s}\to\lambda_{12...s}. \end{aligned}$$

Some references Krishnamoorthy (1951); Wishart (1949).

102. THE MULTIVARIATE POISSON DISTRIBUTION

The s-variate Poisson distribution is defined by the pgf

$$G(z_1, z_2, \ldots, z_s) = \exp\left\{\sum_{i=1}^{s} \lambda_i(z_i-1) + \sum_{i<j} \lambda_{ij}(z_iz_j-1) + \ldots + \lambda_{12\ldots s}(z_1z_2\ldots z_s-1)\right\},$$

where the 2^s-1 parameters $\lambda_{i_1i_2\ldots i_k}$, $\{i_1<i_2<\ldots<i_k\}\subset\{1, 2, \ldots, s\} \equiv J_s$ are all positive. Each component of the s-dimensional random vector can take the values 0, 1, 2, ...

Let $\lambda_{i_1i_2\ldots i_k} = \lambda_{j_1j_2\ldots j_k}$ where $\{j_1, j_2, \ldots, j_k\}$ is a permutation of $\{i_1, i_2, \ldots, i_k\}\subset J_s$.

$$\mu_i = \lambda_i + \sum_j \lambda_{ij} + \ldots + \lambda_{12\ldots s}.$$

$$\sigma_{ij} = \begin{cases} \lambda_i + \sum_j \lambda_{ij} + \ldots + \lambda_{12\ldots s}, & i=j \\ \lambda_{ij} + \sum_k \lambda_{ijk} + \ldots + \lambda_{12\ldots s}, & i \neq j. \end{cases}$$

Some inter-relations

(i) Let $Y_{i_1i_2\ldots i_k}$ be 2^{s-1} independent random variables having Poisson distribution with parameters $\lambda_{i_1i_2\ldots i_k}$ where $\{i_1, i_2, \ldots, i_k\}\subset J_s$, $J_s \equiv \{1, 2, \ldots, s\}$, $Y_{i_1i_2\ldots i_k} = Y_{j_1j_2\ldots j_k}$ and $\lambda_{i_1i_2\ldots i_k} = \lambda_{j_1j_2\ldots j_k}$ if $\{j_1j_2, \ldots, j_k\}$ is a permutation of $\{i_1, i_2, \ldots, i_k\}$. Define for fixed i, $X_i = \Sigma Y_{ij_1j_2\ldots j_k}$, the summation extending over all subsets $\{j_1, j_2, \ldots, j_k\}$ of $J_s-\{i\}$, $i = 1, 2, \ldots, s$. Then $(X_1, X_2, \ldots, X_s)$ has the s-variate Poisson distribution with the pgf

$$G(z_1, z_2, \ldots, z_s) = \exp\{\Sigma\lambda_i(z_i-1) + \sum_{i<j} \lambda_{ij}(z_iz_j-1) + \ldots + \lambda_{12\ldots s}(z_1z_2\ldots z_s-1)\}.$$

(ii) The s-variate binomial distribution with the factorial moment generating function

$$(1+\Sigma p_iz_i+\Sigma p_{ij}z_iz_j+\ldots+p_{12\ldots s}z_1z_2\ldots z_s)^n$$

tends to the s-variate Poisson distribution with the pgf

$$\exp\{\Sigma\lambda_i(z_i-1)+\Sigma\lambda_{ij}(z_iz_j-1)+\ldots+\lambda_{12\ldots s}(z_1z_2\ldots z_s-1)\}$$

as $n\to\infty$, $p_i\to 0$, $i = 1, 2, \ldots, s$ such that

$$n\{p_i - \sum_j p_{ij} + \sum_{j<k} p_{ijk} - \ldots + (-1)^{s-1}p_{12\ldots s}\}\to\lambda_i,$$

$$n\{p_{ij} - \sum_k p_{ijk} + \sum_{k<l} p_{ijkl} - \ldots + (-1)^{s-2}p_{12\ldots s}\}\to\lambda_{ij},$$

$$\vdots$$

$$np_{12\ldots s}\to\lambda_{12\ldots s}.$$

Some references Krishnamoorthy (1951); Teicher (1954*c*).

103. THE MULTIVARIATE NEGATIVE BINOMIAL DISTRIBUTION

The s-variate negative binomial distribution arises as the mixtures on u of the s-variate Poisson distribution with the pgf

$$\exp\{\Sigma\lambda_iu(z_i-1) + \sum_{i<j} \lambda_{ij}u(z_iz_j-1) + \ldots + \lambda_{12\ldots s}u(z_1z_2\ldots z_s-1)\}$$

where u has the gamma distribution. If parameters of the distribution of u are α and k, then the fmgf of the resulting s-variate negative binomial distribution is

$$p_0^k(p_0-\Sigma p_i z_i-\sum_{i<j} p_{ij}z_i z_j-\ldots-p_{12\ldots s}z_1 z_2\ldots z_s)^{-k}$$

where

$$p_0=\frac{\alpha}{\Lambda+\alpha},\ p_i=\frac{\Lambda_i}{\Lambda+\alpha},\ p_{ij}=\frac{\Lambda_{ij}}{\Lambda+\alpha},\ \ldots,\ p_{12\ldots s}=\frac{\Lambda_{12\ldots s}}{\Lambda+\alpha}.$$

and $\Lambda=\Sigma\lambda_i+\sum_{i<j}\lambda_{ij}+\ldots+\lambda_{12\ldots s}$,

$\Lambda_i=$ sum of all λ's each of which has i as one of its subscripts,

$\Lambda_{ij}=$ sum of all λ's each of which has i and j as two of the subscripts

$\vdots$

$\Lambda_{12\ldots s}=\lambda_{12\ldots s}$.

$$\mu_i=\frac{kp_i}{p_0}.$$

$$\sigma_{ij}=\begin{cases}\dfrac{kp_i}{p_0}\left(1+\dfrac{p_i}{p_0}\right)=\mu_i\left(1+\dfrac{p_i}{p_0}\right), & i=j,\\[2ex] k\left(\dfrac{p_ip_j}{p_0^2}\right)+\left(\dfrac{p_{ij}}{p_0}\right), & i\neq j.\end{cases}$$

Some inter-relations

The s-variate Pascal distribution with the fmgf

$$p_0^k(p_0-\Sigma p_i z_i-\sum_{i<j} p_{ij}z_i z_j-\ldots-p_{12\ldots s}z_1 z_2\ldots z_s)^{-k}$$

is a special case of the s-variate negative binomial distribution with the fmgf

$$p_0^k(p_0-\Sigma p_i z_i-\sum_{i<j} p_{ij}z_i z_j-\ldots-p_{12\ldots s}z_1 z_2\ldots z_s)^{-k},$$

when k is an integer.

104. THE MULTIVARIATE PASCAL DISTRIBUTION

Consider a population of individuals each having one or more or none of the s characters A_1, A_2, …, A_s. Let the probabilities of observing different characters on a single individual be as follows.

p_i—probability of A_i,

p_{ij}—probability of A_iA_j,

$\vdots$

$p_{12\ldots s}$—probability of $A_1A_2\ldots A_s$.

Suppose individuals are observed at random one after another, observations being independent of each other, until individuals possessing none of the characters are observed exactly k times. Let X_i denote the number of occurrences of A_i in the process, $i = 1, 2, \ldots, s$. Then $(X_1, X_2, \ldots, X_s)$ has the s-variate Pascal distribution with the factorial moment generating function

$$p_0^k(p_0 - \Sigma p_i z_i - \sum_{i<j} p_{ij} z_i z_j - \ldots - p_{12\ldots s} z_1 z_2 \ldots z_s)^{-k}$$

where

$$p_0 = \text{probability of } \bar{A}_1 \bar{A}_2 \ldots \bar{A}_s, \text{ where } \bar{A}_i \equiv \text{complement of } A_i$$
$$= 1 - \Sigma p_i + \sum_{i<j} p_{ij} - \ldots + (-1)^s p_{12\ldots s}.$$

$$\mu_i = \frac{kp_i}{p_0}.$$

$$\sigma_{ij} = \begin{cases} \dfrac{kp_i}{p_0}\left(1 + \dfrac{p_i}{p_0}\right) = \mu_i\left(1 + \dfrac{p_i}{p_0}\right), & i = j, \\ k\left(\dfrac{p_i p_j}{p_0^2} + \dfrac{p_{ij}}{p_0}\right), & i \neq j. \end{cases}$$

Some inter-relations

The s-variate Pascal distribution with the fmgf

$$p_0^k(p_0 - \Sigma p_i z_i - \sum_{i<j} p_{ij} z_i z_j - \ldots - p_{12\ldots s} z_1 z_2 \ldots z_s)^{-k}$$

is a special case of the s-variate negative binomial distribution with the fmgf

$$p_0^k(p_0 - \Sigma p_i z_i - \sum_{i<j} p_{ij} z_i z_j - \ldots - p_{12\ldots s} z_1 z_2 \ldots z_s)^{-k}$$

when k is an integer.

105. THE MULTIPLE POISSON DISTRIBUTIONS

1. The joint distribution of independent Poisson variables is called the multiple Poisson distribution [Feller (1957), p. 162].

2. The joint distribution of s arbitrary subsums of random variables whose joint distribution is multivariate Poisson is called s-variate multiple Poisson. Since the r-variate Poisson distribution arises as the joint distribution of certain overlapping sums of $2^r - 1$ independent Poisson variables, the multivariate multiple Poisson distribution can be interpreted as the joint distribution of arbitrary sub-sums of non-negative integer multiples of independent Poisson variables [Edwards (1962)].

106. THE BIVECTOR MULTINOMIAL DISTRIBUTION

(The bivariate multinomial distribution)

Consider a sequence of n independent trials each of which has $(s_1+1)(s_2+1)$ mutually exclusive outcomes E_{ij} with probabilities p_{ij}, $0 < p_{ij} < 1$, $i = 0, 1, 2, \ldots, s_1$, $j = 0, 1, 2, \ldots, s_2$, $\sum_i \sum_j p_{ij} = 1$.

Let

$$A_i^{(1)} = \bigcup_{j=0}^{s_2} E_{ij}, \quad i = 1, 2, \ldots, s_1$$

$$A_j^{(2)} = \bigcup_{i=0}^{s_1} E_{ij}, \quad j = 1, 2, \ldots, s_2.$$

Let $X_j^{(i)}$ denote the numbers of occurrences of $A_j^{(i)}$, $i = 1, 2$; $j = 1, 2, \ldots, s_i$, in the above sequence. Then the double vector $(X_1^{(1)}, X_2^{(1)}, \ldots, X_{s_1}^{(1)}; X_1^{(2)}, X_2^{(2)}, \ldots, X_{s_2}^{(2)})$ has the bivector multinomial distribution with parameters n, p_{ij}, $i = 0, 1, 2, \ldots, s_1$; $j = 0, 1, 2, \ldots, s_2$. Its pf is

$$P(x_1^{(1)}, x_2^{(1)}, \ldots, x_{s_1}^{(1)}, x_1^{(2)}, x_2^{(2)}, \ldots, x_{s_2}^{(2)}) = \Sigma n! \prod_{i=0}^{s_1} \prod_{j=0}^{s_2} \frac{p_{ij}^{r_{ij}}}{r_{ij}!}, \quad x_j^{(i)} = 0, 1, 2, \ldots, n,$$

$$i = 1, 2,$$

$$j = 1, 2, \ldots, s_i,$$

$$n = 1, 2, \ldots,$$

$$0 < p_{ij} < 1, \; i = 0, 1, 2, \ldots, s_1,$$

$$j = 0, 1, 2, \ldots, s_2,$$

$$\sum_{i=0}^{s_1} \sum_{j=0}^{s_2} p_{ij} = 1,$$

where the summation extends over all integers r_{ij}'s such that

$$0 \leqq r_{ij}, \; \sum_{j=0}^{s_2} r_{ij} \leqq x_i^{(1)}, \; i = 1, 2, \ldots, s_1 \text{ and } \sum_{i=1}^{s_1} r_{ij} \leqq x_j^{(2)}, j = 1, 2, \ldots, s_2.$$

$$G(z_1^{(1)}, z_2^{(1)}, \ldots, z_{s_1}^{(1)}; z_1^{(2)}, z_2^{(2)}, \ldots, z_{s_2}^{(2)}) = \left(p_{00} + \sum_{i=1}^{s_1} p_{i0} z_i^{(1)} + \sum_{j=1}^{s_2} p_{0j} z_j^{(2)} + \sum_{i=1}^{s_1} \sum_{j=1}^{s_2} p_{ij} z_i^{(1)} z_j^{(2)}\right)^n.$$

Let

$$p_{i\cdot} = \sum_{j=0}^{s_2} p_{ij}, \quad i = 1, 2, \ldots, s_1$$

$$p_{\cdot j} = \sum_{i=0}^{s_1} p_{ij}, \quad j = 1, 2, \ldots, s_2$$

$$\mu_j^{(i)} = E(X_j^{(i)})$$

$$\sigma_{ij}^{(k,l)} = E[(X_i^{(k)} - \mu_i^{(k)})(X_j^{(l)} - \mu_j^{(l)})]$$

$$\mu_i^{(1)} = np_{i\cdot}$$

$$\mu_j^{(2)} = np_{\cdot j}$$

$$\sigma_{ij}^{(1,1)} = \begin{cases} np_{i\cdot}(1 - p_{i\cdot}), & i = j, \\ -np_{i\cdot}\, p_{j\cdot}, & i \neq j, \end{cases}$$

$$\sigma_{ij}^{(2,2)} = \begin{cases} np_{\cdot i}(1 - p_{\cdot i}), & i = j, \\ -np_{\cdot i}\, p_{\cdot j} & i \neq j. \end{cases}$$

$$\sigma_{ij}^{(1,2)} = n(p_{ij} - p_{i\cdot}\, p_{\cdot j}).$$

Some references Krishnamoorthy (1951); Wishart (1949).

107. THE MULTIVECTOR MULTINOMIAL DISTRIBUTION

(The multivariate multinomial distribution)

Consider a sequence of n independent trials each of which has $(s_1+1)(s_2+1)\dots(s_t+1)$ mutually exclusive outcomes $E_{i_1 i_2 \dots i_t}$ with probabilities $p_{i_1 i_2 \dots i_t}$, $0<p_{i_1 i_2 \dots i_t}<1$, $i_j = 0, 1, 2, \dots, s_j$, $j = 1, 2, \dots, t$. $\sum\limits_{i_1}\sum\limits_{i_2}\dots\sum\limits_{i_t} p_{i_1 i_2 \dots i_t} = 1$. Let for each $r = 1, 2, \dots, t$, $A_i^{(r)} = \cup E_{i_1 i_2 \dots i_t}$, where $i_r = i$ is fixed, $i = 1, 2, \dots, s_r$, and the union is over all the subscripts of E except the rth subscript, i_r. Let $X_i^{(r)}$ denote the numbers of occurrences of $A_i^{(r)}$ in the above sequence, $r = 1, 2, \dots, t$; $i = 1, 2, \dots, s_r$. Then the multiple vector $(X^{(1)}, X^{(2)}, \dots, X^{(t)})$ where $X^{(r)} = (X_2^{(r)}, \dots, X_{s_r}^{(r)})$, has the $s_1+s_2+\dots+s_t$-dimensional t-vector multinomial distribution with parameters n, $p_{i_1 i_2 \dots i_t}$, $i_j = 0, 1, 2, \dots, s_j$, $j = 1, 2, \dots, t$. Its gf

$$G(z_i^{(r)}, r = 1, 2, \dots, t, i = 1, 2, \dots, s_r)$$

is given by

$$\Big(p_{000\dots0} + \sum_{i=1}^{s_1} p_{i00\dots0} z_i^{(1)} + \sum_{i=1}^{s_2} p_{0i0\dots0} z_i^{(2)} + \dots + \sum_{i=1}^{s_t} p_{000\dots i} z_i^{(t)}$$

$$+ \sum_{i=1}^{s_1}\sum_{j=1}^{s_2} p_{ij0\dots0} z_i^{(1)} z_j^{(2)} + \dots + \sum_{i_1=1}^{s_1}\sum_{i_2=1}^{s_2}\dots\sum_{i_t=1}^{s_t} p_{i_1 i_2 \dots i_t} z_{i_1}^{(1)} z_{i_2}^{(2)} \dots z_{i_t}^{(t)}\Big)^n.$$

Let $p_i^{(k)} = \Sigma p_{i_1 i_2 \dots i_t}$, where the kth subscript i_k is fixed at i, $i = 1, 2, \dots, s_k$ and the sum is taken over all the other subscripts of p = probability of $A_i^{(k)}$ in a single trial.

Let $p_{ij}^{(kl)} = \Sigma p_{i_1 i_2 \dots i_t}$, where the kth subscript i_k is fixed at i, $i = 1, 2, \dots, s_k$, the lth subscript i_l is fixed at j, $j = 1, 2, \dots, s_l$ and then the sum is taken over the remaining $t-2$ subscripts of p = probability of $A_i^{(k)} A_j^{(l)}$ in a single trial.

$$\mu_i^{(k)} = E(x_i^{(k)}) = np_i^{(k)}.$$

$$\sigma_{ij}^{(k,l)} = \text{covariance}\,(X_i^{(k)}, X_j^{(l)}) = \begin{cases} np_i^{(k)}(1-p_i^{(k)}), & k = l,\ i = j, \\ -np_i^{(k)}p_j^{(k)}, & k = l,\ i \neq j, \\ n(p_{ij}^{(kl)} - p_i^{(k)}p_j^{(l)}), & k \neq l. \end{cases}$$

A reference Wishart (1949).

108. THE MULTINOMIAL CLASS SIZE DISTRIBUTION

Let $(Y_1, Y_2, \dots, Y_m)$ have the m-dimensional multinomial distribution with parameters $s, p_1, p_2, \dots, p_m$ where $p_i = \dfrac{1}{(m+1)}$, $i = 1, 2, \dots, m$. Let $X_0, X_1, X_2, \dots, X_s$ be the numbers of components of $(Y_0, Y_1, Y_2, \dots, Y_m)$, where $Y_0 = s - Y_1 - Y_2 - \dots - Y_m$, which are 0, 1, 2, ..., s respectively. Then $(X_0, X_1, X_2, \dots, X_s)$ has the $(s-1)$-dimensional multinomial class size distribution with pf

$$P(x_0, x_1, x_2, \dots, x_s) = \frac{(m+1)!}{x_0!x_1!x_2!\dots x_s!}\,\frac{s!}{(0!)^{x_0}(1!)^{x_1}(2!)^{x_2}\dots(s!)^{x_s}}\left(\frac{1}{m+1}\right)^s,$$

$$x_i = 0, 1, 2, \dots, m+1;\ i = 0, 1, 2, \dots, s$$

$$\text{such that } \sum_{i=0}^{s} x_i = m+1,\ \sum_{i=1}^{s} ix_i = s,$$

$$m = 1, 2, \dots.$$

$$\mu_i = \mu(s, m, i) = (m+1)\binom{s}{i}\left(\frac{1}{m+1}\right)^i \left(\frac{m}{m+1}\right)^{s-i}.$$

$$\sigma_{ij} = \begin{cases} \mu(s, m, i)[\mu(s-i, m-1, i) - \mu(s, m, i) + 1], & i = j, \\ \mu(s, m, i)[\mu(s-i, m-1, j) - \mu(s, m, j)], & i \neq j. \end{cases}$$

Some references Tukey (1949); Wilks (1962), p. 152.

109. THE HYPERGEOMETRIC CLASS SIZE DISTRIBUTION

Let $(Y_1, Y_2, \ldots, Y_m)$ have the m-dimensional hypergeometric distribution with parameters $N, M_1, M_2, \ldots, M_m, s$ where $N = (m+1)M$, $M_i = M$, 1, 2, ..., m and $s>1$. Let

$$Y_0 = s - Y_1 - Y_2 - \ldots - Y_m, \text{ and } X_0, X_1, X_2, \ldots, X_s$$

be the numbers of components of $(Y_0, Y_1, Y_2, \ldots, Y_m)$ which are 0, 1, 2, ..., s respectively. Then $(X_0, X_1, X_2, \ldots, X_s)$ has the $(s-1)$-dimensional hypergeometric class size distribution with the pf

$$P(x_0, x_1, x_2, \ldots, x_s) = \frac{(m+1)!}{x_0!x_1!x_2!\ldots x_s!} \frac{\binom{M}{0}^{x_0}\binom{M}{1}^{x_1}\binom{M}{2}^{x_2}\cdots\binom{M}{s}^{x_s}}{\binom{N}{s}},$$

$x_i = 0, 1, 2, \ldots, m+1;\ i = 0, 1, 2, \ldots, s,$

such that $\sum_{i=0}^{s} x_i = m+1,\ \sum_{i=0}^{s} ix_i = s,$

$M = 1, 2, \ldots,$

$m = 1, 2, \ldots,$

$N = (m+1)M.$

$$\mu_i = \mu(s, m, i) = (m+1)\frac{\binom{M}{i}\binom{Mm}{s-i}}{\binom{M(m+1)}{s}}, \quad i = j.$$

$$\sigma_{ij} = \begin{cases} \mu(s, m, i)[\mu(s-i, m-1, i) - \mu(s, m, i) + 1], & i = j, \\ \mu(s, m, i)[\mu(s-i, m-1, j) - \mu(s, m, j)], & i \neq j. \end{cases}$$

Some references Tukey (1949); Wilks (1962, p. 152).

110. THE DISTRIBUTION OF RUN LENGTHS

Let n elements of m kinds, n_i being of the ith kind, $i = 1, 2, \ldots, m$, be arranged in a random order along a straight line. A run is defined to be a sequence of objects of the same kind terminated by objects of another kind.

Let X_{ij} = the number of runs of objects of the ith kind and of length j.
The random vector $(X_{ij}, i = 1, 2, \ldots, m; j = 1, 2, \ldots, n_i)$ has the pf

$$P(\{x_{ij}\}) = \frac{n_1!n_2!\ldots n_m!}{n!}\left[\prod_{i=1}^{m} \frac{x_i!}{x_{i1}!x_{i2}!\ldots x_{in_i}!}\right]\gamma(x_1, x_2, \ldots, x_m), \quad x_{ij} = 1, 2, \ldots; \ i = 1, 2, \ldots, m; \quad j = 1, 2, \ldots, n_i, \quad n_i = 1, 2, \ldots; \ i = 1, 2, \ldots, m,$$

where

$$x_i = \sum_{j=1}^{n_i} x_{ij}, \quad n = \sum_{i=1}^{m} n_i,$$

and $\gamma(x_1, x_2, \ldots, x_m)$ = the number of ways x_1 objects of the first kind, x_2 objects of the second kind, ..., x_m objects of the mth kind can be arranged so that no two adjacent objects are of the same kind

= the coefficient of $Y_1^{x_1}Y_2^{x_2}\ldots Y_m^{x_m}$ in the expansion of

$$(Y_1+Y_2+\ldots+Y_m)^m(Y_2+Y_3+\ldots+Y_m)^{x_1-1}(Y_1+Y_3+\ldots+Y_m)^{x_2-1}\ldots(Y_1+Y_2+\ldots+Y_{m-1})^{x_m-1}.$$

Some references Mood (1940); Wilks (1962), p. 150.

111. THE DISTRIBUTION OF NUMBERS OF RUNS

Let X_i be the number of runs (for definition, see the distribution of run lengths) of any length of the ith kind in a random arrangement of n object of m kinds, n_i being of the ith kind, $i = 1, 2, \ldots, m$, along a straight line. The pf of the distribution of $(X_1, X_2, \ldots, X_m)$ is

$$P(x_1, x_2, \ldots, x_m) = \frac{n_1!n_2!\ldots n_m!}{n!}\left[\prod_{i=1}^{m} \binom{n_i-1}{x_i-1}\right]\gamma(x_1, x_2, \ldots, x_m), \quad x_i = 1, 2, \ldots, n_i; \ i = 1, 2, \ldots, m, \quad n_i = 1, 2, \ldots; \ i = 1, 2, \ldots, m,$$

where $n = n_1+n_2+\ldots+n_m$ and $\gamma(x_1, x_2, \ldots, x_m)$ is the coefficient of $Y_1^{x_1}Y_2^{x_2}\ldots Y_m^{x_m}$ in the expansion of $(Y_1+Y_2+\ldots+Y_m)^m(Y_2+Y_3+\ldots+Y_m)^{x_1-1}(Y_1+Y_3+\ldots+Y_m)^{x_2-1}\ldots(Y_1+Y_2+\ldots+Y_{m-1})^{x_m-1}$.

Some references Mood (1940); Wilks (1962), p. 150.

Chapter 2
A BIBLIOGRAPHY OF DISCRETE DISTRIBUTIONS

2.1 Introduction

The present bibliography is an extension of *A selected bibliography of statistical literature on classical and contagious discrete distributions* by Patil (1965e).

An effort has been made to make the bibliography current, as well as to incorporate useful changes. About 1800 relevant entries have been added to the ones listed in Patil (1965e). A prominent change is the use of a citation scheme as discussed and used by I. Richard Savage (1962) in his *Bibliography of nonparametric statistics* published by the Harvard University Press. This technique gives a list of all entries that cite a particular item. The citation lists can serve as a detailed indexing. Attempt has been made to classify individual entries.

A systematic search was made for most of the entries published in English during the period ending 1966. We have also listed later publications that we came across and found relevant. The original objective of achieving comprehensive coverage of non-English publications was partially abandoned due to lack of adequate resources.

Papers in statistics and probability discussing discrete distributions in general or those providing explicit treatment of one or more of them on individual basis were considered relevant. Publications in other fields discussing applications of these distributions have been also included. A few relevant topics such as matching, occupancy, run length and contingency tables are omitted since these have been covered well in the *Bibliography of nonparametric statistics* by Savage. Publications on the general theory of queues and other stochastic processes have been excluded to a large extent. The scope of the bibliography, including topics covered, may be more apparent from the discussion on matters related to the classification scheme as discussed later.

Textbooks as a class have been excluded, but the more intimate ones have been included. Technical reports and dissertations have been treated as ‘ published papers ’.

Some of the promising periodicals could not be scanned either because they were not readily available or because of language-difficulties or some other reason. However, in view of (2), (3), (4) and (5) below, a satisfactory coverage of them may be claimed. These periodicals are given in Section 2.4.

Entries: A complete entry consists of the name(s) of author(s); year of publication; title of article or book or technical report, etc.; name and volume number of the periodical in which the article appeared, or name of the publisher and place of publication of a book or a technical report, etc; page numbers of journal articles or total number of pages of a technical report; review; users; classification; a special note if needed and a serial number. It is possible that some of the entries are incomplete or less accurate on one count or another.

Names: For each entry the last name of an author is given followed by his first name or initials. Entries are arranged alphabetically by last names of first authors of articles or books. In the case of a publication having more than one author, entry is made as many times as the number of the authors with cyclic permutation of the order of names in which they appear in the publication. However, in such a case, details of the entry are provided only once in the ' main ' entry which is listed under the name of the author appearing first on the publication. For a specified author, publications due to him alone are entered first. And they are listed chronologically for him. Then follow publications written by that author jointly with other listed alphabetically with second author, third author etc., and chronologically for the same joint author(s).

While alphabetizing names, prefixes such as ' de ', ' dela ', ' van ', ' von ', etc., have been treated as parts of the last names. A compound name is entered under the initial letter of the first part of the last name, e.g. Tiago de Oliveira is listed under T; G-Rodeya, F. E. under G. Publications under a corporate body which do not give any author are listed under the name of the body, e.g. National Bureau of Standards, Power Apparatus System or under Anonymous.

The bibliography has been prepared by means of a card index made in the following way:

(1) Certain periodicals were scanned for relevant articles, mainly English. These are given in Section 2.3.

(2) A large number of papers were cited in the following reviewing or abstracting journals:
- (a) *Biological Abstracts* (abbreviated as BA)
- (b) *Mathematical Reviews* (abbreviated as MR)
- (c) *Psychological Abstracts* (abbreviated as PA)
- (d) *Science Abstracts*, Sections A and B (abbreviated as SA)
- (e) *Statistical Theory and Method Abstracts* (earlier: *International Journal of Abstracts: Statistical Theory and Method*) (abbreviated as STMA; earlier as IJA).
- (f) *Zentralblatt für Mathematik und ihre Grenzgebiete* (abbreviated as ZBL).

(3) The following bibilography was very useful in locating some of the papers from relatively remote/obscure periodicals.

Bibliography of statistical literature 1950-1958, by M. G. Kendall and A. G. Doig, Oliver and Boyd, Edinburgh (1962).

(4) Most of the papers on applications in the field of biology were traced from the following:
- (a) *An annotated bibliography on the uses of statistics in ecology—A search of 31 periodicals*, by Vincent Schultz, a technical report TID 3908 from Environmental Sciences Branch, Division of Biology and Medicine, Atomic Energy Commission, Washington, D.C. (1961). Later issues of most of the promising periodicals in this group were also examined.
- (b) *Quantitative plant ecology*, by P. Greig-Smith, 2nd edition, Butterworth and Co. Ltd., London (1964).

(5) In response to the requests made for contributions of references and reprints of publications discussing discrete distributions, several research workers in a variety of disciplines brought to our attention a number of relevant articles. In particular, Dr Frank A. Haight provided several entries on processes and chains involving Poisson or Poisson-related distributions. Dr Collin Forrester helped with entries from the field of operations research.

Dates: The year given is the year in which the volume of the periodical in which the article appeared was published or a book (last edition) was published. Hence if a volume spreads over more than a calendar year, a single entry may have more than one year associated with it, for example, J. B. S. Haldane (1947-49). Most of the entries are dated.

If the same group of authors has more than one publication in the same year, the year for such publications is usually suffixed by different letters for easy reference. There may appear some minor irregularities in suffixing as a result of a few last minute alterations.

Titles: Titles are given in the language of publication and/or an English translation is provided. Sometimes the language of publication is indicated in a parenthesis following the title if it is not English. If an article is summarized at the end of the publication then it is also indicated in this parenthesis and the language of the summary is given, for example, the entry of Medgyessy (1954a), in the bibliography. The individual identity of each article has been preserved, perhaps at the cost of some duplication at times. In such cases, titles for some entries may be identical. Instances of this kind can particularly occur with published translations and also with some technical reports appearing in periodicals. Titles of books, dissertations and technical reports are italicized.

Periodical or Publisher and Place: The names of journals are italicized and the volume number which follows is in heavy, bold, type. Volume number along with the date and the starting and end page numbers of the article should enable the reader to locate the article quickly. Sometimes issue numbers are given in the subsequent parenthesis because in these cases either the page numbers start with 1 in each issue or the description of the entry was copied from some secondary source (i.e. not from the original publication). Sometimes place of publication of the journal (although it is not a part of its name), is given because it is well known that way and may help the reader recalling some information, for example, *Ann. Inst. Statist. Math. Tokyo.* Almost in all cases, the names of periodicals are abbreviated to save space. Abbreviations used are fairly standard, either borrowed from the *Mathematical Reviews* or the *International Journal of Abstracts.* In the case of a book, the name of the publishing company and the place are given in short. If the publication is neither a journal-article nor a book, it is indicated whether it is a technical report or a research report or a dissertation, etc. At these places also standard abbreviations are used whenever possible, for example, Expt. for experiment or experimental, Tech. Rep. for technical report, Univ. for university, etc. We hope that all of these abbreviations are quite clear in the context and hence are not listed separately. Locations of American universities are not given.

Reviews: Mathematical Reviews, Statistical Theory and Method Abstracts and *Zentralblatt für Mathematik und ihre Grenzgebiete* through 1966 were scanned for reviews. *Psychological Abstracts, Biological Abstracts* and *Science Abstracts* through 1966 were scanned. In connection with reviews are given abbreviated names of reviewing and/or abstracting journals, their volume number, year (year in parenthesis) and page number in the case of *Mathematical Reviews*, or abstract number in the case of *Psychological Abstracts, Biological Abstracts, or Science Abstracts.* This item is omitted if we have not come across a review for the entry.

Users: If a publication is cited as a reference in another publication then the latter is a user of the former. Thus Feller (1943) is listed as a user under Neyman (1939) because Feller (1943) referred to Neyman (1939). We neither listed users of books nor did we list books as users of any publication. Several articles were not available when users were being listed. Sometimes footnotes may have been overlooked. Thus, lists of users may not be taken to be complete. Users have been listed chronologically. This item is omitted whenever there is no user.

Classification: We did not classify books and dissertations, and in rare cases technical reports, which generally cover vast areas and hence are difficult to classify in a reasonably small number of classes. Also abstracts were not classified. In many other cases we could not see the original publication or language difficulty was apparent and these publications could not be classified. Whenever classification is done it is given in code letters as indicated later in the discussion.

Classification is three ways:

(1) by distribution
(2) by statistical inference and
(3) by fields of application.

Notations for these three classification are separated by colons(:). The first and the third classifications are denoted by capital Roman letters, and the second by small Roman letters. An entry may be classified in more than one category, by each of the three criteria. Codes for different categories in the same classification are separated by dashes (-). For example, B-P:pe-ie:BM tells that the article deals with binomial and Poisson distributions, treats point estimation and interval estimation and discusses application in biological and medical sciences.

The classification scheme may be described as follows:

Classification by Distributions: We give classes and their code in Section 2.2 This classification does not need explanation for each class separately. However, a few comments are necessary. The names of distributions as seemingly accepted at large in the statistical literature are listed. The words ' compound ' and ' generalized ' are used in the sense of Gurland (1957). Of course, well-known distributions, like negative binomial, have been classified by their own name rather than just as a compound or a generalized distribution. Pascal distribution being a special case of negative binomial has been classified as the negative binomial distribution. In a ' related ' class, for example, ' other multinomial related ', distributions obtained by modifying the underlying probabliity model one way or another (not covered otherwise) are included. Thus, for example, see Dandekar (1955), Tallis (1962), Stevens (1937). All bivariate distributions are classified as multivariate. Distributions not included in the above list by any specific name are classified as miscellaneous. Articles on discrete distributions in general or sometimes articles containing several—six or more—of them are also classified as miscellaneous.

Classification by Statistical Inference: This classification is more difficult than the other two and almost each class of it needs some explanation. If at least a section or a paragraph of an article deals with a type of statistical inference, it has been included in that particular class.

(1) *Tabulation and Charts* (*tc*)*:* This includes entries which give tables and/or charts such as probability tables of a discrete distribution, graphs of discrete distribution functions or a table of significance values, etc. For example, Nicholson (1960), Nicholson (1961). However, articles giving goodness of fit table or an analysis of variance table do not fall in this class.

(2) *Moments* (*m*)*:* Publications discussing raw moments, absolute moments, moments about mean or some other origin, incomplete moments, expected values of negative powers of a random variable, factorial moments, cumulants are in this class.

(3) *Approximations, Asymptotics, etc.* (*a*)*:* This includes entries discussing limiting or asymptotic forms of distributions. Publications giving other types of approximations, for example, to an expected value or to a probability function, are also put in this class.

(4) *Other Structural Properties* (*osp*)*:* This class covers structural properties other than those covered in the previous two classes, such as characterizations of distributions, inter-relationships between distributions, sampling distributions, etc.

(5) *Processes and Chains* (*pc*)*:*

(6) *Point Estimation* (*pe*)*:*

(7) *Sequential Estimation* (*se*)*:*

(8) *Interval Estimation* (*ie*)*:*

(9) *Order Statistics* (*os*)*:*

(10) *Test on Parameter* (*tp*)*:* We have restricted ourselves here to tests on parameter(s) of a single distribution. The case of two populations where interest lies in testing equality or inequality of parameters of the two distributions is classified in a different class, ' comparison of two populations '. The case of several populations where equality of parameters of the populations is being tested is included in the class ' homogeneity ' or ' analysis of variance and transformation '.

(11) *Goodness of Fit* (*gf*)*:* An entry falls into this class if it discusses general theory of goodness of fit criteria or if it contains data to which a distribution is fitted.

(12) *Statistical Quality Control and Acceptance Sampling* (*sqc*)*:* There are several publications which discuss statistical quality control using discrete distributions. We have included in this bibliography only those publications from this field which discuss discrete models and methods employed rather than the operational and industrial aspects as such.

(13) *Analysis of Variance and Transformations* (*anovat*)*:* Transformations for discrete distributions have been considered in literature mainly to apply the analysis of variance techniques to the discrete data. Therefore, the publications dealing with analysis of variance of discrete data and also with transformations of discrete distributions are put in a common class.

(14) *Index of Dispersion* (*id*)*:* This is perhaps too specialized a topic to warrant a separate class. But since quite a few papers were observed to discuss this index it was thought worthwhile to prepare a separate class. In fact this is a subclass of ' homogeneity ' !

(15) *Homogeneity* (*h*)*:* This has been used in two, more or less distinct senses. In one case, the hypothesis to be tested is that given k samples come from a common population with a given probability distribution. In the second, the hypothesis to be tested is that there is no contagion, or that the experimental material is homogenous. In some articles this was done by testing the Poisson goodness of fit to the data by using index of dispersion. Such entries are classified as ' gf ' or ' id '. But in some articles the ' no contagion ' or randomness hypothesis was tested using some other statistics and these were put in this class. The randomness hypothesis is of great importance in ecological problems.

(16) *Comparison of Two Populations* (*ctp*)*:* Although this could be looked upon as a special case of ' test on parameters ' of a probability function which is a product of individual probability functions, a separate category for this problem is justified by the number of entries in it. The hypothesis to be tested in such a case could be, for example, H: $\lambda_1=\lambda_2$ or H: $\lambda_1<c\lambda_2$ with c a known constant, where λ_1 and λ_2 are parameters of two independent Poisson populations.

(17) *Comparison of Models* (*cm*)*:* If two models are to be compared, the two distributions based on the models may be fitted to the available data and statements on comparison may be made using chi-square criterion. Such articles can be classified in ' gf '. In the present class we include those papers in which different models are compared by means of other reasoning. Univariate versus bivariate approach to accident statistics may be cited as an example of this. For example, Blum and Mintz (1951).

(18) *Selection and Ranking Problems* (*srp*).

(19) *Computations* (*c*)*:* Articles dealing with computational methods such as discussing programs for preparing tables or providing simplified formulae to evaluate an expression are classified in this class. For example, Birch (1963) and Molina (1929).

(20) *Regression and Prediction* (*rp*).

(21) *Model Building* (*mb*)*:* All publications which derive discrete distributions under suitable assumptions have been classified under ' mb '. For example, Neyman (1939).

(22) *Miscellaneous* (*mi*)*:* Even with the 21 classes described above it was found necessary to make this class. A publication which could not be classified in any of the above 21 classes was put in this class. Increase in the number of classes was not considered very desirable just for a few entries of some kind. It was realized only after a large number of entries were classified that many articles on infinite divisibility were classified as miscellaneous. We are hopeful that in most such cases titles of articles would be self-descriptive.

Classification by Fields of Application: Entries were classified into one or more of the following fields of application.

Biological and Medical Sciences (BM), Physical Sciences (P),
Engineering (E), Social Sciences (S), Accidents, Absenteeism (A),
Linguistics (L), General Theory (G) and Other (O).

Classification into any of BM, P, E, S, L is relatively easy. Papers on accidents data and absenteeism could not be classified in any of these five classes conveniently. Because of this reason, and due to wide interest generated by the theory of accidents and absenteeism, such papers were classified in a separate class. Papers discussing general statistical or probability theory without application to any particular field are put in ' G '. In a number of publications a general theory is developed with a view to applying it to a specific problem in one of the above fields. In such a case the entry is classified as ' G ' as well as the particular field. The same procedure is followed if the general theory is applied in some field as an illustration of the general theory, or even if it was felt that the statistical method under discussion could be of general interest in a particular field. There are not many papers which had to be put in the last class ' O '.

2.2 List of Classes

(1) *Classification by distribution*:

1.	Binomial	B
2.	Borel-Tanner	BT
3.	Compound Binomial	COB
4.	Compound Geometric	COG
5.	Compound Multinomial	COM
6.	Compound Negative Binomial	CONB
7.	Compound Negative Multinomial	CONM
8.	Compound Poisson	COP
9.	Discrete Lognormal	DL
10.	Generalized Binomial	GB
11.	Generalized Binomial Distribution of Poisson	GBDP
12.	Generalized Geometric	GG
13.	Generalized Multinomial	GM
14.	Generalized Negative Binomial	GNB
15.	Generalized Poisson	GP
16.	Geometric	G
17.	Hypergeometric	H
18.	Inverse Factorial Series	IFS
19.	Inverse Hypergeometric	IH
20.	Logarithmic Series	LS
21.	Miscellaneous	MI
22.	Multinomial	M
23.	Multivariate Binomial	MB
24.	Multivariate Hypergeometric	MH
25.	Multivariate Hypergeometric Related	MHR
26.	Multivariate Inverse Hypergeometric	MIH
27.	Multivariate Negative Binomial	MNB
28.	Multivariate Poisson	MP
29.	Multivariate Power Series	MPS
30.	Negative Binomial	NB
31.	Negative Hypergeometric	NH
32.	Negative Multinomial	NM
33.	Neyman's Type A	N
34.	Other Binomial Related	OBR
35.	Other Hypergeometric Related	OHR
36.	Other Multinomial Related	OMR
37.	Other Poisson Related	OPR
38.	Poisson	P
39.	Polya	PO
40.	Power Series	PS
41.	Thomas	T
42.	Truncated or Censored Binomial	TCB
43.	Truncated or Censored Negative Binomial	TCNB
44.	Truncated or Censored Poisson	TCP

(2) *Classification by statistical inference*:

1.	Anova and Transformations	anovat
2.	Approximations, Asymptotics	a
3.	Comparison of Models	cm
4.	Comparison of Two Populations	ctp
5.	Computations	c
6.	Goodness of Fit	gf
7.	Homogeneity	h
8.	Index of Dispersion	id
9.	Interval Estimation	ie
10.	Miscellaneous	mi
11.	Model Building	mb
12.	Moments	m
13.	Order Statistics	os
14.	Other Structural Properties	osp
15.	Point Estimation	pe
16.	Processes and Chains	pc
17.	Regression and Prediction	rp
18.	Selection and Ranking Problems	srp
19.	Sequential Estimation	se
20.	SQC and AS	sqc
21.	Tabulation and Charts	tc
22.	Test on Parameters	tp

(3) *Classification by fields of application cited*:

1.	Accidents, Absenteeism	A
2.	Biology and Medicine	BM
3.	Engineering	E
4.	General Theory	G
5.	Linguistics	L
6.	Physical Sciences	P
7.	Social Sciences	S
8.	Other	O

2.3 List of scanned Periodicals

Acta Math. Acad. Sci. Hungar.
Amer. Math. Monthly
Ann. Eugenics (now *Ann. Human Genetics*)
Ann. Human Genetics (formerly *Ann. Eugenics*)

Ann. Math. Statist.
Ann. of Math.
Appl. Statist.
Austral. J. Statist.
Bell System Tech. J.
Biometrics
Biometrika
Bull. Amer. Math. Soc.
Bull. Calcutta Math. Soc.
Canad. J. Bull.
Canad. Math. Bull.
Duke Math. J.
Ecology
Econometrica
Industrial Quality Control
Information and Control
J. Amer. Statist. Assoc.
J. Appl. Prob.
J. Ecol.
J. Roy. Statist. Soc. Ser. A.
J. Roy. Statist. Soc. Ser. B
J. Roy. Statist. Soc. Suppl.
Math. Nachr.
Metrika (formerly *Mitteilungsbl. Math. Statis.*)
Mitteilungsbl. Math. Statist. (now *Metrika*)
Operations Res. (formerly *J. Operations Res. Society Amer.*)
Pacific J. Math.
Philos. Trans. Roy. Soc. London Ser. A
Population Studies
Proc. Berkeley Symp. Math. Statist and Probab.
Proc. Cambridge Philos. Soc.
Proc. Edinburgh Math. Soc. Ser. 2
Proc. 4*th Berkeley Symp. on Math. Statist. and Probab.*
Proc. London. Math. Soc.
Proc. Roy. Soc. Edinburgh Sect. A
Proc. 2*nd Berkeley Symp. Math. Statist. Probab.*
Proc. 3*rd Berkeley Symp. Math. Statist. Probab.*
Psychometrika
Sankhyā Ser. A and B
Technometrics
Teor. Verojatnost. i Primenen.
Theor. Probability Appl.
Trans. Amer. Math. Soc.

2.4 List of other Periodicals appearing in the Bibliography

Abh. Deutsch. Akad. Wiss. Berlin Kl. Math. Phys. Tech.
Acad. R. P. Romine Bul. Sti. Mat. Fiz.
Acad. Roy. Belg. Bull. Cl. Sci.
Acad. Roy. Belg. Cl. Sci. Mem. Coll. in-8°
Acta Gent. Statist. Med.
Acta XI Congr. Int. Orn.
Acta Math. Sinica
Acta. Pontif. Acad. Sci.
Acta Sci. Math. (*Szeged*)
Acta Theriol.
Actas Acad. Ci. Lima
Actas 2.ª *Reunion Mat. Espanolas*
Actuar. Studiën
Agra Univ. J. Res.
Akad. Nauk. SSR Zhurnal Eksper. Teoret. Fiz.
Akad. Nauk. SSSR Inzenernyi. Sbornik
Aktuar. Vedy
Algorytmy
Allgemein. Statist. Arch.
Amer. Econ. Rev.
Amer. J. Hum. Genet.
Amer. J. Physiology
Amer. J. Publ. Health
Amer. J. Sci.
Amer. J. Sociology
Amer. Midland Nat.
Amer. Naturalist
Amer. Psychologist
Amer. Sociol. Rev.
Amer. Statistician
An. Fac. Ci. Porto
An. Soc. Ci. Argentina
An. Univ. Bucureşti Ser. Sci. Natur. Mat.—Fiz.
Ann. Appl. Biol.
Ann. Bot. Lond.
Ann. Entomol. Soc. America
Ann. Fac. Econ. Com. Palermo
Ann. Fac. Sci. Univ. Toulouse
Ann. Inst. H. Poincaré

Ann. Inst. Statist. Math. (Tokyo)
Ann. Sci. École. Norm. Sup.
Ann. Sci. Textiles Belges
Ann. Soc. Polon. Math.
Ann. Soc. Sci. Bruxelles Ser. I
Ann. Univ. Sci. Budapest Sec. Math.
Ann. Univ. Lyon Sect. A
Apl. Mat.
Appl. Sci. Res. B
Arch. Math.
Arch. Math. Wirtsch.-u Sozialforschg
Arh. hig. rada
Arkiv für Matematik Astronomi och Fysik
Assoc. Roy. Actuairies Belges, Bull.
ASTIN Bull.
Astrophys. J.
Atti Accad. Gioenia Catania
Atti 1st. Veneto Sci. Etc.
AUK
Austral. J. Appl. Sci.
Austral. J. Biol. Sci.
Austral. J. Bot.
Austral. J. Marine and Freshwater Res.
Automobilismo
Behavioral Sci.
Behaviour
Ber. Tagung Wahrsch. Rechnung Math. Statist. Berlin
Ber. Verh. Sachs. Akad. Wiss. Leipzig
Bibliotèca Metron Serie C Rome
Bi.-Deutsch. Ges. Versicherungs-Math.
Biol. Sci. Tokyo
Biom. Zeit.
Biometrie-Praximetrie
Bl. Versich.-Math.
Boll. Centr. Ric. Operat.
Boll. Un. Mat. Ital.
Bot. Rev.
Brit. J. Prev. Soc. Med.
Brit. J. Psychol.
Brit. J. Statist. Psychol.
Bull. Acad. Polon. Sci. Classe III
Bull. Acad. Polon. Sci. Ser. Sci. Math. Astronom. Phys.
Bull. Assoc. Actuaires Belges
Bull. Assoc. Actuaries Diplômes Inst. Sci. Financ. Assuar. Mars
Bull. Assoc. Licencies en Sci Actuarielles, Univ. Libre de Bruxelles (Bruxelles)
Bull. Calcutta Statist. Assoc.
Bull. Coll. Sci. (Bagdad)
Bull. Inst. Internat. Statist.
Bull. Math. Biophys.
Bull. Math. Phys. École Polytechn. Bucarest
Bull. Math. Statist.
Bull. Sci. Math. Biology
Bull. Soc. Math. France
Bull. Soc. Math. Grèce
Bull. Soc. Roy. Sci. Liège
Bull. Soc. Sci. Lettres Lodz.
Bull. Trimest. Inst. Actuaires Franç.
C.R. Acad. Sci. Paris
C.R. Accad. Lincei
C.R. (Doklady) Acad. Sci. URSS N. S.
C. R. 11th Congr. Math. Scand.
California Fish and Game
Canad. Entomologist
Canad. J. Bot.
Canad. J. Res. D.
Canad. J. Zool.
Cashiers Centre Recherche Opérat.
Časopis Mat.
Chiffres
Ciencia (Lisboa)
Cold Spring Harbour Symp. on Quantitative Biol.
Colloq. Math.
Comm. and Electronics
Comm. Math. Helv.
Comm. Pure Appl. Math.
Comment Math. Univ. Carolinae
Commentary
Compositio Math.
Comptes Rendus de l'Académie des Sciences d'URSS
Condor
Contr. Lab. Vertebrate Biol. Univ. Mich.
CORS J.
Czechoslovak. Math. J.
Defence Sci.

Deutsche Math.
Dokl. Akad. Nauk SSSR
Dopovidi Akad. Nauk Ukrain RSR
Ecol. Monogr.
Educ. Psychol. Measmt.
Egyptian Statist. J.
Elec. Comm. Lab. Tech. J.
Ericsson Technics
Estadist. Española
Evaluation Engineering
Food Res.
Forest Sci.
50*th Indian Sci. Congress*
Gac. Mat. (*Madrid*)
Ganita
Genie Civil
Giorn. Economisti
Giorn. Inst. Ital. Attuari.
Growth
Hermes (*Quebec*)
Het. Verzek.—Arch., Actar. Bij.
Highway Res. Board Proceedings
Hum. Biol.
Indian J. Medical Res.
Indian J. Meteo. Geo.
Indian J. Vet. Sci. and Animal Husbandry
Industria
Industritidningen Norden
Internat. Z. Versicherungsmath. Statist. Probl. Soz. Sicherheit
IRE Trans.
Ist. Italiana degli Attuari Giornale
Izv. Akad. Nauk SSSR Ser. Math.
Izv. Akad. Nauk Turkmen SSR Ser. Fiz.—Techn. Him Geol. Nauk
Izv. Akad. Nauk UzSSR Ser. Fiz.—Mat. Nauk
J. Agri. Res.
J. Amer. Soc. Agron.
J. Analyse Math.
J. Animal Ecol.
J. Appl. Psychology
J. Assoc. Comput. Mach.
J. Austral. Math. Soc.
J. Boston Soc. Civil Engrs.
J. Chron. Dis.
J. College Arts Sci. Chiba Univ.
J. Conseil
J. Dental Res.
J. Econ. Ent.
J. Educ. Res.
J. Elisha Mitchell Sci. Soc.
J. Expt. Biol.
J. Expt. Educ.
J. Fisheries Res. Board Canada
J. Forestry
J. Geology
J. Gen. Microbiol.
J. Genetics
J. Helminthology
J. Hyg., Camb.
J. Immunology
J. Indian Soc. Agri. Statist.
J. Insurance
J. Inst. Actuar.
J. Maharaja Sayajirao Univ. Baroda
J. Marine Biol. Assoc.
J. Marine Res.
J. Math. Analysis Appl.
J. Math. Mech.
J. Math. Psychol.
J. Math. Soc. Japan
J. Nat. Inst. Personn. Res. Johannesburg
J. Operations Res. Soc. Amer.
J. Operations Res. Soc. Japan
J. Proc. Roy. Soc. New South Wales
J. Res. Nat. Bur. Standards. Sect B
J. Sci. Res. Banaras Hindu Univ.
J. Soc. Indust. Appl. Math.
J. Soc. Statist. Paris
J. Textile Inst.
J. Washington Acad. Sci.
J. Wildlife Managem.
Jahrbücher für Nationalokonomie und Statistik
Jap. J. Appl. Entomol. Zool.
Jap. J. Ecol.
Jber. Deutsch. Math.—Verein.
Jenaische Zeitschrift fuer Medizin und Naturwissenschaft
Junior Inst. Eng. J.

K. Nederl. Akad.
Kagaku (Science)
Kodai Math. Sem. Rep.
Kungl. Lantbrukshogskolans Annaler
Limnology and Oceanography
Litovsk. Mat. Sb.
Magyar. Tud. Akad. Mat. Kutató Inst. Közl. (Publ. Math. Inst. Hungar. Akad. Sci.)
Magyar. Tud. Akad. Mat. Fiz. Oszt. Közl.
Magyar. Tud. Akad. III Oszt. Közl.
Marketing Res.
Mat.-Fys. Skr. Danske Vid. Selsk.
Mat. Sb.
Math. Ann.
Math. Centrum Amsterdam Rap.
Math. Centrum Amsterdam Statist. Afdeling Rap.
Math. Comp.
Math. Japon.
Math. Mag.
Math. Phys. Semesterber.
Math. Scandinav.
Math. Statist. Prob.
Math. Student
Math., Tech., Wirtschatz
Math. Z.
Med. Verzek. Actuar. Bij.
Meded. Kon. Akad. Belg.
Mem. Amer. Math. Soc.
Mem. College Agri. Kyoto Imperial Univ.
Mem. Fac. Sci. Kyushu Univ. Ser. A
Mem. Fac. Sci. Kyushu Univ. Ser. E
Mem. Real Acad.. Ci. Art. Barcelona
Mém. Sav. Étrangers Acad. Sci. Paris
Mém. Sci. Math.
Mem. Soc. Astronom. Ital. N. S.
Mem. Soc. Ital Sci.
Mém. Soc. Roy. Sci. Liège
Metron
Mh. Math. Phys.
Micropaleontology
Milbank Memorial Fund Quart.
Mitt. Verein. Schweiz. Versich—Math.
Monatsh. Math.
Nature
Natur. Sci. Rep. Ochanomizu Univ.
Naturwissenschaften
Nat. Conv. Trans. Amer. Soc. Qual. Contr.
Naval Res. Logist Quart.
Nederl. Akad. Wetensch. Proc. Ser. A (Indag. Math.)
New Phytologist
New Zealand J. Sci.
Nieuw Arch. Wisk
Nordisk Mat. Tidskr.
Numer. Math.
Nuovo Cimento
Nyt Tidsskrift för Mathematik B
Oesterreichische Revue
Ohio J. Sci.
Oikos
Osaka Math. J.
Ôya-Kontyu
P. O. Elect. Engrs. J.
Papers Michigan Acad. Sci. Arts and Letters
Philos. Mag.
Phil. Trans. Roy. Statist. Soc. B
Phys. Rev.
Pokroky Mat. Fyz. Astronom.
Pont. Acad. Sci. Comment
Prace Lodzkie towarz nauk
Prace Mat.
Proc. Assoc. Pl. Prot.
Proc. Conf. Inter. Population Union
Proc. 11th Internat. Congr. Ent.
Proc. First Pakistan Statist. Conf.
Proc. Harvard Symp. Dig. Computers Appl.
Proc. Indian. Acad. Sci.
Proc. Linn. Soc. Lond.
Proc. Nat. Acad. Sci.
Proc. Nat. Acad. Sci. India Sect. A
Proc. National Inst. Sci. India
Proc. Nat. Acad. Sci. USA
Proc. North-Central Branch, E. S. A.
Proc. Phys. Soc.
Proc. 6th Grassland Congr.
Proc. South Dakota Acad. Sci.
Proc. Symp. Appl. Math.
Proc. Third Int. Con. Oper. Res. Oslo.
Psychol. Bull.

Publ. Inst. Statist. Univ. Paris
Publ. Math.
Publ. Math. Debrecen
Publ. Math. Inst. Hungar. Acad. Sci. (*Magyar Tud. Akad. Mat. Kutató Közl.*)
Qualitätskontrolle
Quart. J. Roy. Meteorol. Soc.
Rend. Mat. Sue Appl. Univ. Roma Ist Naz Alta Mat. V. S.
Rend. Sem. Fac. Sci. Univ. Cagliari
Rep. Statist. Appl. Res. Un. Japan. Sci. Engrs.
Res. Population Ecol.
Rev. Belge Statist. Rech. Opérat.
Rev. Brasil. Estatistica
Rev. Ci. (*Lima*)
Rev. Fac. Ciencias Lisboa, A
Rev. Fac. Sci. Univ. Estabul Ser. A
Rev. Gen. Electr.
Rev. Inst. Internat. Statist.
Rev. Math. Hisp.—Amer.
Rev. Med. Veterinaria
Rev. Modern Phys.
Rev. Sci. Instrum.
Rev. Statist. Appl.
Rev. Suisse Econ. Polit. Statist.
Rev. Un. Mat. Argentina
Rev. Int. Sci. Econ. Comm.
Riv. Ital. Demogr. Statist.
Riv. Ital. Econ. Demogr. Statist.
Schweiz. Arch. Angew. Wiss. Tech.
Sci. Rep. Kagoshima Univ.
Sci. Rep. Tohoku Univ. 4*th Ser.*
Semaine des Hôpitaux
Separata de Rivista
SIAM Review
Simon Stevin
Sitzungsberichte der Berliner Math. Gesellschaft
Skand. Aktuarietidskr.
Soc. Actuar. Trans.
Soviet Math., Doklady
Statistica (*Bologna*)
Statistica Neerlandica
Statist. and Math. in Biology
Statist. Qual. Control
Statist. Vierteljschr.
Statistische Hefte
Studi Univ. Babes-Bolyai Ser. Math.—Phys.
Studia Math.
Studii Cercetări Matematice
Stvh. Math.
Svensk. Botanisk Tidskrift
Symposia Soc. Experimental Biology
Trabajos. Estadist.
Trans. Amer. Fisheries Soc.
Trans. Amer. Inst. Elec. Engrs.
Trans. XVth Inter. Congress Actuaries
Trans. N. Amer. Wildlife Conf.
Trans. Roy. Soc. South Africa
Trans. 17*th Inter. Congress Actuaries*
Travail Hum.
Trudy Inst. Mat. Mekh., Akad. Nauk Uzbek, SSR
Trudy Mat. Inst. Steklov
Trudy Vsesojuzn. Sovešč Teor. Verojatnost Mat. Statist.
Tydskr. Wet. Kuns.
Uchenye Zapiski Moskovskii Gosudarstvennyi Univ. Math.
Ukrain. Mat. Z.
Univ. California Publ. Statist.
Univ. Lisboa Revista Fac. Ci. A (Also abbreviated *Rev. Fac. Cencias Lisboa, A*)
Univ. Washington Publ. Math.
Unternehmensforschung
Uspehi Mat. Nauk
Uspehi Mat. Nauk N. S.
Vestnik Leningrad. Univ.
Vestnik Leningrad. Univ. Ser. Mat. Meh. Astronom.
Vilniaus Valst. Univ. Mokslo Darbai Mat. Fiz.
Virginia J. Sci.
Virginia J. Sci. N. S.
Wahrscheinlichkeitstheorie und Verw
Wiadom. Mat.
William and Mary Quart.
Wiss. Z. Hochschule Schwermachinenbau Magdeburg
Wiss. Z. Humboldt-Univ. Berlin Math.-Nat. Reihe

Yokohama Math. J.
Z. Angew. Math. Mech.
Z. Angew. Math. Phys.
Z. für Philosophie und philosophische Kritik.
Z. Eksper. Teoret. Fiz.
Z. Menschl. Verebungs-u. Konstitutionslehre
Z. Physik.
Z. Versicherungs-Recht und-Wissenchaft
Zastos. Mat.
Zeit. Angew. Entomol.
Zeit. Wahrscheinlichkeitsth.

2.5 INDIVIDUAL ENTRIES

Abbe, E. 1879. 1
Ueber Blutkörper-Zählung. *Jenaische Zeitschrift fuer Medizin und Naturwissenschaft*, **13**, 98–105.

Abdel-Aty, S. H. 1954. 2
Ordered variables in discontinuous distributions (Dutch Summary). *Statistica Neerlandica*, **8**, 61–82.
Review: MR **16** (1955), 729.
Class.: MI-B-P:os:G

Aberdeen, J. E. C. 1958. 3
The effect of quadrat size, plant size, and plant distribution on frequency estimates in plant ecology.
Austral. J. Bot., **6**, 47–58.
User: Morisita, M. (1959).

Abraham, J. K. 1962. 4
Confidence intervals for the reliability of multi-component systems. Memo. Rep. 1404, Ballistic Res. Labs., Maryland, pp. 50.
Class.: B-P:ie:E

Abruzzi, Adam. 1954. 5
The ASN concept in attributes double sampling plans. *Rev. Inst. Internat. Statist.*, **22**, 48–56.
Review: ZBL **57** (1956), 358.

Ackerman, C. L. 1963. 6
Theoretical analysis of a digital correlator. Tech. Memo., Ordnance Res. Lab., Pennsylvania State University, pp. 12.
Class.: B:mi:E

Ackerman, Wolf-Gunter. 1939. 7
Eine Erweiterung des Poissonschen Grenzwertsatzes und ihre Anwendung auf die Risikoprobleme in der Sachversicherung. *Schr. Math. Inst. u Inst. Agnew. Math., Univ. Berlin*, **4**, 211–55.
Review: MR **1** (1940), 251; ZBL **21** (1940), 343.
User: Feller, W. (1943).

Aczel, J. 1952. 8
On composed Poisson distributions III (Russian summary). *Acta Math. Acad. Sci. Hungar.*, **3**, 219–24.
Review: MR **14** (1953), 770; ZBL **49** (1959), 364.
Users: Prekopa, A. (1957a); Philipson, Carl (1963c).
Class.: COP:pc:G

Aczel, J. 1956. 9
Some general methods in the theory of factorial equations of one variable. New applications of functional equations (Russian). *Uspehi Mat. Nauk*, **11**, 3–68.

Aczel, J., Janossy, L., and Renyi, A.
See Janossy, L., Renyi, A., and Aczel, J. (1950).

Aczel, J., and Zubrzycki, S. 1956. 10
Sur un problème de la théorie des nombres lié à la distribution binomiale. *Colloq. Math.*, **4**, 56–67.
Review: MR **17** (1956), 944.
Class.: B:mi:G

Adair, D. 1944a. 11
The authorship of the disputed federalist papers. Part I. *William and Mary Quart.*, **1**, 97–122.
Users: Mosteller, F., and Wallace, D. (1963).

Adair, D. 1944b. 12
The authorship of the disputed federalist papers. Part II. *William and Mary Quart.*, **1**, 235–64.
Users: Mosteller, F., and Wallace, D. (1963).

Adam, A. 1950. 13
Reproduktive Systeme und ihre Anwendungen in der technischen Statistik (German). *Statist. Vierteljschr.*, **3**, 55–70.
Review: MR **13** (1952), 962.

Adam, Adolf. 1953. 14
Klassenstatistik. *Metteilungsbl. Math. Statist.*, **5**, 1–28.
Review: ZBL **50** (1954), 145.

Adams, G. C. 1957. 15
A study of the morphology and variation of some upper Lias Formaminifera. *Micropaleontology*, **3**, 205–26.

Adelman, I. G. 1958. 16
A stochastic analysis of the size distribution of firms. *J. Amer. Statist. Assoc.*, **53**, 893–904.
Class.: MI:pc:S

Adelstein, A. M. 1952. 17
Accident proneness: a criticism of the concept based upon an analysis of shunters' accidents. *J. Roy. Statist. Soc. Ser. A*, **115**, 354–410.
Users: Gurland, J. (1959), Haight, Frank (1964a), Haight, Frank (1965a), Haight, Frank (1965b).
Class.: P-NB:gf:A

Adke, S. R. 1964. 18
A multi-dimensional birth and death process. *Biometrics*, **20**, 212–16.
Review: STMA **6** (1965), 257.
Class.: MI:pc:G

Adke, S. R. 1964. 19
The generalized birth and death process and Gaussian diffusion. *J. Math. Analysis Appl.*, **9**, 336–40.
Review: ZBL **124** (1966), 126.

Adler, Franz. 1951. 20
Yates' correction and the statisticians. *J. Amer. Statist. Assoc.*, **46**, 490–501.
Class.: MI:mi:G

Adler, H. A., and Miller, K. W. 1946. 21
A new approach to probability problems in electrical engineering. *Trans. Amer. Inst. Engrs.*, **65**, 630–32.
User: Goldberg, S. (1954).

Agarwala, S. P., Chakraborty, P. N., and Chandra, S. C.
See Chandra, S. C., Agarwala, S. P., and Chakraborty, P. N. (1955).

Agnew, Ralph Palmer. 1959. 22
Asymptotic expansion in global central limit theorems. *Ann. Math. Statist.*, **30**, 721–37.
Class.: B:a:G

Ahlstorm, E. H., and Sette, O. E.
See Sette, O. E., and Ahlstrom, E. H. (1947).

Ahmed, M. S. 1955. 23
A locally optimal test for the independence of two Poisson variables (Abstract). *Ann. Math. Statist.*, **26**, 157.

Ahmed, M. S. 1959. 24
Poisson type stationary Markov chains I. *Bull. Coll. Sci. (Bagdad)*, **4**, 63–68.
User: Ahmed, M. S. (1960).

Ahmed, M. S. 1960. 25
Binomial type stationary Markov chains. *Bull. Coll. Sci. (Bagdad)*, **5**, 1–9.
Review: MR **24A** (1962), 558.
Class.: B:pc-tp:G

Ahmed, M. S. 1961. 26
On locally most powerful boundary randomized similar test for the independence of two Poisson variables. *Ann. Math. Statist.*, **32**, 809–27.
Review: MR **24A** (1962), 214; STMA **5** (1964), 89.
Users: Kemp, C. D., and Kemp, A. W. (1965).
Class.: MP-MI:tp:G

Aitchison, J. 1955. 27
On the distribution of a positive random variable having a discrete probability mass at the origin. *J. Amer. Statist. Assoc.*, **50**, 901–08.
Review: MR **17** (1956), 169; PA **30** (1956), Ab.No. 3720.
Class.: TCP9pe:G

Aitchison, J. 1962. 28
Large-sample restricted parametric tests. *J. Roy. Statist. Soc. Ser. B.*, **24**, 234–50.
User: Selby, B. (1965).
Class.: P-M:tp:G

Aitchison, J., and Sculthorpe, Diane. 1965. 29
Some problems of statistical prediction. *Biometrika*, **52**, 469–83.
Class.: B-P:rp:G

Aitken, A. C., and Gonin, H. T. 1935. 30
On fourfold sampling with and without replacement. *Proc. Roy. Soc. Edinburgh*, **55**, 114–25.
Users: Shenton, L. R. (1950), Krishnamoorthy, A. S. (1951), Hogben, L., and Cross, K. W. (1952), Lancaster, H. O. (1954), Wiid, A. J. B. (1956–58), Shenton, L. R. (1951), Gonin, H. T. (1961), Gonin, H. T. (1966).

Akaike, H. 1956. 31
On a zero-one process and some of its applications. *Ann. Inst. Statist. Math., Tokyo*, **8**, 87–94.

Akao, Y. 1958. 32
A significance test for fraction-defective by compressed limits. *Statist. Qual. Control*, **19**, 54–62.

Albert, G. E., and Nelson, L. 1953. 33
Contributions to the statistical theory of counter data. *Ann. Math. Statist.*, **24**, 9–22.
Users: Girshick, M. A., Rubin, H., and Sitgreaves (1955).

Alda, V. 1952. 34
A note on Poisson's distribution (Russian, English summary). *Czechoslovak Math. J.*, **2**, (77), 243–46.
Review: MR **15** (1954), 634.

Alda, V. 1953. 35
Completeness of polynomals for Poisson's distribution (Russian, English summary). *Czechoslovak. Math. J.*, **3** (87), 83–85.

Aldanondo, I. 1942. 36
Über eine Wahrscheinlichkeisverteilung (Spanish). *Rev. Mat. Hisp.-Amer.*, **IV 2**, 232–41.
Review: ZBL **27** (1943), 338.

Alfred, F. M. 1952. 37
Modeles de courbes statistiques gouvernant le hasard. Distribution hypergéométrique et système Personien. *Hermes (Quebec)*, **2**, 56–80.

Ali, Mir M., and Kabir, A. B. M. Lutful. 1965. 38
The estimation of the parameters of a mixture of distributions (Abstract). *Ann. Math. Statist.*, **36**, 1592.

Allen, J. L., and Beekman, J. A. 1966. 39
A statistical test involving a random number of random variables. *Ann. Math. Statist.*, **37**, 1305–11.
Class.: P:cm:G

Allen, K. R. 1941. 40
Studies on the biology of the early stages of the salmon (Salmo salar). *J. Animal Ecol.*, **10**(1), 47–76.

Allen, W. R. 1963. 41
Simple inventory models with bunched inputs. *Naval Res. Logist. Quart.*, **9**, 265–73.
Review: ZBL **114** (1965), 117.

Allendoerfer, C. B., and Dunn, E. R.
See Dunn, E. R., and Allendoerfer, C. B. (1949).

Alling, David W. 1966. 42
Closed Sequential tests for binomial probabilities. *Biometrika*, **53**, 73–84.
Class.: B:tp:G

Alma, D. 1965. 43
An efficiency comparison between two different estimators for the fraction p of a normal distribution falling below a fixed limit τ (Dutch). *Statistica Neerlandica*, **19**, 81–91.
Review: STMA **7** (1966), 65.
Class.: B:pe:G

Almond, Joyce. 1954. 44
A note on the χ^2 test applied to epidemic chains. *Biometrics*, **10**, 459–77.
Review: ZBL **58** (1958), 133.

Amaral, Edilberto. 1952. 45
On cálculo das probabilidades e suas aplicações na experimentação agrícola e biológica. *Rev. Brasil. Estadistica.*, **13** (52), 359–421.

Amato, V. 1950. 46
Sui limiti di applicabilita della formula di Poisson. *Statistica (Bologna)*, **10**, 149–52.
Review: MR **12** (1951), 190.

Amato, V. 1959. 47
L'esponenziale di Poisson e la distribuzione del numero dei morti per giorno. *Statistica*, **19**, 20–59.

Amato, V. 1959a. 48
Approximate relationships between the negative binomial and the Γ distribution (Italian). *Industria*, 3–16.
Review: IJA **1** (1960), 368.

Amato, V. 1960. 49
A new theoretical scheme of a factorial type and the distribution of plural births (Italian). *Riv. Ital. Econ. Demogr. Statist.*, **14**, 33–64.
Review: IJA **2** (1961), 459.
Class.: G-MI:mb-a:BM

Amato, V. 1963. 50
On the relation between a random variable and its inverted (Italian). *Statistica*, **23**, 565–73.
Review: STMA **7** (1966), 215.
Class.: B:mi:G

Ambarcumjan, G. A. 1960. 51
Über die Menge der Information über die unbekannte Wahrscheinlickheit im Bernoullischen Schema von Experimenten (Russian). *Trudy Vsesojuzn. Sovešč. Teor. Verojatnost. Mat. Statist, Erevan 1958*, 112–20.
Review: ZBL **93** (1962), 319.

Amerine, M. A., Roessler, E. B., and Baker, G. A.
See Baker, G. A., Amerine, M. A., and Roessler, E. B. (1954).

Amerine, M. A., Roessler, E. B., and Baker, G. A.
See Roessler, E. B., Baker, G. A., and Amerine, M. A. (1956).

Amerine, M. A., Roessler, E. B., Filipelio, F., and Baker, G. A. *See* Baker, G. A., Amerine, M. A., Roessler, E. B., and Filipello, F. (1960).

Ammeter, Hans. 1948. 52
A generalization of the collective theory of risk in regard to fluctuating basic—probabilities. *Skand. Aktuarietidskr.*, **31**, 171–98.
Review: ZBL **32** (1950), 421.
User: Philipson, Carl (1963c).

Ammeter, Hans. 1949. 53
Die elementè der kollektiven Risikotheorie von festen und zufallsartig schwandenden Grundwahrscheinlichkeiten. *Mitt. Verein. Schweiz. Versich-Math.*, **49**, 34–95.

Amster, Sigmund J. 1963. 54
A modified Bayes stopping rule. *Ann. Math. Statist.*, **34**, 1404–13.
Class.: B:se-tp:G

Ancker, C. J., and Gafarian, A. V. 1961. 55
Queuing with multiple Poisson inputs and exponential service times. *Operations Res.*, **9**, 321–27.
Review: ZBL **108** (1964), 314.

Ancker, C. J. Jr., and Gafarian, A. V. 1963. 56
Queueing with reneging and multiple heterogeneous servers. *Naval Res. Logist. Quart.*, **10**, 125–49.
Review: ZBL **117** (1965), 360.

Anderson, E. O., Bliss, C. I., Morgan, M. E., and Macleod, P. *See* Morgan, M. E., MacLeod, P., Anderson, E. O., and Bliss, C. I. (1951).

Anderson, R. L., Binet, F. E., Leslie, R. T., and Weiner, S. *See* Binet, F. E., Leslie, R. T., Weiner, S., and Anderson, R. L. (1956).

Anderson, T. W. 1965. 57
Over estimation of binomial probabilities by Poisson probabilities (Abstract). *Ann, Math. Statist.*, **36**, 1611.

Anderson, T. W., and Friedman, Milton. 1960. 58
A limitation of the optimum property of the sequential probability ratio test. *Contributions to probability and statistics*, Ed. by Olkin, I., *et al.*, Stanford Univ. Press, Stanford, Calif., pp. 57–69.
Review: MR **22** (1961), 1957.

Andrzejewski, R., and Wierzbowska, T. 1961. 59
An attempt at assessing the duration of residence of small rodents in a defined forest area. *Acta Theriol.*, **5**, 153–72.
User: Holgate, P. (1966a).

Anonymous. 1943. 60
Annual meeting in the university botany school, Cambridge. *J. Ecol.*, **31**(1), 97–99.

Anonymous. 1947. 61
Report of the Hon. secretaries for the year 1946. *J. Ecol.*, **34**(1), 236–40.

Anscombe, F. J. 1948. 62
The transformation of Poisson, binomial and negative-binomial data. *Biometrika*, **35**, 246–54.
Users: Anscombe, F. J. (1949a), Anscombe, F. J. (1950a), Freeman, M. F., and Turkey, J. W. (1950), Bailey, N. T. J. (1951), Good, I. J. (1953b), Abdel-Aty, S. H. (1954), Blom, G. (1954), David, F. N. (1955a), Holt, S. J. (1955), Moore, P. G. (1956b), Bliss, C. I., and Owen, A. R. G. (1958), Clemans, K. G. (1959), Ehrenberg, A. S. C. (1959), Hairston, N. G. (1959), Jensen, P. (1959), Harcourt, D. G. (1960), Laubscher, N. F. (1960), Harcourt, D. G. (1961), Laubscher, N. F. (1961), Tsao, C. M. (1962), Berthet, P., and Gerard, G. (1965), Govindarajulu, Z. (1965), Harcourt, D. G. (1965).
Class.: P-B-NB:anovat:G

Anscombe, F. J. 1949a. 63
The statistical analysis of insect counts based on the negative binomial distribution. *Biometrics*, **5**, 165–73.
Users: Anscombe, F. J. (1950a), Anscombe, F. J. (1950b), Oakland, G. B. (1950), Barnes, H., and Stanbury, F. (1951), Hunter, G. C., and Quenouille, M. H. (1952), Bliss, C. I., and Fisher, R. A. (1953), David, F. N., and Moore, P. G. (1954), Robinson, P. (1954), Douglas, J. B. (1955), Waters, W. E. (1955), Putnam. L. G., and Shklov, N. (1956), Smith, J. H. G., and Ker, J. W. (1957), Bliss, C. I. (1958), Bliss, C. I., and Owen, A. R. G. (1958), Kutkuhn, J. H. (1958), Hairston, N. G. (1959), Henson, W. R. (1959), Jensen, P. (1959), Patil, G. P. (1959), Waters, W. (1959), Waters, W. E., and Henson, W. R. (1959), Barton, D. E., David, F. N., and Merrington, M. (1960), Pielou, D. P. (1960a), Forsythe, H. Y., and Gyrisco, G. (1961), Harcourt, D. G. (1961), Taylor, C. J. (1961), Ito, Nakamura, Kondo, Miyashita, and Nakamura (1962), Tsao, C. M. (1962), Kuno, E. (1963), Shiyomi, M., and Nakamura, K. (1964), Berthet P., and Gerard, G. (1965), Harcourt, D. G. (1965), Katti, S. K., and Sly, L. E. (1965), Kobayashi, S. (1965), Shenton, L. R., and Mayers, R. (1965).
Class.: BN:pe-anovat:BM

Anscombe, F. J. 1949b. 64
Large-sample theory of sequential estimation. *Biometrika*, **36**, 455–58.
Review: ZBL **36** (1951), 214.
Users: Cox, D. R. (1952), Tweedie, M. C. K. (1952), Schwarz, G. (1962).
Class.: MI:se:G

Anscombe, F. J. 1950a. 65
Sampling theory of the negative-binomial and logarithmic series distributions. *Biometrika*, **37**, 358–82.
Review: MR **12** (1951), 510; ZBL **39** (1951), 142.
Users: Ramakrishnan, A. (1951), Adelstein, A. M. (1952), Skellam, J. G. (1952), Bliss, C. I., and Fisher, R. A. (1953), Evans, D. A. (1953), Good, I. J. (1953a), Cochran, W. G. (1954), David, F. N., and Moore,

P. G. (1954), Robinson, P. (1954), Douglas, J. B. (1955), Sampford, M. R. (1955), Barton, D. E. (1957), Smith, J. H. G., and Ker, J. W. (1957), Bliss, C. I. (1958), Bliss, C. I., and Owen, A. R. G. (1958), Gurland, J. (1958), Kutkuhn, J. H. (1958), Schaefer, M. B., and Bishop, Y. M. (1958), Shenton, L. R. (1958), Sprott, D. A. (1958), Cohen, A. C. Jr. (1959d), Gurland, J. (1959), Jensen, P. (1959), Waters, W. (1959), Waters, W. E., and Henson, W. R. (1959), Darwin, J. H. (1960), Edwards, A. W. F. (1960a), Harcourt, D. G. (1960), Harcourt, D. G. (1961), Katti, S. K., and Gurland, J. (1961), Martin L. (1961), Cassie, R. M. (1962), Katti, S. K., and Gurland, J. (1962b), Martin, D. C., and Katti, S. K. (1962a), Martin, L. (1962b), Patil, G. P. (1962b), Shenton, L. R., and Wallington, P. A. (1962), Tsao, C. M. (1962), Birch, M. W. (1963), Birch, M. W. (1963), Kuno, E. (1963), Mosimann, J. E. (1963), Nelson, W. C., and David H. A. (1964),, Williamson, E., and Bretherton, M. (1964), Berthert, P., and Gerard, G. (1965), Blischke, W. R. (1965), Bowman, K. O., and Shenton, L. R. (1965b), Chatfield, C., Ehrenberg, A. S. C., and Goodhardt, G. J. (1965), Cohen, A. Clifford, Jr. (1965), Harcourt, D. G. (1965), Kobayashi, S. (1965), Mellinger, G., Sylwester, D., Gaffey, W., and Manheimer, D. (1965), Shenton, L. R., and Myers, R. (1965), Chatfield, C., Ehrenberg, A. S. C., and Goodhardt, G. J. (1966), Holgate, P. (1966a), Katti, S. K. (1966), Patil, G. P., and Shorrock, R. W. (1966), Kobayashi, Shiro (1966).
Class.: NB-LS-N-T-DL-MI:mi:G

Anscombe, F. J. 1950b. 66
Soil sampling for potato root eelworm cysts. *Ann. Appl. Biol.*, **37**, 286–95.
User: Anscombe, F. J. (1950a).

Anscombe, F. J. 1964. 67
Normal likelihood functions. *Ann. Inst. Statist. Math.*, **16**, 1–19.

Anscombe, F. J., and Page, E. S. 1954. 68
Sequential tests for binomial and exponential populations. *Biometrika*, **41**, 252–53.
Review: MR **19** (1958), 1205; ZBL **55** (1955), 131.
Class.: B:tp:G

Anselone, Philip M. 1960. 69
Persistence of an effect of a success in a Bernoulli sequence. *J. Soc. Indust. Appl. Math.*, **8**, 272–79.
Review; MR **22** (1961), 1020; IJA **4** (1963), 129; ZBL **99** (1963), 134.
Class.: B:pc:G-P

Anselone, Philip M., and Porcelli, P. 1960. 70
Oscillatory limiting behavior of a random sequence. *Amer. Math. Monthly*, **67**, 565–66.
Review: MR **22** (1961), 1461.
Class.: MI:pc:G

Antoneas, G., and Papamichail, D.
See Papamichail, D., and Antoneas, G. (1964).

Aoyama, Hirojiro 1957. 71
On a certain statistic in a social group. *Ann. Inst. Statist. Math.*, **9**, 23–30.
Review: ZBL **84** (1960), 363.

Appel, V. 1952. 72
Companion nomographs for testing the significance of the difference between uncorrelated percentages. *Psychometrika*, **17**, 325–30.
User: Peterson, R. L. (1955).
Class.: B:ctp-tc:G

Arbous, A. G., and Kerrich, J. C. 1951. 73
Accident statistics and the concept of accident-proneness. *Biometrics*, **7**, 340–432.
Users: Bliss, C. I., and Fisher, R. A. (1953), Arbous, A. G., and Sichel, H. S. (1954a), Arbous, A. G., and Sichel, H. S. (1954b), Taylor, W. F. (1956), Taylor, W. F. (1956), Johnson, N. L. (1957a), Fitzpatrick, R. (1958), Gurland, J. (1959), Edwards, C. B., and Gurland, J. (1960), Edwards, C. B., and Gurland, J. (1961), Haight, F. (1964a), Subrahamaniam, K. (1964), Chatfield, C. Ehrenberg, A. S. C., and Goodhardt, G. J. (1965), Haight, F. (1965a), Mellinger, G., Sylwester, D., Gaffrey, W., and Manheimer, C. (1965), Neyman, J. (1965), Shenton, L. R., and Myers, R. (1965), Chatfield, C., Ehrenberg, A. S. C., and Goodhardt, G. J. (1966), Subrahamanian, K. (1966a).
Class.: P-NB-N-MI:mi:A

Arbous, A. G., and Sichel, H. S. 1954a 74
The use of estimates of absence-proneness for guiding executive action. *Appl. Statist.*, **3**, 159–73.
Class.: NB:rp:A

Arbous, A. G., and Sichel, H. S. 1954b. 75
New techniques for the analysis of absenteeism data. *Biometrika*, **41**, 77–90.
Review: ZBL **55** (1955), 137.
Users: Arbous, A. G., and Sichel, H. S. (1954a), Johnson, N. L. (1957a.)
Class.: NB-NM:gf:A

Archibald, E. E. A. 1948. 76
Plant populations. I. A new application of Neyman's contagious distribution. *Ann. Bot. Lond., N. S.*, **12**, 221–35.
Users: Archibald, E. E. A. (1949a), Thomas, M. (1949), Bateman, G. I. (1950), Curtis, J., and McIntosh, R. (1950), Barnes, H., and Stanbury, F. (1951), Cain, S. A., and Evans, F. C. (1952), Grieg-Smith, P. (1952a), Skellam, J. G. (1952), Thomson, G. W. (1952), Beall, G., and Rescia, R. R. (1953), Bliss, C. I., and Fisher, R. A. (1953), Evans, D. A. (1953), Moore, P. G. (1953), David, F. N., and Moore, P. G. (1954), Moore, P. G. (1954b), Robinson, P. (1954), Pielou, E. C. (1957), Kono Sugino (1958), Morisita, M. (1959), Pielou, E. C. (1960).

Archibald, E. E. A. 1949a. 77
The specific character of plant communities. I. Herbaceous communities. *J. Ecol.*, **37**(2), 260–75.
Class.: LS:mi:BM

Archibald, E. E. A. 1949b. 78
The specific character of plant communities. II. A quantitative approach. *J. Ecol.*, **37**(2), 274–88.
User: Brian, M. V. (1953).
Class.: LS:mi:BM

Archibald, E. E. A. (1950). 79
Plant populations. II. The estimation of number of

individuals per unit area of species in heterogeneous plant populations. *Am. Bot. Lond., N. S.*, **14**, 7–21.
Users: Skellam, J. G. (1952), Greig-Smith, P. (1952a), Moore, P. G. (1953), Pielou, E. C. (1957), Kono and Sugino (1958), Pielou, E. C. (1960).

Arfwedson, B. 1951. 80
A probability distribution connected with Stirling's second class numbers. *Skand. Aktuartidskr.*, **34**, 121-32.
Review: MR **13** (1952), 956; ZBL **45** (1955), 71.

Arfwedson, G. 1955. 81
Research in collective risk theory, Part II. *Skand. Aktuartidskr.*, **38**, 53–100.

Armitage, P. 1953. 82
A note on the time-homogeneous birth process. *J. Roy. Statist. Soc. Ser. B*, **15**, 90–91.
Review: ZBL **50** (1954), 365.
User: Irwin, J. O. (1953).

Armitage, P. 1957. 83
Studies in the variability of pock counts. *J. Hyg. Camb.*, **55**, 564–581.

Armitage, P. 1958. 84
Numerical studies in the sequential estimation of a binomial parameter. *Biometrika*, **45**, 1–15.
Review: MR **19** (1958), 1096.
Users: Edwards, A. W. F. (1960a), Wasan, M. T. (1964), Wasan, M. T. (1965).
Class.: B:se-tc:G

Armitage, P., and Bartsch, G. E. 1960. 85
The detection of host variability in a dilution series with single observations. *Biometrics*, **16**, 582–92.
Review: IJA **2** (1961), 513.
Class.: B:tp-BM

Armitage, P., and Spicer, C. C. 1956. 86
The detection of variation in host susceptibility in dilution counting experiments. *J. Hyg., Camb.*, **54**, 401–14.

Armsen, P. 1953. 87
On the formula useful in the theory of negative binomial distribution. *J. Nat. Inst. Personn. Res. Johannesburg*, **5**, 169–72.
Review: PA **29** (1955), Ab. No. 1774.

Asai, A., Murakami, M., and Kawamura, M.
See Kawamura, M., Asai, A., and Murakami, M. (1954).

Asano, Chooichiro. 1965. 88
On estimating multinomial probabilities by pooling incomplete samples. *Ann. Inst. Statist. Math.*, **17**, 1–13.
Review: MR **30** (1965), 1004.
Class.: M:pe:G

Ashby, E. 1935. 89
The quantitative analysis of vegetation. *Ann. Bot. Lond.*, **49**, 779–802.
Users: Clapham, A. R. (1936), Singh, B. N., and Chalam, G. (1937), Stevens, W. L. (1937), Singh, B. N., and Chalam, G. (1939), Singh, B. N., and Das, K. (1939), Blackman, G. E. (1942), Fracker, S. B., and Brischle, H. (1944), Archibald, E. E. A. (1948, Dice, L. R. (1948), Archibald, E. E. A. (1950), Curtis, J., and McIntosh, R. (1950), Thomson, G. W. (1952), Clark, P. J., and Evans, F. C. (1954a), Kono and Sugino (1958).

Ashby, E. 1936. 90
Statistical ecology. *Bot. Rev.*, **2**, 221–35.
Users: Ashby, E. (1948), Archibald, E. E. A. (1949a), Curtis, J., and McIntosh, R. (1950), Barnes, H., and Marshall, S. M. (1951), Barnes, H., and Stanbury, F. (1951).

Ashby, E. 1948. 91
Statistical ecology. II. A reassessment. *Bot. Rev.*, **14**, 222–34.
Users: Archibald, E. E. A. (1949a), Barnes, H., and Marshall, S. M. (1951), Barnes, H., and Stanbury, F. (1951), Grieg-Smith, P. (1952b), Thomson, G. W. (1952), Thompson, H. R. (1958).

Atwood, E. L. 1956. 92
Validity of mail survey data on bagged waterfowl. *J. Wildlife Managem*, **20**(1), 1–16.

Auerbach, S. I. 1951. 93
The centipedes of the Chicago area with special reference to their ecology. *Ecol. Monogr.*, **21**(1), 97–124.
Class.: P:gf:BM

Avondo Bondino, G. 1960. 94
On the distribution laws of the number of events happening in a given time interval (Italian). *Bull. Centro Ricerca Operat., Serie metodologica*, **4**, 3–13. *Review*: STMA **6** (1965), 375.

Ayyangar, A. A. K. 1934a. 95
Note on the recurrence formulae for the moments of the point binomial. *Biometrika*, **26**, 262–64.
Users: Riordan, J. (1937), Patil, G. P. (1959).
Class.: B:m:G

Ayyangar, A. A. K. 1934b. 96
A note on the incomplete moments of the hypergeometric series. *Biometrika*, **26**, 264–65.
Class.: H:m:G

Azorin Poch, Francisco. 1963. 97
On estimation of proportions (Spanish). *Estadist. Española No. 20*, 5–14.
Review: MR **30** (1965), 811.
Class.: B-P-M:pe:G

B

Babinin, B. V. 1952. 98
A nomogram of the basic statistical distributions and its application to certain problems of sampling (Russian). *Akad. Nauk SSR. Inzenernyi. Sbornik.*, **11**, 169–80.

Bagenal, T. B. 1951. 99
A note on the papers of Elton and Williams on the generic relations of species in small ecological communities. *J. Animal Ecol.*, **20**(2), 242–45.

Bahadur, R. R. 1958. 100
Examples of inconsistency of maximum likelihood estimates. *Sankyhā*, **20**, 207–10.
User: Rao, C. R. (1958).
Class.: MI:pe:G

Bahadur, R. R. 1960. 101
On the number of distinct values in a large sample from an infinite discrete distribution (Abstract). *Ann. Math. Statist.*, **31**, 1215.

Bahadur, R. R. 1960a. 102
Some approximations to the binomial distribution function. *Ann. Math. Statist.*, **31**, 43–54.
Review: MR **22** (1961), 1954; IJA **1** (1960), 567; ZBL **92** (1962), 352.
Users: Patil, G. P. (1963c), Brockwell, P. (1964), Mott-Smith, J. C. (1964), Govindarajulu, Z. (1965).
Class.: B:a:G

Bahadur, R. R. 1960b. 103
On the number of distinct values in a large sample from an infinite discrete distribution. *Proc. Nat. Inst. Sci. India Part A*, **26**(2), 67–75.
Review; MR **25** (1963), 147.
Class.: MI-P-G:osp:G

Bahn, Rudolf. 1937. 104
Uber den Grenzwert der Wahrscheinlichkeiten seltener Ereignisse. *Deutsche Math.*, **2**, 698–708.

Bailar, John C. III. 1964. 105
Significance factors for the ratio of a Poisson variable to its expectation. *Biometrics*, **20**, 639–43.

Bailey, N. T. J. 1951. 106
On estimating the size of mobile populations from recapture data. *Biometrika*, **38**, 293–306.
Users: Bailey, N. T. J. (1952), Chapman, D. G. (1952b), Craig, C. C. (1953a), Goodman, L. A. (1953), Chapman, D. G. (1954), Wohlschlag, D. E. (1954), Holt, S. J. (1955), Darroch, J. N. (1958), Sen, P. K. (1960), Shenton, L. R., and Bowman, K. (1963), Pathak, P. K. (1964).
Class.: B-IH:pe:BM

Bailey, N. T. J. 1951-2a. 107
The estimation of the frequencies of recessives with incomplete multiple selection. *Ann. Eugenics*, **16**, 215–22.
Users: Bailey, N. T. J. (1951b), Gittelsohn, A. M. (1960).

Bailey, N. T. J. 1951-2b. 108
A classification of methods of ascertainment and analysis in estimating the frequencies of recessives in man. *Ann. Eugenics*, **16**, 223–25.
Users: Bailey, N. T. J. (1951a), Drooth, R. S. (1952).

Bailey, N. T. J. 1952. 109
Improvements in the interpretation of recapture data. *J. Animal Ecol.*, **21**, 120–27.
User: Chapman, D. G. (1954).
Class.: MI:pe:BM

Bailey, N. T. J. 1953. 110
The use of chain-binomials with a variable chance of infection for the analysis of intra-household epidemics. *Biometrika*, **40**, 279–86.
Review: ZBL **51** (1954), 111.
Users: Bailey, N. T. J. (1954), Bailey, N. T. J. (1956a), Bailey, N. T. J. (1956b), Bailey, N. T. J. (1956c), Taylor, W. F. (1956).

Bailey, N. T. J. 1954. 111
Maximum-likelihood estimation of the relative removal rate from the distribution of the total size of an intra-household epidemic. *J. Hyg., Camb.*, **52**, 400–02.

Bailey, N. T. J. 1956a. 112
On estimating the latent and infectious periods of measles. I. Families with two susceptibles only. *Biometrika*, **43**, 15–22.
Class.: COB:pe:BM

Bailey, N. T. J. 1956b. 113
On estimating the latent and infectious periods of measles. II. Families with three or more susceptibles. *Biometrika*, **43**, 322–31.
Users: Bailey, N. T. J. (1956a), Bailey, N.T J. (1956c).
Class.: COB:pe:BM

Bailey, N. T. J. 1956c. 114
Significance tests for a variable chance of infection in chain binomial theory. *Biometrika*, **43**, 332–36.
Review: MR **19** (1958), 932.
Class.: COB:tp:BM

Bailey, N. T. J. 1957. 115
The mathematical theory of epidemics. Charles Griffiths and Co., London.

Bailey, N. T. J. 1963. 116
The simple stochastic epidemic: a complete solution in terms of known functions. *Biometrika*, **50**, 235–40.
Review: STMA **6** (1965), 261.
User: Siskind, V. (1965).
Class.; MI:pc:BM

Baker, George A., and Briggs, Fred N. 1945. 117
Wheat-bunt field trials. *J. Amer. Soc. Agron.*, **37**, 127–33.
Users: Baker, G. A., and Briggs, F. N. (1949).
Class.: B:mb-tp:BM

Baker, G. A., and Briggs, F. N. 1949. 118
Wheat-bunt field trials, II. *Proc. Berkeley Symp. Math. Statist. and Prob.*, 485–91.
Review: ZBL **38** (1951), 100.

Baker, G. A., Amerine, M. A., and Roessler, E. B. 1954. 119
Errors of the second kind in organoleptic difference testing. *Food Res.*, **19** (2), 206–10.
Users: Baker, G. A., Amerine, M. A., Roessler, E. B., and Filipello, F. (1960).
Class.: B:tp:O

Baker, G. A., Amerine, M. A., and Roessler, E. B.
See Roessler, E. B., Baker, G. A., and Amerine, M. A. (1956).

Baker, G. A., Amerine, M. A., Roessler, E. B., and Filipello, F. 1960. 120
The nonspecificity of differences in taste testing for preference. *Food Res.*, **25**(6), 810–16.
Class.: B:mi:S

Baldessari, B. 1962. 121
Le probabilita di concordanza, discordanza e " invarianza " (French, English, Italian and German Summaries). *Giorn. Ist. Ital. Attuari.*, **25**, 114–38.

Baldessari, B. 1963. 122
Tests non parametrici di concordanza e discordanza probalistiche (English Summary). *Statistica*, **23**, 173–223.

Ballantine, D. 1953. 123
Comparison of the different methods of estimating nanoplankton. *J. Marine Biol. Assoc.*, **32**, (1) 129–47.
User: Kutkuhn, H. (1958).

Ballarin, Silvio. 1948. 124
Espressione rigoroso della scarto mediano nel problema delle prove ripetute nello schema di Bernoulli. *Mem. Soc. Astronom. Ital.* (*N. S.*), **19**, 63–65.
Review: MR **9** (1948), 450.

Bamforth, S. S. 1958. 125
Ecological studies on the planktonic protozoa of a small artificial pond. *Limnology and Oceanography*, 3(4), 398–412.

Bancroft, T. A., McGuire, J. U., and Brindley, T. A.
See McGuire, J. U., Brindley, T. A., and Bancroft, T. A. (1957).

Banerjee, D. P. 1951. 126
On some new recurrence formulae for cumulants of multivariate multinomial distributions. *Proc. Indian Acad. Sci.*, **34**, 20–23.
Review: MR **13** (1952), 665; ZBL **43** (1952), 339.
Users: Patil, G. P., and Bildikar, S. (1966a).

Banerjee, D. P. 1959. 127
On some theorems on Poisson distribution. *Proc. Nat. Acad. Sci. India Sect. A*, **28**, 30–33.
Review: MR **26** (1963), 1066.

Barberi, B. 1959. 128
On the theoretical Interpretation of statistical distributions (Italian) *Riv. Ital. Econ. Demogr. Statist.*, **13**, 225–91.
Review: IJA **1** (1960), 508.

Bardwell, G. E. 1960. 129
On certain characteristics of some discrete distributions. *Biometrika*, **47**, 437–75.
Review: ZBL **114** (1965), 96; IJA **2** (1961), 460.
Users: Bardwell, G. E. (1961), Kamat, A. R. (1965).
Class.: B-P-NB-LS-G:osp:G

Bardwell, G. E. 1961. 130
Certain discrete distributions. Ph.D. Thesis, Univ. Colorado.

Bardwell, G. E., and Crow, E. L. 1964. 131
A two-parameter family of hyper-Poisson distributions. *J. Amer. Statist. Assoc.*, **59**, 133–41.
Review: MR **28** (1964), 337; STMA **7** (1966), 40; STMA **6** (1964), 819.
Users; Bardwell, G. E., and Crow, E. L. (1965), Kamat, A. R. (1966).
Class.: OPR:mb-m-pe-gf:G-BM

Bardwell, G. E., and Crow, E. L.
See Crow, E. L., and Bardwell, G. E. (1965).

Bargmann, Rolf E., and Carter, Frederick L. 1960. 132
Group testing in binomial and multinomial situations (Abstract). *Virginia J. Sci.*, **11**(4), 230.
Review: BA **37** (1962), Ab. No. 12490.
Users: Patil, G. P., and Bildikar, S. (1966a).

Barnard, G. A. 1946. 133
Sequential tests in industrial statistics. *J. Roy. Statist. Soc. Suppl.*, **8**, 1–21.
Users: Wise, M. E. (1946), Plackett, R. L. (1948), Walker, A. M. (1950), Cox, D. R. (1953), Vogholkar, M. K. (1959), Vogholkar, M. K., and Wetherill, G. (1960), Bartko, J. J. (1961b), Cox, D. R. (1963), Ellner, H. (1963), Bennett, B. M., and Birch, B. (1964c).
Class.: MI:sqc:E
Notes: Discussion, *J. Roy. Statist. Soc. Suppl.*, **8**, 22–26.

Barnard, G. A. 1954. 134
Sampling inspection and statistical decisions. *J. Roy. Statist. Soc. Ser. B*, **16**, 151–74.
Users: Kemp, C. D., and Kemp, A. W. (1956b), Vohgolkar, M. K. (1959), Cox, D. R. (1960), Hald, A. (1960), Vogholkar, M., and Wetherill, G. (1960), Wetherill, G. (1960a), Samuel, E. (1963), Blischke, W. R. (1965).
Class.: COB:sqc:E

Barnard, G. A. 1959. 135
Control charts and stochastic processes. *J. Roy. Statist. Soc. Ser. B.*, **21**, 239–71.
Review; IJA **2** (1961), 119.
Class.: P:pe-sqc:E

Barndorff-Nielsen, O. 1965. 136
Identifiability of mixtures of exponential families. *J. Math. Analysis Appl.*, **12**, 115–21.
Class.: MI:mi:G

Barnes, H. 1949a. 137
A statistical study of the variation in vertical plankton hauls, with special reference to the loss of the catch with divided hauls. *J. Marine Biol. Assoc.*, **28**(2), 429–46.
Users: Barnes, H., and Marshall, S. M. (1951).

Barnes, H. 1949b. 138
On the volume measurement of water filtered by a plankton pump with some observations on the distribution of planktonic animals. *J. Marine Biol. Assoc.*, **28**(3), 651–62.
Users: Barnes, H., and Marshall, S. M. (1951).

Barnes, H. 1952. 139
The use of transformations in marine biological statistics. *J. Conseil*, **18**(1), 61–71.
Users: Schaefer, M. B., and Bishop, Y. M. (1958), Cassie, R. M. (1962).

Barnes, H., and Marshall, S. M. 1951. 140
On the variability of replicate plankton samples and some applications of "contagious" series to the statistical distribution of catches over restricted periods. *J. Marine Biol. Assoc.*, **30**(2), 233–63.
Users: Bliss, C. L., and Fisher, R. A. (1953), Comita, G. W., and Comita, J. J. (1957), Pielou, E. C. (1957), Jensen, P. (1959), Cassie, R. M. (1962).
Class.: P-N-T:gf:BM

Barnes, H., and Stanbury, F. A. 1951. 141
A statistical study of plant distribution during the colonization and early development of vegetation on china clay residues. *J. Ecol.*, **39**, 171–81.
Users: Barnes, H., and Marshall, S. M. (1951). Goodall, D. W. (1952), Skellam, J. G. (1952), Evans, D. A. (1953), MacFadyen, A. (1933), Pielou, E. C. (1957), Cassie, R. M. (1962).
Class.; P-N-T:cm-gf:BM

Barnett, B. N., and Lindley, D. V.
See Lindley, D. V., and Barnett, B. N. (1965).

Barra, J. R. 1961. 142
A contribution to the study of empirical probability distributions (French). *Publ. Inst. Statist. Univ. Paris*, **10**, 1–90.
Review: IJA **4** (1963), 245.
Class.: MI:a:G

Barratt, T., and Marsden, E.
See Marsden, E., and Barratt, T. (1911).

Bartholomew, D. J. 1963. 143
On Chassan's test for order. *Biometrics*, **19**, 188–91.
Review: STMA **6** (1965), 538.
Class.: B:h:G

Bartholomew, D. J. 1965. 144
A comparison of some Bayesian and frequentist inferences. *Biometrika*, **52**, 19–35.
Review: STMA **6** (1965), 1149.
Users: Bartholomew, D. J. (1963), Plackett, R. L. (1966), Springer, M. D., and Thompson, W. D. (1966).
Class.: P:mi:G

Bartko, J. J. 1961a. 145
The negative binomial distribution. Master's Thesis, Virginia Polytechnic Institute.
Users: Nelson, W. C., and David, H. A. (1964).

Bartko, J. J. 1961b. 146
The negative binomial distribution: a review of properties and applications. *Virginia J. Sci.* (*N.S.*), **12**, 18–37.
Review: MR **23A** (1962), 122; BA **36** (1961), Ab. No. 62674.
Users: Bartko, J. J. (1962), Chew, V. (1964), Chatfield, C., Ehrenberg, A. S. C., and Goodhardt, G. J. (1965), Shenton, L. R., and Myers, R. (1965), Chatfield, C., Ehrenberg, A. S. C., and Goodhardt, G. J. (1966).
Class.; NB:mi:G

Bartko, J. J. 1962. 147
A note on the negative binomial distribution. *Technometrics*, **4**, 609–10.
Class.: NB:mi:G

Bartko, J. J. 1966. 148
Approximating the negative binomial. *Technometrics*, **8**, 345–50.
Class.: NB-P:a:G

Bartlett, M. S. 1937. 149
Sub-sampling for attributes. *J. Roy. Statist. Soc. Suppli.*, **4**, 131–35.
Review: ZBL **19** (1939), 35.
Class.: B:ie:G

Bartlett, M. S. 1947. 150
The use of transformations. *Biometrics*, **3**, 39–52.
Users: Anscombe, F. J. (1948), Mosteller, F., and Tukey, J. W. (1949), Freeman, M. F., and Tukey, J. W. (1950), David, F. N. (1955a), Bennett, B. M. (1957b), Schultz, V., and Byrd, M. A. (1957), Forsythe, H. Y., and Gyrisco, G. (1961).
Class.: B-P-MI:anovat:G

Bartlett, M. S. 1959. 151
The impact of stochastic process theory on statistics. *Probability and Statistics; Harald Cramer Vol.* pp. 39–49.
Review: STMA **5** (1964), 205.
Class.: P-NB-LS-MI:mb-pc:G

Bartlett, M. S. 1960. 152
Stochastic population models. John Wiley and Sons, Inc.

Bartlett, M. S. 1960. 153
The spectral analysis of point processes. *J. Roy. Statist. Soc. Ser. B*, **25**, 264–96.

Barton, D. E. 1955. 154
A form of Neyman's ψ_k^2 test of goodness of fit applicable to grouped and discrete data. *Skand. Aktuarietidskr.*, **38**, 1–16.
User: Watson, G. S. (1959).
Class.: MI:gf:G

Barton, D. E. 1957. 155
The modality of Neyman's contagious distribution of type A (Spanish summary). *Trabajos. Estadist.*, **8**, 13–22.
Review: MR **19** (1958), 188; ZBL **78** (1959), 313.
User: Douglas, J. B. (1965).
Class.: N:osp:G

Barton, D. E. 1958. 156
The matching distributions: Poisson limiting forms and derived methods of approximation. *J. Roy. Statist. Soc. Ser. B*, **20**, 73–92.
Review: MR **20** (1959), 1200; ZBL **87** (1961), 138.
User: Govinadarajulu, Z. (1965).
Class.: MI-P:m-a:G

Barton, D. E. 1961. 157
Unbiased estimation of a set of probabilities. *Biometrika*, **48**, 227–29.
Review: ZBL **112** (1965), 105; STMA **5** (1964), 62.
User: Glasser, G. J. (1962b).
Class.: B-P:pe:G

Barton, D. E., and David, F. N. 1956. 158
Spearman's rho and the matching distribution. *Brit. J. Statist. Psychol.*, **9**, 69–73.

Barton, D. E., and David, F. N. 1957. 159
Multiple runs. *Biometrika*, **44**, 168–78.
Review: MR **19** (1958), 70.
Users: Barton, D. E. (1958), Barton, D. E., and David, F. N. (1959c), Patil, G. P., and Bildikar, S. (1966a).
Class.: P-B:a:G
Notes: Corrigenda, *Biometrika*, **44**, 534.

Barton, D. E., and David, F. N. 1958. 160
Runs in a ring. *Biometrika*, **45**, 572–78.
Review: ZBL **86** (1961), 356.

Barton, D. E., and David, F. N. 1959a. 161
Contagious Occupancy. *J. Roy. Statist. Soc. Ser. B*, **21**, 120–33.
Review: IJA **1** (1960), 345.
Users: Barton, D., and David, F. N. (1959), Blischke, W. R. (1964), Blischke, W. R. (1965).
Class.: B-P-NB:mb-a-h:g

Barton, D. E., and David, F. N. 1959b. 162
The dispersion of a number of species. *J. Roy. Statist. Soc. Ser. B*, **21**, 190–94.
Review: IJA **1** (1960), 346; ZBL **92** (1962), 355.
Class.: MI:mb:G

Barton, D. E., and David, F. N. 1959c. 163
Combinatorial extreme value distributions. *Mathematika*, **6**, 63–76.
Review: MR **22** (1961), 1020.
User: Young, D. H. (1962).
Class.: M:os:G

Barton, D. E., and David, F. N. 1959d. 164
A collector's problem. *Trabajos Estadist.*, **10**, 75–88.
Review: MR **22** (1961), 179; ZBL **94** (1962), 333.
Users: MacArthur, R. (1957), Barton, D. E. (1958), Blyth, C. R., and Curme, G. L. (1960).
Class.: MI:m-a-osp.:G

Barton, D. E., and David, F. N. 1959e. 165
Sequential occupancy. *Biometrika*, **46**, 218–23.
Review: IJA **1** (1959), 165.
Users: Barton, D. E., and David, F. N. (1959), Bekessy, A. (1964).
Class.: MI:mb:G

Barton, D. E., and David, F. N. 1959f. 166
Haemocytometer counts and occupancy problems. *Trabajos Estadist.*, **10**, 13–18.
Review: IJA **1** (1960), 569.

Barton, D. E., and David, F. N.
See David, F. N., and Barton, D. E. (1962).

Barton, D. E., David, F. N., and Merrington, M. 167
1960.
Tables for thc solution of the exponential equation, exp $(-a)+ka = 1$. *Biometrika*, **47**, 439–45.
Class.: TCP-OMR:tc:G

Barton, D. E., David, F. N., and Merrington, M. 168
1965–66.
A criterion for testing contagion in time and space. *Ann. Human Genetics*, **29**, 97–102.

Bartsch, G. E., and Armitage, P.
See Armitage, P., and Bartsch, G. E. (1960).

Bašarin, G. P. 1959. 169
On statistical estimate for the entropy of a sequence of independent random variables. *Teor. Verojatnost. i Primenen.*, **4**, 361–64.
Review: MR **22** (1961), 528; ZBL **92** (1962), 367.

Basu, D. 1951. 170
On the limit points of relative frequencies. *Sankhyā*, **11**, 379–82.
Class.: B:mi:G

Basu, D. 1952. 171
On the minimax approach to the problem of estimation. *Proc. Nat. Inst. Sci. India*, **18**, 287–99.
Review: MR **14** (1953), 666.
Class.: P:pe:G

Basu, D. 1955. 172
A note on the structure of a stochastic model considered by V. M. Dandekar. *Sankhyā*, **15**, 251–52.
Review: ZBL **67** (1958), 105.
Class.: OBR:pc:G

Basu, D. 1958. 173
On sampling with and without replacement. *Sankhyā*, **20**, 287–94.
Review: MR **21** (1960), 837; IJA **1** (1959), 42; ZBL **88** (1961), 126.
User: Pathak, P. K. (1961).
Class.: MI:mi:O

Bateman, G. I. 1950. 174
Power of the χ^2 index of dispersion test when Neyman's contagious distribution is the alternative hypothesis. *Biometrika*, **37**, 59–63.
Review: ZBL **37** (1951), 91.
Users: Thomas, M. (1951), David, F. N., and Moore, P. G. (1954), Darwin, J. H. (1957), Bennett, B. M. (1959).
Class.: P-N:id:G

Bateman, H. 1910. 175
On the probability distribution of particles. *Philos. Mag.*, **20**, 704–07.

Baten, W. D. 1933. 176
A statistical study of the Daucus Carota L. I. *Biometrika*, **25**, 186–95.
Class.: MI:mi:BM

Baten, W. D. 1934. 177
A statistical study of the Daucus Carota L. II. *Biometrika*, **26**, 443–68.
Class.: MI:mi:BM

Baten, W. D. 1935a. 178
Constancy in the number of Ligulate flowers of Chrysanthemum Leucanthemum, variety Pinnatifidum, during the flowering season. *Biometrika*, **27**, 260–66.
Class.: MI:mi:BM

Beten, W. D. 1935b. 179
Influence of position on structure of inflorescences of zizia aurea. *Papers Michigan Acad. Sci. Arts and Letters*, **21**, 33–58.
Class.: MI:mi:BM

Baten, W. D. 1936. 180
Influence of position on structure of inflorescences of cicuta maculata. *Biometrika*, **28**, 64–85.
Class.: MI:mi:BM

Bates, G. E. 1955. 181
Joint distributions of time intervals for the occurrence of successive accidents in a generalized Polya scheme. *Ann. Math. Statist.*, **26**, 705–20.
Users: Taylor, W. F. (1956), Fitzpatrick, R. (1958), Haight, F. (1965a), Neyman, J. (1965).
Class.: MI:tp:A

Bates, G. E., and Neyman, J. 1952a. 182
Contributions to the theory of accident proneness. I. An optimistic model of the correlation between light and severe accidents. *Univ. California Publ. Statist.*, **1**, 215–53.
Review: MR **14** (1953), 389; ZBL **47** (1953), 134.
Users: Bates, G. E., and Neyman, J. (1952b), Goldberg, S. (1954), Bates, G. E. (1955), Johnson, N. L. (1957a), Fitzpatrick, R. (1958), Edwards, C. B., and Gurland, J. (1960), Mosimann, J. E. (1963), Subrahmaniam, K. (1964), Haight, F. (1965a), Mellinger, G., Sylvester, D., Gaffey, W., and Manheimer, D. (1965), Neyman, J. (1965), Patil, G. P., and Bildikar, S. (1966a).

Bates, G. E., and Neyman, J. 1952b. 183
Contributions to the theory of accident proneness. II. True or false contagion. *Univ. California Publ. Statist.* **1**, 255–75.
Review: MR **14** (1953), 390; ZBL **47** (1953), 135.
Users: Fitzpatrick, R. (1958), Gurland, J. (1959), Subrahmaniam, K. (1964), Neyman, J. (1965), Patil, G. P., and Buildikar, S. (1966a), Subrahmaniam, K. (1966a).
Class.: PO-MI:mb-cm:A

Baticle, E. 1949. 184
Sur une loi de probabilité a priori pour l'interprétation des résultats de tirages dans une urne. *C. R. Acad. Sci. Paris*, **228**, 902–04.
Review: ZBL **34** (1950), 224.

Baticle, E. 1950. 185
Interpretation of results of tests of a sample. *Génie Civil*, **77**, 246–48.
Review: MR **4** (1951), Rev. No. 3446.

Baticle, E. 1951. 186
Sur la probabilité des iterations dans le schéma de Bernoulli. *C. R. Acad. Sci. Paris*, **232**, 472–73.
Review: ZBL **42** (1952), 136.

Batschelet, E. 1960. 187
Uber eine Kontingenztafel mit fehlenden Daten. *Biom. Zeit.*, **2**, 236–43.
Class.: M:pe:G

Baum, Leonard E., and Billingsley, Patrick. 1965. 188
Asymptotic distributions for the coupon collector's problem. *Ann. Math. Statist.*, **36**, 1835–39.
Class.: MI-P-G:a:G

Baumhover, A. H., Graham, A. J., Bitter, B. A., Hopkins, D. E., New, W. D., Dudley, F. H., and Bushland, R. C. 1955. 189
Wire-worm control through release of sterilized flies. *J. Econ. Ent.*, **48**, 462–66.

Bavli, G. M. 1935. 190
Eine Verallgemeinerung des Poissonschen Grenzwertsatzes. (Russian. German summary). *Comptes Rendus de l'Academie des Sciences d'URSS*, **2**, 508–11.
User: Govindarajulu, Z. (1965).

Beall, G. 1935. 191
Study of arthropod populations by the method of sweeping. *Ecology*, **16**, 216–25.
Users: Beall, G. (1940), Beall, G. (1942), Cole, L. C. (1946), Auerbach, S. I. (1951), Dice, L. R. (1952), Whittaker, R. H. (1952), Jensen, P. (1959).
Class.: P:h:BM

Beall, G. 1940. 192
The fit and significance of contagious distributions when applied to observations on larval insects. *Ecology*, **21**, 460–74.
Users: Fracker, S. B., and Brischle, H. (1944), Cole, L. C. (1946a), Cole, L. C. (1946b), Bowen, M. F. (1947), Anscombe, F. J. (1950a), Wadley, F. M. (1950), Barnes, H., and Marshall, S. M. (1951), Kono, Utida, Yosida, and Watanabe (1952), Skellam, J. (1952), Utida, Kono, Watanabe, and Yosida (1952), Beall, G., and Rescia, R. R. (1953), Bliss, C. I. and Fisher, R. A. (1953), Evans, D. A. (1953), Barton, D. E. (1957), McGuire, J., Brindley, T. A., and Bancroft, T. A. (1957), Gurland, J. (1958), Cohen, A. C., Jr. (1959c), Gurland, J. (1959), Jenson, P. (1959), Morisita, M. (1959), Cohen, A. C. (1960c), Katti, S. K. (1960d), Pielou, D. P. (1960a), Martin, D. C., and Katti, S. K. (1962a), Martin, D. C., and Katti, S. K. (1962b), Blischke, W. R. (1965), Blischke, W. R. (1965), Blischke, W. R. (1965), Katti, S. K., and Sly, L. E. (1965), Martin, D., and Katti, S. (1965).

Beall, G. 1942. 193
The transformation of data from entomological field experiments so that the analysis of variance becomes applicable. *Biometrika*, **32**, 243–62.
Users: Curtiss, J. H. (1943a), Fracker, S. B., and Brischle, H. (1944), Wadley, F. M. (1945), Bartlett, M. S. (1947), Anscombe, F. J. (1948), Anscombe, F. J. (1949a), Wadley, F. M. (1950), Hunter, G. C., and Quenouille, M. H. O. (1952), Utida, Kōno, Watanabe, and Yosida (1952), Beall, G., and Rescia, R. R. (1953), David, F. N. (1955a), Bliss, C. I. (1958), Bliss, C. I., and Owen, A. R. G. (1958), Pruess, K. P., and Weaver, C. R. (1959), Forsythe, H. Y., and Byrisco, G. (1961), Hayman, B. I., and Lowe, A. D. (1961).
Class.: P-MI:anovat:BM

Beall, G. 1954. 194
Data in binomial or near-binomial distribution: with particular application to problems in entomological research. *Statist. and Math. in Biology*, Ed. by Kempthorne, Bancroft, Gowen, and Lush, Iowa State College Press, Ames, Iowa. pp. 295–302.
Class.: B-P-NB-N-MI:gf-anovat:BM

Beall, Geoffrey, and Cobb, Sidney. 1961. 195
The frequency distribution of episodes of rheumatoid arthritis as shown by periodic examination. *J. Chron. Dis.*, **14**, 291–310.
Class.: MI:gf:BM

Beall, G., and Rescia, R. R. 1953. 196
A generalization of Neyman's contagious distributions. *Biometrics*, **9**, 354–86.
Review: MR **15** (1954), 239.
Users: Douglas, J. B. (1955), Bliss, C. I. (1958), Gurland, J. (1958), Cohen, A. C., Jr. (1959c), Jesen, P. (1959), Nef, L. (1959), Waters, W. E., and Henson, W. R. (1959), Cohen, A. C. (1960c), Shumway, R., and Gurland, J. (1960c), Katti, S. K. (1960d), Shumway, R., and Gurland, J. (1960b), Bardwell, G. E. (1961), Katti, S. K., and Gurland, J. (1961), Khatri, C. G., and Patel, I. R. (1961), Yassky, D. (1962), Subrahmaniam, K. (1964), Blischke, W. R. (1965), Blischke, W. R. (1965), Crow, E. L., and Bardwell, G. E. (1965), Gurland, J. (1965), Kemp, C. D., and Kemp, A. W. (1965).
Class.: N-MI:gf:BM

Beard, R. E., and Perks, Wilfred. 1949. 197
The relation between the distribution of sickness and the effect of duplicates on the distribution of deaths. *J. Inst. Actuar.*, **75**, 75–86.
Review: ZBL **39** (1951), 154.
User: Seal, H. L. (1953).

Beazley, R., and Shiue, C. J.
See Shiue, C. J., and Beazley, R. (1957).

Bechhofer, R. E., and Sobel, M. 1956. 198
A sequential multiple decision procedure for selecting the multinormal event with the largest probability (Preliminary report) (Abstract). *Ann. Math. Statist.*, **27**, 861.
Users: Bechhofer, R. E., Elmaghraby, S., and Morse, N. (1959).

Bechhofer, R. E., Elmaghraby, Salah, and Morse, Norman. 1959. 199
A single-sample multiple-decision procedure for selecting the multinomial event which has the highest probability. *Ann. Math. Statist.*, **30**, 102–19.
Review: IJA **1** (1959), 276.
Users: Kesten, H., and Morse, N. (1959).
Class.: M:mi:G

Becker, R. M., and Hazen, W. S. Jr. 1961. 200
Particle statistics of infinite populations as applied to mine sampling. Rep. of Investigations, 5669 U.S. Dept. of the Interior, Bureau of Mines, pp. 79.

Beckmann, M. J., and Bobkoski, F. 1958. 201
Airline demand: An analysis of some frequency disbributions. *Naval Res. Logist. Quart.*, **5**, 43–51.

Beekman, J. A., and Allen, J. L.
See Allen, J. L., and Beekman, J. A. (1966).

Beevey, E. S., Jr. 1941. 202
Limnological studies in Connecticut. IV. The quantity and composition of the bottom fauna of thirty-six Connecticut and New York lakes. *Ecol. Monogr.*, **11** (4), 413–55.
Class.: P:mi:BM

Behrens, D. J. 1952. 203
The fitting of a distribution curve of known shape, but arbitrary origin and scale. Tech. Rep. 629, Atomic Energy Res. Establ. H.M.S.O., London, pp. 42.

Békéssy, A. 1963. 204
On classical occupancy problems I. (Russian summary). *Magyar Tud. Akad. Mat. Kutató Int. Közl.*, **8**, 59–71.
Class.: P-MI:mi-a:G

Békéssy, A. 1964. 205
On classical occupancy problems. II. (Sequential occupancy). (Russian summary). *Magyar. Tud. Akad. Mat. Kutató Int. Közl.*, **9**, 133–41.
Review: MR **30** (1965), 303.
Class.: MI:a:G

Békéssy, A. 1964. 206
On classical occupancy problems, II (Sequential occupancy). *Magyar Tud. Akad. Mat. Kutató Int. Közl.*, **9**, 133–40.
Review: STMA **7** (1966), 455.
Class.: MI:a:G

Belevitch, V. 1959. 207
On the statistical laws of linguistic distributions. *Ann. Soc. Sci. Bruxelles*, **73**, 310–26.
Review: IJA **3** (1962), 396.
Class.: MI:mi:L

Bell, C. B., and Doksum, K. A. 1966. 208
Distribution-free independence tests. (Abstract). *Ann. Math. Statist.*, **37**, 539–40.
Class.: MI:mi:G

Beljaew, P. F.
See Belyaev, P. F. (1964).

Belyaev, P. F. 1964a. 209
The probability of non-occurrence of a given number of outcomes. *Teor. Verojatnost. i Primenen.*, **9**, 541–47.
Review: MR **31** (1966), 309.
Class.: MI-P:pc-a:G

Belyaev, P. F. 1964b. 210
The probability of non-occurrence of a given number of outcomes. *Theor. Probability Appl.*, **9**, 489–96.
Class.: MI-P:pc-a:G

Benayoun, R. 1964. 211
On a stochastic model in use in the mathematical theory of epidemics (French). *C.R. Acad. Sci. Paris*, **258**, 5789–91.
Review: STMA **6** (1965), 1196.
Class.: MI:pc:BM

Bender, B. K., and Goldman, A. J.
See Goldman, A. J., and Bender, B. K. (1962).

Bendersky, L. 1948. 212
Sur Quelques problèmes concernant les épreuves répétées. *Bull. Sci. Math., II S.*, **72**, 99–107.
Review: ZBL 33 (1950), 192.

Benedetti, C. 1956. 213
Sulla rappresentabilità di una distribuzione binomiale mediante una distribuzione B e viceversa. *Metron*, **18** (1–2), 121-31.
Review: MR **18** (1957), 606; ZBL **74** (1962), 342.
User: Johnson, N. L. (1960).
Class.: B:a:G

Beneš, V. E. 1957. 214
On queues with Poisson arrivals. *Ann. Math. Statist.*, **28**, 670–77.
Review: ZBL **85** (1961), 347; ZBL **74** (1962), 141.
Users: Prabhu, N. U. (1960), Prabhu, N. U. (1964).

Beneš, V. E. 1965a. 215
Properties of random traffic in non-blocking telephone connecting networks. *Bell System Tech. J.*, **44**, 509–25.

Beneš, V. E. 1965b. 216
Some inequalities in the theory of telephone traffic. *Bell System Tech. J.*, **44**, 1941–75.

Ben-Israel, A., and Naor, P. 1960a. 217
A problem of delayed service. I. *J. Roy. Statist. Soc. Ser. B.*, **22**, 245–69.
Review: IJA **2** (1961), 549.
Users: Ben-Israel, A., and Naor, P. (1960b).
Class.: MI:pc:G

Ben-Israel, A., and Naor, P. 1960b. 218
A problem of delayed service II. *J. Roy. Statist. Soc. Ser. B.*, **22**, 270–76.
Review: IJA **2** (1961), 550.
Class.: MI:pc:G

Bennett, B. M. 1955. 219
The cumulants of a sample mean from a finite population of first n integers. *Trabajos Estadist.*, **6**, 31–32.
Review: ZBL **66** (1957), 119.

Bennett, B. M. 1956. 220
Note on the Poisson index of dispersion. *Trabajos Estadist.*, **7**, 183–85.
Review: MR **18** (1957), 522.
Class.: P:id:G

Bennett, B. M. 1957a. 221
Note on the method of inverse sampling (Spanish summary). *Trabajos Estadist.*, **8**, 29–31.
Review: MR **19** (1958), 89; ZBL **80** (1959), 356.
Class.: NB:m:G

Bennett, B. M. 1957b. 222
On the variance stabilizing properties of certain logarithmic transformations. *Trabajos Estadist.*, **8**, 69–74.
Review: ZBL **80** (1959), 124.
Class.: B:anovat:G

Bennett, B. M. 1959. 223
A sampling study on the power function of the χ^2 'index of dispersion' test. *J. Hyg. Camb.*, **57**, 360–65.
Users: Bennett, B. M., and Hsu, P. (1961), Bennett, B. M. (1962a), Nicholson, W. L. (1963).
Class.: P:id:G

Bennett, B. M. 1962a. 224
On a heuristic treatment of the 'indices of dispersion'. *Ann. Inst. Statist. Math. Tokyo*, **14**, 151–57.
Class.: B-P:anovat-id-h:G

Bennett, B. M. 1962b. 225
On an exact test for trend in binomial trials and its power function. *Metrika*, **5**, 49–53.
Review: MR **24A** (1962), 704; IJA **4** (1963), 65; ZBL **104** (1964), 376.
User: Bennett, B. M. (1964b).
Class.: B:h:G

Bennett, B. M. 1962c. 226
Note on an exact test for the 2×2 contingency table using the negative binomial model. *Metrika*, **5**, 154–57.
Review: STMA **6** (1965), 1150; ZBL **118** (1965), 143.
Users: Bennett, B. M. (1964b), Bennett, B. M., and Birch, B. (1964c).
Class.: NB:tp:G

Bennett, B. M. 1963. 227
On a test for birth-order effect when several abnormalities are present. *Ann. Human Genetics*, **27**, 11–15.
Class.: MI:mi:BM

Bennett, B. M. 1963a. 228
Note on an approximation to the distribution of the logarithm of the relative risk. *Trabajos Estadist.*, **14**, 11–15.
Review: STMA **5** (1964), 557.
Class.: B:a-pe:G

Bennett, B. M. 1963. 229
Optimum moving averages for the estimation of median effective dose in bioassay. *J. Hyg. Camb.*, **61**, 401–06.

Bennett, B. M. 1964a. 230
A non-parametric test for randomness in a sequence of multinomial trials. *Biometrics*, **20**, 182–90.
Review: MR **28** (1964), 1068; ZBL **124** (1966), 350.

Bennett, B. M. 1964b. 231
On a test of homogeneity for samples from a negative binomial distribution. *Metrika*, **8**, 1–4.
Review: MR **31** (1966), 1137; STMA **6** (1965), 135.
Class.: NB:h:G

Bennett, B. M., and Birch, B. 1964. 232
Sampling inspection tables for comparison of two groups using the negative binomial model. *Trabajos Estadist.*, **15**, 1–12.
Review: MR **31** (1966), 749; STMA **6** (1965), 336.
Class.: NB:ctp-tc:G

Bennett, B. M., and Horst, C. 1966. 233
Tables for testing significance in a 2×2 contingency table. Cambridge Univ. Press.

Bennett, B. M., and Hsu, P. 1960. 234
On the power function of the exact test for the 2×2 contingency table. *Biometrika*, **47**, 393–98.
Review: ZBL **104** (1964), 376; IJA **2** (1961), 496.
Users: Bennett, B. M. (1962), Bennett, B. M. (1962b), Bennett, B. M. (1964b), Harkness, W. L. (1965).
Class.: OHR:h:G

Bennett, B. M., and Hsu, P. 1961. 235
A sampling study of the power function of the binomial χ^2 'index of dispersion' test. *J. Hyg. Camb.*, **59**, 449–55.
Review: BA **38** (1962), Ab. No. 12623; IJA **4** (1963), 420.
Users: Bennett, B. M., and Hsu, P. (1962).

Bennett, B. M., and Hsu. P. 1962. 236
Sampling studies on a test against trend in binomial data. *Metrika*, **5**, 96–104.
Review: ZBL **105** (1964), 338; IJA **4** (1963), 452.
Class.: B:h:G

Bennett, Carl A., and Franklin, Norman L. 1954. 237
Statistical analysis in chemistry and the chemical industry. John Wiley and Sons, New York.

Bennett, R. W. 1962. 238
Sizes of the χ^2 test in the multinomial distribution. *Austral. J. Statist.*, **4**, 86–88.
Review: MR **32** (1966), 309; ZBL **111** (1964–65), 337; IJA **4** (1963), 453.

Benson, F., and Cox, D. R. 1951. 239
The productivity of machines requiring attention at random intervals. *J. Roy. Statist. Soc. Ser. B.*, **13**, 65–82.
Review: ZBL **45** (1955), 81.
User: Bharucha-Reid, A. T. (1958).
Class.: MI-TCP:mb:E

Benson, F., and Gregory, G. 1961. 240
Some properties of a disguised Poisson distribution. *Operations Res.*, **9**, 901–03.
Class.: MI:pc-osp:G

Benson, Robert, and Gutzman, Wayne. 1961. 241
Variation of the parameters of the sequential test as applied to the binomial distribution (Abstract). *Proc. South Dakota Acad. Sci.*, **40**, 235–38.
Review: BA **38** (1962), Ab. No. 12624.

Berger, Alfred. 1938. 242
Über eine Funktionalgleichung der Wahrscheinlichkeitstheorie. *Mh. Math. Phys.*, **46**, 366–76.

Review: ZBL **20** (1939), 242.
Class.: B-P:mi:G

Berkson, Joseph. 1938. 243
Some difficulties of interpretation encountered in the application of the chi-square test. *J. Amer. Statist. Assoc.*, **33**, 526–36.
Users: Cochran, W. G. (1954), Hoel, P. G. (1947), Turner, M. E., and Eadie, G. S. (1957).
Class.: P:gf:P

Berkson, Joseph. 1940. 244
A note on the Chi-square test, the Poisson and the binomial. *J. Amer. Statist. Assoc.*, **35**, 362–67.
Review: ZBL **23** (1941), 246.
Users: Cochran, W. .G (1954), Hoel, P. G. (1947), Schilling, W. (1947), Turner, M. E., and Eadie, S. (1957).
Class.: P:gf:P

Berkson, Joseph, Dews, Peter B., and Higgins, George M.
See Dews, Peter B., Higgins, George M., and Berkson, Joseph (1954).

Berkson, Joseph, Magath, Thomas B., and Hurn, Margaret. 1935. 245
Laboratory standards in relation to chance fluctuations of the Erythrocyte count as estimated with the hemocytometer. *J. Amer. Statist. Assoc.*, **30**, 414–26.
Users: Birnbaum, A. (1954b), Turner, M. E., and Eadie, S. (1957).

Berkson, J., Magath, T. B., and Hurn, M. 1940. 246
The error of estimate of blood cell count as made with the hemacytometer. *Amer. J. Physiology*, **128**, 309–23.
Users: Schneiderman, M., and Brecher, G. (1950), Chamberlain, A. C., and Turner, F. (1952).

Berljand, O. S., Nazarov, I. M., and Pressman, A. Ja. 1962. 247
i^n distribution or mixed Gauss-Poisson distribution (Russian). *Dokl. Akad. Nauk SSSR.*, **147**, 1005–07.
Review: MR **26** (1963), 1077.

Berndt, G. D. 1963. 248
Tables of the negative binomial distribution (computer output). *Oper. Analysis HQSAC, Offuti AFB Nebraska.*

Berthet, P., and Gerard, G. 1965. 249
A statistical study of microdistribution of Oribatei (Acari) (Russian abstract). *Oikos*, **16**, 214–27.
Class.: NB-GP-COP:gf:BM

Bertram, G. 1960. 250
Sequential analysis for two binomial experiments (German). *Z. Angew. Math.*, **40**, 185–89.
Review: IJA **2** (1961), 308.
Class.: B:tp:G

Beyer, Otfried. 1957. 251
Eine Anwendung des Poissonschen Gesetzes der Wahrscheinlichkeitsrechung. *Wiss. Zeit. Hochschule Schwermaschinenbau Magdeburg*, **1**, 81–83.
Review: ZBL **88** (1961), 115.

Bezem, J. J. 1954. 252
A sequential method for testing interaction of two factors producing the same all-or-none effect. *Nederl. Akad. Wetensch. Proc. Ser. A*, **57**, (Indag. Math. 16), 424–31.
Review: MR **16** (1955), 273; ZBL **56** (1955), 378.

Bhapkar, V. P. 1966. 253
A note on the equivalence of two test criteria for hypotheses in categorical data. *J. Amer. Statist.*, **61**, 228–35.
Class.: M:tp:G

Bharucha-Reid, A. T. 1958. 254
Comparison of populations whose growth can be described by a branching stochastic process—with special reference to a problem in epidemiology. *Sankhyā*, **19**, 1–14.
Class.: MI:pc-ctp:G-BM-P-E-S

Bhat, B. R., and Kulkarni, N. V. 1966b. 255
On efficient multinomial estimation. *J. Roy. Statist. Soc. Ser. B.*, **28**, 45–52.
Class.: M-NM:pe:G

Bhat, B. R., and Kulkarni, S. R. 1966a. 256
Lamp tests of linear and loglinear hypotheses in multinomial experiments. *J. Amer. Statist. Assoc.*, **61**, 236–45. Correction **61**, 1240.
Class.: B-M:tp:G

Bhat, B. R., and Nagnur, B. N. 1965. 257
Locally asymptotically most stringent tests and Lagrangian multiplier tests of linear hypotheses. *Biometrika*, **52**, 459–68.
Class.: M:tp:G

Bhat, U. N. 1965. 258
On a stochastic process occurring in queueing systems. *J. Appl. Prob.*, **2**, 467–69.
Review: MR **32** (1966), 307.

Bhat, U. N., and Prabhu, N. U.
See Prabhu, N. U., and Bhat, U. N. (1963).

Bhattacharji, A. K. 1959. 259
A note on a stochastic model for dependent binomial events (Abstract). *Math. Student*, **27**, 276.

Bhattacharyya, A. 1943. 260
On a measure of divergence between two statistical populations defined by their probability distributions. *Bull. Calcutta Math. Soc.*, **45**, 99–109.
Review: MR **6** (1945), 7.
User: Rényi, A. (1960b).
Class.: M:ctp:G

Bhattacharyya, A. 1946. 261
On a measure of divergence between two multinomial populations. *Sankhyā*, **7**, 401–06.
Users: Poti, S. J. (1955), Patil, G. P., and Bildikar, S. (1966a).
Class.: M:ctp:G

Bhattacharyya, A. 1954. 262
Notes on the use of unbiased statistics in the binomial population. *Bull. Calcutta Statist. Assoc.*, **5**, 149–64.
Users: Haldane, J. B. S. (1956b), Sirazhdinov, S. H. (1956).

Bhattacharyya, M. N., and Iyer, P. V. K.
See Iyer, P. V. K., and Bhattacharyya, M. N. (1955).

Bhattacharya, S. K., and Holla, M. S. 1965. 263
On a discrete distribution with special reference to the theory of accident proneness. *J. Amer. Statist. Assoc.*, **60**, 1060–66.
Class.: COP:mb-osp:G-A

Bienaymé, J. 1838. 264
Sur la probabilité des résultats moyens des observations; demonstration directe de la règle de Laplace. *Mem. Sav. Etrangers Acad. Sci. Paris*, **5**, 513–58.

Bildikar, Sheela. 1966. 265
Certain contributions to multivariate distributions theory. Ph.D. Thesis, McGill Univ.

Bildikar, Sheela, and Patil, G. P.
See Patil, G. P., and Bildikar, Sheela (1965).

Bildikar, Sheela, and Patil, G. P.
See Patil, G. P., and Bildkar, Sheela (1966a).

Bildikar, Sheela, and Patil, G. P.
See Patil, G. P., and Bildkar, Sheela (1966b).

Billewicz, W. Z. 1956. 266
Matched pairs in sequential trials for significance of a difference between proportions. *Biometrics*, **12**, 282–300.
Class.: B:ctp:G

Billingsley, Patrick, and Baum, Leonard E.
See Baum, Leonard E., and Billingsley, Patrick (1965).

Bílý, Josef. 1959a. 267
Eine Markoffsche Kette, die zu einer Faltung zweier binomischen Verteilungen und zu kumulierten binomischen Verteilungen führt (Czech. Russian and German Summary). *Casopis Mat.*, **84**, 327–34.
Review: ZBL **88** (1961), 350.

Bílý, Josef 1959b. 268
Zusammengesetzte Poissonsche Verteilungen (Czech. Russian and German summaries). *Casopis Mat.*, **84**, 424–32.
Review: ZBL **117** (1965), 133.

Binet, F. E. 1953. 269
The fitting of the positive binomial distribution when both parameters are estimated from the sample. *Ann. Eugenics*, **18**, 117–19.
Review: MR **15** (1954), 240; PA **28** (1954), Ab. No. 3559; ZBL **51** (1954), 372.
Class.: B-P:gf:G-E

Binet, F. E., and Leslie, R. T. 1960. 270
The coefficienti of inbreeding in case of repeated full-sib-matings. *J. Genetics*, **57**, 127–30.
Class.: MI:mi:BM

Binet, F. E., Lesile, R. T., Weiner, S., and Anderson, R. L. 1956. 271
A comparison of discrete and continuous models in agricultural production analysis. *Methodological Procedures in the Economic Analysis of Fertilizer use data*, Iowa State College Press, pp. 39–61.

Birch, B., and Bennett, B. M.
See Bennett, B. M., and Birch, B. (1964).

Birch, M. W. 1963. 272
An algorithm for the logarithmic series distribution. *Biometrics*, **19**, 651–52.
Review: STMA **6** (1965), 47.

Birch, M. W. 1964. 273
A new proof of the Pearson-Fisher theorem. *Ann. Math. Statist.*, **35**, 817–23.
Class : M:gf:G

Birnbaum, Allan. 1952. 274
Sequential tests and estimates for comparing Poisson populations (Abstract). *Ann. Math. Statist.*, **23**, 645.

Birnbaum, Allan. 1953. 275
Some procedures for comparing Poisson processes or populations. *Biometrika*, **40**, 447–49.
Review: MR **15** (1954), 331; ZBL **51** (1954), 108.
User: Birnbaum, Allan (1954b).
Class.: P-B:ctp:G

Birnbaum, Allan. 1954a. 276
Confidence intervals of fixed length for the Poisson mean and the difference between two Poisson means (Abstract). *Ann. Math. Statist.*, **25**, 171.

Birnbaum, Allan. 1954b. 277
Statistical methods for Poisson processes and exponential populations. *J. Amer. Statist. Assoc.*, **49**, 254–66.
Review: ZBL **55** (1955), 375.
Users: Ellner, H. (1963), Singh, Rajinder (1963), Knight, W. (1965).
Class.: P:pc-tp-ctp:G

Birnbaum, Allan. 1961. 278
Confidence curves: an omnibus technique for estimation and testing statistical hypotheses. *J. Amer. Statist. Assoc.*, **56**, 246–49.
Review: IJA **4** (1963), 557.
Class.: B:ie-tp:G

Birnbaum, Allan, and Guthrie, D. 1956. 279
Tables for estimating a proportion or a Poisson mean with prescribed precision. Tech. Rep. 25, Applied Math. and Statist. Lab., Stanford Univ.
User: Cacoullos, T. (1962).

Birnbaum, Allan, and Healy, William C., Jr. 1960. 280
Estimates with prescribed variance based on two-stage sampling. *Ann. Math. Statist.*, **31**, 662–76.
Review: MR **22** (1961), 1464.
User: Samuel, E. (1966).
Class.: H-B-P:pe-mi:G

Bisbee, E. F., and Oliver, R. M.
See Oliver, R. M., and Bisbee, E. F. (1962).

Bishir, John. 1962. 281
Maximum population size in a branching process. *Biometrics*, **18**, 394–403.
Review: IJA **4** (1963), 736.
Class.: MI:pc:BM-P

Bishop, Y. M. M., and Margolis, L. 1955. 282
A statistical examination of Anisakis larvae (Nematoda) in herring (*Cupea pallasi*) of the British Columbia coast. *J. Fisheries Res. Board Canada*, **12** (4), 571–92.

Bishop, Y. M. M., and Schaefer, M. B.
See Schaefer, M. B., Bishop, Y.M.M. (1958).

Bisi, W. 1963. 283
The queues in a pre-stationary phase (Italian). *Statistica*, **23**, 363–77.
Review: STMA **7** (1966), 385.
Class.: MI:pc:G

Bissinger, Barnard H. 1965. 284
A type-resisting distribution generated from considerations of an inventory decision model. *Classical and Contagious Discrete Distributions*, Ed. by G. P. Patil,

Statistical Publishing Society, Calcutta and Pergamon Press, pp. 15–17.
Class.: MI:m:G

Bitter, B. A., Hopkins, D. E., New, W. D., Dudley, F. H., Bushland, R. C., Baumhover, A. H., and Graham, A. J.
See Baumhover, A. H., Graham, A. J., Bitter, B. A., Hopkins, D. E., New, W. D., Dudley, F. H., and Bushland, R. C. (1955).

Bizard, G., and Seguinot, J. 1954. 285
Solution of statistical problems of Poisson accumulation by electronic analogy. *Nuovo Cimento*, **20**, 836–44.
Review: SA **64A** (1961), Ab. No. 15396.

Blackman, G. E. 1935. 286
A study of statistical methods of the distribution of species in grassland associations. *Ann. Bot. Lond.*, **49**, 749–77.
Users: Ashby, E. (1936), Clapham, A. R. (1936), Singh, B. N., and Chalam, G. (1937), Singh, B. N., and Chalam, G. (1938), Singh, B. N., and Das, K. (1939), Fracker, S. B., and Brischle, H. (1944), Williams, C. B. (1944b), Archibald, E. E. A. (1948), Dice, L. R. (1948), Whitford, P. B. (1949), Archibald, E. E. A. (1950), Curtis, J. and McIntosh, R. (1950, Barnes, H., and Stansbury, F. (1951), Grieg-Smith, P. (1952a), Skellam, J. G. (1952), Thomson, G. W. (1952), Bliss, C. I., and Fisher, R. A. (1953), David, F. N., and Moore, P. G. (1954), Robinson, P. (1954), Kemp, C. D., and Kemp, A. W. (1956a), Smith, J. H. G., and Ker, J. W. (1957), Kono, and Sugino (1958).

Blackman, G. E. 1942. 287
Statistical and ecological studies in the distribution of species in plant communities. I. Dispersion as a factor in the study of changes in plant populations. *Ann. Bot., Lond. (N.S.)*, **6**, 351–70.
Users: Archibald, E. E. A. (1950), Barnes, H., and Marshall, S. M. (1951), Barnes, H., and Stanbury, F. (1951), Grieg-Smith, P. (1952b), Thompson, G. W. (1952), Morisita, M. (1959).

Blackwell, D. 1951a. 288
On the translation parameter problem for discrete variables (Abstract). *Ann. Math. Statist.*, **22**, 138.

Blackwell, D. 1951b. 289
On the translation parameter problem for discrete variables. *Ann. Math. Statist.*, **22**, 393–99.
Review: MR **13** (1952), 260; ZBL **43** (1952), 138.
Class.: MI:pe:G

Blackwell, D. 1957. 290
On discrete variables whose sum is absolutely continuous. *Ann. Math. Statist.*, **28**, 520–21.
Review: MR **19** (1958), 467.
Class.: MI:pc-a:G

Blackwell, D., and Hodges, J. L., Jr. 1959. 291
The probability in the extreme tail of a convolution. *Ann. Math. Statist.*, **30**, 1113–20.
Users: Williams, E. J. (1961b), Hald, A. (1962), Hald, A., and Kousgaard, E. (1963), Hald, A. (1964), Govindarajulu, Z. (1965), Hald, A. (1965).
Class.: MI:mi:G

Blackwell, D., and Kendall, D. 1964. 292
The Martin boundary for Polya's urn scheme, and an application to stochastic population growth. *J. Appl. Prob.*, **1**, 284–96.
Review: MR **31** (1966), 143.

Blanc-Lapierre, A. 1951. 293
Notes on Poisson distributions (Spanish). *Rev. Un. Mat. Argentina*, **15**, 3–6.
Review: MR **15** (1954), 138; ZBL **44** (1952), 336.

Blanc-Lapierre, A., and Fortet, R. 1955. 294
Sur les répartitions de Poisson. *C. R. Acad. Sci. Paris*, **240**, 1045–46.
Review: ZBL **64** (1956), 128.
User: Prekopa, A. (1957a).
Class.: OPR:osp:G

Blank, A. A. 1956. 295
Existence and uniqueness of a uniformly most powerful randomized unbiased test for the binomial. *Biometrika*, **43**, 465–66.
Review: MR **18** (1957), 426.
Class.: B:tp:G

Blench, T., and Latif, Abul. 1950. 296
Poisson frequency distribution applied to peak floods. *Proc. First Pakistan Statist. Conf.*, 59–61.

Blenk, H. 1951. 297
Poissonsche Verteilungskurven bei Versuchen mit Drill maschinen. *Z. Angew. Math. Mech.*, **31**, 257–58.
Review: ZBL **42** (1952), 379.

Blischke, W. R. 1962. 298
Moment estimators for the parameters of a mixture of two binomial distributions. *Ann. Math. Statist.*, **33**, 444–54.
Review: MR **25** (1962), 141; IJA **4** (1963), 620.
Users: Teicher, H. (1963), Blischke, W. R. (1965).
Class.: COB:pe:G

Blischke, W. R. 1964. 299
Estimating the parameters of mixtures of binomial distributions. *J. Amer. Statist. Assoc.*, **59**, 510–28.
Review: MR **28** (1964), 1065; STMA **7** (1966), 68.
User: Blischke, W. R. (1965).
Class.: COB:pe:G

Blischke, W. R. 1965. 300
Mixtures of discrete distributions. *Classical and Contagious Discrete Distributions*, Ed. G. P. Patil, Statistical Publishing Society, Calcutta, and Pergamon Press, pp. 351–72.
Class.: MI:mi-tp-pe:G

Bliss, C. I. 1937. 301
The analysis of field experimental data expressed in percentages (Russian). *Plant Protection No.* 12, *Leningrad.*
User: Bliss, C. I. (1938).
Class.:

Bliss, C. I. 1938. 302
The transformation of percentages for use in the analysis of variance. *Ohio J. Sci.*, **38**, 9–12.
Users: Cochran, W. G. (1940), Curtiss, J. H. (1943a).
Class.: B:anovat:G

Bliss, C. I. 1941. 303
Statistical problems in estimating populations of Japanese beetle larvae. *J. Econ. Ent.*, **34** (2), 221–31.
Review: BA **17** (1943), Ab. No. 360.
Users: Bowen, M. F. (1947), Fracker, S. B., and Brischle, H. (1944), Kono, Utida, Yosida, and Watanabe (1952), Utida, Kono, Watanabe, and Yosida (1952).

Bliss, C. I. 1948. 304
Estimation of the mean and its error from incomplete Poisson distributions. Bull. 513, Connecticut Agri. Expt. Stn., New Haven, pp. 1–12.
Users: Hartley, H. O. (1958), Moore, P. G. (1952), Murakami, M., Asai, A., and Kawamura, M. (1954).
Class.: P:pe-gf-tc:G-BM

Bliss, C. I. 1958. 305
The analysis of insect counts as negative binomial distributions. *Proc. 10th Int. Congr. Entom.*, **2**, 1015–32.
Users: Waters, W. (1959), Kuno, E. (1963), Shiyomi, M., and Nakamura, K. (1964), Kobayashi, S. (1965), Kobayashi, S. (1966).
Class.: NB-P:pe-gf-anovat:BM

Bliss, C. I. 1965. 306
An analysis of some insect trap records. *Classical and Contagious Discrete Distributions*, Ed. G. P. Patil, Statistical Publishing Society, Calcutta, and Pergamon Press, pp. 385–97. Reprinted in *Sankhyā A*, **28** (1966).
Class.: LS-COP:gf:BM

Bliss, C. I., and Calhoun, D. W. 1954. 307
An outline of biometry. Yale Co-op. Corp. New Haven.

Bliss, C. I., and Fisher, R. A. 1953. 308
Fitting the negative binomial distribution to biological data, note on the efficient fitting of the negative binomial. *Biometrics*, **9**, 176–200.
Review: MR **14** (1953), 1102 and 1104.
Users: Sampford, M. R. (1955), Waters, W. E. (1955), Kemp, C. D., and Kemp, A. W. (1956a), McGuire, J., Brindley, T. A., and Bancroft, T. A. (1957), Comita, G. W., and Comita, J. J. (1957), Smith, J. H. G., and Ker, J. (1957), Bliss, C. I. (1958), Bliss, C. I., and Owen, A. R. G. (1958), Gurland, J. (1958), Hartley, H. O. (1958), Kutkuhn, J. H. (1958), Nef, L. (1958), Sprott, D. A. (1958), Cohen, A. C. (1959d), Gurland, J. (1959), Hairston, N. G. (1959), Jensen, P. (1959), Nef, L. (1959), Patil, G. P. (1959), Waters, W. (1959), Edwards, A. W. F. (1960a), Griffiths, J. C. (1960), Harcourt, D. G. (1960), Katti, S. K. (1960d), Patil, G. P. (1960c), Pielou, D. P. (1960a), Pielou, E. C. (1960), Harcourt, D. G. (1961), Hayman, B. I., and Lowe, A. D. (1961), Katti, S. K., and Gurland, J. (1961), Khatri, C. G. (1961), Martin, L. (1961), Cassie, R. M. (1962), Katti, S. K., and Gurland, J. (1962b), Itô, Nakamura, Kondo, Miyashita, and Nakamura (1962), Pielou, E. C. (1962), Rider, P. R. (1962a), Rider, P. R. (1962b), Kuno, E. (1963), Kuno, Yamamoto, Satomi, Outi, and Okada (1963), Shiyomi, M., and Nakamura, K. (1964), Berthet, P., and Gerard, G. (1965), Blischke, W. R. (1965), Cohen, A. Clifford Jr. (1965), Harcourt, D. G. (1965), Katti, S. K., and Sly, L. E. (1965), Kemp, D. C., and Kemp, A. W. (1965), Kobayashi, S. (1965), Shenton, L. R., and Myers, R. (1965), Sprott, D. A. (1965b), Kobayashi, S. (1966).
Class.: NB-N-T-LS:pe:BM-G

Bliss, C. I., and Fleischer, David A. 1964. 309
The analysis in angles of a microbial experiment. *Biometrics*, **20**, 883–91.
Class.: B:anovat:BM

Bliss, C. I., and Owen, A. R. G. 1958. 310
Negative binomial distributions with a common k. *Biometrika*, **45**, 37–58.
Review: ZBL **96** (1962), 134.
Users: Bliss, C. I. (1958), Waters, W. (1959), Waters, W. E., and Henson, W. R. (1959), Edwards, A. W. F. (1960a), Pielou, D. P. (1960a), Grimm, H., and Maly, V. (1961), Harcourt, D. G. (1961), Cassie, R. M. (1962), Kuno, E. (1963), Kuno, Yamamoto, Satomi, Outi, and Okada (1963), Berthet, P., and Gerard, G. (1965), Harcourt, D. G. (1965), Kobayashi, S. (1965).
Class.: NB:pe-anovat:G-BM

Bliss, C. I., Morgan, M. E., MacLeod, P., and Anderson, E. O.
See Morgan, M. E., MacLeod, P., Anderson, E. O., and Bliss, C. I. (1951).

Bloch, D. A., and Watson, G. S. 1966. 311
A Bayesian study of the multinomial distribution (Abstract). *Ann. Math. Statist.*, **37**, 760.

Bloemena, A. R. 1960a. 312
On Queueing processes with a certain type of bulk service. *Bull. Inst. Internat. Statist.*, **37** (3), 219–27.
Review: IJA **2** (1961), 553.
Class.: MI:pc:G

Bloemena, A. R. 1960b. 313
On Probability distributions arising from points on a graph. *Report S266A Statist. Dept. Math. Cent. Amsterdam.*
Review: IJA **1** (1960), 540.
Class.: MI:mi:G

Bloemena, A. R., and Van Eeden, Constance.
See Van Eeden, Constance, and Bloemena, A. R. (1960).

Blom, Gunnar. 1954. 314
Transformations of the binomial, negative binomial, Poisson and χ^2 distributions. *Biometrika*, **41**, 302–16.
Review: MR **16** (1955), 940; ZBL **56** (1955), 365.
Users: Wishart, J. (1956), Govindarajulu, Z. (1965).
Class.: B-P-NB:anovat-a-ie:G
Notes: Correction, *Biometrika*, **43**, 235.

Blomqvist, Nils. 1951. 315
Some tests based on dichotomization. *Ann. Math. Statist.*, **22**, 362–71.
User: Govindarajulu, Z. (1965).

Blomqvist, N. 1952. 316
On an exhaustion process. *Skand. Aktuarietidskr.*, 201–10.
Review: ZBL **48** (1953), 111.

Blum, J. R., and Rosenblatt, M. 1959. 317
On the structure of infinitely divisible distributions. *Pacific J. Math.*, **9**, 1–7.
Review: ZBL **85** (1961), 129.
Class.: MI:osp:G

Blum, M. L., and Mintz, A.
See Mintz, A., and Blum, M. L. (1949).

Blum, M. L., and Mintz, Alexander. 1951. 318
Correlation versus curve fitting in research on accident proneness: reply to Maritz. *Psychol. Bull.*, **48**, 413–18.
Users: Webb, W., and Jones, E. R. (1953, Arbous, A. G., and Sichel, H. S. (1954b), Fitzpatrick, R. (1948), Edwards, C. B., and Gurland, J. ((1960), Teicher, H. (1960a), Edwards, C. B., and Gurland, J. (1961).
Class.: NB-NM-P:cm:A

Blyth, C. R. 1959. 319
Note on estimating information. *Ann. Math. Statist.*, **30**, 71–79.
Review: MR **21** (1960), 362; IJA **1** (1959), 296.
Class.: MI:pe:G

Blyth, C. R., and Curme, G. L. 1960. 320
Estimation of a parameter in the classical occupancy problem. *Biometrika*, **47**, 180–85.
Review: ZBL **95** (1962), 141; STMA **5** (1964), 11.
User: Barton, D. E. (1961).
Class.: MI:pe:G

Blyth, C. R., and Hutchinson, D. W. 1960. 321
Table of Neyman-shortest unbiased confidence intervals for the binomial parameter. *Biometrika*, **47**, 381–91.
Review: ZBL **104** (1964), 130.
Users: Blyth, C. R., and Hutchinson, D. W. (1961), Pratt, J. W. (1961), Thatcher, A. C. (1964).
Class.: B:ie-tc:G

Blyth, C. R., and Hutchinson, D. W. 1961. 322
Table of Neyman-shortest unbiased confidence intervals for the Poisson parameter. *Biometrika*, **48**, 191–94.
Review: IJA **4** (1963), 622.
Class.: P:ie-tc:G

Boas, R. P., Jr. 1949. 323
Representation of probability distributions by Charlier series. *Ann. Math. Statist.*, **20**, 376–92.
Review: ZBL **41** (1952), 248.

Bobkoski, F., and Beckmann, M. J.
See Beckmann, M. J., and Bobkoski, F. (1958).

Bochner, S. 1936. 324
A converse of Poisson's theorem in the theory of probability. *Ann. of Math.*, **37**, 816–22.

Bodmer, W. F. 1959. 325
A significantly extreme deviate in data with a non-significant heterogeneity (Chi-square). *Biometrics*, **15**, 538–42.
Review: BA **35** (1960), Ab. No. 12197; IJA **1** (1960), 409.

Bodmer, W. F. 1960. 326
Discrete stochastic processes in population genetics. *J. Roy. Statist. Soc. Ser. B.*, **22**, 218–44.

Boen, J. R., and Sylwester, David. 1966. 327
A quantitative discussion of the effectiveness of voiding as a defense against bladder infection. *Biometrics*, **22**, 53–57.
Class.: G:mb:BM

Bofinger, Eve. 1965. 328
Sufficiency for multinomial and transition probabilities. *J. Appl. Prob.*, **2**, 470–74.
Review: MR **32** (1966), 1454.

Boge, W. 1964. 329
Characterization of successively infinitely divisible probability distributions on locally compact groups (German). *Zeit. Wahrscheinlichkeitsth.*, **2**, 380–94.
Review: STMA **6** (1965), 382.
Class.: P:mi:G

Böhm, Friedrich. 1941. 330
Über eine Aufgabe aus der Theorie der Momente. *Bl. Versich.-Math.*, **5**, 245–59.
Review: ZBL **25** (1942), 416.
Class.: MI:m:G

Bohman, Harold. 1962. 331
Two inequalities for Poisson distributions. *Skand. Aktuarietidskr.*, **45**, 47–52.

Bohman, Harold. 1963. 332
Two inequalities for Poisson distributions. *Skand. Aktuarietidskr.*, **46**, 47–52.
Review: MR **32** (1966), 302; STMA **6** (1965), 48; ZBL **123** (1965), 360.
Class.: P:osp:G

Boldrini, Marcello. 1941. 333
Sulla dispersione dei caratteri mendeliani. *Acta. Pontif. Acad. Sci.*, **5**, 85–101.
Review: ZBL **28** (1944), 172.

Bolger, Edward M. 1961. 334
Exponential distributions. Master's Thesis, Pennsylvania State University.

Bolger, Edward M. 1964. 335
Exponential distributions. Ph.D. Thesis, Pennsylvania State University, pp. 93.

Bolger, Edward M. 1966. 336
The sum of two independent exponential-type random variables. *Pacific J. Math.*, **18**, 31–35.
Class.: OPR:osp:G

Bolger, Edward M., and Harkness, W. L. 1965. 337
Characterizations of some distributions by conditional moments. *Ann. Math. Statist.*, **36**, 703–05.
Review: MR **30** (1965), 993.
Class.: B-P-NB:m:G

Bolger, Edward M., and Harkness, W. L. 1966. 338
Some characterizations of exponential-type distributions. *Pacific J. Math.*, **16**, 5–11.
Class.: P-OPR:osp:G

Bol'shev, L. N. 1960. 339
On estimates of probabilities. *Theor. Probability Appl.*, **5**, 411–15.
Users: Bol'shev, L., Gladkov, B., and Shcheglova, M. (1961), Bol'shev, L. N. (1962), Bol'shev, L. N. (1962a).

Bol'shev, L. N. 1961. 340
A refinement of the Cramer-Rao inequality. *Theor. Probability Appl.*, **6**, 295–301.
Review: ZBL **109** (1964), 376.

Bol'shev, L. N. 1962a. 341
On comparison of means of Poisson distributions. *Theor. Probability Appl.*, **7**, 113–14.
Class.: P:ctp-ie:G

Bol'shev. L. N. 1962b. 342
On comparison of means of Poisson distributions (Russian, English Summary). *Teor. Verojatnost i Primenen*, **7**, 119–20.
Review: STMA **6** (1965), 497.
Class.: P:ctp-ie:G

Bol'shev, L. N. 1964. 343
Distributions similar to the hypergeometric one (Russian, English Summary). *Teor. Verojatnost i Primenen*, **9**, 687–92.
Review: MR **30** (1965), 121.

Bol'shev, L. N. 1964a. 344
Distributions related to the hypergeometric distributions. *Theor. Probability Appl.*, **9**, 619–24.
Class.: NH:mi-a:G

Bol'shev, L. N. 1965a. 345
On a characterization of the Poisson distribution and its statistical appliations (Russian, English Summary). *Teor. Verojatnost. i Primenen.*, **10**, 489–99.

Bol'shev, L. N. 1965b. 346
On a characterization of the Poisson distribution and its statistical applications. *Theor. Probability Appl.*, **10**, 446–56.
Class.: P:osp:G

Bol'shev, L. N., Gladkov, B. V., and Shcheglova, M. V. 1961. 347
Tables for the calculation of B and Z-distribution functions. *Theor. Probability Appl.*, **6**, 410–19.
Class.: B:a-tc:G

Bombara, E. L. 1955. 348
A study of the lower moments of order statistics of discrete uniform distributions. Master's Thesis, Virginia Polytechnic Institute.

Bonferroni, C. E. 1933. 349
Sulla probabilità massima nello schèma di Poisson. *Ist. Italiana degli Attuari Giornale*, **4**, 109–15.

Bonini, C. P., and Simon, H. A.
See Simon, H. A., and Bonini, C. P. (1958).

Borel, Émile. 1939. 350
Sur une interprétation des probabilités virtuelles. *C. R. Acad. Sci. Paris*, **208**, 1369–71.
Review: ZBL **21** (1940), 40.

Borel, E. 1942. 351
Sur l'emploi du théorème de Bernoulli pour faciliter le calcul d'un infinité de coefficients. Application au problème de l'attente à un guichet. *C. R. Acad. Sci. Paris*, **214**, 452–56.
Users: Haight, F. A. (1959), Haight, F. (1960), Haight, F., and Breuer, M. (1960), Takacs, L. (1962), Takacs, L. (1963a), Haight, F. (1965b), Morgenstern, D. (1966).

Borel, Émile. 1951. 352
Sur la transmission d'un caractère héréditaire dans les générations successives. *C. R. Acad. Sci. Paris*, **233**, 1241–43.
Review: ZBL **43** (1952), 341.

Borenius, G. 1953. 353
On the statistical distribution of mine explosions. *Skand. Aktuartidskr.*, **36**, 151–57.
Review: ZBL **52** (1955), 145. .
Class.: MI:a:O

Borovkov, A. A. 1960. 354
Limit theorems on the distribution of maximum of sums of bounded, lattice random variables II (Russian). *Teor. Verojatnost. i Primenen.*, **5**, 377–92.
Review: ZBL **94** (1962), 324.

Borovkov, A. A. 1960a. 355
Limit theorems on the distributions of maxima of sums of bounded, lattice random variables I. *Theor. Probability Appl.*, **5**, 125–55.
Review: ZBL **106** (1964) 126.
Class.: MI:a:G

Borovkov, A. A. 1960b. 356
Limit theorems on the distributions of maximum of sums of bounded latticed random variables. I (Russian, English Summary). *Teor. Verojatnost. i Primenen.*, **5**, 137–71.
Review: ZBL **90** (1961) 355.

Borovkov, A. A. 1960c. 357
Limit theorems on the distributions of maxima of sums of bounded lattice random variables II. *Theor. Probability Appl.*, **5**, 341–55.
Review: ZBL **94** (1964), 324.
Class.: MI:a:G

Borovkov, A. A. 1961. 358
Limit theorems and moments for maxima of sums of bounded lattice components. *Theor. Probability Appl.*, **6**, 99–101.
Review: ZBL **106** (1964) 340.
Class.: MI:a:G

Borovkov, A. A. 1961a. 359
Local theorems and moments for maximums of sums of lattice restricted components (Russian). *Teor. Verojatnost. i Primenen.*, **6**, 108–10.
Review: IJA **2** (1961), 461.

Borror, D. J. 1948. 360
Analysis of repeat records of banded white-throated sparrows. *Ecol. Monogr.*, **18** (3), 411–30.

Bortkiewicz, L. 1894. 361
Kritische Betrachtungen zur Theoretischen Statistik. *Jahrbücher für Nationalökonomie und Statistik.*, **8**, 641–80.
User: Seal, H. L. (1949).

Bortkiewicz, L. 1898. 362
Das Gesetz der kleinen Zahlen Teubner, Leipzig.

Bortkiewizc, L. 1899a. 363
Über die Sterblichkeit der Empfänger von Invalidenrenten vom statistischen und versicherungstechnischen Standpunkt. *Z. Versicherungs-Recht und-Wissenschaft*, **5**, 563–605.

Bortkiewicz, L. 1899b. 364
Die erkenntnistheoretischen Grundlagen der Wahrscheinlichkeitsrechnung. *Jahrbücher für Nationalökonomie und Statistik*, **17**, 230–44.

Bortkiewicz, L. 1901. 365
Anwendungen der Wahrscheinlichkeitsrechnung auf Statistik. *Encyklopädie der Math. Wissenschafter mit Einschluss ihrer Anwendungen*, **1**, 821–51.

Bortkiewicz, L. 1903. 366
Wahrscheinlichkeitstheorie und Erfahrung. *Z. für Philosophie und philosophische Kritik.*, **121**, 71–86.

Bortkiewicz, L. 1906. 367
Der wahrscheninlichkeitstheoretische Standpunkt im Lebensversicherungswesen. *Oesterreichische Revue*, **31**, 24–28.

Bortkiewicz, L. 1908. 368
La légge dei piccoli numeri. Chiarimenti. *Giorn. Economisti*, **37**, 415–427.

Bortkiewicz, L. 1909. 369
Ancora la légge dei piccoli numeri. *Giorn. Economisti*, **39**, 395–415.

Bortkiewicz, L. 1910. 370
Zur Verteiligung des Gesetzes der kleinen Zahlen. *Jahrbücher für Nationalökonomie und Statistik*, **39**, 218–36.

Bortkiewicz, L. 1915a. 371
Über die Zeitfolge zufalliger ereignisse. *Bull. Inst. Internat. Statist.*, **20** (2), 30–111.

Bortkiewicz, L. 1915b. 372
Realismus und Formalismus in der mathematischen Statistik. *Allgemein. Statist. Arch.*, **9**, 225–56.

Bortkiewicz, L. 1917. 373
Homogenität und Stabilität in der Statistik. *Skand. Aktuarietidskr.*, **1**, 1–81.

Bortkiewicz, L. 1920. 374
Das Laplacesche Erganzungsglied und Eggenbergers Grenzberichtigung zum Wahrscheinlichkeitsintegral. *Sitzungsberichte der Berliner Math. Gesellschaft*, **18**, 37–42.

Bortkiewicz, L. 1922. 375
Das Helmertsche Verteilungsgesetz. *Z. Angew. Math. Mech.*, **2**, 358–75.
User: Lancaster, H. O. (1960).

Bortkiewicz, L. 1927. 376
Zum Markoffschen Lemma. *Skand. Aktuarietidskr.*, **10**, 13–16.

Bortkiewicz, L. 1931. 377
The relations between stability and homogeneity. *Ann. Math. Statist.*, **2**, 1–22.

Bortkewitsch, L. 1917. 378
Die Iterationen, ein Beitrag zur Wahrscheinlichkeitsrechnung. Springer, Berlin.

Bosch, A. J. 1963. 379
The Pólya distribution. *Statistica Neerlandica*, **17**, 201–13.
Review: STMA **5** (1964), 326.
Class.: PO-MI:mi:g

Boswell, M. T. 1966. 380
Estimating and testing trend in a stochastic process of Poisson type. *Ann. Math. Statist.*, **37**, 1564-73.
Class.: P:pc-tp:G

Bottema, O., and Van Veen, S. C. 1943. 381
Calculation of probabilities in the game of billiards. *Nieuw Arch. Wisk.*, **22** (2), 15–33.
Review: MR **7** (1946), 209.

Boulter, E. A., Westwood, J. C. N., and Phipps, P. H.
See Westwood, J. C. N., Phipps, P. H., and Boulter, E. A. (1957).

Bouzitat, J. 1947. 382
Note sur un problème de sondage. *Office National d'Études et de Recherche Aéronautiques, Paris*, pp. 57.
Review: ZBL **31** (1949), 170.

Bowden, J. 1965. 383
The distribution of catch of some species of insects and its relationship to their biology (Abstract). *Biometrics*, **21**, 263.

Bowen, M. F. 1947. 384
Population distribution of the leafhopper in relation to experimental field plot layout. *J. Agri. Res.*, **75**, 259–78.
Users: Bliss, C. I., and Fisher, R. A. (1953), Bliss, C. I. (1958), Jenson, P. (1959), Kuno, E. (1963).

Bower, Gordon H. 1961. 385
Application of a model to paired-associate learning. *Psychometrika*, **26**, 255–80.

Bowman, K. O. 1963. 386
Moments of higher orders for maximum likelihood estimators, with an application to the negative binomial distribution. Ph.D. Thesis, Virginia Polytechnic Institute.
Users: Bowman, K. O., and Shenton, L. R. (1965a), Shenton, L. R., and Myers, R. (1965), Bowman, K. O., and Shenton, L. R. (1966).

Bowman, K., and Shenton, L. R.
See Shenton, L. R., and Bowman, K. (1963).

Bowman, K. O., and Shenton, L. R.
See Shenton, L. R., and Bowman, K. O. (1965).

Bowman, K. O., and Shenton, L. R. 1965a. 387
Biases and covariances of maximum likelihood estimators. Tech. Rep. K-1633, Union Carbide Corp., Oak Ridge, Tenn., pp. 21.
Users: Bowman, K. O., and Shenton, L. R. (1966).
Class.: NB:pe:G

Bowman, K. O., and Shenton, L. R. 1965b. 388
Asymptotic covariances for the maximum likelihood estimators of the parameters of a negative binomial distribution. Tech. Rep. K-1643, Union Carbide Corp., Oak Ridge, Tenn.
Users: Bowman, K. O., and Shenton, L. R. (1966).
Class.: NB:pe-tc:G

Bowman, K. O., and Shenton, L. R. 1966. 389
Biases of estimators for the negative binomial distributions. Tech. Rep. ORNL-4005, Oak Ridge National Lab., Oak Ridge, Tennessee.
Class.: NB:pe-tc:G

Bowman, K. O., and Uppuluri, V. R. Rao.
See Uppuluri, V. R. Rao, and Bowman, K. O. (1966).

Boyd, W, C. 1954. 390
Maximum likelihood method for estimation of gene frequencies from M.N.S. data. *Amer. J. Hum. Genet.*, **6**, 1–10.
Users: Ceppellini, R., Siniscalco, M., and Smith, C. A. B. (1955).

Bracewell, R. N. 1953. 391
The sunspot number series. *Nature*, **171**, 649–50.
Class.: MI:pc:P

Brachman, M. K. 1955. 392
Notes on the summation of series. *J. Soc. Indust. Appl. Math.*, **3**, 254–58.
Class.: MI:mi:G

Bracken, Jerome. 1966. 393
Percentage points of the beta distribution for use in Bayesian analysis of Bernoulli process. *Technometrics*, **8**.

Bradley, James V. 1959a. 394
Studies in research methodology: I. Compatibility of psychological measurements with parametric assumptions. Tech. Rep. 58–574 (1), Wright Air Develop. Center, Air Res. and Develop. Com., U.S.A.F., Wright-Patterson, A. F. B., Ohio.
Class.: B-P:a:G

Bradley, James V. 1959b. 395
Studies in research methodology: II. Consequences of violating parametric assumptions—fact and fallacy. Tech. Rep. 58–574 (II), Wright Air Develop. Centre, Air Res. and Develop. Com. U.S.A.F., Wright-Patterson, A.F.B., Ohio.

Bradley, James V. 1963. 396
Studies in research methodology: IV. A sampling study of the central limit theorem and the robustness of one-sample parametric tests. Rep. AMRL-TDR-63–29, Behavioral Sci. Lab., Wright-Patterson A.F.B., Ohio, pp. 305.

Brainerd, B., and Narayana, T. V. 1961. 397
A note on simple binomial sampling plans. *Ann. Math. Statist.*, **32**, 906–08.
Review: MR **23A** (1962), 575; ZBL **106** (1964), 128; STMA **5** (1964), 63.
User: Narayana, T. V. (1962).
Class.: B:sqc:G

Brass, W. 1958a. 398
Simplified methods of fitting the truncated negative binomial distribution. *Biometrika*, **45**, 59–68.
Review: ZBL **95** (1962), 135.
Users: Bartko, J. J. (1961b), Cohen, A. C. Jr. (1961c), Khatri, C. G. (1961), Cassie, R. M. (1962), Khatri, C. G. (1962d), Tauton, (1965), Cohen, A. Clifford Jr. (1965), Cohen, A. C. Jr. (1966).
Class.: TCNB:pe-gf:G

Brass, W. 1958b. 399
Models of birth distributions in human populations. *Bull. Inst. Internat. Statist.*, **36** (2), 165–78.
User: Martin, L. (1961).

Brass, W. 1958c. 400
The distribution of the births in human populations. *Population Studies*, **12**, 51–72.
Class.: NB:mb-gf:BM

Braumann, P. 1959a. 401
Symmetric infinitely divisible probability law (German). *Rev. Fac. Ciencias Lisboa, A*, **7**, 255–62.
Review: IJA **1** (1960), 349.

Braumann, P. 1959b. 402
Comments on the representation of infinitely divisible probability laws. *Rev. Fac. Ciencias Lisboa, A*, **7**, 347–57.
Review: IJA **2** (1961), 15.

Bray, J. R. 1962. 403
Use of non-area analytic data to determine species dispersion. *Ecology*, **43**, 328–33.
Class.: B:h:BM

Breakwell, J. V. 1954. 404
The problem of testing for the fraction of defectives. *Operations Res.*, **2**, 59–69.
User: Breakwell, J. V. (1956).
Class.: B:sqc:E

Breakwell, J. V. 1955. 405
Minimax test for the parameter of a Poisson process (Abstract). *Ann. Math. Statist.*, **26**, 768.

Breakwell, J. V. 1956. 406
Economically optimum acceptance tests. *J. Amer. Statist. Assoc.*, **51**, 243–56.
Users: Somerville, P. N. (1957).
Class.: B:tp:E

Breakwell, John V. 1961. 407
Minimax tests for the rate of a Poisson process and the bias rate of a normal process. *Sankyā*, **23A**, 161–82.
Review: STMA **6** (1965), 539.

Breiman, Leo. 1962. 408
On some probability distributions in traffic flow (French Summary). *Bull. Inst. Internat. Statist.*, **39** (4), 155–61.
Review: MR **29** (1965), 135; STMA **5** (1964), 470; ZBL **115** (1965), 134.

Breiman, Leo. 1963. 409
The Poisson tendency in traffic distribution. *Ann. Math. Statis.*, **34**, 308–11.
Review: STMA **6** (1965), 677; ZBL **117** (1965), 137.
User: Thedeen, T. (1964).
Class.: P:pc:E

Breny, H. 1953. 410
Sur une classe de fonctions aléatoires liees à la loi de Poisson. *Bull. Soc. Roy. Sci., Liège*, **22**, 405–16.
Review: MR **15** (1954), 541.

Breny, H. 1957. 411
Sur quelques problèmes d'analyse statistique posés par la physique des microcorpuscles. I. Distributions de Poisson et mesures relatives. *Ann. Soc. Sci. Bruxelles Ser. I*, **71**, 135–60.
Review: ZBL **95** (1962), 133.

Breny, H. 1961. 412
Quelques propriétés des files d'attente où les clients arrivent en grappes. *Mém. Soc. Roy. Sci. Liège Coll. in* 8° (5) 6, No. 4, 65 pp.
Review: MR **32** (1966), 307.
Class.: MI:pc:G

Breny, H. 1962. 413
Poisson models in bunches for bundles of fibres (French). *Ann. Sci. Textiles Belges*, **2**, 124–146.
Review: IJA **4** (1963), 396.

Bresciani, Costantino. 1908. 414
A proposito della 'légge dei piccoli numeri.' *Giorn. Economisti*, **36**, 357–80.

Bretagnolle, Jean, and Dacunha-Castelle, Didier 415
1964.
Convergence de la nieme convoluée d'une loi de probabilité. *C. R. Acad. Sci. Paris*, **258**, 4910–13.

Bretherton, Michael H., and Williamson, Eric.
See Williamson, Eric, and Bretherton, Michael H. (1963).

Bretherton, Michael H., and Williamson, Eric.
See Williamson, Eric, and Bretherton, Michael H. (1964).

Brett, D. W., and Dormer, K. J. 1960. 416
Observations on a cyclic fluctuation in the leaf serrations of Spiraea salicifolia, and on the asymmetry of the leaf. *New Phytologist*, **59**, 104–08.

Breuer, M., and Haight, F. A.
See Haight, F .A., and Breuer, M. (1960).

Brian, A.D. 1957. 417
Differences in the flowers visited by four species of bumblebees and their causes. *J. Animal Ecol.*, **26** (1), 71–98.

Brian, M. V. 1953. 418
Species frequencies in random samples from animal populations. *J. Aninml Ecol.*, **22**, 57–64.
Users: Clark, P. J., Eckstrom, P. T., and Linden, L. C. (1964), Hairston, N. G. (1959), MacArthur, R. (1957).
Class.: LS-NB:gf:BM

Brian, M. V. 1956. 419
Segregation of species of the ant genus Myrmica. *J. Animal Ecol.*, **25** (2), 319–37.
Class.: B:gf:BM

Briggs, Fred N., and Baker, George A.
See Baker, Geroge A., and Briggs, Fred N. (1945).

Briggs, Fred N., and Briggs, George A.
See Baker, George A., and Briggs, Fred N. (1949).

Brindley, T. A. Bancroft, T. A., and McGuire, J. U.
See McGuire, J. U., Brindley, T. A., and Bancroft, T. A. (1957).

Brinegar, Claude. 1963. 420
Mark Twain and the Quintus Curtius Snodgrass letters: A statistical test of authorship. *J. Amer. Statist. Assoc.*, **58**, 85–96.
Users: Mosteller, F., and Wallace, D. (1963).

Brischle, H. A., and Fracker, S. B.
See Fracker, S. B., and Brischle, H. A. (1944).

Brockmeyer, E. 1952. 421
The use of probability calculations in telephone engineering on the basis of the research of Erlang and Moe. *Teleteknik*, **3**, 1–95.

Brockwell, P. J. 1963. 422
The transient behaviour of a single server queue with batch arrivals. *J. Austral. Math. Soc.*, **3**, 241–48.
Review: STMA **5** (1964), 719.

Brockwell, P. J. 1964. 423
As asymptotic expansion for the tail of a binomial distribution and its application in queueing theory. *J. Appl. Prob.*, **1**, 161–67.
Review: MR **31** (1965), 958.
User: Hald, A. (1965).
Class.: B:a:G

Bross, I. D. 1954a. 424
A confidence interval for a percentage increase. *Biometrics*, **10**, 245–250.
Users: Noether, G. E. (1957), Pfanzagl, E. J. (1960).

Bross, I. D. 1954b. 425
Misclassification in 2×2 tables. *Biometrics*, **10**, 478–86.
Review: MR **16** (1955), 942.
User: Chew, V. (1964).

Brown, B. 1965. 426
Some tables of the negative binomial distribution and their use. Memo. RM-4577-PR, RAND Corporation.

Brown, Bancroft H. 1919. 427
Probabilities in the game of 'Shooting Craps'. *Amer. Math. Monthly*, **26**, 351–52.
Class.: G:gf:O

Brown, B. H. 1941. 428
Simple examples of limiting processes in probability. *Amer. Math. Monthly*, **48**, 98–102.
Class.: MI:mi:G

Brown, B. W., and Johnson, E. A.
See Johnson, E. A., and Brown, B. W. (1961).

Brown, David T. 1959. 429
A note on approximations to discrete probability distributions. *Information and Control*, **2**, 386–92.
Review: ZBL **117** (1965), 148.
User: Good, I. J. (1963).
Class.: MI:a:G

Brown, J. 1956. 430
A modified sigificance test for the difference between two observed proportions. *J. Occup. Psychol.*, **30**, 69–74.

Brown, J. L., Jr. 1962. 431
Analysis of clipping cross-correlator for arbitrary deterministic signal. Tech. Memo., Ordnance Res. Lab., Pennsylvania State Univ., pp. 7.
Class.: B:mi:E

Brown, J. L., Jr. 1963. 432
Analysis of digital polarity-coincidence correlator independent samples and arbitrary signal-to-noise ratios. Tech. Memo., Ordnance Res. Lab., Pennyslvania State Univ., pp. 11.
Class.: B:mi:E

Brown, J. L., Jr., and Patil, G. P. 1966. 433
On the statistical independence of infinitely clipped Gaussian random variables. Tech. Rep., Ordnance Res., Lab., Pennsylvania State Univ., pp. 6.
Class.: B:mi:E

Brown, Richard H. 1963. 434
Theory of combat: The probability of winning. *Operations Res.*, **11**, 418–25.

Brown, T. A. I., and Lancaster, H. O.
See Lancaster, H. O., and Brown, T. A. I. (1965).

Brunel, Pierre, and Jolicoeur, Pierre.
See Jolicoeur, Pierre, and Brund, Pierre. (1966).

Brunk, H. D. 1955. 435
Maximum likelihood estimates of monotone parameters. *Ann. Math. Statist.*, **26**, 607–16.
Class.: B-P-MI:pe:G

Brunk, H. D. 1961. 436
On a theorem of E. Sparre Andersen and its application to tests against trend. *Math. Scandinav.*, **8**, 305–26.
Review: ZBL **97** (1962), 134.

Bruno, Angelo. 1964. 437
Sugli eventi parzialmente scambiabili (French, English, Spanish and German Summaries). *Giorn. Ist. Ital. Attuari*, **27**, 174–96.
Review: MR **31** (1966), 501.

Bryson, Marion R. 1965. 438
Errors of classification in a binomial population. *J. Amer. Statist. Assoc.*, **60**, 217–24.
Review: MR **31** (1966), 313.
Class.: B:pe:G

Buch, K. R. 1949. 439
A note on sentence-length as random variable. *Compte rendu du onzième congrès de mathématiciens scandinaves tenu à Trondheim.*, 272–75.
Class.: MI:a-h:L

Buch, K. R. 1952. 440
A note on sentence-length as random variable, *C. R. 11th Congr. Math. Scand.*, 272–75.
Class.: MI:a-h:L

Buchanan-Wollaston, H. J. 1935. 441
The philosophic basis of statistical analysis. *J. Conseil* **10** (3), 249–63.

Buchanan-Wollaston, H. J. 1936. 442
The philosophic basis of statistical analysis. *J. Conseil*, **11** (1), 7–26.

Buchanan-Wollaston, H. J. 1938. 443
On the application of the statistical theory of space distribution to hydrographic and fishery problems. *J. Conseil*, **13** (2), 173–86.

Buchanan-Wollaston, H. J. 1958. 444
Statistical tests for significance applicable to distribution in space. *J. Conseil*, **23** (2), 161–72.

Buchner, P. 1952. 445
Bemerkungen zum Satz von Bernoulli. *Elem. Math.*, **7**, 8–11.

Buehler, R. J. 1957. 446
Confidence intervals for the product of two binomial parameters. *J. Amer. Statist. Assoc.*, **52**, 482–93.
Review: MR **19** (1958), 1204; ZBL **80** (1959), 127.
Class.: B-P:ie:G

Buhler, W., Fein, H., Goldsmith, D., Neyman, J., and Puri, P. S. 1965. 447
Locally optimal test for homogeneity with respect to very rare events. *Proc. Nat. Acad. Sci.*, **54**, 673–80.

Bulmer, M. G. 1958a. 448
The repeat frequency of twinning. *Ann. Hum. Genet.*, **23**, 31–35.

Bulmer, M. G. 1958b. 449
The numbers of human multiple births. *Ann. Hum. Genet.*, **22**, 158–64.
Review: PA **33** (1959), Ab. No. 8639.

Bunke, O. 1959–60. 450
New confidence intervals for the parameter of the binomial distribution (German). *Wiss. Zeit. Humboldt-Univ. Math.-Nat. Riehe*, **9**, 335–63.
Review: STMA **6** (1965) 99; ZBL **108** (1964), 158.
Class.: B-P:ie:G

Burkholder, D. L. 1960. 451
Effect on the minimal complete class of tests of changes in the testing problem. *Ann. Math. Statist.*, **31**, 325–31.
Class.: MI:tp:G

Burnett, T. 1958. 452
Effect of host distribution on the reproduction of Encarsia formosa Gahan, (Hymenoptera: Chalcidoidea). *Canad. Entomologist*, **90**, 179–91.
Users: Burnett, T. (1958a), Henson, W. R. (1959), Waters, W. (1959), Lyons, L. A. (1962).

Burnett, T. 1958a. 453
Effect of area of search on reproduction of Encarsia formosa Gahan (Hymenoptera: Chalcidoidea). *Canad. Entomologist*, **90**, 225–29.
User: Henson, W. R. (1959).

Burr, E. J. 1961. 454
Length of the longest run of consecutive successes (Abstract). *Ann. Math. Statist.*, **32**, 917.

Burr, I. W. 1952a. 455
Distributions of ranges from an arbitrary discrete population (Abstract). *Ann. Math. Statist.*, **23**, 145.
Review: ZBL **65** (1956), 119.
Users: Govindarajulu, Z. (1965).

Burr, I. W. 1952b. 456
Formulas for approximating the hypergeometric and binomial by the Poisson distribution (Abstract). *Ann. Math. Statist.*, **23**, 145.

Burr, I. W. 1955. 457
Calculation of exact sampling distribution of ranges from a discrete population. *Ann. Math. Statist.*, **26**, 530–32.
User: Leti, G. (1961).
Class.: MI:mi:G

Burrows, C. 1960. 458
Some numerical results for the waiting times in the queue $E_k/M/1$. *Biometrika*, **47**, 202–03. Correction 47, 484.
Review: IJA **2** (1961), 180.
Class.: G:pc:G

Burrows, G. L., and Halperin, M.
See Halperin, M., and Burrows, G. L. (1960).

Burrows, G. L., and Halperin, M.
See Halperin, M., and Burrows, G. L. (1961).

Bushland, R. C., Baumhover, A. H., Graham, A. J., Bitter, B. A., Hopkins, D. E., New, W. D., and Dudley, F. H.
See Baumhover, A. H., Graham, A. J., Bitter, B. A., Hopkins, D. E., New, W. D., Dudley, F. H., and Bushland, R. (1955).

Butler, Margaret Sauls. 1966. 459
On Neyman's Type A contagious distribution. Master's Thesis Oklahoma State Univ., Still Water, 50 pp.

Byers, G. W., and Hairston, N. G.
See Hairston, N. G., and Byers, G. W. (1954).

Byrd, M. A., and Schultz, V.
See Schultz, V., and Byrd, M. A. (1957).

Byron, Frank H. 1935. 460
The point binomial and probability paper. *Ann Math. Statist*, **6**, 21–26.
Class.: B:a-tc:G

Bystrov, N. V. 1956. 461
On some unbiased estimates (Russian). *Vestnik Leningrad. Univ. Ser. Mat. Meh. Astronom.* 1956 (1), 169–75.
User: Bol'shev, L. N. (1961).

C

Cacoullos, T. 1960. 462
Two problems of estimation: binomial variance and truncated Poisson mean. Res. Rep., Columbia Univ. (sponsored by ONR), pp. 34.
Class.: B-TCP:pe-ie-se:G

Cacoullos, T. 1961. 463
A combinatorial derivation of the distribution of the truncated Poisson sufficient statistic. *Ann. Math. Stast.* **32**, 904–05.
Review: MR **23A** (1962), 685; STMA **5** (1964), 45.
Class.: TCP:mi:G

Cacoullos, T., and Sobel, Milton. 1966. 464
An inverse sampling procedure for selecting the most probable event in a multinomial distribution. *Multivariate Analysis*, Ed. P. R. Krishnaiah, Academic Press, pp. 423–55.

Cain, S. A., and Evans, F. C. 1952. 465
The distribution patterns of three plant species in an old-field community in southeastern Michigan. *Contri. Lab. Vertebrate Biol. Univ. Mich.*, **52**, 1–11.
Users: Clark, P. J., and Evans, F. C. (1954a), Hairston, N. G. (1959), Morisita, M. (1959).
Class.: MI:mi:BM

Cain, S. A., and Evans, F. C.
See Evans, F. C., and Cain, S. A. (1952).

Camargo, M. C., and Isabel, M. M. 1956. 466
The logarithmic correlation (Spanish). *An. Real. Soc. Espan. Fis. Quim. Ser. A*, **52** (5–6), 117–36.
Review: SA **59A** (1956), Ab. No. 7103.
Users: Patil, G. P., and Bildikar, S. (1966a).

Camp, B. H. 1924. 467
Probability integrals for the point binomial. *Biometrika*, **16**, 163–71.
Users: Burton, G., and Camp (1925), Govindarajulu, Z. (1965).
Class.: B:a-c:G

Camp, B. H. 1925. 468
Probability integrals for a hypergeometric series. *Biometrika*, **17**, 61–67.
User: Govindarajulu, Z. (1965).
Class.: H:a-c:G

Camp, B. H. 1940. 469
Further comments on Berkson's problem. *J. Amer. Statist. Assoc.*, **35**, 368–76.
Review: ZBL **23** (1941), 247.
Class.: P-MI:gf:G

Camp, B. H. 1951. 470
Approximation to the point binomial. *Ann. Math. Statist.*, **22**, 130–31.
Review: ZBL **42** (1952), 140.
Users: Raff, M. S. (1956), Mott-Smith, J. C. (1964), Govindarajulu, Z. (1965), Bartko, J. J. (1966).
Class.: B:a-c:G

Campagne, C. 1959a. 471
Some considerations concerning probability of ' Ruin ' from the point of view of discontinuity (French). *Het Verzek.-Arch., Actuar. Bij.*, **36**, 1–5.
Review: IJA **3** (1962), 629.
Class.: P:pc:G

Campagne, C. 1959b. 472
An aspect of solvency in non-life insurance business (Dutch). *Het Verzek.-Arch., Actuar. Bij.*, **36**, 4–30.
Review: IJA **3** (1962), 628.
Class.: P:mi:S

Campbell, George A. 1923. 473
Probability curves showing Poisson's exponential summation. *Bell System Tech. J.*, **2**, 95–113.
Users: Thorndike, F. (1926), Wise, M. E. (1946), Blom, G. (1954), Wise, M. E. (1954), Teichrow, G. (1955), Martin, L. (1961), Govindarajulu, Z. (1965).
Class.: B-P:tc:G

Campbell, J. T. 1934. 474
The Poisson correlation function. *Proc. Edinburgh Math. Soc. Ser.*, **24**, 18–26.
Users: Krishnamoorthy, .A S. (1951), Edwards, C. B. (1962), Fuchs, C., and David, H. (1965a), Patil, G. P., and Bildikar, S. (1966a), Holgate, P. (1966b).
Class.: MP:m-rp-gf-mi:G

Campbell, R. C., Hancock, J. L., and Rothschild, Lord. 1953. 475
Counting live and dead bull spermatozoa. *J. Expt. Biol.*, **30**, 44–49.
Class.: B:gf-h-ie:BM

Cansado, Maceda E. 1947. 476
Über die faktorielle charakteristische Funktion (Spanish). *Rev. Mat. Hisp.-Amer.*, **IV** 7, 159–64.
Review: ZBL **33** (1950), 193.
Class.: B-P-NB-MI:mi-m-G

Cansado, E. 1951a. 477
A study of bivariate distributions (Spanish, English Summary). *Trabajos Estadist.*, **2**, 149–78.

Cansado, E. 1951b. 478
An example of a bivariate distribution (Spanish, English Summary). *Trabajos Estadist.*, **2**, 261–272.

Cansado, E. 1957. 479
About moments and factorial coefficients (Spanish and French Summaries). *Bull. Inst. Internat. Statist.*, **35** (2), 77–84.
Review: MR **22** (1961), 1719.

Cansado. E. 1957a. 480
Sampling without replacement from finite populations (Spanish Summary). *Trabajos Estadist.*, **8**, 3–11.
Review: ZBL **87** (1961), 141.

Capobianco, Michael F. 1964. 481
On the bivariate binomial distribution and related problems. Ph.D. Thesis, Polytechnic Inst., Brooklyn, pp. 55.
Users: Patil, G. P., and Bildikar, S. (1966a).

Caregradskii, I. P. 1958. 482
On uniform approximation to a binomial distribution by infinitely divisible laws (Russian, English Summary). *Teor. Verojatnost. i Primenen.*, **3**, 470–74.
Review: MR **21** (1960), 175; IJA **1** (1960), 581; ZBL **100** (1963), 141.
Class.: B:a:G

Carlitz, L. 1964. 483
Comment on the paper ' Some probability distributions and their associated structures '. *Math. Mag.*, **37**, 51–52.
Class.: MI:mi:G

Carnal, H. 1962. 484
Indefinitely divisible random distributions on compact groups (French). *C. R. Acad. Sci. Paris*, **255**, 1179–80.
Review: STMA **5** (1964), 535.
Class.: P:mi:G

Carroll, John B. 1961. 485
The nature of the data, or how to choose a correlation coefficient. *Psychometrika*, **26**, 347–72.

Carter, F. L., Jr. 1960. 486
Group testing in binomial and multinomial situations. Master's Thesis, Virginia Polytechnic Institute.
Users: Patil, G. P., and Bildikar, S. (1966a).

Carter, Frederick L., and Bargmann, Rolf E.
See Bargmann, Rolf E., and Carter, Frederick, L. (1960).

Cartwright, Desmond S. 1956. 487
A rapid non-parametric estimate of multi-judge reliability. *Psychometrika*, **21**, 17–29.
Review: ZBL **74** (1962), 135.

Cassie, R. M. 1954. 488
Some uses of probability paper in the analysis of size frequency distributions. *Austral. J. Marine and Freshwater Res.*, **5**, 513–22.
Users: Cassie, R. M. (1962), Blischke, W. R. (1965).

Cassie, R. M. 1959a. 489
An experimental study of factors inducing aggregation in marine plankton. *New Zealand J. Sci.*, **2**, 339–65.
User: Cassie, R. M. (1962)..

Cassie, R. M. 1959b. 490
Microdistribution of Plankton. *New Zealand J. Sci.*, **2**, 398–409.
User: Cassie, R. M. (1962).

Cassie, R. M. 1959c. 491
Some correlations in replicate plankton samples. *New Zealand J. Sci.*, **2**, 436–48.
User: Cassie, R. M. (1962).

Cassie, R. M. 1960. 492
Factors influencing the distribution pattern of plankton in the mixing zone between oceanic and harbour water. *New Zealand J. Sci.*, **3**, 26–50.

Cassie, R. M. 1962. 493
Frequency distribution models in the ecology of Plankton and other organism. *J. Animal Ecol.*, **31**, 65–92.
Users: Kuno, E. (1963), Berthet, P., and Gerard, G. (1965).
Class.: B-NB-DL-N-T-MI:mb-cm-mi:BM

Castagnetto, Louis, and Cernuschi, Felix.
See Cernuschi, Felix, and Castagnetto, Louis (1946).

Casteliani, M. 1950. 494
On multinomial distributions with limited freedom: a stochastic genesis of Pareto's and Perason's curves. *Ann. Math. Statist.*, **21**, 289–93.
Review: MR **11** (1950), 673; ZBL **41** (1952), 461.
Users: Patil, G. P., and Bildikar, S. (1966a).
Class.: M:pc:G

Castellano, V. 1960. 495
Sull'universo dei campioni come fondamento della teoría dei campioni (Italian, French and English Summaries). *Bull. Inst. Internat. Statist.*, **37** (3).

Castoldi, L. 1954. 496
Un problèma generale di prove ripetute. *Rend. Sem. Fac. Sci. Univ. Cagliari*, **24**, 21–27.
Review: ZBL **56** (1955), 358.

Castoldi. L. 1958. 497
A variant of a probability problem by Banach (Italian). *Rend. Sem. Fac. Sci. Univ. Cagliari*, **28**, 175–80.
Review: IJA **3** (1962), 1.
Class.: MI:mi:G

Castoldi, L. 1959. 498
A property of Poissonian distributions. *Rend. Sem. Fac. Sci. Univ. Cagliari*, **29**, 220–22.
Review: MR **22** (1961), 1457; ZBL **95** (1962), 128; IJA **3** (1962), 516.
Class.: P:osp:G

Castoldi, L. 1959a. 499
A variant to a probabilistic problem by Banach (Italian). *Boll. Cent. Ric. Operat.*, **3**, 24–28.
Review: IJA **1** (1960), 541.
Class.: MI:mi:G

Castoldi, L. 1959b. 500
The distribution of the number of births and probability problems related thereto (Italian). *Rend. Sem. Fac. Sci. Univ. Cagliari*, **29**, 14–25.
Review: IJA **3** (1962), 2; ZBL **98** (1963), 333.
Class.: MI:mb-a:BM

Castoldi, L. 1959c. 501
Considerations on the reckoning of Poissonian events (Italian). *Rend. Sem. Fac. Sci. Univ. Cagliari*, **29**, 32–36.
Review: IJA **4** (1963), 597.
Class.: P-MI:pc:G

Castoldi, L. 1963a. 502
A continuous analogue of Poisson's distribution. *Rend. Sem. Fac. Sci. Univ. Cagliari*, **33**, 245–49.
Review: STMA **6** (1965), 1116.

Castoldi, L. 1963b. 503
Poisson processes with events in clusters. *Rend. Sem. Fac. Sci. Univ. Cagliari*, **33**, 433–37.
Review: STMA **6** (1965), 1117.

Catana, A. J., Jr., Cottam, G., and Curtis, J. T.
See Cottam, G., Curtis, J. T., and Catana, A. J., Jr. (1957).

Catcheside, D. G., Lea, D. E., and Thoday, J. M. 504
1945–6.
Types of chromosome structural change induced by the irridiation of Tradescantia microspores. *J. Genetics*, **47**, 113–36.

Ceppellini, R., Siniscalco, M., and Smith, C. A. B. 505
1955.
The estimation of gene frequencies in a random-mating population. *Ann. Hum. Genet.*, **20**, 97–115.

Cernuschi, Felix, and Castagnetto, Louis. 1946. 506
Chains of rare events. *Ann. Math. Statist.*, **17**, 53–61.
Review: MR **7** (1946), 457.
Users: Anscombe, F. J. (1950a), Douglas, J. B. (1965).
Class.: P-NB-MI:pc-a:G

Cernuschi, Felix, and Saleme, Ernesto. 1944. 507
A new scheme of contagion in probability (Spanish). *An. Soc. Ci. Argentina*, **138**, 201–13.
Review: MR **6** (1945), 233; ZBL **60** (1957), 289.
Class.: OBR:pc:G

Chacko, V. J., and Negi, G. S. 1966. 508
A statistical study of the spatial distribution of dead trees in a Casuarina plantation. *Res. Notes of the Second Conf. Internat. Advisrory Group of Forest Statist. Royal College of Forestry, Stockholm*, pp. 45–66.
Class.: B-TCB-G-COB:gf:BM

Chaddha, R. L. 1962. 509
Binomial distribution: when both parameters are random variables (Abstract). *Ann. Math. Statist.*, **33**, 816.

Chaddha, Roshan L. 1965. 510
A case of contagion in binomial distribution. *Classical and Contagious Discrete Distributions.* Ed. G. P. Patil, Statistical Publishing Society, Calcutta and Pergamon Press, pp. 273–90.
Class.: OBR:mb-tc-pe:G

Chakraborty, P. N., Chandrasekar, C., and Agarwala, S. P.
See Chandra Sekar, C., Agarwala, S. P., and Chakraborty, P. N. (1955).

Chakravarti, I. M., and Rao, C. R.
See Rao, C. R., and Chakravarti, I. M. (1956).

Chakravarti, I. M., and Rao, C. R. 1959. 511
Tables for some small sample tests of significance for Poisson distributions and 2×3 contingency tables. *Sankhyā*, **21**, 315–26.
Review: MR **22** (1961), 184; ZBL **91** (1962), 317; IJA **2** (1961), 66.
Class.: P-TCP-B:h-gf-mi-tc:G

Chakravorty, P. N., and Mathen, K. K.
See Mathen, K. K., and Chakravorty, P. N. (1950).

Chalam, G. V., and Singh, B. N.
See Singh, B. N., and Chalam, G. V. (1937).

Chalkley, Herald W., and Cornfield, Jerome.
See Cornfield, Jerome, and Chalkley, Herald W. (1951).

Chamberlain, A. C., and Turner, F. M. 1952. 512
Errors and variations in white-cell counts. *Biometrics*, **8**, 55–65.
Class.: P-B:anovat-gf:BM

Chambers, E. G., and Farmer, E.
See Farmer, E., and Chambers, E. G. (1939).

Chambers, E. G., and Yule, G. U. 1941. 513
Theory and observation in the investigation of accident causation. *J. Roy. Statist. Soc. Suppl.*, **7**, 89–101.
Review: MR **4** (1943), 28.
Users: Greenwood, M. (1950), Sichel, H. S. (1951), Good, I. J. (1953b), Barton, D. E., and David, F. N. (1959a), Bartko, J. J. (1961b).
Class.: COP-NB:mi:A
Notes: Discussion, *J. Roy. Statist. Soc. Suppl.*, **7**, 101–09.

Chambers, M. L., and Jarratt, P. 1964. 514
Use of double sampling for selecting best population. *Biometrika*, **51**, 49–64.

Champernowne, D. G. 1956. 515
An elementary method of solution of the queueing problem with a single server and constant parameters. *J. Roy. Statist. Soc. Ser. B.*, **18**, 125–28.
Review: ZBL **73** (1960), 130.
User: Darroch, J. N. (1958).

Champernowne, D. G. 1960. 516
An experimental investigation of the robustness of certain procedures for estimating means and regression coefficients. *J. Roy. Statist. Soc. Ser. A*, **123**, 398–412.
Review: IJA **4** (1963), 623.
Class.: MI:pe-rp:G

Chanda, K. C. 1963. 517
On the efficiency of two-sample Mann-Whitney test for discrete populations. *Ann. Math. Statist.*, **34**, 612–17.
Review: MR **26** (1963), 1347; STMA **6** (1965), 542.
Class.: MI-P-B:ctp:G

Chandra Sekar C., Agarwala, S. P., and Chakraborty, P. N. 1955. 518
On a power function of a test of significance for the difference between two proportions. *Sankhyā*, **15**, 381–90.
Review: ZBL **68** (1957), 372.
Class.: B:ctp-tc:G

Chandrasekahr, S. 1943. 519
Stochastic problems in physics and astronomy. *Rev. Modern Phys.*, **15**, 2–89.

Chandrasekhar, S. 1949. 520
On a class of probability distributions. *Proc. Cambridge Philos. Soc.*, **45**, 219–24.
Review: MR **10** (1949), 464.

Chapanis, Alphonse. 1962. 521
An exact multinomial one-sample test of significance. *Psychol. Bull.*, **59**, 306–10.
Review: PA **37** (1963, 5921.)
Users: Patil, G. P., and Bildikar, S. (1966a).
Class.: M:tp:G

Chapman, Douglas. 1949. 522
The application of the hypergeometric distribution to problems of estimating and comparing zoological population sizes (Abstract). *Ann. Math. Statist.*, **20**, 134.

Chapman, D. D. 1951. 523
Some properties of the hyper-geometric distribution with applications to zoological sample censuses. *Univ. California Publ. Statist.*, **1**, 131–59.
Review: ZBL **145** (1955), 91.
Users: Chapman, D. G. (1952b), Goodman, L. A. (1953), Chapman, D. G. (1954), Cooper, G., and Lagler, K. (1956), Gilbert, N. E. G. (1956), Darroch, J. N. (1958), Govindarajulu, A. (1962c).
Class.: H:pe-tp-BM

Chapman, D. G. 1952a. 524
On tests and estimates for the ratio of Poisson means. *Ann. Inst. Statist. Math. Tokyo*, **4**, 45–49.
Review: MR **14** (1953), 488; ZBL **49** (1959), 220.
Users: Cox, D. R. (1953).

Chapman, D. G. 1952b. 525
Inverse, multiple and sequential sample censuses. *Biometrics*, **8**, 286–306.
Review: MR **14** (1953), 777.
Users: Chapman, D. G. (1954), Darroch, J. N. (1958), Nadler, J. (1960), Sen, P. K. (1960), Bartko, J. J. (1961b), Pathak, P. K. (1964), Knight, W. (1965).
Class.: MI:pe:BM

Chapman, D. G. 1952c. 526
Sufficient statistics for 'selected distributions'. *Univ. Washington Publ. Math.*, **3**, 59–64.

Chapman, D. G. 1954. 527
The estimation of biological populations. *Ann. Math. Statist.*, **25**, 1–15.
Users: Darroch, J. N. (1958), Parker, R. A. (1963).
Class.: MI:pe-ie:G

Chapman, D. G., and Birnbaum, Z. W.
See Birnbaum, Z. W., and Chapman, D. G. (1950).

Chapman, D. G., and Junge, C. O. 1956. 528
The estimation of the size of a stratified animal population. *Ann. Math. Statist.*, **27**, 375–89.
Class.: H:pe:B

Chapman, D. G., and Robson, D. S. 1960. 529
The analysis of a catch curve. *Biometrics*, **16**, 354–68.
Review: IJA **2** (1961), 483.
Users: Williams, E. J. (1961b).
Class.: G-B-P-MI:pe:BM

Chapman, R. A. 1938. 530
Applicability of the z test to a Poisson distribution. *Biometrika*, **30**, 188–90.
User: Bowen, M. F. (1947).
Class.: P:anovat:G

Chapman, V. J., and Myers, E.
See Myers, E., and Chapman, V. J. (1953).

Charlier, C. V. L. 1905a. 531
Die zweite Form des Fehlergesetzes. *Arkiv fur Matematik Astronomi och Fysik*, **2**, 1–8.

Charlier, C. V. L. 1905b. 532
Über die Darstellung willkurlicher Funktionen. *Arkiv für Matematik Astronomi och Fysik*, **2**, 1–35.

Chassan, J. B. 1960. 533
On a test for order. *Biometrics*, **16**, 119–21.
Review.: ZBL **92** (1962), 363.

Chatfield, C., Ehrenberg, A. S. C., and Goodhardt, G. J. 1965. 534
Progress on a simplified model of stationary purchasing behaviour. ASKE Research Limited.
Class.: NB-NM-LS:mb-gf:S

Chatfield, C., Ehrenberg, A. S. C., and Goodhardt, G. J. 1966. 535
Progress on a simplified model of stationary purchasing behaviour. *J. Roy. Statist. Soc. Ser. A.*, **129**, 317–67.
Class.: NB-NM-LS:mb-gf:S

Chatterji, S. D. 1963. 536
Some elementary characterizations of the Poisson distribution. *Amer. Math. Monthly*, **70**, 858–64.
Review: MR **70** (1964), 130; ZBL **117** (1965), 365.
Users: Bolger, E. M., and Harkness, W. L. (1965).
Class.: P:osp:G

Cheng, Tseng Tung. 1949. 537
The normal approximation to the Poisson distribution and a proof of a conjecture of Ramanujan. *Bull. Amer. Math. Soc.*, **55**, 396–491.
Review: MR **10** (1949), 613.
User: Govindarajulu, Z. (1965).
Class.: P:a:G

Chernoff, H. 1960. 538
A comprise between bias and variance in the use of nonrepresentative samples. *Contributions Prob. Statist.*, 153–67.
Review: IJA **3** (1962), 85.
Class.: B:pe:G

Chernoff, H., and Daly, J. F. 1957. 539
The distribution of shadows. *J. Math. Mech.* **6**, 567–84.
Review: ZBL **85** (1961), 348.

Chernoff, Herman, and Ray, S. N. 1965. 540
A Bayes sequential sampling inspection plan. *Ann. Math. Statist.*, **36**, 1387–1407.
Class.: B:sqc:G

Chew, Victor. 1964. 541
Application of the negative binomial distribution with probability of misclassification. *Virginia J. Sci.*, **15**, 34–40.
Class.: NB:ie:O

Chiang, Chin Long. 1951. 542
On the design of mass medical surveys. *Hum. Biol.*, **23**, 242–71.
Users: Freund, J. E. (1956), Samuel, E. (1963), Blischke, W. R. (1964).
Class.: MI:gf:BM

Chiang, Chin Long. 1957. 543
An application of stochastic processes to experimental studies on flour beetles. *Biometrics*, **13**, 79–97.
Users: Chiang, C. L. (1960a), Chiang, C. L. (1960).
Class.: B-P:pc-pe-gf-mi:BM
Notes: Correction, *Biometrics*, **13**, 543.

Chiang, Chin Long. 1960a. 544
A stochastic study of the life table and its applications: I. Probability distributions of the biometric functions. *Biometrics*, **16**, 618–35. Correction, **17**, 669.
Review: ZBL **104** (1964), 139; IJA **4** (1963), 738.
User: Chiang, C. L. (1960b).
Class.: M:mi-pe:BM

Chiang, Chin Long. 1960b. 545
On the probability of death from specific causes in the presence of competing risks. *Proc. 4th Berkeley Symp. on Math. Statist. Probab.*, **4**, 169–80.
Review: ZBL **104** (1964), 139; IJA **4** (1963), 134.
Class.: MI:mi-pe:BM

Chibosov, D. M. 1958. 546
Limit distributions for the number of runs in Bernoulli trial (Summary). *Theor. Probability Appl.*, **3**, 199.

Chipperfield, P. N. J. 1951. 547
The breeding of Crepidula Fornicata (L.) in the River Blackwater, Essex. *J. Marine Biol. Assoc.*, **30** (1), 49–71.

Chistyakov, V. P. 1964a. 548
On the foundation of the calculation of the power of

the test for empty boxes (Russian; English summary). *Teor. Verojatnost. i Primenen.*, **9**, 718–24.
Review: MR **31** (1966), 1137.
Class.: M:gf:G

Chistyakov, V. P. 1964b. 549
On the foundation of the calculation of the power of the test for empty boxes. *Theor. Probability Appl.*, **9**, 648–53.
Review: MR **31** (1966), 1137.
Class.: M:gf:G

Chistyakov, V. P., and Viktorova, I. I. 1965. 550
Asymptotic normality in a problem of balls when the probabilities of falling into different boxes are different. *Theor. Probability Appl.*, **10**, 149–54.
Class.: MI-P:m-a:G

Chistyakov, V. P., and Sevast'yanov, B. A.
See Sevast'yanov, B. A., and Chistyakov, V. P. (1964).

Chistyakov, V. P., and Victorova, I. I.
See Victorova, I. I., and Chistyakov, V. P. (1965).

Chistyakov, V. P., and Viktorova, I. I.
See Viktorova, I. I., and Chistyakov, V. P. (1966).

Chorneyko, I., Sathe, Y. S., and Narayana, T. V.
See Narayana, T. V., Chorneyko, I., and Sathe, Y. S. (1960).

Choudhary, Nazir Ahmad. 1947–48. 551
A generalization of the binomial, Lexian and Poisson distributions. *Math. Student*, **15**, 8.
Review: MR **10** (1949), 386; ZBL **35** (1950), 85.
Class.: GBDP:m:G

Chow, Bryant, and Maynard, James M.
See Maynard, James M., and Chow, Bryant. (1966).

Chu, J. T., Topp, C. W., Leone, F. C., and Hayman, G. E.
See Leone, F. C., Hayman, G. E., Chu, J. T., and Topp, C. W. (1960).

Chung, J. H. 1950. 552
Sequential sampling from finite lots when the proportion defective is small. *J. Amer. Statist. Assoc.*, **45**, 557–69.
Review: MR **12** (1951), 510; ZBL **39** (1951), 359.
Class.: P:pe-gf:BM-P

Chung, J. H., and Delury, D. B. 1950. 553
Confidence limits for the hypergeometric distribution. Univ. Toronto Press.
Review: ZBL **41** (1952), 466.

Chung, Kai Lai. 1941. 554
On the probability of the occurrence of at least *m*-events among *n* arbitrary events. *Ann. Math. Statist.*, **12**, 328–38.
Review: MR **3** (1942), 168; ZBL **26** (1942), 329.
User: Takacs, L. (1958b).
Class.: MI:mi:G

Chung, K. L., and Erdös, P. 1951. 555
Probability limit theorems assuming only the first moment. I. *Mem. Amer. Math. Soc. No.* 6, 19 pp.
Review: ZBL **42** (1952), 376.
Class.: MI:a:G

Chung, Kai Lai, and Feller, W. 1949. 556
On fluctuations in cointossing. *Proc. Nat. Acad. Sci. USA*, **35**, 605–08.
Review: MR **11** (1950), 444; ZBL **37** (1951), 363.
Users: Csaki, E., and Vincze, I. (1963), Newell, D. (1965).
Class.: MI:pc:G

Clapham, A. R. 1932. 557
The form of the observational unit in quantitative ecology. *J. Ecol.*, **20** (1), 192–97.
Users: Ashby, E. (1936), Archibald, E. E. A. (1949a), Myers, E., and Chapman, V. (1953).

Clapham, A. R. 1936. 558
Over-dispersion in grassland communities and the use of statistical methods in plant ecology. *J. Ecol.*, **24** (1), 232–51.
Users: Ashby, E. (1936), Singh, B. N., and Chalam, G. (1938), Singh, B. N., and Das, K. (1939), Beall, G. (1942), Fracker, S. B., and Brischle, H. (1944), Jones, E. W. (1945), Cole, L. C. (1946a), Whitford, P. B. (1949), Curtis, J., and McIntosh, R. (1950), Barnes, H., and Marshall, S. M. (1951), Barnes, H., and Stanbury, F. (1951), Dawson, G. W. P. (1951), Cain, S. A., and Evans, F. C. (1952), Dice, L. R. (1952), Grieg-Smith, P. (1952a), Skellam, J. G. (1952), Thomson, G. W. (1952), Evans, D. A. (1953), David, F. N., and Moore, P. G. (1954), Thompson, H. R. (1958), Pielou, D. P. (1960a).

Claringbold, P. J. 1955. 559
Matrices in quantal analysis. *Biometrics*, **11**, 481–501.
Review: BA **30** (1956), Ab. No. 18600.
Users: Naylor, A. F. (1964).

Clark, Andrew, and Leonard, W. H. 1939. 560
The analysis of variance with special reference to data expressed as percentages. *J. Amer. Soc. Agron.*, **31**, 55–66.
Users: Cochran, W. G. (1940), Curtiss, J. H. (1943 a Cochran, W. G. (1943).
Class.: B:anovat:G

Clark, C. E. 1937. 561
Note on the binomial distribution. *Ann. Math. Statist.*, **8**, 116–17.
Class.: B:osp:G

Clark, C. R., and Koopmans, L. R. 1959. 562
Graphs of the hypergeometric O.C. and A.O.Q. functions for lot sizes 10 *to* 225. Monograph SCR–121, Sandia Corp.
Class.: H:sqc-tc:E

Clark, Frank E. 1957. 563
Truncation to meet requirements on means. *J. Amer. Statist. Assoc.*, **52**, 527–536.
Review: PA **33** (1959), Ab. No. 2508.
Class.: TCB-TCP:mi:G

Clark, P. J. 1956. 564
Grouping in spatial distributions. *Science*, **123**, 373–74.

Clark, P. J., and Evans, F. C. 1954a. 565
Distance to nearest neighbor as a measure of spatial relationships in populations. *Ecology*, **35**, 445–53.
Users: Clark, P. J., and Evans, F. C. (1954b), Thompson, H. R. (1956), Hairston, N. G. (1959), Morisita, M. (1959), Waters, W. E., and Henson, W. R. (1959), Bray, J. R. (1962).
Class.: P:mi:BM

Clark, P. J., and Evans, F. C. 1955. 566
On some aspects of spatial pattern in biological populations. *Science*, **121**, 397–98.
Users: Clark, P. J. (1956), Smith, J. H. G., and Ker, J. W. (1957), Pielou, E. C. (1960).

Clark, P. J., Eckstrom, P. T., and Linden, L. C. 567
1964.
On the number of individuals per occupation in a human society. *Ecology*, **45**, 367–72.
Users: Patil, G. P., and Bildikar, S. (1966a), Patil, G. P., and Shorrock, R. W. (1966).
Class.: LS:gf:S

Clark, R. B., and Milne, A. 1955. 568
The sublittoral fauna of two sandy bays on the Isle of Cumbrae, Firth of Clyde. *J. Marine Biol. Assoc.*, **34** (1), 161–80.

Clark, Robert E. 1953. 569
Percentage points of the incomplete beta function. *J. Amer. Statist. Assoc.*, **48**, 831–43.
Users: Crow, E. L. (1956), Fabian, V. (1959), Pachares, J. (1960).
Class.: B:tc:G

Clark, R. D. 1946. 570
An application of the Poisson distribution. *J. Inst. Actuar.*, **72**, 481.

Clemans, K. G. 1959. 571
Confidence limits in the case of the geometric distribution. *Biometrika*, **46**, 260–64.
Review: IJA **1** (1960), 396.
Class.: G:ie-tc:G

Clopper, C. J., and Pearson, E. S. 1934. 572
The use of confidence or fiducial limits illustrated in the case of the binomial. *Biometrika*, **26**, 404–13.
Review: BA **10** (1936), Ab. No. 7706.
Users: Bartlett, M. S. (1937), Ludwig, W. (1937), Ricker, W. E. (1937), Von Schelling, H. (1937), Rietz, H. L. (1938), Scheffe, H. (1943–6), Eudey, M. W. (1949), Pearson, E. S. (1950), Stevens, W. L. (1950), Young, H., Neess, J., and Emlen, J. T. (1952), Birnbaum, A. (1953), Katz, L. (1953), Birnbaum, A. (1954b), Sterne, T. E. (1954), Walsh, J. E. (1955c), Crow, E. L. (1956), Buehler, R. J. (1957), Stevens, W. L. (1957), Clunies-Ross, C. W. (1958), Clemans, K. G. (1959), Blyth, C. R., and Hutchinson, D. W. (1960), Bol'shev, L. N. (1960), Leon, F. C., Hayman, G. E., Chu, J. T., and Topp, C. W. (1960), Pratt, J. W. (1961).
Class.: B:ie-tp-tc:G

Clunies-Ross, C. W. 1958. 573
Interval estimation for the parameter of a binomial distribution. *Biometrika*, **45**, 275–79.
Review: ZBL **88** (1961), 355.
Users: Crow, E. L., and Gardner, R. S. (1959), Govindarajulu, Z. (1965).
Class.: B:ie:G

Cobb, E. B., and Harris, Bernard. 1966. 574
As asymptotic lower bound for the entropy of discrete populations with application to the estimation of entropy for approximately uniform populations. *Ann. Inst. Statist. Math.*, **18**, 289–97.
Class.: MI:mi-pe:G

Cobb, Sidney, and Beall, Geoffrey.
See Beall, Geoffrey, and Cobb, Sidney (1961).

Cochran, W. G. 1936–37. 575
The χ^2 distribution for the binomial and Poisson series, with small expectations. *Ann. Eugenics*, **7**, 207–17.
Users: Haldane, J. B. S. (1937), Cochran, W. G. (1940), Hoel, Paul G. (1943), Tweedie, M. C. K. (1946), Tweedie, M. C. K. (1947), Seal, H. L. (1949), Robertson, A. (1951), Lancaster, H. O. (1952), Kathirgamatamby, N. (1953), Cochran, W. G. (1954), Rao, C. R., and Chakravarti, I. (1956), Tweedie, M. C. K. (1956), Chakravarti, I., and Rao, C. R. (1959), Bennett, B. M., and Hsu, P. (1961), Bennett, B. M. (1962a), Nicholson, W. L. (1963), Tweedie, M. C. K. (1965).
Class.: B-P:h:G

Cochran, W. G. 1940. 576
The analysis of variance when experimental errors follow the Poisson or binomial laws. *Ann. Math. Statist.*, **11**, 335–47.
Review: MR **2** (1941), 111; ZBL **23** (1941), 341.
Users: Cochran, W. G. (1943), Curtiss, J. H. (1943a), Bhattacharyya, A. (1946), Bartlett, M. W. (1947), Bliss, C. I., and Fleischer, D. A. (1964).
Class.: B-P:anovat:G

Cochran, W. G. 1943. 577
Analysis of variance for percentages based on un equal numbers. *J. Amer. Statist. Assoc.*, **38**, 287–301.
User: Cochran, W. G. (1954).
Class.: B:anovat:G

Cochran, William G. 1950a. 578
Estimation of bacterial densities by means of the 'most probable number'. *Biometrics*, **6**, 105–16.
Users: Wadley, F. M. (1954), Harris, E. K. (1958), Johnson, E. A., and Brown, B. W. (1961), Thompson, K. H. (1962).
Class.: B:pe-ctp:BM

Cochran, W. G. 1950b. 579
The comparison of percentages in matched samples. *Biometrika*, **37**, 256–66.
Review: PA **25** (1951), Ab. No. 5835; ZBL **40** (1951), 222.
Users: Iyer, P. V. K. (1955), Cox, D. R. (1958).
Class.: B:tp:G

Cochran, W. G. 1954. 580
Some methods for strengthening the common χ^2 tests. *Biometrics*, **10**, 417–51.
Users: Cochran, W. G. (1955), Rao, C. R., and Chakravarti, I. (1956), Owens, G. H., and Leslie, P. (1958), Patil, G. P. (1959), Watson, G. S. (1959), Chassen, J. B. (1960), Edwards, C. B., and Gurland, J. (1960), Griffiths, J. C. (1960), Bennett, B. M., and Hsu, P. (1961), Edwards, C. B., and Gurland, J. (1961), Rao, C. R. (1961), Williams, E. J. (1961b), Bennett, B. M. (1962b), Bennett, B. M., and Hsu, P. (1962), Bray, J. R. (1962), Keats, J. (1963), Keats, J. (1964), Lord, F. M. (1964), Quesenberry, C. P., and Hurst, D. C. (1964), Potthoff, Richard F., and Whittinghill, Maurice (1966).
Class.: P-B-NB:mi:G

Cochran, W. G. 1955. 581
A test of a linear function of the deviations between observed and expected numbers. *J. Amer. Statist. Assoc.*, **50**, 377–97.
Class.: MI-B-P:gf:G

Cochran, W. G., and Hopkins, C. E. 1960. 582
Some problems in multivariate classification using discrete variables (Summary). *J. Amer. Statist. Assoc.*, **55**, 357–58.
Class.: MI:tp:G

Cochran, W. G., and Hopkins, C. E. 1961. 583
Some classification problems with multivariate qualitative data. *Biometrics*, **17**, 10–32.
Review: IJA **3** (1962), 586.
User: Hills, M. (1966).

Cohen, A. C. 1954. 584
Estimation of the Poisson parameter from truncated samples and from censored samples. *J. Amer. Statist. Assoc.*, **49**, 158–68.
Review: MR **15** (1954), 637; ZBL **55** (1955), 375.
Users: Moore, P. G. (1954a), Finney, D. J., and Varley, G. C. (1955), Rider, P. R. (1955b), Clark, F. E. (1957), Tate, R. F., and Goen, R. L. (1958), Patil, G. P. (1959), Cohen, A. C. Jr. (1959e), Cohen, A. C. (1960b), Cohen, A. C. (1960d), Cohen, A. C. (1960e), Singh, N. (1960), Cohen, A. C. (1961a), Cacoullos, T. (1962), Hughes, E. J. (1962), Doss, S. A. D. C. (1963), Doss, S. A. D. C. (1963).
Class.: TCP:pe:G

Cohen, A. C., Jr. 1959a. 585
Estimation in the Poisson distribution when sample values of c+1 are sometimes erroneously reported as c. *Ann. Inst. Statist. Math. Toyko*, **11**, 189–93.
Review: ZBL **113** (1965), 136; IJA **2** (1961, 47.
Class.: P:pe:G

Cohen, A. C., Jr. 1959b. 586
Tables for the sign test when observations are estimates of binomial parameters. *J. Amer. Statist. Assoc.*, **54**, 784–93.
Review: PA **35** (1961), Ab. No. 56; ZBL **90** (1961), 364; IJA **3** (1962) 190.
Class.: B:ctp:G

Cohen, A. C., Jr. 1959c. 587
Maximum likelihood estimation in a contagious distribution. Tech. Rep. **5**, Institute Statist., Univ. Georgia, pp. 10.
User: Cohen, A. C., Jr. (1959d).
Class.: G:m-pe-gf-tc:BM

Cohen, A. C., Jr. 1959d. 588
Estimation in a binomial-negative binomial contagious distribution. Tech. Rep., **6**, Institute Statist., Univ. Georgia, pp. 12.
Class.: GB:pe-gf-tc:BM

Cohen, A. C., Jr. 1959e. 589
Estimating the Poisson parameter from truncated samples with missing zero observations. Tech. Rep. **15** (rev.), Institute Statist., Univ. Georgia, pp. 11.
Class.: TCP:pe-tc:G

Cohen, A. C., Jr. 1960a. 590
Extensions of the Poisson and the negative binomial distribution (Abstract). *Ann. Math. Statist.*, **31**, 532.
User: Cohen, A. C., Jr. (1961c).

Cohen, A. C., Jr. 1960b. 591
Estimating the parameter in a conditional Poisson distribution. *Biometrics*, **16**, 203–11.
Review: MR **22** (1961), 2177; ZBL **117** (1965), 369; IJA **2** (1961), 46.
Users: Cohen, A. C., Jr. (1960c), Cohen, A. C., Jr. (1960f), Cohen, A. C., Jr. (1961a), Cohen, A. C., Jr. (1961c), Martin, D. C., and Katti, S. K. (1962a), Martin, D. C., and Katti, S. K. (1962b), Patil, G. P. (1962e), Doss, S. A. D. C. (1963), Bardwell, G. E., and Crow, E. L. (1964), Cohen, A. C., Jr. (1965), Crow, E. L., and Bardwell, G. E. (1965), Katti, S. K., and Sly, L. E. (1965), Martin, D., and Katti, S. (1965), Tweedie, M. C. K. (1965).
Class.: TCP:pe-tc:G

Cohen, A. C., Jr. 1960c. 592
An extension of a truncated Poisson distribution. *Biometrics*, **16**, 446–50.
Review: MR **22** (1961), 2177; ZBL **114** (1965), 358; IJA **2** (1961), 463.
Users: Cohen, A. C., Jr. (1961c), Khatri, C. G. (1961), Khatri, C. G., and Patel, I. R. (1961), Cohen, A. C., Jr. (1965), Cohen, A. C., Jr. (1966).
Class.: TCP:pe-gf:BM

Cohen, A. C., Jr. 1960d. 593
Estimating the parameters of modified Poisson Distribution. *J. Amer. Statist. Assoc.*, **55**, 139–43.
Review: ZBL **104** (1964), 377; IJA **2** (1961), 287.
Users: Shah, and Venkataraman (1963).
Class.: OPR:pe:G

Cohen, A. C., Jr. 1960e. 594
Estimation in truncated Poisson distribution when zeros and some ones are missing. *J. Amer. Statist. Assoc.*, **55**, 343–48.
Review: ZBL **93** (1962), 322; IJA **2** (1961), 651.
Class.: TCP:pe-tc-gf:G-BM

Cohen, A. C., Jr. 1960f. 595
Misclassified data from a binomial population. *Technometrics*, **2**, 109–13.
Review: MR **22** (1961), 51; IJA **3** (1962), 405.
Class.: B:pe:G

Cohen, A. C., Jr. 1961a. 596
Estimating the Poisson parameter from samples that are truncated on the right. *Technometrics*, **3**, 433–38.
Review: MR **23A** (1962), 685; ZBL **117** (1965), 369; STMA **5** (1963), 26.
User: Doss, S. A. D. C. (1963).
Class : TCP:pe-tc:G

Cohen, A. C., Jr. 1961b. 597
Tables for maximum likelihood estimates: singly truncated and singly censored samples. *Technometrics*, **3**, 535–41.
Review: MR **26** (1963), 1082.
User: Bliss, C. I. (1965).

Cohen, A. C., Jr. 1961c. 598
On a class of pseudo-contagious distributions. Tech. Rep. **11**, Institute Statis., Univ. Georgia, pp. 12.

User: Cohen, A. C., Jr. (1965).
Class.: GB:pe-gf:G-BM

Cohen, A. C., Jr. 1965. 599
Estimation in mixtures of discrete distributions. *Classical and Contagious Discrete Distributions*, Ed. G. P. Patil, Statistical Publishing Society, Calcutta and Pergamon Press, pp. 373–78.
Class.: P-B:pe-gf:G-E

Cohen, A. C., Jr. 1965. 600
Estimation in the negative binomial distribution. *NASA Tech. Memo. No. X*-53372, George C. Marshall Space Flight Centre, Huntsville, Alabama.
Class.: NB-TCNB:pe:G

Cohen, A. C., Jr. 1965a. 601
A hybrid problem on the exponential family. *Ann. Math. Statist.*, **36**, 1185–1206.
Class.: B-P-MI:tp-pe:G

Cohen, A. C., Jr. 1966. 602
A note on certain discrete mixed distributions. *Biometrics*, **22**, 566–72.
Class.: NB-MI:pe-gf:G-BM

Cohen, B. H., and Sakoda, J. M.
See Sakoda, J. M., and Cohen, B. H. (1957).

Cohen, Leonard. 1958. 603
On mixed single sample experiments. *Ann. Math. Statist.*, **29**, 947–71.
Class.: B:tp:G

Cole, L. C. 1945. 604
A simple test of the hypothesis that alternative events are equally probable. *Ecology*, **26** (2), 202–05.
User: Cole, L. C. (1946).
Class.: B:ctp-tp:G

Cole, L. C. 1946a. 605
A theory for analyzing contagiously distributed populations. *Ecology*, **27**, 329–41.
Users: Ashby, E. (1948), Dice, L. R. (1948), Whitford, P. B. (1949), Curtis, J., and McIntosh, R. (1950), Kono, Utida, Yosida, and Watanabe (1952), Thomson, G. W. (1952), Bliss, C. I., and Fisher, R. A. (1953), Pearson, P. G. (1955), Hairston, N. G. (1959), Jensen, P. (1959), Pielou, E. C. (1962), Kobayashi, S. (1966).
Class.: P-N-MI:mi-gf:BM

Cole, L. C. 1946b. 606
A study of the cryptozoa of an Illinois woodland. *Ecol. Monogr.*, **16**, 49–86.
Users: Cole, L. C. (1946a), Cole, L. C. (1949), Dice, L. R. (1952), Kuenzler, E. J. (1958), Hairston, N. G. (1959), Jensen, P. (1959).
Class.: P-N:gf:BM

Cole, L. C. 1949. 607
The measurement of interspecific association. *Ecology*, **30**, 411–24.
Users: Dice, L. R. (1952), Grieg-Smith, P. (1952b), Bliss, C. I., and Fisher, R. A. (1953), Fager, E. W. (1957).
Class.: MI:mi:BM

Collis-Bird, J. J. 1963. 608
The Poisson distribution. Master's Thesis, McGill Univ.

Comita, G. W., and Comita, J. J. 1957 609
The interval distribution patterns of a colanoid copepod population and a description of a modified Clarke-Bumpus plankton sampler. *Limnology and Oceanography*, **2**, 321–32.

Comita, J. J., and Comita, G. W. 610
See Comita, G. W., and Comita, J. J. (1957).

Comstock, R. E., and Robinson, G. F. 1952. 611
Genetic parameters, their estimation and significance. *Proc. 6th Int. Grassland Congr.*, **1**, 284–91.

Conolly, B. W. 1960. 612
Queueing at a single serving point with group arrival. *J. Roy. Statist. Soc. Ser. B*, **22**, 285–98.
Review: IJA **2** (1961), 557.
User: Shanbhag, D. N. (1965).

Consael, R. 1949. 613
Sur le schéma de Pólya-Eggenberger à deux variables aléatoires. *Bull. Assoc. Actuaires Belges*, **55**, 11–23.
Review: ZBL **41** (1952), 250.

Consael, R. 1952a. 614
Sur les processus composés de Poisson à deux variables aléatoires. *Acad. Roy. Belg. Cl. Sci. Mém. Coll. in-8°*, **27** (6), 1–44.
Review: MR **15** (1953), 138; ZBL **48** (1953), 362.

Consael, R. 1952b. 615
Sur les processus de Poisson du type composé. *Acad. Roy. Belg. Bull. Cl. Sci.*, **38** (5), 442–61.
Review: MR **14** (1953), 293; ZBL **47** (1953), 125.

Consael, R. 1959. 616
Concerning the mathematical theory of populations according to J. Neymann and E. L. Scott, (French). *Meded. Kon. Acad. Belg. Wetensch.*, **45**, 845–58.
Review: IJA **3** (1962), 435.

Consoli, T. 1940. 617
Généralization d'un théorème sur la probabilité de la somme d'un nombre infini de variables aléatoires. *Rev. Fac. Sci. Univ. Istabul. Ser. A.*, **5**, 1–17.
Review: MR **2** (1941), 228.
User: Govindarajulu, Z. (1965).

Cooper, G. P., and Lagler, K. F. 1956. 618
The measurement of the fish population size. *Trans. N. Amer. Wildlife Conf.*, **21**, 281–97.

Cooper, L. H. N. 1947. 619
The distribution of iron in the waters of the western English channel. *J. Marine Biol. Assoc.*, **27**, 279–359.

Copeland, Arthur H., and Regan, Francis. 1936. 620
A postulational treatment of the Poisson law. *Ann. of Math.*, **37**, 357–62.

Corbet, A. S., Williams, C. B., and Fisher, R. A.
See Fisher, R. A., Corbet, A. S., and Williams, C. B. (1943).

Cormack, R. M. 1966. 621
A test for equal catchability. *Biometrics*, **22**, 330–42.

Cornfield, Jerome, and Chalkley, Herald W. 1951. 622
A problem in geometric probability. *J. Washington Acad. Sci.*, **41**, 226–29.
Review: BA **26** (1952), Ab. No. 5412.
Class.: MI:mi:G

Cornfield, Jerome, Mosimann, James E., and Gurian, Joan M.
See Gurian, Joan M., Cornfield, Jerome, and Mosimann, James E. (1964).

Corsten, L. C. A. 1957. 623
Partition of experimental vectors connected with multinomial distributions. *Biometrics*, **13**, 451–84.
Review: MR **19** (1958), 1095.
Users: Patil, G. P., and Bildikar, S. (1966a).
Class.: M:mi:G

Costa, M. A. F. 1953. 624
On the graduation of discrete frequency distributions. *Bol. Inst. Actuar. Port.*, **8**, 21–27.

Cota, P., and Patil, G. P,
See Patil, G. P., and Cota, P. (1962).

Cote, L., and Skibinsky, M.
See Skibinsky, M., and Cote, L. (1963).

Cottam, G., Curtis, J. T., and Catana, A. J., Jr. 625
1957.
Some sampling characteristics of a series of aggregated populations. *Ecology*, **38**, 610–22.

Cottam, G., Curtis, J. T., and Hale, B. W. 1953. 626
Some sampling characteristics of a population of randomly dispersed individuals. *Ecology*, **34**, 741–57.
Class.: P:gf-h:BM

Coumetou, M. 1950. 627
Abaques pour la determination des erreurs sur les proportions et sur leur difference (English summary). *Travail Hum.*, **13**, 104–08
Review: PA **25** (1951), Ab. No. 2746.
Class.: B:tc-ctp:G

Cover, Thomas M. 1966. 628
The probability that a random game is unfair. *Ann. Math. Statist*, **37**, 1796–99.
Class.: MI:mi:G

Coward, L. E. 1949. 629
The distribution of sickness. *J. Inst. Actuar.*, **75**, 12–38.
Review: ZBL **39** (1951), 153.

Cowden, Dudley J. 1946. 630
An application of sequential sampling to testing students. *J. Amer. Statist. Assoc.*, **41**, 547–56.
Class.: B:tp:S

Cox, D. R. 1952. 631
Estimation by double sampling. *Biometrika*, **39**, 217–27.
Review: PA **28** (1954), Ab. No. 118.
Users: Birnbaum, A., and Healy, W. C. (1960).
Class.: MI-B:se:G

Cox, D, R. 1953. 632
Some simple approximate tests for Poisson variates. *Biometrika*, **40**, 354–60.
Review: MR **15** (1954), 332; ZBL **51** (1954), 108.
Users: Birnbaum, A. (1954b), Ellner, H. (1963).
Class.: P:a-ctp-ie-anovat:E

Cox, D. R. 1955. 633
Some statistical methods connected with series of events. *J. Roy. Statist. Soc. Ser. B.*, **17**, 129–64.
Users: Cox, D. R. (1958b), Johnson, N. L. (1959), Miller, A. J. (1961).

Cox, D. R. 1958a. 634
Two further applications of a model for binary regression. *Biometrika*, **45**, 562–65.
Review: IJA **1** (1959), 79.
User: Cox, D. R. (1963).

Cox, D. R. 1958b. 635
The regression analysis of binary sequences. *J. Roy. Statist. Soc. Ser. B.*, **20**, 215–42.
Review: IJA **1** (1959), 80.
Users: Gart, J. J. (1962), Bennett, B. M. (1964).

Cox, D. R. 1960. 636
Serial sampling acceptance schemes derived from Bayes' theorem. *Technometrics*, **2**, 353–60.
Review: MR **22** (1961), 1032; ZBL **97** (1962), 137; IJA **2** (1961), 701.
Class.: MI:sqc:E

Cox, D. R. 1960a. 637
A note on tests of homogeneity applied after sequential sampling. *J. Roy. Statist. Soc. Ser. B.*, **22**, 368–71.
Review: IJA **2** (1961), 310.
Class.: B:h:G

Cox, D. R. 1962. 638
Further results on tests of separate families of hypotheses. *J. Roy. Statist. Soc. Ser. B.*, **24**, 406–24.
Review: IJA **4** (1963), 455.
User: Good, I. J. (1963).
Class.: MI-G-P:cm:G

Cox, D. R. 1963. 639
Large sample sequential tests for composite hypotheses. *Sankhyā*, **25**A, 5–12.
Review: STMA **5** (1964), 384.
Class.: B:ctp:G

Cox, D. R., and Benson, F.
See Benson, F., and Cox, D. R. (1951).

Cox, D. R., and Smith, W. L. 1957. 640
On the distribution of Tribolium confusum in a container. *Biometrika*, **44**, 328–35.
Class.: MI:m:BM

Crabb, W. D., and Emik, L. O. 1946. 641
Evaluating rat baits by field acceptance trials on Guam. *J. Wildlife Managem.*, **10**, 162–71.

Craig, C. C. 1936. 642
Sheppard's corrections for a discrete variable. *Ann. Math. Statist.*, **7**, 55–61.

Craig, C. C. 1953a. 643
On the utilization of marked specimens in estimating populations of flying insects. *Biometrika*, **40**, 170–76.
Users: Darroch, J. N. (1958), Sen, P. K. (1960), Pathak, P. K. (1964).
Class.: TCP-OMR:pe:BM

Craig, C. C. 1953b. 644
On a method of estimating biological populations in the field. *Biometrika*, **40**, 216–18.
Review: ZBL **50** (1954), 365.
Class.: B:pe:BM

Craig, C. C. 1953c. 645
Note on the use of fixed number of defectives and variable sample sizes in sampling by attributes. *Industrial Quality Control*, **9** (7), 83–86.
Class.: NB:sqc:E

Craig, C. C. 1962. 646
On the mean and variance of the smaller of two drawings from a binomial population. *Biometrika*, **49**, 566–69.
Review: MR **28** (1964), 338; IJA **4** (1963), 614.
User: Shah, B. K. (1966).
Class.: B:os-m:G

Craig, R., and Upholt, W. M.
See Upholt, W. M., and Craig, R. (1940).

Cramer, G. F. 1948. 647
An approximation to the binomial summation. *Ann. Math. Statist.*, **19**, 592–94.
Review: MR **10** (1949), 465; ZBL **37** (1951), 363.
User: Govindarajulu, Z. (1965).
Class.: B:a:G

Cramér, H. 1946. 648
Mathematical Methods of Statistics. Princeton Univ. Press.

Cramér, Harald 1956. 649
Ein Satz über geordnete Mengen von Wahrscheinlichkeitsverteilungen (Russian summary). *Teor. Verojatnost i Primenen*, **1**, 19–24.
Review: ZBL **73** (1960), 124.

Cramér, H. 1962. 650
Model building with the aid of stochastic processes. *Bull. Inst. Internat. Statist.*, **39** (2), 3–30.
Review: STMA **5** (1964), 475.
Class.: MI:pc-mb:G-BM-E

Cramér, Harald. 1964. 651
Model building with the aid of stochastic processes. *Technometrics*, **6**, 133–59.
Review: MR **32** (1966), 315.

Craven, B. D. 1963. 652
Some results for the bulk service queue. *Austral. J. Statist.*, **5**, 49–56.
Review: STMA **5** (1964), 721.
Class.: MI:pc:G

Crawford, G. B. 1965. 653
Characterization of Geometric and exponential distributions. Math. Note No. 421, Math. Res. Lab. Boeing Scientific Res. Lab.

Crawford, Gordon B. 1966. 654
Characterization of geometric and exponential distributions. *Ann. Math. Statist.*, **37**, 1790–95.
Class.: G:osp:G

Crawley, J. F., Rhodes, A. J., and Reid, D. B. W.
See Reid, D. B. W., Crawley, J. F.,and Rhodes, A. J. (1949).

Cresswell, W. L., and Froggatt, P. 1963. 655
The causation of bus drive accidents: An epidemiological study. Oxford Univ. Press.
Review: *J. Roy. Statist Soc. Ser. A*, **127** (1964), 483–51.

Cronholm, James N. 1963. 656
A general method of obtaining exact sampling probabilities of the Shannon-Wiener measure of information H. *Psychometrika*, **28**, 405–13.
Review: ZBL **118** (1965–1966), 141.
Class.: M:mi:G

Cronholm, James N. 1964. 657
A two-variable generating function for computing the sampling probabilities of a class of widely used statistics. *J. Amer. Statist. Assoc.*, **59**, 487–91.
Review: STMA **7** (1966), 28.
Class.: MI:mi:G

Cross, K. W., and Hogben, L.
See Hogben, L., and Cross, K. W. (1952).

Crow, E. L. 1956. 658
Confidence intervals for a proportion. *Biometrika*, **43**, 423–35.
Review: MR **19** (1958), 1204; ZBL **74** (1962), 140.
Users: Stevens, W. L. (1957), Clemans, K. G. (1959), Crow, E. L., and Gardner, R. W. (1959), Blyth, C. R., and Hutchinson, D. W. (1960), Pachares, J. (1960), Pratt, J. W. (1961).
Class.: B:ie-tc:G
Notes: Correction, *Biometrika*, **45**, 291.

Crow, E. L. 1958a. 659
A property of additively closed families of distributions. *Ann. Math. Statist.*, **29**, 892–97.
Review: ZBL **88** (1961), 115.
Class.: G-P-MI:osp:G

Crow, E. L. 1958b. 660
The mean deviation of the Poisson distribution. *Biometrika*, **45**, 556–559.
Review: IJA **1** (1959) 23; ZBL **88** (1961), 360.
Users: Katti, S. K. (1960a), Bardwell, G. E. (1961), Kamat, A. R. (1965).
Class.: P-B:m:G

Crow, E. L., and Bardwell, George E.
See Bardwell, George E., and Crow, E. L. (1964).

Crow, E. L., and Bardwell, George E. 1965. 661
Estimation of the parameters of the hyper-Poisson distributions. *Classical and contagious discrete distributions*. Ed. G. P. Patil, Statistical Publishing Society, Calcutta and Pergamon Press, pp. 127–40.
Class.: OPR:pe:G-BM-P

Crow, E. L., and Gardner, R. S. 1959. 662
Confidence intervals for the expectation of a Poisson variable. *Biometrika*, **46**, 441–53.
Review: IJA **1** (1960), 397; ZBL **93** (1962), 322.
Users: Bardwell, G. E., and Crow, E. L. (1964).
Class.: P:ie-tc:G

Cruon, R., and Kaufman, A.
See Kaufman, A., and Cruon, R. (1961).

Csáki, Péter, and Fischer, János. 1963. 663
On the general notion of maximal correlation (Russian summary). *Magyar. Tud. Akad. Mat. Kutató Int. Közl.*, **8**, 27–51.
Class.: M-MH-MI:mi:G

Csáki, E., and Vincze, I. 1963. 664
On some distributions connected with the arcsine law (Russian summary). *Magyar Tud. Akad. Mat. Kutató Int. Közl.*, **8**, 281–91.

Csiszar, I. 1961. 665
Some remarks on the dimension and entropy of random variables. *Second Hungarian Math. Congress*.
Review: IJA **2** (1961), 558.
User: Csiszar, I. (1962).
Class.: MI:mi:G

Csiszar, I. 1962. 666
On the dimension and entropy of order α and of the mixture of probability distributions. *Acta Math. Acad. Sci. Hungar.*, **13**, 245–55.
Review: MR **27** (1964), 166.
Class.: MI:mi:G

Csorgo, M., and Guttman, I. 1962. 667
On the empty cell test. *Technometrics*, **4**, 235–57.
Review: IJA **4** (1963), 277.
Class: MI:cm:G

Curme, G. L., and Blyth, C. R.
See Blyth, C. R., and Curme, G. L. (1960).

Curtis, J. T. 1955. 668
A note on recent work dealing with the spatial distribution of plants. *J. Ecol.*, **43**, 309.

Curtis, J. T., and McIntosh, R. P. 1950. 669
The interrelations of certain analytic and synthetic phytosocialogical characters. *Ecology*, **31**, 434–55.
Users: Thomson, G. W. (1952), Clark, P. J., and Evans, F. C. (1954a).
Class.: P-MI:mi:BM

Curtis, J. T., Catana, A. J., Jr., and Cottam, G.
See Cottam, G., Curtis, J. T., and Catana, A. J., Jr. (1957).

Curtis, J. T., Hale, B. W., and Cottam, G.
See Cottam, G., Curtis, J. T., and Hale, B. W. (1953).

Curtiss, J. H. 1941. 670
Generating functions in the theory of statistics. *Amer. Math. Monthly*, **48**, 374–86.

Curtiss, J. H. 1943a. 671
On transformations used in the analysis of variance. *Ann. Math. Statist.*, **14**, 107–22.
Review: MR **5** (1944), 128; ZBL **60** (1957), 312.
Users: Ghurye, S. G. (1949), Freeman, M. F., and Tukey, J. W. (1950), Blom, G. (1954), Laubscher, N. F. (1960).
Class.: P-B:anovat:G

Curtiss, J. H. 1943b. 672
Convergent sequences of probability distributions. *Amer. Math. Monthly*, **50**, 94–105.
Users: Curtiss, J. H. (1943a), Govindarajulu, Z. (1965).
Class.: OBP-P:a:G

Czen, Pin, and Dzan Dzo. 1961. 673
On estimating the size of population by capture-mark method. *Zastos. Mat.*, **6**, 51–63.
Review: IJA **4** (1963), 249.
User: Pathak, P. K. (1964).

D

Daboni, L. 1959. 674
A property of Poisson distributions (Italian). *Boll. Un. Mat. Ital.*, **14**, 318–20.
Review: IJA **2** (1961), 251.
User: Castoldi, L. (1959).

Dacunha-Castelle, Didier, and Bretagnolle, Jean.
See Bretagnolle, Jean, and Dacunha-Castelle, Didier (1964).

Dahl, E. 1960. 675
Some measures of uniformity in vegetation analysis. *Ecology*, **41**, 805–08.
Class.: LS:mi:BM

Dahlberg, Gunnar. 1942. 676
Methodik zur Unterscheidung von Erblicheits-und Milieuvariationen mit Hilfe von Zwillingen. *Hereditas (Lund)*, **28**, 409–28.
Review: ZBL **26** (1942), 337.

Dall'aglio, G. 1961. 677
On bivariate distributions with assigned margins, subject to some limitations (Italian). *Giorn. Ist. Ital. Attuari*, **24**, 94–108.
Review: IJA **4** (1963), 80.
Class.: MI:mi:G

Dall'aglio, Georgia. 1965. 678
Comportamento asintòtico delle stime della differénza media e del rapporto di concentraziόne (Italian, French, English, German and Spanish summaries). *Metron*, **24**, 379–414.

Dalenius, Tore. 1950. 679
Glücksrad statt Urnenschema. *Mitteilungsbl. Math. Statist.*, **2**, 127–29.
Review: ZBL **37** (1951), 365.

Daly, J. F., and Chernoff, H.
See Chernoff, H., and Daly, J. F. (1957).

Daly, J. F., and Wilks, S. S.
See Wilks, S. S., and Daly, J. F. (1939).

Dandekar, V. M. 1955. 680
Certain modified forms of binomial and Poisson distributions. *Sankhyā*, **15**, 237–50.
Review: MR **17** (1956), 278; ZBL **67** (1958), 105
Users: Basu, D. (1955), Martin, L. (1961), Dharmadhikari (1964).
Class.: OBR-OPR:mb-pe-gf:G-BM

Daniels, H. E. 1954. 681
Saddlepoint approximations in statistics. *Ann. Math. Statist.*, **25**, 631–50.
users: Good, I. J. (1957a), Good, I. J. (1958), Good, I. J. (1961b), Good, I. J. (1961c), Williams, E. J. (1961b), Douglas, J. B. (1965).
Class.: MI:a:G

Daniels, H. E. 1961. 682
Mixtures of geometric distributions. *J. Roy Statist. Soc. Ser. B*, **23**, 409–13.
Review: MR **24A** (1962), 452; STMA **5** (1964), 948.
User: Govindarajulu, Z. (1965).
Class.: G-MI:pc:G

Darroch, J. N. 1958. 683
The multiple recapture census. I. Estimation of a closed population. *Biometrika*, **45**, 343–59.
Review: IJA **1** (1959), 99.
Class.: MI:mb-pe:G

Darroch, J. N. 1959. 684
The multiple-recapture census. II. Estimation where there is immigration or death. *Biometrika*, **46**, 336–51.
Review: IJA **1** (1960), 675
User: Parker, D. A. (1963).
Class.: MI:mb-pe:G

Darroch, J. N. 1964. 685
On the distribution of the number of successes in independent trials. *Ann. Math. Statist.*, **35**, 1317–21.
Review: MR **29** (1965), 330.
Users: Samuels, S. M. (1964), Samuels, S. (1965).
Class.: GBDP:osp:G

Darwin, J. H. 1957. 686
The power of the Poisson index of dispersion. *Biometrika*, **44**, 286–89.
Review: ZBL **78** (1959) 333.
Users: Bennett, B. M. (1959), Bennett, B. M. (1962a), Nicholson, W. L. (1963), Selby, B. (1965).
Class.: P:id:G

Darwin, J. H. 1959. 687
Note on a three-decision test for comparing two binomial populations. *Biometrika*, **46**, 106–13.
Review: IJA **1** (1959), 225; ZBL **90** (1961), 113.
Class.: B:ctp:G

Darwin, J. H. 1960. 688
An ecological distribution akin to Fisher's logarithmic distribution. *Biometrics*, **16**, 51–60.
Review: IJA **2** (1961), 21
Users: Mosimann, J. E. (1962), Bosch, A. J. (1963), Nelson, W. C., and David, H. A. (1964), Potthoft, Richard F., and Whittinghill, Maurice (1966).
Class.: MI:mb-pe-gf:BM

Das Gupta, P. 1964. 689
On the estimation of the total number of events and of the probabilities of detecting an event from information supplied by several agencies. *Bull. Calcutta Statist. Assoc.*, **13**, 89–100.
Review: STMA **7** (1966), 70.
Class.: B:pe:G

Das, K., and Singh, B. N.
See Singh, B. N., and Das, K. (1939).

Das, S. R. 1953a. 690
A Mathematical analysis of the phenomena of human twins and higher plural births. I. Twins. *Metron*, **17** (1-2), 65–88.
Class.: MI:mb-mi:BM

Das, S. R. 1953b. 691
A mathematical analysis of the phenomena of human twins and higher plural births. II. Triplets and the application of the analysis in the interpretation of the twin and triplet data. *Metron*, **17** (3-4), 67–92.
Class.: MI:mb:BM

Das, S. R. 1956. 692
A mathematical analysis of the phenomena of human twins and higher plural births. III. *Metron*, **18** (1–2), 219–62.
Class.: MI:mi:BM

Davenport, W. B. Jr. 1950. 693
A study of speech probability distributions. Tech. Rep. 148, Res. Lab. Electronics, Msss. Institute Tech.

David, F. N. 1950. 694
Two combinatorial tests of whether a sample has come from a given population. *Biometrika*, **37**, 97–110.
Review: ZBL **37** (1951), 366.
Users: Barton, D. E., and David, F. N. (1959a), Jones, H. L. (1959), Nicholson, W. L. (1960), Nicholson, W. L. (1961), Chistjakov, V. P. (1964b), Laurent, A. G. (1965).
Class.: OMR:gf:G

David, F. N. 1950a. 695
An alternative form of χ^2. *Biometrika*, **37**, 448–51.
Review: ZBL **39** (1951), 114.

David, F. N. 1955a. 696
The transformation of discrete variables. *Ann. Human Genetics*, **19**, 174–82.
Review: MR **16** (1955), 940; ZBL **64** (1956), 136.
User: Govindarajulu, Z. (1965).
Class.: P-B-NB:anovat-tp-ctp:G

David, F. N. 1955b. 697
Studies in the history of probability and statistics. I. Dicing and gaming (a note on the history of probability). *Biometrika*, **42**, 1-15.
Class.: MI:mi:G

David, F. N., and Barton, D. E.
See Barton, D. E., and David, F. N. (1956).

David, F. N., and Barton, D. E.
See Barton, D. E., and David, F. N. (1957).

David, F. N., and Barton, D. E.
See Barton, D. E., and David, F. N. (1958).

David, F. N., and Barton, D. E.
See Barton, D. E., and David, F. N. (1959).

David, F. N., and Barton, D. E. 1962. 698
Combinatorial chance. Charles Griffin, London.

David, F. N., and Johnson, N. L. 1950. 699
The probability integral transformation when the variable is discontinuous. *Biometrika*, **37**, 42–49.
Review: ZBL **38** (1951), 288.
Users: Pearson, E. S. (1950), Barton, D. E. (1955), Ellner, H. (1963).
Class.: MI-B-P:mi:G

David, F. N., and Johnson, N. L. 1952. 700
The truncated Poisson. *Biometrics*, **8**, 275–85.
Review: MR **14** (1953), 65.
Users: Plackett, R. L. (1953), Cohen, A. C. Jr. (1954), Moore, P. G. (1954a), Aitchison, J. (1955), Finney, D. J., and Varley, G. C. (1955), Rider, P. R. (1955b), Sampford, M. R. (1955), Rao, C. R., and Chakravarti, I. (1956), Roy, J., and Mitra, S. K. (1957), Brass, W. (1958a), Hartley, H. O. (1958), Tate, R. F., and Goen, R. L. (1958), Chakravarti, I. M., and Rao, C. R. (1959), Cohen, A. C. (1959d), Cohen, A. C. (1959e), Irwin, J. O. (1959), Patil, G. P. (1959), Barton, D. E., David, F. N., and Merrington, M. (1960), Cohen, A. C. (1960b), Cohen, A. C. (1960d), Cohen, A. C. (1960e), Bardwell, G. E. (1961), Cohen, A. C. (1961a), Cohen, A. C. (1961c), Patil, G. P. (1961b), Hughes, E. J. (1962), Patil, G. P. (1962b), Patil, G. P. (1962e), Rider, P. R. (1962b), Doss, S. A. D. C. (1963), Chatfield, C., Ehrenberg, A. S. C., and Goodhardt, G. J. (1966).
Class.: TCP-NB:id-pe:G

David, F. N., and Johnson, N. L. 1953. 701
Reciprocal Bernoulli and Poisson variables. *An. Fac. Ci. Porto*, **37**, 200–203.
Review: MR **19** (1958), 187; ZBL **53** (1961), 409.
Users: Johnson, N. L. (1960), Govindarajulu, Z. (1962b), Govindarajulu, Z. (1962c), Govindarajulu, Z. (1962d), Govindarajulu, Z. (1962e), Govindarajulu, Z. (1963).
Class.: OBR-OPR:m:G

David, F. N., and Johnson, N. L. 1956. 702
Reciprocal Bernoulli and Poisson variables. *Metron*, **18** (1–2), 77–81.
Review: MR **18** (1957), 520.
Users: Johnson, N. L. (1960) Govindarajulu, Z. (1962b), Govindarajulu, Z. (1962c), Govindarajulu, Z. (1963d), Govindarajulu, Z. (1963).
Class.: OBR-OPR:m:G

David, F. N., and Moore, P. G. 1954. 703
Notes on contagious distributions in plant populations. *Ann. Bot. Lond. N. S.*, **18**, 47-53.
Review: BA **29** (1955), Ab.No. 15577.
Users: David, F. N. (1955a), Barton, D. E. (1957), Pielou, E. C. (1962).
Class.: MI:h:BM

David, F. N., and Moore, P. G. 1957. 704
A bivariate test for the clumping of supposedly random individuals. *Ann. Bot. Lond. N.S.*, **21**, 315–20.
Users: Barton, D. E., and David, F. N. (1959a).

David, F. N. Merrington, M., and Barton, D. E.
See Barton, D. E., David, F. N., and Merrington, M. (1960).

David, F. N. Merrington, M., and Barton, D. E.
See Barton, D. E., David, F. N., and Merrington, M. (1965).

David, H. A. 1952. 705
An operational method for derivation of relations between moments and cumulants. *Metron*, **16** (3–4), 41–47.
Review: BA **27** (1953), Ab.No. 31362.
Class.: MI:m:G

David, H. A. 1960. 706
A conservative property of binomial tests. *Ann. Math. Statist.*, **31**, 1205–07. Correction: **32**, 1343.
Review: MR **22** (1961), 1465; IJA **3** (1962), 192; ZBL **97** (1962), 141.
Users: Darroch, J. N. (1964), Samuels, S. M. (1964).
Class.: B:tp:G

David, H. A., and Nelson, W. C.
See Nelson, W. C., and David, H. A. (1964).

David, H. A., and Norman, J. E. Jr. 1965. 707
Exact distribution of the sum of independent identically distributed discrete random variables. *J. Amer. Statist. Assoc.*, **60**, 837–42. Correction: **61**, 1246.
Review: MR **32** (1966), 525.
Class.: MI:mi:G

David, H. A., and Taylor, R. J.
See Taylor, R. J., and David, H. A. (1962).

David, H. T., and Edwards, C. B.
See Edwards, C. B., and David, H. T. (1962).

David, H. T., and Mengido, R. M. 1963. 708
Asymptotically exact truncation in binomial sequential analysis (Abstract). *Ann. Math. Statist.*, **34**, 1617–18.

Davies, O. L. 1933. 709
On asymptotic formulae for the hypergeometric series. I. Hypergeometric series in which the fourth element, *x*, is unity. *Biometrika*, **25**, 295–322.
Review: BA **9** (1935), Ab.No. 2371.
Users: Davies, O. L. (1934), Wise, M. E. (1954), Johnson, N. L. (1960), Govindarajulu, Z. (1965).
Class.: H:a:G

Davies, O. L. 1934. 710
On asymptotic formula for the hypergeometric series. II. Hypergeometric series in which the fourth element, *x*, is not necessarily unity. *Biometrika*, **26**, 59–107.
Review: BA **9** (1935), 15359.
Class.: H:a:G

Davies, R. O., and Leech, J. W. 1954. 711
The statistics of scaled random events. *Proc. Cambridge Philos. Soc.*, **50**, 575–80.
Review: MR **16** (1955), 272.
Class.: OPR:pc-m:P

Davis, D. E., Komarek, E. V., and Wood, J. E.
See Wood, J. E., Davis, D. E., and Komarek, E. V. (1958).

Davis, D. H. S., and Leslie, P. H.
See Leslie, P. H., and Davis, D. H. S. (1939).

Davis, D. J. 1952. 712
An analysis of some failure data. *J. Amer. Statist. Assoc.*, **47**, 113–50.
Users: Birnbaum, A. (1954b), Duker, S. (1955).
Class.: P:gf-ie:E-G

Davis, E. G., and Wadley, F. M. 1949. 713
Grasshopper egg-pod distribution in the northern Great Plains and its relation to egg-survey methods. *Circular* 816 U.S. *Dept. Agri.*, 1–15.
Class.: P:gf:BM

Davis, Harold T. 1927. 714
Elementary derivation of the fundamental constants in the Poisson and Lexis frequency distributions. *Amer. Math. Monthly*, **34**, 183–88.
Class.: OBR:m:G

Davis, S. A. Jacobsen, A. C., and Marino, R. T. 715
1965
B and H Acceptance Sampling Plans. *Ann. Tech. Conf. Trans., Amer. Soc. Qual. Contr.* pp. 257–69.
Review: STMA **7** (1966), 367.
Class.: H-B:sqc:G

Dawkins, C. J. 1939. 716
Tussock formation by Schoenus nigricans: the action of fire and water erosion. *J. Ecol.*, **27**, 78-88.

Dawson, G. W. P. 1951. 717
A method for investigating the relationship between the distribution of individuals of different species in a plant community. *Ecology*, **32**, 332–34.
Class.: MI:h:BM

Deboer, J. 1953. 718
Sequential test with three possible decisions for testing an unknown probability. *Appl. Sci. Res. B*, **3**, 249–59.
Review: MR **15** (1954), 727.
Class.: B:tp:G

Decastro, G. 1952a. 719
Note on differences of Bernoulli and Poisson variables. *Portugal. Math.*, **11**, 173–75.
Review: MR **14** (1953), 566; ZBL **49** (1959), 214.
User: Shanfield, F. (1964).
Class.: B-P:mi:G

Decastro, G. 1952b. 720
Elementos da estatística da distribuisão binomial. *Rev. Med. Veterinaria*, **47**, 153–68.

Deevey, E. S. Jr., and Potzger, J. E. 1951. 721
Peat samples for radio carbon analysis. *Amer. J. Sci.*, **249**, 473–511.

Definetti, B. 1953-4. 722
Una légge riguardante l'estinzióne nei processi di eliminazione. *Giorn. Ist. Ital. Attuari.*, **16**, 94–99.
Review: MR **16** (1955), 147; ZBL **53** (1961), 278.

Definetti, Bruno. 1957. 723
Sul numero di elementi al di là dell'ultimo osservato. *Scritti Mat. in Onore di F. Sibirani*, pp. 95–107.
Review: ZBL **77** (1958), 127.

Definetti, Bruno. 1964. 724
Alcune osservazioni in tema di ' suddivisione casuale.' (French, English, Spanish and German Summaries). *Giorn. Ist. Ital. Attuari*, **27**, 151–73.
Review: MR **30** (1965), 993.

deGroot, Morris H. 1959. 725
Unbiased sequential estimation for binomial populations. *Ann. Math. Statist.*, **30**, 80–101.
Review: IJA **1** (1959), 209; ZBL **91** (1962), 309.
Users: Nadler, J. (1960), Brainerd, B., and Narayana, T. V. (1961), Narayana, T. V., and Sathe, Y. S. (1961), Nelson, A. C. Jr., Williams, J. S., and Fletcher, N. T. (1963), Wasan, M. T. (1964), Knight, W. (1965), Wasan, M. T. (1965).
Class.: B–NB:se:G

deGroot, M. H. 1966. 726
Optimal allocation of observations. *Ann. Inst. Statist. Math.*, **18**, 13–28.
Class.: B:pe:G

Dehalu, M. 1942. 727
Sur la démonstration de la formule de K. Pearson dans le cas du schéma simple des urnes. *Bul. Soc. Roy. Sci. Liege*, **11**, 146-51.
Review: ZBL **27** (1943), 337.

Dejong, A. J., Dewolff, P., and Sittig, J.
See deWolff, P., Sittig, J., and deJong, A. J. (1950).

Dejongh, B. H. 1961. 728
A short note on extended Poisson processes. *Skand. Aktuarietidskr.*, **44**, 210–11.

Delaporte, Pierre. 1959. 729
Quelques problèmes de statistique mathématique posés par l'assurance automobile et le bonus pour non sinistre. *Bull. Trimest. Inst. Actuaires Franc.*, **70**, 87–102.
Users: Dickerson, O. D., Katti, S. K., and Hofflander, A. E. (1961).

Delury, D. B. 1947. 730
On estimation of biological populations. *Biometrics*, 3, 145–67.
Users: Bailey, N. T. J. (1951), MacLulich, D. A. (1951), Chapman, D. G. (1952b), Chapman, D. G. (1954), Cooper, G., and Lagler, K. (1956), Pathak, P. K. (1964).
Class.: MI:mb-pe:BM

Delury, D. B. 1951. 731
On the planning of experiments for the estimation of fish populations. *J. Fisheries Res. Board Canada*, **8**, 281–307.
Users: Chapman, D. (1952b), Chapman, D. G. (1954), Geiss, A. D. (1955), Cooper, G., and Lagler, K. (1956).

Delury, D. B. 1958. 732
The estimation of population size by a marking and recapture procedure. *J. Fisheries Res. Board Canada*, **15**, 19–25.

Delury, D. B., and Chung, J. H.
See Chung, J. H., and DeLury, D. B. (1950).

Demars, Clarence John, Jr. 1953. 733
An analysis of within-tree changes in distribution of the Western pine beetle, Dendroctonus brevicomis Leconte, during development. Ph.D. Thesis, Univ. California, Berkeley, 197 pp.

Deming, Lola S. 1960. 734
Selected bibliography of statistical literature, 1930 to 1957. III. Limit theorems. *J. Res. Nat. Bur. Standards Sect. B*, **64**, 175–92.
Review: IJA **2** (1961), 773.

Deming, Lola S. 1963. 735
Selected bibliography of statistical literature: supplement, 1958–60. *J. Res. Nat. Bur. Standards Sect. B*, **67**, 91–133.
Review: STMA **5** (1964), 993.

Deming, Lola S., and Gupta, D. 1961. 736
Selected bibliography of statistical literature, 1930 to 1957. IV. Markoff chains and stochastic processes. *J. Res. Nat. Bur. Standards Sect. B*, **65**, 61–93.
Review: IJA **2** (1961), 774.

Deming, W. E., and Glasser, G. G. 1959. 737
On the problem of matching lists by sampling. *J. Amer. Statist. Assoc.*, **54**, 403–15.
Review: IJA **1** (1960), 595.
Class.: MI:mb-pe-tp:G

Demoivre, Abraham. 1718. 738
The doctrine of chance.

Dempster, A. P. 1966. 739
New methods for reasoning towards posterior distributions based on sample data. *Ann.* Math. Statist., **37**, 355–74.
Review: MR **32** (1966), 809.
Class.: MI:mi:G

Devol, L., and Ruark, A.
See Ruark, A., and Devol, L. (1936).

Dewit, C. T., Keuls, M., and Over, H. J.
See Keuls, M., Over, H. J., and DeWit, C. T. (1962).

Dewolff, P., Sittig, J., and Dejong, A. J. 1950. 740
Frequentie verdelingen. *Statistica Neerlandica*, **4**, 89–120.

Dews, Peter B., Higgins, George M., and Berkson, Joseph. 1954. 741
Error of the determination of the Eosinophil count in the peritoneal fluid of the rat. *Biometrics*, **10**, 221–26.

Dharmadhikari, S. W. 1964. 742
A generalisation of a stochastic model considered by V. M. Dandekar. *Sankhyā Sec. A*, **26**, 31–38.
Review: STMA **6** (1965), 1027; MR **30** (1965), 664.
Class.: MI:pc:G-BM

Dhruvarajan, P. S., and Singal, M. K. 1957. 743
A note on moments and cumulants. *Math. Student*, **25**, 27–32.
Review: MR **20** (1959), 719.
Class.: MI:m:G

Diamantides, N. D. 1956. 744
Analogue computer generation of probability distributions for operations research. *Comm. Electronics*, 23, 86–91.

Diamond, E. L., Mitra, S. K., and Roy, S. N. 1960. 745
Asymptotic power and asymptotic independence in the statistical analysis of categorical data. *Bull. Inst. Internat. Statist.*, **37** (3), 309–29.
Review: IJA **2** (1961), 67.
Class.: M:tp-anovat:G

Dice, L. R. 1948. 746
Relationship between frequency index and population density. *Ecology*, **29**, 389–91.
Class.: P:h:BM

Dice, L. R. 1952. 747
Measure of spacing between individuals within a population. *Contri. Lab. Vert. Biol. Univ. Mich.*, **55**, 1–23.
Users: Cain, S. A., and Evans, F. C. (1952), Clark, P. J., and Evans, F. C. (1954a), Clark, P. J., and Evans, F. C. (1954b), Johnston, R. F. (1956a), Hairston, N. G. (1959).

Dick, Ronald S. 1966. 748
On the queue length distribution with balking in a single server queuing process (Abstract). *Ann. Math. Statist.*, **37**, 762–63.

Dickerson, O. D., Katti, S. K., and Hofflander, A. E. 1961. 749
Loss distribution in non-life insurance. *J. Insurance*, **28**, 45–54.
Class.: P-NB-COB:mi:S

Dickson, Charlotte. 1920. 750
On a theorem in the theory of probabilities. *Amer. Math. Monthly*, **27**, 166–167.
Class.: G:m:O

Dieulefait, C. E., 1939a. 751
Bestimmung der Momente der gewöhnlichen hypergeometrischen Wahrscheinlichkeiten und Bestimmung der Momente für den Fall der Ansteckung (Spanish). *An. Soc. Ci. Argentina*, **127**, 108–17.
Review: ZBL **21** (1940), 40.

Dieulefait, C. E. 1939b. 752
On incomplete moments (Spanish). *Rev. Ci. (Lima)*, **41**, 543–47.
Review: MR **1** (1940), 245.

Dieulefait, C. E. 1939c. 753
Sui momenti delle distribuzioni ipergeometriche. *Giorn. Inst. Ital. Attuari.*, **10**, 221–24.
Review: ZBL **22** (1940), 245.

Dijkman, J. G. 1965. 754
Probability theory and intuitionism. Discrete state-space. *Compositio Math.*, **17**, 72–95.
Review: MR **31** (1966), 953.
Class.: MI:mi:G.

Dilley, N. R. 1963a. 755
Some probability distributions and their associated structures. Part I. *Math. Mag.*, **36**, 175–79.
User: Carlitz, L. (1963).
Class.: B:mi:G

Dilley, N. R. 1963b. 756
Some probability distributions and their associated structures. Part II. *Math. Mag.*, **36**, 227–31.
Class.: MI:mi:G

Diveev, R. Ch. 1959. 757
Eine wesentlich vollständige Klasse von Entscheidungregeln für die Bestimmung der Warscheinlickheit eines Zustandes. *Izv. Akad. Nauk UzSSr, Ser. Fiz.-Mat., No.*, **1**, 45–53.
Review: ZBL **92** (1962), 361.

Dobrushin, R. L. 1956. 758
On Poisson's law for distribution of particles in space. *Ukrain. Math. Z.*, **8**, 127–134.
Review: ZBL **73** (1960), 352

Dodd, Edward L. 1913. 759
The probability integral deduced by means of developments in finite form. *Amer. Math. Monthly*, **20**, 123–27.
Class.: B:a:G

Dodd, S. C. 1952. 760
On all-or-none elements and mathematical models for sociologists. *Amer. Sociol. Rev.*, **17**, 167–77.

Dodge, H. F. 1963. 761
A general procedure for sampling inspection by attributes based on the AQL concept. *Nat. Conv. Trans., Amer. Soc. Qual. Contr.*, pp. 7–19.
Review: STMA **5** (1964), 435.
User: Hald, A. (1965).
Class.: B:sqc:E

Dodge, Harold F., and Romig, Harry G. 1959. 762
Sampling inspection tables. John Wiley and Sons, New York and Chapman and Hall, London, 224 pp.
Review: ZBL **86** (1961), 124.

Dodge, H. F., and Stephens, K. S. 1965. 763
Some new chain sampling inspection plans. *Annual Techn. Conf. Trans., Amer. Soc. Qual. Contr.*, pp. 8–17.
Review: STMA **7** (1966), 368.
Class.: B:sqc:G

Doig, Alison. 1957. 764
A bibliography on the theory of queues. *Biometrika*, **44**, 490–514.
Users: Miller, R. G., Jr. (1959), Winsten, C. B. (1959), Prabhu, N. U. (1960).

Doig, Alison G., and Kendall, Maurice G.
See Kendall, Maurice G., and Doig, Alison G. (1965).

Doksum, K. A., and Bell, C. B.
See Bell, C. B., and Doksum, K. A. (1966).

Domb, C. 1948. 765
Some probability distributions connected with recording apparatus. *Proc. Cambridge Philos. Soc.*, **44**, 335–41.

Review: ZBL **30** (1949), 403.
User: Domb, C. (1950).
Class.: P:pc:G

Domb, C. 1950. 766
Some probability distributions connected with recording apparatus. II. *Proc. Cambridge Philos. Soc.*, **46**, 429–35.
Review: ZBL **38** (1951), 89
Class.: P:pc:G

Domb, C. 1952. 767
On the use of a random parameter in combinatorial problems. *Proc. Phys. Soc.*, **65**, 305–09.
Review: SA **55**A (1952), Ab.No. 4854.

Domb, C. 1954. 768
On multiple returns in the random walk problem. *Proc. Cambridge Philos. Soc.*, **50**, 586–91.
User: Good, I. J. (1958).
Class.: MI:pc-a:G

Donnelly, J., and Macleod, J.
See MacLeod, J., and Donnelly, J. (1958).

Doornbos, R., and Prins, H. J. 1958a. 769
On slippage tests. I. A general type of slippage test and a slippage test for normal variates. *Nederl. Akad. Wetensch. Proc. Ser. A*, **61** (Indag. Math. 20), 38–46.
Class.: MI:tp:G

Doornbos, R., and Prins, H. J. 1958b. 770
On slippage tests. II. Slippage tests for discrete variates. *Nederl. Akad. Wetensch. Proc. Ser. A*, **61** (Indag. Math. 20), 47–55.
Class.: P-B-NB-MI:tp-osp:G

Doornbos, R., and Prins, H. J. 1958c. 771
On slippage tests. III. Two distribution-free slippage tests and two tables. *Nederl. Akad. Wetensch. Proc. Ser. A*, **61** (Indag. Math. 20), 438–47.
Review: IJA **1** (1959), 67.
Class.: P:tp-tc:G

Dormer, K. J., and Brett, D. W.
See Brett, D. W., and Dormer, K. J. (1960).

Doss, S. A. D. C. 1963. 772
On the efficiency of best asymptotically normal estimates of the Poisson parameter based on singly and doubly truncated or censored samples. *Biometrics*, **19**, 588–94.
Review: MR **28** (1964), 511; STMA **6** (1965), 433.
Class.: TCP:pe:G

Douglas, J. B. 1955. 773
Fitting the Neyman type A (two parameter) contagious distribution. *Biometrics*, **11**, 149–73.
Review: MR **16** (1955), 1039; MR **20** (1959), 600.
Users: Barton, D. E. (1957), Sprott, D. A. (1958), Nef, L. (1959), Shumway, R., and Gurland, J. (1960a), Shumway, R., and Gurland, J. (1960b), Shumway, R., and Gurland, J. (1960c), Martin, D. C., and Katti, S. K. (1962a), Martin, D. C., and Katti, S. K. (1962b), Martin, D. C., and Katti, S. K. (1962c), Blischke, W. R. (1965), Gurland, J. (1965), Katti, S. K., and Sly, L. E. (1965), Martin, D., and Katti, S. K. (1965), Sprott, D. A. (1965b), Butter, Margaret Sauls (1966).
Class.: N:pe-gf-tc:G

Douglas, J. B. 1956. 774
Tables of Poisson power moments. *Biometrika*, **43**, 489.
Review: MR **18** (1957), 238.
Class.: P:m-tc:G

Douglas, J. B. 1965. 775
Asymptotic expansions for some contagious distributions. *Classical and Contagious Discrete Distributions.* Ed. G. P. Patil, Statistical Publishing Society, Calcutta and Pergamon Press, pp. 291–302.
Class.: MI-N:a:G

Douglass, R. D., and Rutledge, George.
See Rutledge, George and Douglass, R. D. (1936).

Doust, J. F., and Josephs, K. J. 1941. 776
A simple introduction to the use of statistics in telecommunications engineering. III. The Poisson probability law. *P. O. Elect. Engrs. J.*, **34**, 139–44.
Review: SA **45**A (1952), Ab.No. 9.

Downton, F. 1961. 777
A note on vacancies on a line. *J. Roy. Statist. Soc. Ser. B*, **23**, 207–14.
Review: STMA **5** (1964), 793.
Class.: MI:mi:G

Dreze, J. P. 1962. 778
Queues with non-preemptive priorities. *Cahiers Centre Recherche Opérat.*, **4**, 20-51.
Review: IJA **4** (1963), 140.
Class.: MI:pc:G

Dubey, S. D. 1965a. 779
Statistical solutions of a combinatorial problem. *Amer. Statistician*, **19**, 30–31.

Dubey, Satya D. 1965. 780
The compound Pascal distributions (Abstract). *Ann. Math. Statist.*, **36**, 734.

Dubey, Satya D. 1966. 781
Graphical tests for discrete distributions. *Amer. Statistician*, **20**, 23-24.
Class.: B-P-NB-MI:gf-osp:G

Dubey, S. D. 1966a. 782
Compound Pascal distributions. *Ann. Inst. Statist. Math.*, **18**, 357–65.
Class.: NH-CONB:m-a:G

Dubins, L. E. 1957. 783
A discrete evasion game. *Contri. to the theory of games III, Ann. Math. Stud.* 39, *Princeton Univ.*, pp. 231–55.

Dubourdieu, Jules. 1938a. 784
Remarques relatives à la théorie de l'assurance-accidents. *C. R. Acad. Sci. Paris*, **206**, 303–05.
Users: Feller, W. (1943), Consael, R. (1952b).

Dubourdieu, Jules. 1938b. 785
Les fonctions absolument monotones et la theorie mathématique de l'assurance-accidents. *C. R. Acad. Paris*, **206**, 556–57.

Dudley, F. H., Bushland, R. C., Baumhover, A. H., Graham, A. J., Bitter, B. A., Hopkins, D. E., and NEW, W. D.
See Baumhover, A. H., Graham, A. J., Bitter, B. A., Hopkins, D. E., New, W. D., Dudley, F. H., and Bushland, R. C. (1955).

Dogué, Daniel. 1941. 786
Sur quelques exemples de factorisation de variables aléatoires. *C. R. Acad. Sci. Paris*, **212**, 838–40.
Review: MR 3 (1942), 2.
Class.: P:osp:G

Dukar, Sam. 1952. 787
The Poisson distribution and its significance to educational research. Ph.D. Thesis, Columbia Univ.
Review: PA **27** (1953), Ab.No. 7498.

Dukar, Sam. 1955. 788
The Poisson distribution in educational research. *J. Expt. Educ.*, **23**, 263–69.
Review: PA **30** (1956), Ab.No. 1895.
Class.:

Dumas, M. 1951. 789
L'évaluation des probabilités des lois binômiales. *J. Soc. Statist. Paris*, **92** (7-8-9), 227–33.

Dumas, M. 1952. 790
L'interprétation des séries de résultats blancs et noirs. *Mém. Artillérie Française*, **26**, 589-624.
Review: MR **14** (1953), 1102.

Duncan, D. C. 1965. 791
The diagonals of Pascal's triangle. *Amer. Math. Monthly*, **72**, 1115.

Dunn, E. R., and Allendoerfer, C. B. 1949. 792
The application of Fisher's formula to collections of Panamanian snakes. *Ecology*, **39**, 533–36.
Class.: LS:gf:BM

Durbin, J. 1961. 793
Some methods of constructing exact tests. *Biometrika*, **48**, 41–55.

Dwass, Meyer, and Teicher, Henry. 1957. 794
On infinitely divisible random vectors. *Ann. Math. Statist.*, **28**, 461–70.
Review: ZBL **78** (1959), 313.
Users: Edwards, C. B. (1962), Fuchs, C. E., and David, H. (1965a).
Class.: P-MP:pe-mi:G

Dvoretsky, W. 1953. 795
Sequential decision problems for processes with continuous time parameter. *Ann. Math. Statist.*, **24** 403–15.

Dynkin, E. B. 1950. 796
On sufficient and necessary statistics for families of probability distributions (Russian). *Dokl. Akad. Nauk SSSR*, **75**, 161–64.

Dynkin, E. B. 1951. 797
Necessary and sufficient statistics for a family of probability distributions (Russian). *Uspehi Mat. Nauk*, **6** (41), 68–90.

Dzan-Dzo, I. 1961. 798
On minimax estimation of the parameter in binomial distribution (Polish). *Zastos. Mat.*, **6**, 31–42.
Review: ZBL **102** (1963), 143; IJA **4**(1963), 433.
Class.: B:Pe:G

E

Eadie, G. S., and Turner, M. E.
See Turner, M. E., and Eadie, G. S. (1957).

Earles, D., Massa, J., and Sirull, R. 1963. 799
Some ' plain talk ' about confidence and sample size Parts I and II. *Evaluation Engineer*, **2**, 14–15; 26–27.
Review: STMA **5** (1964), 598.
Class.: B:ie-sqc:G-E

Eberhardt, L., and Whitlock, S. C.
See Whitlock, S. C., and Eberhardt, L. (1956).

Eckstrom, P. T., Linden, L. C., and Clark, P. J.
See Clark, P. J., Eckstrom, P. T., and Linden, L. C. (1964).

Ederer, Fred, Myers, Max H., and Mantel, Nathan. 800
1964.
A statistical problem in space and time: do leukemia cases come in clusters? *Biometrics*, **20**, 626–38.

Edie, Leslie, C. 1954. 801
Traffic delays at toll booths. *J. Operations Res. Soc. America*, **2**, 107–38.

Edie, Leslie C. 1955. 802
Expectancy of multiple vehicular breakdowns in a tunnel. *J. Operations Res. Soc. America*, **3**, 513–22.

Edwards, A. W. F. 1958. 803
An analysis of Geissler's data on the human sex ratio. *Ann. Human Genetics*, **23**, 6–15.
Users: Edwards, A. W. F. (1960), Edwards, A. W. F., and Fraccaro, M. (1960).
Class.: OBR:mb-gf:BM

Edwards, A. W. F. 1960a. 804
On a method of estimating frequencies using the negative binomial distribution. *Ann. Human Genetics*, **24**, 313–18.
Review: BA **36** (1961), ab.No. 29327; ZBL **95** (1962), 334.
Class.: NB:pe–m:BM

Edwards, A. W. F. 1960. 805
The meaning of binomial distribution. *Nature*, **186**, 1074.
Review: IJA **1** (1960), 571; ZBL **91** (1962), 305.
Class.: OBR:mb:G

Edwards, A. W. F., and Fraccaro, M. 1958. 806
The sex distribution in the offspring of 5,477 Swedish ministers of religion, 1585–1920. *Hereditas*, **44**, 447–50.
Users: Edward, A. W. F., and Fraccaro, M. (1960), Potthoff, Richard F., and Whittinghill, Maurice (1966).

Edwards, A. W. F., and Fraccaro, M. 1960. 807
Distribution and sequences of sexes in a selected sample of Swedish families. *Ann. Human Genetics*, **24**, 245–52.

Edwards, Allen L. 1964. 808
Expected values of discrete random variables and elementary statistics. Wiley, New York.
Review: P. G. Moore, 1964, *J. Amer. Statist. Assoc.*, **59**, 1317–18.

Edwards, C. B. 1962. 809
Multivariate and multiple Poisson distributions. Ph.D. Thesis, Iowa State Univ.
Users: Patil, G. P., and Bildikar S. (1966a).

Edwards, C. B. and Gurland, J. 1960. 810
A class of distributions applicable to accidents. MRC Tech. Summary Rep. 172, Univ. Wisconsin, pp. 22.
Class.: MNB–NM:gf:A

Edwards, C. B. and Gurland, J. 1961. 811
A class of distributions applicable to accidents. *J. Amer. Statist. Assoc.*, **56**, 503–17.
Review: IJA **4** (1963), 221.
Users: Blischke, W. R. (1965), Haight, F. (1965a).
Class.: MNB-NM:gf:A

Edwards, C. B., and David, H. T. 1962. 812
Poisson limits of bivariate run distributions (Abstract). *Ann. Math. Statist.*, **33**, 1489.

Efron, Bradley. 1965. 813
Increasing properties of Polya frequency functions. *Ann. Math. Statist.*, **36**, 272–79.
Review: MR **30** (1965), 311.
Class.: MI:mi:G

Egert, P. 1954. 814
The mathematical principles of traffic statistics (German.) Tech. und Volkswirt. Ber. Wirtschaftsund Verkehrsminist. Nordrhein-Westfalen, pp. 34.

Egg, K., Rüst, H., and Van der Waerden, B. L. 815
1965.
The level of significance of the χ^2-test for the limiting case of the Poisson distribution. *Zeit. Wahrscheinlichkeitsth.*, **4**, 260–64.
Review: STMA **7** (1966), 517.
Class.: P:ctp:G

Eggenberger, F. and Polya, G. 1923. 816
Über die Statistik verketteter Vorgange. *Z. Angew. Math. Mech.*, **3**, 279–89.
Users: Kitagawa, T. (1940), Kitagawa, T. (1941b), Kitagawa, T., Huruya, S., and Yazima, T. (1942), Feller, W. (1943), von Schelling, H. (1949a), Janossy, L., Renyi, A., and Aczel, J. (1950), Johnson, N. L. (1951), McFadden, J. A. (1955), Martin, L. (1961), Martin, L. (1962b), Bosch, A. J. (1963), Blackwell, D., and Kendall, D. (1964), Kitagawa, T. (1965).

Ehrenberg, A. S. C. 1959. 817
The pattern of consumer purchases. *Appl. Statist.*, **8**, 26–41.
Users: Ehrenberg, A. S. C. (1963), Ehrenberg, A. S. C. (1964), Chatfield, C., Ehrenberg, A. S. C., and Goodhardt, G. J. (1965), Chatfield, C., Ehrenberg, A. S. C., and Goodhardt, G. J. (1966).
Class.: NB:gf-pe-pc-ctp-mb:S

Ehrenberg, A. S. C. 1963. 818
Verified predictions of consumer purchasing patterns. *Commentary*, **10**, 1–6.
Users: Chatfield, C., Ehrenberg, A. S. C., and Goodhardt, G. J. (1965), Chatfield, C., Ehrenberg, A. S. C., and Goodhardt, G. J. (1966).
Class.: NB:gf-mb-pe-rp:S

Ehrenberg, A. S. C. 1964. 819
Estimating the proportion of loyal buyers. *J. Marketing Res.*, **1**, 56–59.
Users: Chatfield, C., Ehrenberg, A. S. C., and Goodhardt, G. J. (1965), Chatfield, C., Ehrenberg, A. S. C., and Goodhardt, G. J. (1966).
Class.: NB:rp:S

Ehrenberg, A. S. C., Goodhardt, G. J., and Chatfield, C.
See Chatfield, C., Ehrenberg, A. S. C., and Goodhardt, G. J. (1965).

Ehrenberg, A. S. C., Goodhardt, G. J., and Chatfield, G.
See Chatfield, C., Ehrenberg, A. S. C., and Goodhardt, G. J. (1966).

Eidelnant, M. I. 1958. 820
Approximate formulas for hypergeometric distribution. *Izv. Akad. Nauk. UzSSR. Ser. Fiz.-Mat. Nauk.*, **5**, 79–92.
Review: MR **22** (1961), 1458; ZBL **84** (1960), 139.

Eisenberg, H. B., Geoghagen, R. R. M., and 821
Walsh, J. E. 1962.
A general probability model for binomial events, with application to surgical mortality. SP-612, System Develop. Corp., California, pp. 11.

Eisenberg, Herbert B., Geoghagen, R. R. M., and 822
Walsh, J. E. 1963.
A general probability model for binomial events, with application to surgical mortality. *Biometrics*, **19**, 152–57.
Review: ZBL 114 (1965), 117.
Class.: B:pe:BM

Eisenberg, Herbert B., Geoghagen, Randolph, R. M., 823
and Walsh, John E. 1966.
A general use of the Poisson approximation for binomial events with application to bacterial endocarditis data. *Biometrics*, **22**, 74–82.

Elashoff, Robert M. 1963. 824
Multivariate two sample problems with discrete and continuous variables. Ph.D. Thesis, Harvard Univ.

Elashoff, Robert M. 1964. 825
Discriminatory analysis with binomial variables. (Abstract). *Biometrics*, **20**, 902–03.

Elfving, G. 1947. 826
On a class of elementary Markoff processes. 10 *Skand. Mat. Kongr. Kobenhavn*, 149–59.
Review: ZBL **30** (1949), 402.

Elias, N. J. 1964. 827
Spare parts reliability nomograph (reliability mathematics corner). *Evaluation Engineering*, **3**, 16.
Review: STMA **6** (1965), 1079.
Class.: P:rp-tc:E

Elkin, J. M. 1953. 828
Estimating the ratio between the proportions of two classes when one is a sub-class of the other. *J. Amer. Statist. Assoc.*, **48**, 128–30.
Review: ZBL **50** (1954), 363.
Class.: MI:pe:G

Elliott, E. O. 1963. 829
Estimates of error rates for codes on burst-noise channels. *Bell System Tech. J.*, **42**, 1977–97.

Ellner, Henry. 1963. 830
Validating results of sampling inspection by attributes. *Technometrics*, **5**, 23–46.
Review: STMA **5** (1964), 152
Class.: B-P-H:a-tp:E

Elmaghraby, Salah, Morse, Norman, and Bechhofer, Robert E.
See Bechhofer, Robert E., Elmaghraby, Salah, and Morse, Norman (1959).

Elveback, L. 1958. 831
Actuarial estimation of survivorship in chronic disease. *J. Amer. Statist. Assoc.*, **53**, 420–40.
Class.: B:pe:BM

Embree, D. G. and Sisojevic, P. 1965. 832
The binomics and population density of *Cyzenis albicans* (Fall.) (Tachinidae: Diptera). *Canad. Entomologist*, **97**, 631–39.

Emik, L. O. 1947. 833
Statistical treatment of counts of trichostronglid eggs. *Biometrics*, **3**, 89–93.

Emik, L. O., and Crabb, W. D.
See Crabb, W. D., and Emik, L. O. (1946).

Emlen, J. T. Jr., Young, H., and Neess, J.
See Young, H., Neess, J., and Emlen, J. T. Jr. (1952).

Engelberg, O. 1965. 834
On some problems concerning a restricted random walk. *J. Appl. Prob.*, **2**, 396–404.

Epstein, B. 1954. 835
Tables for the distribution of the number of exceedances. *Ann. Math. Statist.*, **25**, 762–68.
Review: ZBL **56** (1955), 375.
User: Sarkadi, K. (1957a).
Class.: MI:tc:G

Epstein, J. L. 1966. 836
An upper confidence limit on the product of two binomial parameters (Abstract). *Biometrics*, **22**, 645.

Erdös, P., and Chung, K. L.
See Chung, K. L., and Erdös, P. (1951).

Erdös, P., and Rényi, A. 1959a. 837
On the central limit theorem for samples from a finite population (Hungarian and Russian summaries). *Magyar. Tud. Akad. Mat. Kutató Int. Közl.*, **4**, 49–61.
Review: IJA **1** (1959), 171; MR **21** (1960), 1121.
User: Govindarajulu, Z. (1965).
Class.: MI:a:G

Erdös, P., and Rényi, A. 1959b. 838
On random Graphs I. *Publ. Math. Debrecen*, **6**, 290–97.
Review: IJA **2** (1961), 238.
Class.: P-MI:a:G

Erdös, P. and Rényi, A. 1961. 839
On a classical problem of probability theory (Russian summary). *Magyar Tud. Akad. Mat. Kutató Int. Közl.*, **6**, 215–20.
Review: MR **27** (1964), 167; IJA **3** (1962), 311.
Users: Békéssy, A. (1963), Sevast'Janov, B. A., and Christyakov, V. P. (1966), Békéssy, A. (1964), Baum, Leonard E., and Billingsley, Patrick (1965).
Class.: MI-P:a:G

Erlang, A. K. 1909. 840
Probability calculus and telephone conversations (Danish). *Nyt Tidsskrift for Matematik B*, **20**, 33.

Erlang, A. K. 1917. 841
Solutions of some problems of probability calculus of significance for automatic telephone switchboards (Danish). *Elektroteknikeren*, **13**, 5.

Erlang, A. K. 1918. 842
Solution of some problems in the theory of probabilities of significance in automatic telephone exchanges. *P. O. Elect. Engrs. J.*, **10**, 189-197.
User: Takacs, L. (1956b).

Erickson, R. O., and Stehn, J. R. 1945. 843
A technique for analysis of population density data. *Amer. Midland Nat.*, **33**, 781–87.

Erokhin, U. 1958a. 844
ϵ-Entropy of a discrete random variable (Russian; English summary). *Teor. Verjatnost. i Primenen.*, **3** 103–07.
Review: MR **20** (1959), 126; ZBL **83** (1960), 138.
Class.: MI:mi:G

Erokhin, U. 1958b. 845
ϵ-Entropy of a discrete random variable. *Theor. Probability Appl.*, **3**, 97–100.
Class.: MI:mi:G

Esscher, F. 1932. 846
On the probability function in the collective theory of risk. *Skand. Aktuarietidskr.*, **15**, 175–195.

Eudey, Mark W. 1949. 847
On the treatment of discontinuous random variables. Tech. Rep. 13, Statist. Lab., Univ. California, pp. 52.
Users: Pearson, E. S. (1950), Abdel-Aty, S. H. (1954), David, F. N. (1955a), Buehler, R. J. (1957), Stevens, W. L. (1957), Crow, E. L., and Gardner, R. S. (1959), Blyth, C. R., and Hutchinson, D. W. (1960).
Class.: B-P:pe-ie-tp:G

Evans, D. A. 1953. 848
Experimental evidence concerning contagious distributions in ecology. *Biometrika*, **40**, 186–211.
Review: ZBL **50** (1954), 365.
Users: Moore, P. G. (1954b), Pielou, E. C. (1957), Bliss, C. I., and Owen, A. R. G. (1958), Gurland, J. (1958), Ehrenberg, A. S. C. (1959), Gurland, J. (1959), Jensen, P. (1959), Waters, W. (1959), Waters, W., and Henson, W. R. (1959), Pielou, D. P. (1960a), Katti, S. K. (1960d), Katti, S. K., and Gurland, J. (1961), Khatri, C. G. (1961), Khatri, C. G., and Patel, I. R. (1961), Katti, S. K., and Gurland, J. (1962b), Martin, D. C., and Katti, S. K. (1962a), Pielou, E. C. (1962), Kuno, E. (1963), Ehrenberg, A. S. C. (1963), Berthet, P., and Gerard, G. (1965), Blischke, W. R. (1965), Bowman, K. O., and Shenton, L. R. (1965b), Gurland, J. (1965), Katti, S. K., and Sly, L. E. (1965), Kobayashi, S. (1965), Shenton, L. R., and Myers, R. (1965), Butler, Margaret Sauls (1966).

Evans, D. A., and Philip, U. 1964. 849
On the distribution of Mendelian ratios. *Biometrics*, **20**, 794–817.
Review: STMA **7** (1966), 41.
Class.: B-OBR:mb-gf:BM

Evans, F. C. 1942. 850
Studies of a small mammal population in Baley Wood, Berkshire. *J. Animal Ecol.*, **11**, 182–97.
Users: Hayne, D. W. (1949), Tanton (1965).

Evans, F. C. 1952. 851
The influence of size of quadrat on the distributional patterns of plant populations. *Contr. Lab. Vertebrate Biol. Univ. Mich.*, **54**, 1–15.
Users: Skellam, J. G. (1952), Dice, L. R. (1952), Comita, G. W., and Comita, J. J. (1957), Bliss, C. I.,

and Owen, A. R. G. (1958), Hairston, N. G. (1959), Morisita, M. (1959), Harcourt, D. G. (1961).

Evans, F. C. 1960. 852
Population dispersion. *McGraw-Hill Encyclopedia of Science and Technology Publ.* McGraw-Hill Book Co. pp. 500–03.

Evans, F. C., and Cain, S. A.
See Cain, S. A., and Evans, F. C. (1952).

Evans, F. C., and Cain, S. A. 1952. 853
Preliminary studies on the vegetation of an old-field community in south-eastern Michigan. *Contr. Lab. Vertebrate Biol. Univ. Mich.*, **51**, 1–17.
Users: Dice, L. R. (1952), Berthet, P., and Gerard, G. (1965),

Evans, F. C., and Clark, P. J.
See Clark, P. J., and Evans, F. C. (1954).

Evans, W. D. 1940. 854
Note on the moments of a binomially distributed variate. *Ann. Math. Statist.*, **11**, 106–07.
Review, MR **1** (1940), 247.
Class.: B:m:G

F

Fabian, Vaclav. 1959. 855
Some modifications of interval estimation and choice of number of observations for a binomial random variable. *Appl. Mat.*, **4**, 35–52.
Review: MR **21** (1960), 581; ZBL **83** (1960), 151.
Class.: B:tp-ie-a:G

Faegri, K., and Ottestad, P. 1949. 856
Statistical problem in pollen analysis. *Univ. Bergen, Arbok. Naturvitensk, rekka*, 1948, *No.* 3, 1–29.
Users: Deevey, E. J. Jr., and Potzger, J. E. (1951).

Fairbanks, C. W., and Whittaker, R. H.
See Whittaker, R. H., and Fairbanks, C. W. (1958).

Faleschini, L. 1949. 857
Sullo schèma generale del problèma delle prove ripetute con probabilità indipendenti. *Riv. Italiana di Demografia e Statist.*, **3**, 37–59.

Faleschini, L. 1954. 858
Sullo schèma generale del problèma delle prove ripetute con probabilità dipendenti secondo lo schèma di contagio (o immunità). *Bull. Inst. Internat. Statist.*, **34** (2), 274–82.
Review: ZBL **58** (1958), 119.

Farmer, E., and Chambers, E. G. 1939. 859
A study of accident proneness among motor drivers. Rep. 84, Industrial Health Res. Board, London: H. M. Stationary Office.
Users: Greenwood, M. (1950), Arbous, A., and Kerrich, J. (1951), Bates, G. E. (1955), Edwards, C. B., and Gurland, J. (1960), Edwards, C. B., and Gurland, J. (1961), Haight, F. (1965a).

Farris, D. A. 1958. 860
Diet-induced variation in the free amino acid complex of Sardinops caerules. *J. Conseil*, **23** (2), 235–44.

Feather, N. 1957. 861
Note on the population of segments of a random series which include the origin. *Proc. Phys. Soc. Sect. A*, **70**, 226–27.
Review: ZBL **90** (1961), 349.
Class.: OPR:mb:G

Fein, H., Goldsmith, D., Neyman, J., Prui, P. S., and Buhler, W.
See Buhler, W., Fein, H., Goldsmith, D., Neyman, J., and Puri, P. S. (1965).

Feldman, Dorian and Fox, Martin. 1966. 862
Estimation of the number of trials in the binomial case when several observations are available (Abstract). *Ann. Math. Statist.*, **37**, 541.
Class.: B:pe-a:G

Feller, W. 1943. 863
On a general class of 'contagious' distributions. *Ann. Math. Statist.*, **14**, 389–400.
Review: MR **5** (1944), 209.
Users: Maceda, E. C. (1948), Robbins, H. (1948), Kendall, D. G. (1949), Feller, W. (1949), Shenton, L. R. (1949), Anscombe, F. J. (1950a), Janossy, L., Renyi, A., and Aczel, J. (1950), Johnson, N. L. (1951), Morgan, M. E., MacLeod, P., Anderson, E. O., and Bliss, C. I. (1951), Consael, R. (1952a), Consael, R. (1952b), Schmetterer, L. (1952), Bliss, C. I., and Fisher, R. A. (1953), David, F. N., and Moore, P. G. (1954), Taylor, W. F. (1956), Gurland, J. (1957), McGuire, J., Brindley, T. A., and Bancroft, T. A. (1957), Bliss, C. I., (1958), Crow, E. L. (1958a), Gurland, J. (1958), Skellam, J. G. (1958b), Cohen, A. C. (1959c), Gurland, J. (1959), Patil, G. P. (1959), Waters, W. E., and Hensen, W. R. (1959), Cohen, A. C. (1960c), Edwards, C. B., and Gurland, J. (1960), Katti, S. K. (1960d), Pielou, D. P. (1960a), Pielou, E. C. (1960), Shumway, R., and Gurland, J. (1960b), Shumway, R., and Gurland, J. (1960c), Teicher, H. (1960a), Bardwell, G. E. (1961), Dickerson, O. D., Katti, S. K., and Hofflander, A. E. (1961), Edwards, C. B., and Gurland, J. (1961), Khatri, C. G., and Patel, I. R. (1961), Teicher, H. (1961), Cassie, R. M. (1962), Martin, D. C., and Katti, S. K. (1962a), Martin, D. C., and Katti, S. K. (1962b), Pielou, E. C. (1962), Tsao, C. M. (1962), Yassky, D. (1962), Bosch, A. J. (1963), Blischke, W. R. (1964), Barndorff-Nielson, O. (1965), Blischke, W. R. (1965), Gurland, J. (1965), Katti, S. K., and Sly, L. E. (1965), Sprott, D. A. (1965b), Butler, Margaret Sauls (1966).
Class.: N-NB-COP-GP-MI:mi-mb:G

Feller, W. 1945. 864
On the normal approximation to the binomial distribution. *Ann. Math. Statist.*, **16**, 319–29.
Review: MR **7** (1946), 459; ZBL **60** (1957), 287.
Users: Cheng, T. T. (1949), Walsh, J. E. (1952), Nicholson, W. L. (1956), Brown, R. (1963), Hannan, J., and Harkness, W. (1963), Mott-Smith, J. C. (1964), Govindarajulu, Z. (1965).
Class.: B:a:G

Feller, W. 1948. 865
On probability problems in the theory of counters. *Courant Anniversary Volume*, pp. 105–15.
Users: Domb, C. (1950), Haight, F. (1959a), Jewell W. S. (1960).

Feller, W. 1949. 866
On the theory of stochastic processes, with particular reference to applications. *Proc. Berkeley Symp. Math. Statist. Probab.* pp. 403–32.
Review: ZBL **39** (1951), 137.
Users: Bates, G. E., and Neyman, J. (1952b), Redheffer, R. M. (1953), Goldberg, S. (1954), Fitzpatrick, R. (1958), Martin, L. (1961), Martin, L. (1962b), Mellinger, G., Sylwester, D., Gaffey, W., and Manheimer, D. (1965).

Feller, W. 1951. 867
The problem of *n* liars and Markov chains. *Amer. Math. Monthly*, **58**, 606–08.

Feller, W. 1957. 868
An introduction to probability theory and its applications. John Wiley and Sons, Inc., New York, Vol. 1. 2nd edition.

Feller, W., and Chung, Kai Lai.
See Chung, Kai, Lai, and Feller, W. (1949).

Ferguson, T. S. 1962. 869
A characterization of the geometric distribution. (Abstract). *Ann. Math. Statist.*, **33**, 1207.
Users: Patil, G. P., and Seshadri, V. (1964), Shanfield, F. (1964).

Ferguson, T. S. 1965. 870
A characterization of the geometric distribution. *Amer. Math. Monthly*, **72**, 256–260.
Class.: G:osp:G

Fernandez de Troconiz, A. 1964. 871
A study of the efficiency and design of quality control in industrial processes (Spanish). *Estadist. Española No.*, **25**, 5–19.
Review: STMA **7** (1966), 369.

Ferreri, C.. 1958. 872
An important new model of dependence (Italian). *Ann. Fac. Econ. Com. Palermo*, **12**, 221–38.
Review: IJA **1** (1959), 25.
Class.: MI:mb:G

Ferreri, C. 1959. 873
On some properties and on the application of a new stochastic model of dependence (Italian). *Statistica*, **19**, 63–71.
Review: IJA **1** (1959), 189.

Ferreri, C. 1962. 874
Some statistical aspects and applications of a probability distribution (Italian). *Ann. Fac. Econ. Comm. Palermo*, **16**, 207–27.
Review: IJA **4** (1963), 496.
Users: Crawford, Gordon B. (1966).
Class.: B-NB:sqc:G

Ferschl, F. 1959. 875
Queues models (German). *Math., Tech., Wirtschaft*, **3**. 98–102.
Review: IJA **1** (1959), 299.

Ferschl, F. 1961. 876
Queues with grouped input (German). *Unternehmensforschung*, **5**, 185–96.
Review: IJA **3** (1962), 635.
Class.: P-MI:pc:G

Fidler, J. E. 1961. 877
Properties of the multivariate Poisson distribution. Master's Thesis, Pennsylvania State Univ.

Fieller, E. C., and Pearson, K.
See Pearson, K., and Fieller, E. C. (1933).

Figa'talamanca, M, 1960. 878
On a new criterion for the addition of binomial coefficients (Italian). *Atti. Riun. Sci., Soc. Ital. Statist.*, 253–63.
Review: STMA **6** (1965), 435.
Class.: B:mi:G

Filipello, F., Baker, G. A., Amerine, M. A., and Roessler, E. B.
See Baker, G. A., Amerine, M. A., Roessler, E. B., and Filipello, F. (1960).

Finch, P. D. 1959a. 879
Balking in the queueing system GI/M/1 (Russian Summary). *Acta Math. Acad. Sci. Hungar.*, **10**, 241-47.
Review: ZBL **86** (1961), 338; IJA **2** (1961), 385.

Finch, P. D. 1959b. 880
On the distribution of queue size in queueing problems. *Acta Math. Acad. Sci. Hungar.*, **10**, 327–36.
Review: IJA **2** (1961), 386.
User: Finch, P. D. (1960b).
Class.: MI:pc:G

Finch, P. D. 1960a. 881
Deterministic customer impatience in the queuing system. GI/M/I. *Biometrika*, **47**, 45-52.
Review: IJA **2** (1961), 748.
Class.: MI:pc:G

Finch, P. D. 1960b. 882
On the transient behaviour of a simple queue. *J. Roy. Statist. Soc. Ser. B.*, **22**, 277–84.
Review: IJA **2** (1961), 561.
Class.: MI:pc:G

Finch, P. D. 1961-62. 883
On the busy period in the queuing system GI/G/I. *J. Austral. Math. Soc.*, **2**, 217–28.
Review: IJA **3** (1962), 636.
User: Shanbhag, D. N. (1965).
Class.: MI:pc:G

Finney, D. J. 1947a. 884
Errors of estimation in inverse sampling. *Nature*, **160**, 195–96.

Finney, D. J. 1947–49. 885
The truncated binomial distribution. *Ann. Eugenics*, **14**, 319–28.
Review: MR **11** (1950), 42.
Users: Moore, P. G. (1954a), Rider, P. R. (1955b), Roy. J., and Mitra, S. K. (1957), Hartley, H. O. (1958), Patil, G. P. (1959), Patil, G. P. (1962d), Patil, G. P. (1962e), Rider, P. R. (1962b), Newell, D. (1965).
Class.: TCB:pe-c-tc:G-BM

Finney, D. J. 1948. 886
Transformation of frequency distributions. *Nature*, **162**, 898.
Review: PA **23** (1949), Ab.No. 5867; ZBL **32** (1950), 292.
Class.: MI:anovat:G

Finney, D. J. 1949. 887
On a method of estimating frequencies. *Biometrika*, **36**, 233–34.
Users: Finney, D. J. (1947-9), Anscombe, F. J. (1949b), Sandelius, M. (1950), Sandelius, M. (1951a), Sandelius, M. (1951b), Sandelius, M. (1951c), Nadler, J. (1960), Bartko, J. J. (1961b), Bartko, J. J. (1962), Patil, G. P. (1963c), Knight, W. (1965).
Class.: NB:pe:G

Finney, D. J., and Leander, E. K.
See Leander, E. K., and Finney, D. J. (1956)

Finney, D. J., and Varley, G. C. 1955. 888
An example of the truncated Poisson distribution. *Biometrics*, **11**, 387–94.
Review: PA **30** (1956), Ab. No. 3738.
Users: Roy, J., and Mitra, S. K. (1957), Cohen, A. C. (1959e), Patil, G. P. (1959), Irwin, J. O. (1959), Cohen, A. C. (1960b), Cohen, A. C. (1960e), Cohen, A. C. (1961a), Patil, G. P. (1962e), Ito, Nakamura, Kundo, Miyashita and Nakamura (1962).
Class.: TCP:pe-gf:BM

Finucan, H. M. 1964. 889
The mode of multinomial distribution. *Biometrika*, **51**, 513–17.
Review: MR **30** (1965), 804; ZBL **124** (1966), 111; STMA **6** (1965), 436.

Fischer, Carl H. 1942. 890
A sequence of discrete variables exhibiting correlation due to common elements. *Ann. Math. Statist.*, **13**, 97–101.
Review: MR **4** (1943), 23; ZBL **60** (1957), 309.
Class.: OBR:pc-m:G

Fischer, János, and Csáki, Peter
See Csáki, Peter, and Fischer, János (1963).

Fisher, G. R., and Haight, Frank A.
See Haight, Frank A., and Fisher, G. R. (1966).

Fisher, R. A. (1923-25). 891
Theory of statistical estimation. *Proc. Cambridge Philos. Soc.*, **22**, 700–25.
Users: Johnson, N. L. (1951), Claringbold, P. J. (1955), Edwards, A. W. F. (1960), Bardwell, G. E. (1961), Rao, C. R. (1961), Rao, C. R. (1963), Patil, G. P., and Wani, J. K. (1965), Sprott and Kalbfleish (1965), Tweedie, M. C. K. (1965), Plackett, R. L. (1966).
Class.: B-P-MI:pe:G

Fisher, R. A. 1939. 892
Stage of development as a factor influencing the variance in the number of offspring, frequency of mutants and related quantities. *Ann. Eugenics*, **9**, 406–08.
Review: ZBL **22** (1940), 252.
Users: Moran and Watterson (1959).

Fisher, R. A. 1941. 893
The negative binomial distribution. *Ann. Eugenics*, **11**, 182-87.
Review: MR **4** (1943), 26; ZBL **60** (1957), 296.
Users: Fisher, R. A., Corbet, A. S., and Williams, C. B. (1943), Wise, M. E. (1946), Bowen, M. F. (1947), Anscombe, F. J. (1949a), Kendall, D. G. (1949), Shenton, L. R. (1949), Anscombe, F. J. (1950a), Oakland, G. B. (1950), Shenton, L. R. (1950), Wadley, F. M. (1950), Morgan, M. E., MacLeod, P., Anderson, E., and Bliss, C. I. (1951), Sichel, H. S. (1951), Binet F. E. (1953), Bliss, C. I., and Fisher, R. A. (1953), David, F. N., and Moore, P. G. (1954), Moore, P. G. (1954b), Rider, P. R. (1955b), Sampford, M. R. (1955), Smith, J. H. G., and Ker, J. W. (1957), Bliss, C. I., and Owen, A. R. G. (1958), Brass, W. (1958a), Shenton, L. R. (1958), Sprott, D. A. (1958), Cohen, A. C. (1959d), Gurland, J. (1959), Patil, G. P. (1959), Waters, W. (1959), Edwards, A. W. F. (1960a), Katti, S. K. (1960d), Patil, G. P. (1960c), Harcourt, D. G. (1960), Bartko, J. J. (1961b), Harcourt, D. G. (1961), Rider, P. R. (1961), Katti, S. K., and Gurland, J. (1962b), Martin, D. C., and Katti, S. K. (1962b), Martin, D. C., and Katti, S. K. (1962c), O'Carroll, F. M. (1962), Rider, P. R. (1962a), Tsao, C. M. (1962), Blischke, W. R., (1965), Cohen, A. Clifford, Jr. (1965), Berthet, P., and Gerard, G. (1965), Bowman, K. O., and Shenton, L. R., (1965a), Bowman, K. O. and Shenton, L. R. (1965b), Harcourt, D. G. (1965), Katti, S. K., and Sly, L. E. (1965), Martin, D., and Katti, S. (1965), Patil, G. P., and Wani, J. K. (1965), Shenton, L. R., and Myers, R. (1965).
Class.: NB:pe-gf:G

Fisher, R. A. 1950. 894
The significance of deviations from expectation in a Poisson series. *Biometrics*, **6**, 17–24.
Users: Skellam, J. G. (1952), Binet, F. E. (1953), Rao, C. R., and Chakravarti, I. (1956), Fitzpatrick, R. (1958), Williams, E. J. (1961b), Cassie, R. M. (1962) Cox, D. R. (1962), Patil, G. P., and Wani, J. K. (1965), Potthoff, Richard, and Whittinghill, Maurice (1966).
Class.: P:id-gf:G
Notes: Reprinted, *Biometrics*, **20**, 265–72.

Fisher, R. A. 1954. 895
The analysis of variance with various binomial transformations. *Biometrics*, **10**, 130–39.
Review: MR **16** (1955), 271; ZBL **58** (1958), 127.
Users: Claringbold, P. J. (1955), Naylor, A. F. (1964), Patil, G. P., and Wani, J. K. (1965).
Class.: B:anovat:G
Notes: Discussion, *Biometrics*, **10**, 140–51.

Fisher, R. A. 1964. 896
The significance of deviations from expectation in a Poisson series (Reprinted from *Biometrics*, **6**, 17–24). *Biometrics*, **20**, 265–72.
Review: ZBL **124** (1966), 99.

Fisher, R. A., and Bliss, C. I.
See Bliss, C. I., and Fisher, R. A. (1953).

Fisher, R. A., Corbet, A. S., and Williams, C. B. 897
1943.
The relation between the number of species and the number of individuals in a random sample from an animal population. *J. Animal Ecol.*, **12**, 42–58.
Users: Harrison, J. L. (1943). Riddell, W. J. B. (1944), Williams, C. B. (1944), Williams, C. B. (1944a), Williams, C. B. (1944b), Williams, C. B. (1946), Williams, C. B. (1947), Williams, C. B. (1947b), Williams, C. B. (1947c), Jones, P. C. T., Mollison, J. E.,

and Quenouille, M. H. (1948), Kendall, D. G. (1948), Preston, F. W. (1948b), Anscombe, F. J. (1949a), Archibald, E. E. A. (1949a), Dunn, E. R. (1949), Kendall, D. G. (1949), Simpson, E. H. (1949), Quenouille, M. H. (1949), Williams, C. B. (1949), Anscombe, F. J. (1950a), Williams, C. B. (1952), Bliss, C. I., and Fisher, R. A. (1953), Brian, M. V. (1953), Good, I. J. (1953), Myers, E., and Chapman, V. (1953), Williams, C. B. (1953), Good, I. J., and Toulmin, G. (1956), Good, I. J. (1957b), Herdan, G. (1957), MacArthur, R. (1957), Bliss, C. I. (1958), Gurland, J. (1959), Hairston, N. G. (1959), Waters, W. (1959), Dahl, E. (1960), Darwin, J. H. (1960), Harcourt, D. G. (1960), Pielou, D. P. (1960a), Bartko, J. J. (1961b) Gower, J. C. (1961), Harcourt, D. G. (1961), Taylor, C. J. (1961), Patil, G. P. (1962b), Sironomey, G. (1962), Tsao, C. M. (1962), Clark, P. J., Eckstrom, P. T., and Linden, L. C. (1964), Nelson, W. C., and David, H. A. (1964), Williamson, E., and Bretherton, M. (1964), Berthet, P., and Gerard, G. (1965), Chatfield, C., Ehrenberg, A. S. C., and Goodhardt, G. J. (1965), Gurland, J. (1965), Haight, F. (1965b), Haight, F., and Reirchenbach, H. (1965), Chatfield, C., Ehrenberg, A. S. C., and Goodhardt, G. J. (1966), Patil, G. P., and Shorrock, R. W. (1966), Van Heerden, D. F. I., Gonin, H. T. (1966).
Class.: LS:mb-pe-gf:BM

Fisher, R. A., Thornton, H. G., and Mackenzie, W. A. 1922. 898
The accuracy of the plating method of estimating the density of bacterial populations. *Ann. Appl. Biol.*, **9**, 325–59.
Users: Lancaster, H. O. (1952), Tippett, L. H. C. (1955), Turner, M. E., and Eadie, G. S. (1957), Bennett, B. M. (1959), Lancaster, H. O. (1961), Bennett, B. M. (1962a), Thomson, K. H. (1962), Nicholson, W. L. (1963), Potthoff, Richard, and Whittinghill, Maurice (1966b).

Fisz, M. 1953a. 899
The limiting distributions of sums of arbitrary independent and equally distributed r-point random variables. *Studia Math.*, **14**, 111–23.
Class.: MI-OPR:pc-a:G

Fisz, M. 1953b. 900
The limiting distribution of the difference of two Poisson random variables (Polish; English and Russian summaries). *Z. Mat.*, **1**, 41–45.
Review: MR **15** (1954), 138; ZBL **52** (1955), 140.
Users: Fisz, M. (1955a), Shanfield, F. (1964), Govindarajulu, Z. (1965).

Fisz, M..1954a 901
The limiting distributions of the multinomial distribution. *Studia Math.*, **14**, 272–75.
Review: MR **16** (1955), 834; ZBL **64** (1956), 131.
Users: Fisz, M. (1955b), Govindarajulu, Z. (1965).
Class.: M-OPR:a:G

Fisz, M. 1954b. 902
Accuracy of an asymptotical formula (Polish; English and Russian summaries). *Z. Mat.*, **2**, 62–66.
Review: MR **16** (1955), 1034; ZBL **57** (1956), 111.
User: Govindarajulu, Z. (1965).

Fisz, M. 1955a. 903
Refinement of a probability limit theorem and its application to Bessel functions (Russian summary). *Acta. Math. Acad. Sci. Hungar.*, **6**, 199–202.
Review: MR **19** (1958), 184.
User: Shanfield, F. (1964).
Class.: P:a;G

Fisz, M. 1955. 904
The limiting distribution of a function of two independent random variables and its statistical application. *Colloq. Math.*, **3**, 138–46.
Review: ZBL **64** (1956), 382.
User: Fisz, M. (1955a).

Fisz, M. 1955b. 905
A limit theorem for a modified Bernoulli scheme. *Studia Math.*, **15**, 80-83.
Review: MR **17** (1956), 634.
User: Govindarajulu, Z. (1965).
Class.: M-OPR:a:G

Fisz, M. 1956. 906
Die Grenzverteilungen der Multinomialverteilung. *Ber. Tagung Wahrschrechnung Math. Statist., Berlin*, pp. 51–53.
Review: MR **18** (1957), 605; ZBL **70** (1957), 138.

Fisz, M. 1958. 907
Characterization of some probability distributions. *Skand. Aktuartidskr.*, **41**, 65–67.

Fisz, M. 1962. 908
On an inequality due to Gatti and Birnbaum. *Metron*, **22** (1-2), 110–12.
Review: STMA **5** (1964), 330; ZBL **115** (1965), 138.
Class.: MI:m:G

Fisz, M., and Urbanik, K. 1955. 909
The analytical characterization of the composed non-homogeneous Poisson process. *Bull. Acad. Polon. Sci. Classe III*, **3**, 149–50.
Review: ZBL **67** (1958), 366.

Fisz, M., and Urbanik, K. 1956. 910
Analytical characterization of a composed, non-homogeneous Poisson process. *Studia Math.*, **15**, 328–36.
Review: ZBL **70** (1957), 365.

Fitch, E. R., and Hartley, H. O.
See Hartley, H. O., and Fitch, E. R. (1951).

Fitzpatrick, R. 1958. 911
The detection of individual differences in accident susceptibility. *Biometrics*, **14**, 50–68.
Users: Edwards, C. B., and Gurland, J. (1960), Edwards, C. B., and Gurland, J. (1961), Subrahmaniam, K. (1964), Subrahmaniam, K. (1966a).
Class.: P-NB-MP-NM:cm:A

Fitzpatrick, R., Vasilas, J. N., and Paterson, R. O. 912
1953.
Personnel and training factors in fighter aircraft accidents. Res. Lab. Rep. 37, Hum. Factors Operation, U.S.A.F.

Fleischer, David A., and Bliss, C. I.
See Bliss, C. I., and Fleischer, David A. (1964).

Fletcher, N. T., Nelson, A. C. Jr., and Williams, J. S.
See Nelson, A. C. Jr., Williams, J. S., and Fletcher, N. T. (1963).

Florek, K., Marczewski, E., and Ryll-Nardzewski, C. 1953. 913
Remarks on the Poisson stochastic process. *Studia Math.*, **13**, 122–29.
User: Prekopa, A. (1957a).

Fluckiger, Fritz A., Tripp, Clarence, A., and Weinberg, George H.
See Weinberg, George H., Fluckiger, Fritz A., and Tripp, Clarence A. (1960).

Foerster, R. D. 1936. 914
An investigation of the life history and propagation of the sockeye salmon (*Oncorhynchus nerka*) at Cultus Lake, British Columbia, No. 5. The life history cycle of the 1926 year class with artificial propagation involving the liberation of free-swimming fry. *J. Fisheries Res. Board Canada*, **2** (3), 311–33.
User: Ricker, W. E. (1937).

Folger, J. 1953. 915
Confidence limits tables for small samples of binomially distributed data. Memo. 6, Hum. Resources Res. Inst., Maxwell A. F. Base, Alabama.

Folks, J. L., Pierce, D. A., and Stewart, C. 1965. 916
Estimating the fraction of acceptable product. *Technometrics*, 7, 43–50.
Review: STMA **7** (1966), 73.
Class.: H-B-P:pe-sqc:G

Forbes, T. W. 1951. 917
Statistical techniques in the field of traffic engineering and traffic research. *Proc. 2nd Berkeley Symp. Math. Statist. Prob.* pp. 603–25.

Forsyth, C. H. 1924. 918
Simple derivations of the formulas for the dispersion of a statistical series. *Amer. Math. Monthly*, **31**, 190–96.
Class.: B:anovat:G

Forsythe, H. Y. Jr., and Gyrisco, George C. 1961. 919
Determining the appropriate transformation of data from insect control experiments for use in the analysis of variance. *J. Econ. Ent.*, **54**, 859–61.

Fortet, R. 1950. 920
Calcul des probabilités. Centre National de la Recherche Scientifique, Paris.

Fortet, R. 1956. 921
Random distributions with an application to telephone engineering. *Proc. 3rd Berkeley Symp. Math. Statist. Prob.*, **2**, 81–88.

Fortet, R., and Blanc-Lapierre, A.
See Blanc-Lapierre, A., and Fortet, R. (1955).

Foster, F. G. 1952a. 922
A Markov chain derivation of discrete distributions. *Ann. Math. Statist.*, **23**, 624–27.
Review: MR **14** (1953), 663; ZBL **47** (1953), 376.
Class.: P-NB-MI:pc-mi:G

Foster, F. G., and Nyunt, K. M. 1961. 923
Queues with batch departures. I. *Ann. Math. Statist.*, **32**, 1324–32.
Review: ZBL **117** (1965), 135.

Foster, F. G., and Perera, A. G. A. D. 1965. 924
Queues with batch arrivals. II (Russian summary). *Acta Math. Acad. Sci. Hungar.*, **16**, 275–87.
Review: MR **32** (1966), 802.
Class.: MI:pc:G

Foster, R. E., and Johnson, A. L. S. 1963. 925
Assessments of pattern, frequency distribution, and sampling of forest disease in Douglas Fir plantations. Publ. 1011, Canada Dept. Forestry, Studies in Forest Pathology, pp. 52.

Foster, R. E., and Pielou, E. C.
See Pielou, E. C., and Foster, R. E. (1962).

Fournier, G. 1952. 926
Sur la distribution moyenne des nombres premiers. *C. R. Acad. Sci. Paris*, **234**, 2411–13.

Fox, Martin, and Feldman, Dorian.
See Feldman, Dorian, and Fox, Martin (1966).

Fraccaro, M., and Edwards, A. W. F.
See Edwards, A. W. F., and Fraccaro, M. (1958).

Fraccaro, M., and Edwards, A. W. F.
See Edwards, A. W. F., and Fraccaro, M. (1960).

Fracker, S. B., and Brischle, H. A. 1944. 927
Measuring the local distribution of Ribes. *Ecology*, **25**, 283–303.
Users: Cole, L. C. (1946a), Whitford, P. B. (1949), Curtis, J., and McIntosh, R. (1950), Wadley, F. M. (1950), Cain, S. A., and Evans, F. C. (1952), Thomson, G. W. (1952), Beall, G., and Rescia, R. R. (1953), Bliss, C. I., and Fisher, R. A. (1953), Waters, W. E., and Henson, W. R. (1959), Morisita, M. (1959).
Class.: P-N:h-gf:BM

Frame, J. S. 1945. 928
Mean deviation of the binomial distribution. *Amer. Math. Monthly*, **52**, 377–79.
Review: MR **7** (1946), 128.
Users: Crow, E. L. (1958b), Bardwell, G. E. (1961), Kamat, A. R. (1965).
Class.: B:m:G

Franchetti, S. 1943. 929
Probabilità di errore nelle distribuzioni di Poisson. *Pont. Acad. Sci. Comment.*, **7**, 697–708.
Review: MR **10** (1949), 200.

Frankel, E. T. 1950. 930
A calculus of figurate numbers and finite differences. *Amer. Math. Monthly*, **57**, 14–25.
Class.: MI:mi:G

Franken, Peter. 1963. 931
Approximation durch Poissonsche Prozesse. *Math. Nachr.*, **26**, 101–14.
Review: ZBL **123** (1966), 354; STMA **5** (1964), 952.

Franken, P. 1964a. 932
Zur Approximation durch Poissonsche Verteilungen. *Abh Deutsch. Akad. Wiss. Berlin Kl. Math. Phys. Tech. No.* 4, 59–61.
Review: MR **32** (1966), 303.
Class.: P:a:G

Franken, Peter. 1964. 933
Approximation der Verteilungen von Summen unabhängiger nichtnegativer ganzzahliger Zufallsgrössen

durch Poissonsche Verteilungen. *Math. Nachr.*, **27**, 303–40.
Review: MR **30** (1965), 303; STMA **6** (1965), 438.
Class.: MI-P:ctp:G

Franklin, Norman L., and Bennett, Carl A.
See Bennett, Carl A., and Franklin, Norman L. (1954).

Fraser, D. A. S. 1963. 934
On the sufficiency and likelihood principles. *J. Amer. Statist. Assoc.*, **58**, 641–47.
Class.: MI:pe:G

Fraser, D. A. S., and Guttman, Irwin. 1952. 935
Bhattacharyyá bounds without regularity assumptions. *Ann. Math. Statist.*, **23**, 629–32.
Class.: B:pe:G

Fraser, D. A. S., Thomas, J. B., and Ghent, A. W.
See Ghent, A. W., Fraser, D. A. S., and Thomas, J. B. (1957).

Fraser, G. R. 1963. 936
Parental origin of the sea chromosomes in the XO and XXY karyotypes in man. *Ann. Human Genetics*, **36**, 297–304.

Frechet, Maurice. 1939. 937
Événements compatibles et probabilités fictives. *C. R. Acad. Sci. Paris*, **208**, 1703–05.
Review: ZBL **21** (1940), 421.

Frechet, Maurice. 1944. 938
Les systèmes d'événements et le jeu des rencontres. *Rev. Mat. Hisp.-Amer.*, **4** (4), 95–126.
Review: MR **6** (1945), 231.

Frechet, Maurice. 1947. 939
The general relation between the mean and the mode for a discontinuous variate. *Ann. Math. Statist.*, **18**, 290–93.
Review: MR **9** (1948), 45.
Class.: MI:mi:G

Frechet, Maurice. 1948-49. 940
Sur l'estimation statistique. *Ann. Soc. Polon. Math.*, 207–13.
Review: MR **11** (1950), 42.
Class.: B:ie:G

Freedman, David D. 1962. 941
Poisson processes with random arrival rate. *Ann. Math. Statist.*, **33**, 924–29.
Review: STMA **6** (1965), 281.
User: Freedman, David A. (1965).
Class.: COP:mi-pe:G

Freedman, D. A. 1962b. 942
Asymptotic properties of Bayes solutions: discrete case (French). *C. R. Acad. Sci. Paris*, **255**, 1851–52.
Review: STMA **6** (1965), 393
User: Freedman, D. A. (1963a).

Freedman, David A. 1963a. 943
On the asymptotic behavior of Bayes' estimates in the discrete case. *Ann. Math. Statist.*, **34**, 1386–1403.
Review: MR **28** (1964), 341; STMA **5** (1964), 842.
Users: Freedman, David A. (1965).

Freedman, David A. 1963b. 944
L'urne de Bernard Friedman. *C. R. Acad. Sci. Paris*, **257**, 3809.
Review: MR **28** (1964), 506; ZBL **117** (1965), 140.

Freedman, David A. 1965a. 945
On the asymptotic behavior of Bayes estimates in the discrete case II. *Ann. Math. Statist.*, **36**, 454–56.
Class.: MI:pe-a:G

Freedman, David A. 1965b. 946
Bernard Friedman's urn. *Ann. Math. Statist.*, **36**, 956–70.
Review: MR **31** (1966), 307.
Class.: MI-PO:mi:G

Freeman, M. F., and Tukey, J. W. 1950. 947
Transformations related to the angular and the square root. *Ann. Math. Statist.*, **21**, 607–11.
Review: MR **12** (1951), 344; ZBL **39** (1951), 353.
Users: Blom, G. (1954), David, F. N. (1955a), Raff, M. S. (1956), Wishart, J. (1956), Gupta, S. S., and Sobel, M. (1958), Birnbaum, A., and Healy, W. C. (1960), Laubscher, N. F. (1960), Laubscher, N. F. (1961).
Class.: B-P:anovat:G

Freeman, D., and Weiss, L. 1964. 948
Sampling plans which approximately minimise the maximum expected sample size. *J. Amer. Statist. Assoc.*, **59**, 67–88.
Review: STMA **7** (1966), 157: ZBL **117** (1965), 147.
Class.: B:tp:G

Freimer, M., Gold, B., and Tritter, A. L. 1959. 949
The Morse distribution. *IRE Trans.*, **5**, 25–31.
Review: MR **22** (1961), 1711.

Freudenberg, K 1951. 950
Die Grenzen für die Anwendbarkeit des Gesetzes der kleinen Zahlen. *Metron*, **16** (1–2), 285–310.

Freune, J. E. 1950. 951
The degree of stereotypy. *J. Amer. Statist. Assoc.*, **45**, 265–69.
Review: ZBL **37** (1951), 365.
Class.: M:tp:G

Freund, J. E. 1951a. 952
Some observations on Laplace's rule of succession and Perk's rule of indifference. *J. Inst. Actuar.*, **77**, 439–44.

Freund, J. E. 1951b. 953
The transfer distribution. *Math. Mag.*, **25**, 63–66.
Class.: OBR:mb-mi:G

Freund, J. E. 1956. 954
Some methods of estimating the parameters of discrete heterogeneous populations. *J. Roy. Statist. Soc. Ser. B*, **18**, 222–26.
Review: MR **18** (1957), 606; ZBL **73** (1960), 149.
Class.: MI:pe:G

Freund, J. E. 1957. 955
Some results on recurrent events. *Amer. Math. Monthly*, **64**, 718–20.

Friede, G. 1950a. 956
Pascalische Verteilungen, Confidence-und Fiduzialschluss. *Mitteilungsbl. Math. Statist.*, **2**, 171–83.
Review: ZBL **39** (1951), 149.

Friede, G. 1950b. 957
Über reziprozitatsbezierhungen in der Wahrscheinlichkeitsrechung (German; English, French and Russian Summaries). *Z. Angew. Math. Mech.*, **30**, 65–72.
Review: MR **11** (1950), 604; ZBL **36** (1951), 207.

Friedman, Bernard. 1949. 958
A simple urn model. *Comm. Pure Appl. Math.*, **2**, 59–70.
Review: ZBL 33 (1950), 71.

Friedman, E. A. Miller, G. A., and Norman, E. B.
See Miller, G. A., Newman, E. B., and Friedman, E. A. (1958).

Friedman, L. 1956. 959
A competitive-bidding strategy. *Operations Res.*, **4**, 104–12.

Friedman, Milton, and Anderson, T. W.
See Anderson, T. W., and Friedman, Milton (1960).

Frisch, Ragnar. 1925. 960
Recurrence formulae for the moments of the point binomial. *Biometrika*, **17**, 165–71.
User: Patil, G. P. (1959).
Class.: B:m:G

Frisch, R. 1932. 961
On the use of difference equations in the study of frequency distributions. *Metron*, **10**, 35–59.
Users: Linder, A. (1935), Katz, L. (1945).

Froggatt, P. 1965. 962
Accident proneness: The concept and its development (Abstract). *Biometrics*, **21**, 776.

Froggatt, P., and Cresswell, W. L.
See Cresswell, W. L., and Froggatt, P. (1963).

Fu, Yumin, Srinath, Mandyam D., and Yen, Andrew T. 1965. 963
A note on multivariate generating functions and applications to discrete stochastic process. *SIAM Review*, **7**, 31–41.
Review: MR **30** (1965), 807.

Fuchs, A., and Roby, N. 1960. 964
Sur le domaine d'attraction de la loi de Poisson. *Publ. Inst. Statist. Univ. Paris*, **9**, 392–94.
Review: MR **24**A (1962), 447; ZBL **103** (1964), 119.
Class.: P:a:G

Fuchs, C. F., and Dàvid, H. T. 1965a. 965
Poisson limits of multivariate run distributions. *Ann. Math. Statist.*, **36**, 215–25.
Review: MR **30** (1965), 668.
Class.: P-MP:a:G

Fuchs, C. F., and David, H. T.
See David, H. T., and Fuchs, C. F. (1965).

Fucks, W. 1965. 966
Statistical distributions with bounded components (German). *Z. Physik.*, **45** (4), 520–33.
Review: SA **59**A (1956), Ab.No. 6290.

Fujii, Koichi. 1965. 967
A statistical model of the competition curve (Japanese summary). *Res. Population Ecol.*, **7**, 118–25.
Class.: P-B:mb:BM

Fujisawa, Takehisa, and Homma, Tsuruchiuo.
See Homma, Tsuruchiuo, and Fujisawa, Takehisa (1964).

Fukao, Kichiji. 1963. 968
Several problems in designing sequential sampling inspection plans by control point ($p_{0.5}$) and relative slope ($h_{0.5}$). *Rep. Statist. Appl. Res. Un. Japan. Sci. Engrs.*, **10**, 257–80.
Review: ZBL **117** (1965), 148.

Furry, W. H. 1939. 969
On fluctuation phenomena in the passage of high energy electrons through lead. *Phys. Rev. 2nd Ser.*, **52**, 569–81.
User: Bartko, J. J. (1961b).
Class.: NB:pc:P

Furukawa, N., and Kudo, A.
See Kudo, A., and Furukawa, N. (1958).

G

Gabriel, K. R. 1959. 970
The distribution of the number of successes in a sequence of dependent trials. *Biometrika*, **46**, 454–60.
Review: IJA **1** (1960), 492.
Class.: OBR:pc-m-osp:G-P

Gafarian, A. V., and Ancker, C. J.
See Ancker, C. J., and Gafarian, A. V. (1961).

Gafarian, A. V., and Ancker, C. J., Jr.
See Ancker, C. J., Jr., and Gafarian, A. V. (1963).

Gaffey, W. R., Manheimer, D. I., Mellinger, G. D., and Sylwester, D. L.
See Mellinger, G. D., Sylwester, D. L., Gaffey, W. R., and Manheimer, D. I. (1965).

Galliher, H. P., Morse, Philip M., and Simond, M. 1959. 971
Dynamics of two classes of continuous-review inventory systems. *Operations Res.*, **7**, 362–84.
User: Jewell, W. S. (1960).

Gallot, S. 1966. 972
Assymptotic absorption probabilities for a Poisson process. *J. Appl. Prob.*, **3**, 445–52.

Gamkrelidze, N. G. 1964a. 973
On a local limit theorem for lattice random variables. *Theor. Probability Appl.*, **9**, 662–64.
Class.: MI:a:G

Gamkrelidze, N. G. 1964. 974
On the local limit theorem for lattice distributions (Russian, English summary). *Teor. Verojatnost. i Primenen*, **9**, 733–46.
Review: MR **31** (1966), 307.
Class.: MI:a:G

Gamkrelidze, N. G. 1966. 975
On the speed of convergence in the local limit theorem for lattice distributions (Russian, English summary). *Teor. Verojatnost. i Primenen.*, **11**, 129–40.
Class.: MI:a:G

Gani, J. 1962. 976
The extinction of a bacterial colony by phages: a branching process with deterministic removals. *Biometrika*, **49**, 272–76.
Review: ZBL **108** (1964), 327.

Gani, J. 1963. 977
Models for a bacterial growth process with removals. *J. Roy. Statist. Soc. Ser. B.*, **25**, 140–419.
Review: MR **30** (1965), 499; STMA **6** (1965), 285; ZBL **118** (1965/66), 350.
User: Gani, J. (1963).

Gani, J. 1965. 978
On a partial differential equation of epidemic theory. *Biometrika*, **52**, 617–22.
Review: STMA **7** (1966), 239.
User: Siskind, V. (1965).
Class.: MI:pc:BM

Gani, J., and Prabhu, N. U. 1959. 979
The time-dependent solution for a storage model with Poisson input. *J. Math. Mech.*, **8**, 653–63.
Users: Phatarfod, R. M. (1963), Takács, L. (1963a).

Gardner, R. S., and Crow, E. L.
See Crow, E. L., and Gardner, R. S. (1959).

Gart, J. J. 1959. 980
An extension of the Cramér-Rao inequality. *Ann. Math. Statist.*, **30**, 367–80.
Review: IJA **1** (1959), 214.
User: Mosimann, J. E. (1962).
Class.: MI-B-NB:pe:G

Gart, J. J. 1962. 981
Approximate confidence intervals for the relative risk. *J. Roy. Statist. Soc. Ser. B.*, **24**, 454–63.
Review: IJA **4** (1963), 434.
Class.: B:ie:G

Gart, J. J. 1963a. 982
Poisson regression: confidence limits and tests of the model (abstract). *Ann. Math. Statist.*, **34**, 1618.

Gart, J. J. 1963b. 983
A median test with sequential application. *Biometrika*, **50**, 55–62.
User: Bol'shev, L. N. (1964a).
Class.: NH:ctp:G

Gart, J. J. 1964a. 984
Approximate distribution theory for some common discrete distributions (Abstract). *Ann. Math. Statist.*, **35**, 927.

Gart, J. J. 1964b. 985
The analysis of Poisson regression with an application in virology. *Biometrika*, **51**, 517–21.
Review: STMA **6** (1965), 1140; MR **32** (1966), 1133.
Class.: P:rp:G-BM

Garwood, F. 1936. 986
Fiducial limits for the Poisson distribution. *Biometrika*, **28**, 437–42.
Users: Sandelius, M. (1951a), Cox, D. R. (1953), Birnbaum, A. (1954b), Crow, E. L., and Gardner, R. S. (1959), Blyth, C. R., and Hutchinson, D. W. (1961).
Class.: P:ie:G

Garwood, F. 1940. 987
The application of the theory of probability to the operation of vehicular controlled traffic signals. *J. Roy. Statist. Soc. Suppl.*, **7**, 65–77.
Users: Haight, F. (1959), Miller A. J. (1961), Weiss, G. H. (1963).

Garwood, F. 1947. 988
The variance of the overlap of geometrical figures with reference to a bombing problem. *Biometrika*, **34**, 1–17.
Class.: B-P-N-MI:mi:O

Garwood, F., and Tanner, J. C. 1956. 989
Accident studies before and after road changes. Final Rep., Publ. Works Municip. Services Congr., pp. 329–54.
User: Tanner, J. C. (1958).

Gause, G. F. 1930. 990
Studies on the ecology of the Orthoptera. *Ecology*, **11**, 307–25.
Class.: B:gf:BM

Geiringer, Hilda. 1937. 991
Sur les variables aléatoires arbitrairement liées (convergence vers la loi de Poisson). *C. R. Acad. Sci. Paris*, **204**, 1856–57.
Review: ZBL **16** (1937), 364.
User: Geiringer, H. (1938).

Geiringer, Hilda. 1938. 992
On the probability theory of arbitrarily linked events. *Ann. Math. Statist.*, **9**, 260–71.
Review: ZBL **20** (1939), 380.
Class.: MI:a:G

Geiringer, Hilda. 1959–60. 993
On a limit theorem leading to a compound Poisson distribution. *Math. Z.*, **72**, 229–34.
Review: MR **22** (1961), 1458; IJA **2** (1961), 16; ZBL **91** (1962), 307.
Class.: COP:a:G

Geis, A. D. 1955. 994
Trap response of the cottontail rabbit and its effect on censusing. *J. Wildlike Managem.*, **19**, 466–72.

Geisler, M. A., Mirkovich, A. R., and Young, J. W. T.
See Youngs, J, W. T., Geisler, M. A., and Mirkovich, A. R. (1954).

General Electric Company. 1962. 995
Tables of the individual and cumulative terms of Poisson distribution. Defense Systems Dept., Princeton, N.J.

Gengerelli, J. A. 1947. 996
A method of binomial analysis (Abstract). *Amer. Psychologist*, **2**, 405.
Review: PA **22** (1948), Ab. No. 936.

Gengerelli, J. A. 1948. 997
A binomial method for analyzing psychological functions. *Psychometrika*, **13**, 69–77.
Class.: B:mi:BM

Gentry, C. J. 1957. 998
On a negative binomial distribution. Master's thesis Univ. Florida Gainesville, Florida.

Geohagen, R. R. M., Walsh, J. E., and Eisenberg, H. B.
See Eisenberg, H. B., Geoghagen, R. R. M., and Walsh, J. E. (1962).

Geohagen, R. R. M., Walsh, J. E., and Eisenberg, H. B.
See Eisenberg, H. B., Geoghagen, R. R. M., and Walsh, J. E. (1963).

Geoghagen, Randolph R. M., Walsh, John E., and Eisenberg, Herbert B.
See Eisenberg, Herbert B., Geoghagen, Randolph, and Walsh, John E. (1966).

Georgescu-Roegen, Nicholas. 1959. 999
On the extrema of some statistical coefficients. *Metron*, **19** (3–4), 38–45.
Review: MR **22** (1961), 348.
Class.: MI:osp:G

Geppert, Maria-Pia. 1942. 1000
Die Anwendung der Pascalschen Verteilung auf das Bayessehe Ruckschlub problem. *Arch. Math. Wirtsch. u. Sozialforschg.*, **8**, 129–37.
Review: ZBL **27** (1943), 234.

Gerard, G., and Berthet, P.
See Berthet, P., and Gerard, G. (1965).

Gerlough, D. L., and Schuhl, A. 1955. 1001
Poisson and Traffic. I. The use of the Poisson distribution in highway traffic. D.L.G. II. The calculation of probabilities and the distribution of traffic on two lane roads. ENO Foundation for Highway Control, Saugatuck Conn.
Notes: Schuhl's paper is a revised translation from *Travaux*, **39**, (1955), pp. 24.

Gerrits, H. J. M., and Van Elteren, Ph.
See Van Elteren, Ph., Gerrits, H. J. M. (1961).

Gerstenkorn, Tadeusz. 1960. 1002
On the formula for the moments of the Poisson distribution. *Bull. Soc. Sci. Lettres Lodz*, **11** (10), 1–6.
Review: MR **25** (1963), 1081.

Gerstenkorn, T. 1962. 1003
On the generalized Poisson distributions. *Prace Lodzkie towarz nauk Wydz* 3 *No.* 85.

Ghent, A. W. 1960. 1004
A study of the group-geeding behaviour of larvae of the Jack Pine Sawfly, *Neodiprion Pratti Banksianae Roh* (German summary). *Behaviour*, **16**, 110–48.
Class.: B-P:gf:BM

Ghent, A. W. 1963. 1005
Studies of regeneration in forest stands devastated by the Spruce budworm. *Forest Sci.*, **9**, 295–310.

Ghent, A. W., Fraser, D. A., and Thomas, J. B. 1006
1957.
Studies of regeneration in forest stands devastated by the spruce budworm. I. Evidence of trends in forest succession during the first decade following budworm devastation. *Forest Sci.*, **3**, 184–208.
Class.: MI:mk:BM

Ghizzetti, Aldo and Roma, Maria Sofia. 1964. 1007
Calcolo di alcune probabilità relative alla subdividione casuale di un segmento in n parti. (French, English, Spanish and German summaries). *Giorn. Ist. Ital. Attuari*, **27**, 140–50.
Class.: MI:mb-mi:G

Ghosal, A. 1962. 1008
Finite dam with negative binomial input. *Austral. J. Appl. Sci.*, **13**, 71–74.

Ghosh, S. P. 1966. 1009
Polychotomy sampling. *Ann. Math. Statist.*, **37**, 657–65.

Ghurye, S. G. 1949. 1010
Transformations of a binomial variate for the analysis of variance. *J. Indian Soc. Agri. Statist.*, **2**, 94–109.
Review: MR **11** (1950), 528; BA **25** (1951), Ab. No. 22539.
Class.: B:anovat:G

Gilbert, E. J., Steck, G. P., Young, D. A., and Owen, D. B.
See Owen, D. B., Gilbert, E. J., Steck, G. P., and Young, D. A. (1959).

Gilbert, E. N. 1959. 1011
Random graphs. *Ann. Math. Statist.*, **30**, 1141–44.
Review: IJA **1** (1960), 354.
Class.: MI:mi:G

Gilbert, E. N. 1965. 1012
The probability of covering a sphere with N circular. *Biometrika*, **52**, 323–30.
Review: STMA **7** (1966), 30.
Class.: MI:mi:G

Gilbert, E. N., and Pollak, H. O. 1957. 1013
Coincidences in Poisson patterns. *Bell System Tech. J.*, **36**, 1005–33.

Gilbert, John P. 1966. 1014
Recognizing the maximum of a sequence. *J. Amer. Statist. Assoc.*, **61**, 35–73.
Class.: P-MI:srp:G

Gilbert, N. E. G. 1956. 1015
Likelihood function for capture-recapture samples. *Biometrika*, **43**, 488–89.
Review: MR **18** (1957), 426.
Class.: H-B:pe:BM

Gilchrist, H., Greenberg, B. G., and Larsh, J. E.
See Greenberg, B. G., Larsh, J. E., and Gilchrist, H. (1952).

Gini, Corrado. 1907. 1016
La legge dei piccoli numeri. *Giorn. Economisti*, **35**, 758–75.

Gini, Corrado. 1908. 1017
Su la legge die piccoli numeri e la regolarità dei fenomeni rari. *Giorn. Economisti*, **37**, 649–92.

Gini, C. 1942. 1018
Su la misura della dispersione e la sua relazione con l'indice di connessione. *Atti. Soc. Ital. Statist.* pp. 467–70.
Review: ZBL **27** (1943), 412.

Girshick, M. A. 1946a. 1019
Contributions to the theory of sequential analysis. *Ann. Math. Statist.*, **17**, 123–43.
Users: Birnbaum, A. (1954b), Bharucha-Reid, A. T. (1958), Cox, D. R. (1963).
Class.: B-P:tp:G

Girshick, M. A. 1946b. 1020
Contributions to the theory of sequential analysis, II, III. *Ann. Math. Statist.*, **17**, 282–98.
Users: Waler, A. M. (1950), Breakwell, J. V. (1954).
Class.: B:tp:G

Girshick, M. A., Mosteller, F., and Savage, L. J. 1021
1946.
Unbiased estimates for certain binomial sampling problems with applications. *Ann. Math. Statist.*, **17**, 13–23.
Users: Girshick, M. A. (1946b), Wolfowitz, J. (1946), Savage, I. R. (1947), McGarthy, P. J. (1947), Wolfowitz, J. (1947), Plackett, R. L. (1948), Craig, C. C. (1953c), Dvoretsky, W. (1953), Sirazhdinov, S. H. (1956), Bennett, B. M. (1957a), Armitage, P. (1958), Guttman, I. (1958), deGroot, M. H. (1959), Patil, G. P. (1959), Edwards, A. W. F. (1960a), Nadler, J. (1960), Nelson, A. C., Jr., Williams, J. S., and Fletcher, N. T. (1963), Patil, G. P. (1963a), Wasan, M. T. (1964), Knight, W.

(1965), Tweedie, M. C. K. (1965), Wasan, M. T. (1965), Bhat, B. R., and Kulkarni, N. V. (1966b).
Class.: B:pe:G-E

Girshick, M. A., Rubin, H., and Sitgreaves, R. 1955. 1022
Estimates of bounded relative error in particle counting. *Ann. Math. Statist.*, **26**, 276–85.
Class.: P:ie-se:G-P

Gittelsohn, A. M. 1960. 1023
A model for the analysis of the distribution of qualitative characters in sibships. *Biometrics*, **16**, 534–46.
Review: IJA **2** (1961), 487.
Class.: B-NB-P:mb-pe:BM

Giveen, Samuel M. 1963. 1024
A taxicab problem with time-dependent arrival rates. *SIAM Review*, **5**, 119–27.

Gladkov, B. V., Shcheglova, M. V., and Bol'shev, L. N.
See Bol'shev, L. N., Gladkov, B. V., and Shcheglova, M. V. (1961).

Glasgow, J. P. 1953. 1025
The estimation of animal populations by artificial predation and the estimation of populations. *J. Animal Ecol.*, **22**, 32–46.
Class.: G:gf:BM

Glasgow, M. O., and Greenwood, R. E.
See Greenwood, R. E., and Glasgow, M. O. (1950).

Glasser, G. G., and Deming, W. E.
See Deming ,W. E., and Glasser, G. G. (1959).

Glasser, Gerald J. 1961. 1026
An unbiased estimator for powers of the arithmetic mean. *J. Roy. Statist. Soc. Ser. B.*, **23**, 154–59.
Class.: B:pe:G

Glasser, Gerald J. 1962a. 1027
On estimators for variances and covariances. *Biometrika*, **49**, 259–62.

Glasser, Gerald J. 1962b. 1028
Minimum variance unbiased estimators for Poisson probabilities. *Technometrics*, **4**, 409–18.
Review: MR **26** (1963), 1078; ZBL **102** (1963), 357; STMA **5** (1964), 66.
Class.: P:pe:G

Glaven, Frederik. 1961a. 1029
Pointsgivning i spil med Poissonfordeiling (Danish, English summary). *Nordisk Math. Tidskr.*, **9**, 167–68.
Review: ZBL **107** (1964), 373.

Glaven, Frederik. 1961b. 1030
Points in games with binomial distribution (Danish; English summary on p. 144). *Nordisk Mat. Tidskr.*, **9**, 109–16.
Review: ZBL **107** (1964), 373.

Gnedenko, B. 1939. 1031
To the theory of the domains of attraction of stable laws (Russian). *Uchenye Zapiski Moskovskii Gosudarstvennyi Univ. Math.*, **30**, 61–81.

Gnedenko, B. V. 1941. 1032
On the theory of Geiger-Muller counters. *Akad. Nauk. SSR. Zhurnal Eksper. Teoret. Fiz.*, **11**, 101–06.
Review: MR **7** (1946), 18.
Users: Malmquist, S. (1947), Neyman, J. (1949).

Gnedenko, B. V. 1959. 1033
On a generalization of Erlang's formula. *Dopovidi Akad. Nauk Ukrain, RSR*, 347–50.
Review: ZBL **113** (1965), 340; IJA **2** (1961), 566.
Class.: MI:pc:G
Notes: English translation in *Selected translations in Math. Statist. and Prob.*, **3**, 337–40 (1962).

Godfrey, G. K. 1954. 1034
Tracing field voles (*Microtus agretis*) with a Geiger-Muller counter. *Ecology*, **35**, 5–10.
Class.: B:gf:BM

Godini, G. 1965. 1035
A queueing problem with a single service station, (Rumanian). *Studii cercetări matematice*, **17**, 777–80.
Review: STMA **7** (1966), 624.
Class.: MI:pc:G

Godini, G. 1965a. 1036
A queueing problem with heterogeneous service (Rumanian). *Studii cercetări matematice*, **7**, 765–75.
Review: STMA **7** (1966), 623.
Class.: MI:pc:G

Goen, R. L., and Tate, R. F.
See Tate, R. F., and Goen, R. L. (1958).

Gold, B., Tritter, A. L., and Freimer, M.
See Freimer, M., Gold, B., and Tritter, A. L. (1959).

Gold, L. 1957. 1037
Generalized Poisson distributions. *Ann. Inst. Statist. Math. Tokyo*, **9**, 43–47.
Review: ZBL **80** (1959), 119.
User: Khatri, C. G. (1960).

Gold, R. Z. 1962. 1038
On comparing multinomial probabilities. SAM Rep. 62–81, U.S.A.F. pp. 13.
Review: PA **37** (1963), Ab. No. 5926; STMA **5** (1964), 357.
Users: Patil, G. P., and Bildikar, S. (1966a).
Class.: M:ie:G-BM

Goldberg, S. 1954. 1039
Probability models in biology and engineering. *J. Soc. Indust. Appl. Math.*, **2**, 10–19.
Review: ZBL **59** (1965/66), 135.
Class.: MI:mb:BM-E-A

Goldman, A. J. 1957. 1040
The probability of a saddlepoint. *Amer. Math. Monthly*, **64**, 729–30.
Users: Cover, Thomas M. (1966).
Class.: B:mi:G

Goldman, A. J., and Bender, B. K. 1962. 1041
The first run preceded by a quota. *J. Res. Nat. Bur. Standards Sect. B.*, **66**, 77–89.
Review: IJA **4** (1963), 403.
Class.: GBDP:pc:G

Goldsmith, D., Neyman, J., Puri, P. S., Buhler, W., and Fein, H.
See Buhler, W., Fein, H., Goldsmith, D., Nayman, J., and Puri, P. S. (1965).

Gonin, H. T. 1944. 1042
Curve fitting by means of the orthogonal polynomials in binomial statistical distributions. *Trans. Roy. Soc. South Africa*, **30**, 207–15.

Review: MR **6** (1945), 234.
Class.: B:mi:G

Gonin, H. T. 1961. 1043
The use of orthogonal polynomials of the positive and negative binomial frequency functions in curve fitting by Aitken's method. *Biometrika*, **48**, 115–23. Correction: **48**, 476.
Users: Van Heerden, D. F. I., and Gonin, H. T. (1966).
Class.: B-NB:mi:G

Gonin, H. T. 1966. 1044
Poisson and binomial frequency surfaces. *Biometrika*, **53**, 617–19.

Gonin, H. T., and Aitken, A. C.
See Aitken, A. C., and Gonin, H. T. (1935).

Gonin, H. T., and Van Heerden, D. F. I.
See Van Heerden, D. F. I., and Gonin, H. T. (1966).

Good, I. J. 1949. 1045
The number of individuals in a cascade process. *Proc. Cambridge Philos. Soc.*, **45**, 360–63.
User: Good, I. J. (1951), Good, I. J. (1955a), Good, I. J. (1960).
Class.: MI:pc:G

Good, I. J. 1951. 1046
Random motion on a finite Abelian group. *Proc. Cambridge Philos. Soc.*, **47**, 756–762.
User: Good, I. J. (1953).

Good, I. J. 1953a. 1047
The population frequencies of species and the estimation of population parameters. *Biometrika*, **40**, 237–64.
Review: ZBL **51** (1954), 371.
Users: Simon, H. A. (1955), Good, I. J. (1956), Good, I. J., and Toulmin, G. (1956), Good, I. J. (1957a), Good, I. J. (1957b), Herdan, G. (1957), Simon, H. (1960), Williamson, E., and Bretherton, M. (1964), Martiz, J. S. (1966).
Class.: MI:pe:G

Good, I. J. 1953. 1048
The serial test and other tests for randomness. *Proc. Cambridge Philos. Soc.*, **49**, 276–84.
Review: ZBL **51** (1954), 362.
Users: Good, I. J. (1955).

Good, I .J. 1955. 1049
The likelihood ratio test for Markoff chains. *Biometrika*, **42**, 531–33.
Users: Cox, D. R. (1958b), Good, I. J. (1963).
Class.: MI:pc:G
Notes: Correction; *Biometrika*, **44**, 301.

Good, II. J. 1955a. 1050
The joint distribution for the sizes of the generations in a cascade process. *Proc. Cambridge Philos, Soc.*, **51**, 240–42.
Users: Good, I. J. (1960), Good, I. J. (1965).

Good, I. J. 1955b. 1051
Conjectures concerning the Mersenne numbers. *MTAC*, **9**, 120–21.

Good, I. J. 1956. 1052
On the estimation of small frequencies in contingency tables. *J. Roy. Statist. Soc. Ser. B*, **18**, 113–24.
Class.: MI:pe:G

Good, I. J. 1957a. 1053
Saddle-point methods for the multinomial distribution. *Ann. Math. Statist.*, **28**, 861–81.
Review: ZBL **91** (1962), 143.
Users: Watson, G. S. (1959), Good, I. J. (1961a), Good, I. J. (1961b), Good, I. J. (1961c), Williams, E. J. (1961b), Kullback, S., Kupperman, M., and Ku, H. H. (1962), Young, D. H. (1962), Good, I. J. (1963), Good, I. J. (1965), Plackett, R. L. (1966).
Class.: M-OMR:c-a:G

Good, I. J. 1957b. 1054
Distribution of work frequencies. *Nature*, **179**, 595.
Users: Herdan, G. (1958), Rider, P. R. (1965).
Class.: MI:mb:L

Good, I. J. 1957c. 1055
On the serial test for random sequencies. *Ann. Math. Statist.*, **28**, 262–64.
Class.: MI:mi:G

Good, I. J. 1958. 1056
Legendre polynomials and trinomial random walks. *Proc. Cambridge Soc.*, **54**, 39–42.
User: Good, I. J. (1965).

Good, I. J. 1960. 1057
Generalizations to several variables of Lagrange's expansion, with applications to stochastic processes. *Proc. Cambridge Philos. Soc.*, **56**, 367–80.
Users: Good, I. J. (1961a), Good, I. J. (1965).

Good, I. J. 1961a. 1058
The frequency count of a Markov chain and the transition to continuous time. *Ann. Math. Statist.*, **32**, 41–48.
User: Iyer, P. V. K. (1963).
Class.: MI:pc-m:G

Good, I. J. 1961b. 1059
The multivariate saddlepoint method and chi-squared for the multinomial distribution. *Ann. Math. Statist.*, **32**, 535–48.
Review: MR **25** (1963), 143; IJA **4** (1963), 225.
Users: Good, I. J. (1963), Good, I. J. (1965).
Class.: MI-M:c-a:G

Good, I. J. 1961c. 1060
An asymptotic formula for the differences of the powers at zero. *Ann. Math. Statist.*, **32**, 249–56.
Class.: MI:pc-m:G

Good, I. J. 1962. 1061
Proofs of some 'binomial' identities by means of MacMahon's 'master theorem.' *Proc. Cambridge Philos. Soc.*, **58**, 161–62.

Good, I. J. 1963. 1062
Maximum entropy for hypothesis formulation, especially for multinomial contingency tables. *Ann. Math. Statist.*, **34**, 911–34.
Class.: MI-P-B:mi-ctp:G

Good, I. J. 1965. 1063
The generalization of Lagrange's expansion, and the enumeration of trees. *Proc. Cambridge Philos. Soc.*, **61**, 499–517.

Good, I. J., and Toulmin, G. H. 1956. 1064
The number of new species, and the increase in population coverage, when a sample is increased. *Biometrika*, **43**, 45–63.
Review: ZBL **70** (1957), 144.
Class.: MI:mi:G

Goodall, D. W. 1952. 1065
Quantitative aspects of plant distribution. *Biol. Rev.*, **27**, 194–245.
Users: MacFadyen, A. (1953), Clark, P. H., and Evans, F. C. (1954a), Pielou, E. C. (1957), Waters, W. E., and Henson, W. R. (1959).

Goodall, D. W. 1962. 1066
Bibliography of statistical plant sociology. *Excerpta Bot. Sec. B*, **4**, 253–322.

Goodhardt, G. J., Chatfield, C., and Ehrenberg, A. S. C.
See Chatfield, C., Ehrenberg, A. S. C., and Goodhardt, G. J. (1965).

Goodhardt, G. J., Chatfield, C., and Ehrenberg, A. S. C.
See Chatfield, C., Ehrenberg, A. S. C., and Goodhardt, G. J. (1966).

Goodman, L. A. 1949. 1067
On the estimation of the number of classes in a population. *Ann. Math. Statist.*, **20**, 572–79.
Users: Good, I. J. (1953b), Goodman, L. A. (1953), Deming, W. E., and Glasser, G. G. (1959), Goodman, L. A. (1961).
Class.: MI:pe:G

Goodman, L. A. 1952a. 1068
On the Poisson-Gamma distribution problem. *Ann. Inst. Statist. Math. Tokyo*, **3**, 123–25.
Review: MR **14** (1943), 189; ZBL **48** (1953), 107.
Users: Haight, F. (1959a), Haight, F. (1965b).

Goodman, L. A. 1952b. 1069
Serial number analysis. *J. Amer. Statist. Assoc.*, **47**, 622–34.
Review: MR **14** (1953), 777.
Users: Goodman, L. A. (1954), Goodman, L. A. (1961).
Class.: MI:tp-ie:S-E

Goodman, L. A. 1953. 1070
Sequential sampling tagging for population size problems. *Ann. Math. Statist.*, **24**, 56–69.
Review: MR **14** (1953), 776; ZBL **50** (1954), 148.
Users: Chapman, D. G. (1954), Darroch, J. N. (1958).
Class.: MI:pe-ie:G

Goodman, L. A. 1954. 1071
Some practical techniques in serial number analysis. *J. Amer. Statist. Assoc.*, **49**, 97–112.
Class.: MI:mi:G

Goodman, L. A. 1961. 1072
Snowball sampling. *Ann. Math. Statist.*, **32**, 148–70.
Review: MR **23A** (1962), 259; ZBL **99** (1963), 142; IJA **3** (1962), 29.
Class.: MI:mb-pe:G

Goodman, Leo A. 1961. 1073
Some possible effects of birth control on the human sex ratio. *Ann. Human Genetics*, **25**, 75–81.
Class.: OBR:mi:BM

Goodman, Leo A. 1963. 1074
Some possible effects of birth control on the incidence of disorders and on the influence of birth order. *Ann. Human Genetics*, **27**, 41–52.
Class.: MI:mi:BM

Goodman, L. A. 1964. 1075
Simultaneous confidcnce intervals for contrasts among multinomial populations. *Ann. Math. Statist.*, **35** 716–25.
Review: MR **28** (1964), 1066.
User: Goodman, L. A. (1965).
Class.: M:ie-pc:G

Goodman, L. A. 1965. 1076
On simultaneous confidence intervals for multinomial proportions. *Technometrics.*, **7**, 247–54.
Review: STMA **7** (1966), 505.

Gordon, R. D. 1939. 1077
Estimating bacterial populations by the dilution method. *Biometrika*, **31**, 167–80.
Review: ZBL **23** (1941), 147.

Gotoh, A., Miyashita, K., and Ito, Y.
See Ito, Y., Gotoh, A., and Miyashita, K. (1960).

Goudriaan, J. 1962. 1078
Maxima for the one-sided probability of exceedance and for the magnitude of the mean exceedance of discrete and continuous unimodal distributions (Dutch). *Statistica Neerlandica*, **16**, 215–20.
Review: IJA **4** (1963), 55.
Class.: MI:mi:G

Gould, H. W. 1963. 1079
Note on two binomial coefficient sums found by Riordan. *Ann. Math. Statist.*, **34**, 333–35.
Review: MR **26** (1963), 10.
Users: Govindarajulu, Z., and Suzuki, Y. (1963).
Class.: B:mi:G

Govindarajulu, Z. 1962a. 1080
Inverse moments (Abstract). *Ann. Math. Statist.*, **33**, 1208.

Govindarajulu, Z. 1962b. 1081
Approximations to the first two inverse moments of the decapitated negative binomial variable. Statist. Lab. Rep. 1058, Case Inst. Tech., pp. 7.
Class.: TCNB:m-a:G

Govindarajulu, Z. 1962c. 1082
First two moments of the reciprocal of a positive hypergeometric variable. Statist. Lab. Rep. 1061, Case. Inst. Tech., pp. 44.
Class.: H:m:G

Govindarajulu, Z. 1962d. 1083
Recurrence relations for the first two inverse moments of the positive binomial variable. Statist. Lab. Rep. 1063, Case Inst. Tech. pp. 6.
Class.: TCB:m:G

Govindarajulu, Z. 1962e. 1084
The reciprocal of the decapitated negative binomial variable. *J. Amer. Statist. Assoc.*, **57**, 906–13.
Review: STMA **7** (1966), 43; STMA **6** (1965), 827; ZBL **112** (1965), 110.
Class.: TCNB:m-a:G
Notes: Corrigenda; *J. Amer. Statist. Assoc.*, **58**, 1162.

Govindarajulu, Z. 1963. 1085
Recurrence relations for the inverse moments of the positive binomial variable. *J. Amer. Statist. Assoc.*, **58**, 468–73.
Review: ZBL **115** (1965), 370; STMA **6** (1965), 828.
Class.: TCB:m:G

Govindarajulu, Z. 1964. 1086
The first two moments of the reciprocal of the positive hypergeometric variable. *Sankhyā Ser. B*, **26**, 217–36.
Review: MR **32** (1966), 309.
Class.: H:m:G

Govindarajulu, Zakkula. 1965a. 1087
Normal approximations to the classical discrete distributions. *Sankyhā Ser. A*, **27**.

Govindarajulu, Z. 1965. 1088
Normal approximations to the classical discrete distributions. *Classical and Contagious Discrete Distributions*, Ed. G. P. Patil, Statistical Publishing Society, Calcutta and Pergamon Press, pp. 79–108. Reprinted in *Sankhyā A*, **27** (1966)
Class.: B-P-NB-H-GBDP-GNB-PS:a:G

Govindarajulu, Z., and Suzuki, Yukio. 1963. 1089
A note on an identity involving binomial coefficients. *Ann. Statist. Math. Toyko*, **15**, 83–85.
Review: STMA **6** (1965), 396; ZBL **117** (1965), 373.
Class.: B:mi:G

Gower, J. C. 1961. 1090
A note on some asymptotic properties of the logarithmic series distribution. *Biometrika*, **48**, 212–15.
Review: STMA **5** (1964), 29.
Users: Nelson, W. C., and David, H. A. (1964), Bliss, C. I. (1965).
Class.: LS:c:G

Gower, J. C., and Leslie, P. H.
See Leslie, P. H., and Gower, J. C. (1958).

Grab, E. L., and Savage, I. R. 1954. 1091
Tables of the expected value of 1/x for positive Bernoulli and Poisson variables. *J. Amer. Statist. Assoc.* **49**, 169–77.
Review: MR **15** (1954), 636.
Users: David, F. N., and Johnson, N. L. (1956), Patil, G. P. (1959), Mendenhall, W., and Lehman, E. H. (1960), Patil, G. P. (1961b), Govindarajulu, Z. (1962c) Govindarajulu, Z. (1962d), Rider, P. R. (1962b), Govindarajulu, Z. (1963), Shanfield, F. (1964).
Review: ZBL **55** (1955), 127.
Class.: TCB-TCP:tc:G
Notes: Addenda: *J. Amer. Statist. Assoc.*, **49**, 906.

Graca, J. G., and White, R. F.
See White, R. F., and Graca, J. G. (1958).

Graham, A. J., Bitter, B. A., Hopkins, D. E., New, W. D., Dudley, F. H., Bushland, R. C., and Baumhover, A. H.
See Baumhover, A. H., Graham, A. J., Bitter, B. A., Hopkins, D. E., New, W. D., Dudley, F. H., and Bushland, R. C. (1955).

Grainger, R. M., and Reid, D. B. W. 1954. 1092
Distribution of dental caries in children. *J. Dental Res.*, **33**, 613–23.
Users: Katti, S. K. (1960c), Katti, S. K. (1960d).
Class.: B-COB-P-NB:gf:BM

Green, C. V. 1933. 1093
Differential growth in the crania of mature mice. *J. Mammalogy*, **14**, 122–31.

Greenberg, B. G., Larsh, J. E., and Gilchrist, H. 1094
1952.
A study of the distribution and longevity of adult T. Spiralis in immunized and non-immunized mice. *J. Elisha Mitchell Sci. Soc.*, **68**, 1–11.
Class.: MI:mi:BM

Greenhood, E. Russell, Jr. 1940. 1095
A detailed proof of the chi-square test of goodness fit. Harvard Univ. Press, Oxford Univ. Press, 61 pp.
Review: ZBL **25** (1942), 199.

Greenwood, J. A. 1962. 1096
Guide to tables in mathematical statistics. Princeton Univ. Press.
Review: MR **27** (1964), 827.

Greenwood, J. A., and Greville, T. N. E. 1939. 1097
On the probability of attaining a given standard deviation ratio in an infinite series of trials. *Ann. Math. Statist.*, **10**, 297–98.
Review: ZBL **21** (1940), 338.
Class.: B:a:G

Greenwood, M. 1931. 1098
On the statistical measure of infectiousness. *J. Hyg. Camb.*, **31**, 336–51.
Users: Bailey, N. T. J. (1953), Bailey N. T. J. (1954), Irwin, J. O. (1954), Bailey, N. T. J. (1955c), Taylor, W. F. (1956).

Greenwood, M. 1946. 1099
The statistical study of infectious diseases. *J. Roy. Statist. Soc. Ser. A*, **109**, 85–110.

Greenwood, M. 1950. 1100
Accident proneness. *Biometrika*, **37**, 24–29.
User: Adelstein, A. M. (1952).
Class.: NB:mb:A

Greenwood, M., and Yule, G. U. 1920. 1101
An inquiry into the nature of frequency distributions representative of multiple happenings. *J. Roy. Statist. Soc. Ser. A*, **83**, 255–79.
Users: Muench, H, (1936), Haldane, J. B. S. (1941), Kitagawa, T., Huruya, S., and Yazima, T. (1942), Feller, W. (1943), Cernuschi, F., and Castagnetto, (1946), Cole, L. C. (1946a), Kendall, D. G. (1948), Skellam, J. G. (1948), Blum, Mintz (1949), Feller, W. (1949), Kendall, D. G. (1949), Skellam, J. G. (1949), Anscombe, F. J. (1950a), Greenwood, M. (1950), Arbous, A., and Kerrich, J. (1951), Blum, M. L., and Mintz, A. (1951), Johnson, N. L. (1951), Sichel, H. S. (1951), Consael, R. (1952b), Maguire, B. A., Pearson, E. S., and Wynn, A. H. A. (1952), Maguire, B. A., Pearson, E. S., and Wynn, A. H. A. (1952), Skellam, J. G. (1952), Bliss, C. I., and Fisher, R. A. (1953), Good, I. J. (1953b), Irwin, J. O. (1953), Arbous, A. G., and Sichel, H. S. (1954b), Goldberg, S. (1954), Grainger, R. M., and Reid, D. B. (1954), Rutherford, R. S. G. (1954), Bates, G. E. (1955), Dandekar, V. M. (1955), Duker, S. (1955), Okamoto, M. (1955), Sampford, M. R. (1955), Taylor, W. (1956), Taylor, W. F. (1956),

Gurland, J. (1957), Johnson, N. L. (1957a), Fitzpatrick, R. (1958), Cohen, A. C. (1959d), Gurland, J. (1959), Patil, G. P. (1959), Edwards, C. B., and Gurland, J. (1960), Katti, S. K. (1960d), Patil, G. P. (1960c), Pielou, E. C. (1960), Pielou, D. P. (1960a), Bartko, J. J. (1961b), Edwards, C. B., and Gurland, J. (1961), Martin, L. (1961), Martin, C. D., and Katti, S. K. (1962a), Yassky, D. (1962), Subrahmaniam, K. (1964), Williamson, E., and Bretherton, M. (1964), Blischke, W. R. (1965), Cohen, A Clifford, Jr., (1965), Chatfield, C., Ehrenberg, A S. C., and Goodhardt, G. J. (1965), Mellinger, G., Sylwester, D., Gaffey, W., and Manheimer, D. (1965), Neyman, J. (1965), Chatfield, C., Ehrenberg, A. S. C., and Goodhardt, G. J. (1966), Kemp, A. W., and Kemp, C. D. (1966), Dubey, S. D. (1966a), Patil, G. P., and Bildikar, S. (1966a), Potthoff, Richard, and Whittinghill, Maurice (1966b), Subrahmaniam, K. (1966a).

Greenwood, M., and Woods, H. M. 1919. 1102
The incidence of industrial accidents upon individuals with special reference to multiple accidents. Rep. 4, Indust. Fatigue Res. Board, London: H. M. Stationery Office.
Users: Blum, Mintz (1949), Arbous, A., and Kerrich, J. (1951), Fitzpatrick, R. (1958), Edwards, C. B., and Gurland, J. (1960,) Edwards, C. B., and Gurland, J. (1961), Haight, F. (1965a), Haight F. (1965b).

Greenwood, R. E., and Glasgow, M. O. 1950. 1103
Distribution of maximum and minimum frequencies in a sample drawn from a multinomial distribution. *Ann. Math. Statist.*, **21**, 416–24.
Review: MR **12** (1951), 428; ZBL 38 (1951), 90.
Users: Sandelius, M. (1952a), Kozelka, R. M. (1956), Good, I. J. (1957a), Patil, G. P., and Bildikar, S. (1966a).
Class.: B-M:mi:G

Gregory, G,, and Benson, F.
See Benson, F., and Gregory, G. (1961).

Grenander, U. 1957. 1104
On heterogeneity in non-life insurance. Part I. *Skand. Aktuarietidskr.*, **40**, 71–84.

Grenander, U. 1957. 1105
On heterogeneity in non-life insurance. Part II. *Skand. Aktuarietidskr.*, **40**, 153–79.

Greville, T. N. E., and Greenwood, J. A.
See Greenwood, Ja A., and Greville, T. N. E. (1939).

Greville, T. N. E., and White, R. P.
See White, R. P., and Greville, T. N. E. (1959).

Grey, Peter. 1941. 1106
A simplified method for computing the theoretical class frequencies of a binomial expansion. *Growth*, **5**, 267–71.
Review: BA **17** (1943), Ab. No. 375.

Grieg-Smith, P. 1952a. 1107
The use of random and contiguous quadrats in the study of the structure of plant communities. *Ann. Bot. Lond., N. S.*, **16**, 293–316.
Users: Grieg-Smith, P. (1952b), Pielou, E. C. (1957), Thompson, H. R. (1958), Morisita, M. (1959), Grieg-Smith, P. (1961).

Grieg-Smith, P. 1952b. 1108
Ecological observations on degraded and secondary forest in Trinidad, British West Indies. *J. Ecol.*, **40**, 316–30.
User: Grieg-Smith, P. (1961).
Class.: P:h:BM

Grieg-Smith, P. 1961a. 1109
Data on pattern within plant communities. I. The analysis of pattern. *J. Ecol.*, **49**, 695–702.

Grieg-Smith, P. 1961b. 1110
Data on pattern within plant communities. II. *Ammophila arenaria* (L.) Link. *J. Ecol.*, **49**, 703–08.

Grieg-Smith, P. 1964. 1111
Quantitative plant ecology. Butterworths, London 2nd edition.

Griffiths, B., and Rao, A. G. 1965. 1112
An application of least cost acceptance sampling schemes. *Unternehmensforschung*, **9**, 8–17.
Review: STMA **7** (1966), 159.
Class.: B:sqc:G

Griffiths, John C. 1960. 1113
Frequency distributions in accessory mineral analysis. *J. Geology*, **68**, 353–65.

Griffiths, John C. 1962. 1114
Frequency distributions in some natural resource materials. Circular 63, Mineral Indust. Expt. Station, College Mineral Indust., Pennsylvania State Univ.

Grigelionis, B. 1962a. 1115
On the asymptotic expansion of the remainder term in the case of convergence to the Poisson law (Russian, Lithuanian and English Summaries). *Litovsk. Mat. Sb.*, **2** (1), 35–48.
Review: MR **27** (1964), 168.

Grigelionis, B. 1962b. 1116
Sharpening of a higher-dimensional limit theorem on convergence to the Poisson law (Russian; Lithuanian and English summaries). *Litovsk. Mat. Sb.*, **2** (2), 127–33.
Review: MR **2** (1964), 332.

Grigelionis, B. 1962. 1117
On the degree of approximation of the composition of renewal processes by a Poisson process. *Litovsk. Mat. Sb.*, **2** (2), 135–43.

Grigelionis, B. 1963. 1118
On the convergence of sums of random step processes to a Poisson process. *Theor. Probability Appl.*, **8**, 177–82.

Grimm, H. 1960. 1119
Transformation of variates (German). *Biom. Zeit.*, **2**, 164–82.
Review: IJA **2** (1961), 466; ZBL **117** (1965), 145.
Class.: MI:anovat:G

Grimm, H. 1961. 1120
Some problems in bacterial counting (French). *Biometrie-Proximetrie*, **2**, I, 1–18.
Review: IJA **4** (1963), 253.
Class.: B-NB-P-N-MI:gf-tc:BM-P

Grimm, H. 1962. 1121
Tables of the negative binomial distribution (German). *Biom. Zeit.*, **4**, 240–62.

Review: IJA **4** (1963), 777; ZBL **104** (1964), 379.
Class.: NB:tc:G

Grimm, H. 1964. 1122
Tables of the Neyman distribution, type A. *Biom. Zeit.*, **6**, 10–23.
Review: STMA **5** (1964), 995; ZBL **116** (1965), 378.
Class.: N:tc:G

Grimm, H., and Maly, V. 1961. 1123
Application of the sequential analysis to the negative binomial distribution. *Bull. Inst. Internat. Statist.*, **39**, 1–11.

Grimm, H., and Maly, V. 1962. 1124
Application of sequential analysis to the negative binomial distribution (German). *Biom. Zeit.*,**4**, 182–92.
Review: IJA **4** (1963), 602; ZBL **108** (1964), 317.
Class.: NB:tp:G

Grizzle, James E. 1961. 1125
A new method of testing hypotheses and estimating parameters for the logistic model. *Biometrics*, **17**, 372–85.
Review: BA **37** (1962), Ab. No. 257.

Grizzle, James E., and Novick, Melvin R.
See Novick, Melvin R., and Grizzle, James E. (1965).

G.-Rodeya F. E. 1961. 1126
On a relation between the binomial coefficients (Spanish). *Gac. Mat.* (*Madrid*), **13**, 3–5.
Review: MR **24A** (1962), 576.

Groll, P. A., and Sobel, M.
See Sobel, M., and Groll, P. A. (1959).

Groll, Phyllis, A., and Sobel, Milton.
See Sobel, Milton, and Groll, Phyllis A. (1966).

Groshev, A. 1941. 1127
Sur le domaine d'attraction de la loi de Poisson. *Izv. Akad. Nauk SSSR Sèr. Mat.*, **5**, 165–72.
Review: MR **3** (1942), 2.

Grubbs, Frank E. 1949. 1128
On designing single sampling inspection plans. *Ann. Math. Statist.*, **20**, 242–56.
Users: Clark, R. E. (1953), Tippett, L. H. C. (1958), Hald. A. (1960), Pachares, J. (1960), Hald. A. (1964), Hald, A. (1965).
Class.: B-P:sqc-tc:E

Grunberg, H. 1955. 1129
Genetical studies on the skeleton of the mouse. XV. The interaction between major and minor variants. *J. Genetics*, **53**, 515–35.
Notes: Appendix by C. A. B. Smith.

Grundy, P. M. 1951. 1130
The expected frequencies in a sample of an animal population in which the abundances of species are lognormally distributed Part I. *Biometrika*, **38**, 427–34.
Review: ZBL **43** 1952), 340.
Users: Williams, C. B. (1953), Cassie, R. M. (1962), Nelson, W. C., and David, H. A. (1964), Bliss, C. I. (1965).

Guiasu, Silviu. 1964. 1131
Sur la répartition asymptotique pour les suites alèatoires de variables aléatoires. *C. R. Acad. Sci. Paris*, **259**, 973–76.
Class.: MI:a:G

Gulberg, Alf. 1931a. 1132
On Poisson's frequency function. *Skand. Aktuarietidskr.*, **14**, 43–48.

Gulberg, Alf. 1931b. 1133
On discontinuous frequency functions and statistical series. *Skand. Aktuarietidskr.*, **14**, 167–87.
Users: Linder, A. (1935), Katz, L. (1945).

Gulberg, Alf. 1934. 1134
On discontinuous frequency functions of two variables. *Skand. Aktuarietidskr.*, **17**, 89–117.
User: Tsao, C. M. (1962).

Gulberg, S. 1935. 1135
Sui momenti della legge di distribuzione del Pólya. *Inst. Italiano degli Attuari Giornale*, **6**, 394–498.
User: Bosch, A. J. (1963).

Gulberg, S. 1936. 1136
A remark on Pólya's law. *Aktuar. Vedy*, **5**, 182–84.

Gumbel, E. J. 1939. 1137
La dissection d'une répartition. *Ann. Univ. Lyon Sect. A.*, **11**, 11 (3), 39–51.
Review: MR **1** (1940), 247.
User: Blischke, W. R. (1965).

Gumbel, E. J. 1941. 1138
The limiting form of Poisson's distribution. *Phys. Rev. 2nd Ser.*, **60**, 689.
Review: ZBL **27** (1943), 338.

Gumbel, E. J. 1963. 1139
Deux lois limites pour la distribution des dépassements. *Bull. Assoc. Actuaires Diplomés Inst. Sci. Financ. Assuar. Mars* 1963, 1–10.

Gumbel, E. J., and Von Schelling, H. 1950. 1140
The distribution of the number of exceedances. *Ann. Math. Statist.*, **21**, 247–62.
Review: MR **11** (1950), 732; ZBL **38** (1951), 90.
Users: Sarkadi, K. (1957), Sarkadi, K. (1957a), Sarkadi, K. (1960), Sarndal, C. E. (1964), Govindarajulu, Z. (1965).
Class.: MI:mb-a-m-osp:G

Gupta, D., and Deming, Lola S.
See Deming, Lola S., and Gupta, D. (1961).

Gupta, S. K. 1965. 1141
Some queues with hyper-general service time distributions. *CORS J.*, **3**, 90–95.
Review: MR **32** (1966), 307.
Class.: MI:pc:G

Gupta, S. S. 1960. 1142
Order statistics from the gamma distribution. *Technometrics.*, **2**, 243–62.
Review: IJA **2** (1961), 640.
Class.: P:os:G

Gupta, S. S. 19 . 1143
Order statistics from the binomial distribution (Tabulation). Tech. Rep., Bell Telephone Lab., Allentown, Penna., pp. 124.

Gupta, S. S. 1965. 1144
Selection and ranking procedures and order statistics for the binomal distribution. *Classical and Contagious Discrete Distributions*, Ed. G. P. Patil, Statistical Publishing Society, Calcutta and Pergamon Press, pp. 219–30.
Class.: B:srp-os:G

Gupta, S. S., and Sobel, M. 1958. 1145
On selecting a subset which contains all populations better than a standard. *Ann. Math. Statist.*, **29**, 235–44.
Users: Mosteller, F., and Yountz, C. (1961), Gupta, S. S. (1965).
Class.: B:srp:G

Gupta, S. S., and Sobel, M. 1960. 1146
Selecting a subset containing the best of several binomial populations. *Contributions to Probability and Statistics*, Ed. Olkin, Stanford Univ. Press, Stanford, Calif. pp. 224–48.
Review: MR **22** (1961), 1964; ZBL **99** (1963), 142.

Gupta, S. S., Huyett, J., and Sobel, M. 1957. 1147
Selection and ranking problems with binomial populations. *Nat. Conv. Trans. Amer. Soc. Qual. Contr.*, 635–43.
Class.: B:srp:G

Gurian, Joan M., Cornfield, Jerome, and Mosimann, James E. 1964. 1148
Comparisons of power for some exact multinomial significance tests. *Psychometrika*, **29**, 409–19.
Review: MR **30** (1965), 813.

Gurland, J. 1955. 1149
Extension of certain classes of contagious distributions (Abstract). *Ann. Math. Statist.*, **26**, 152.

Gurland, J. 1957. 1150
Some interrelations among compound and generalized distributions. *Biometrika*, **44**, 265–68.
Review: ZBL **84** (1960), 140.
Users: Gurland, J. (1958), Gurland, J. (1959), Shumway, R., and Gurland, J. (1960b), Shumway, R., and Gurland, J. (1960c), Katti, S. K., and Gurland, J. (1961), Khatri, C. G., and Patel, I. R. (1961), Martin, D. C., and Katti, S. K. (1962a), Martin, D. C., and Katti, S. K. (1962b), Subrahmaniam, K. (1964), Blischke, W. R. (1965), Douglas, J. B. (1965), Gurland, J. (1965), Katti, S. K., and Sly, L. E. (1965), Kemp, C. D., and Kemp, A. W. (1965), Martin, D. C., and Katti, S. K. (1965), Katti, S. K. (1966), Kemp, A. W., and Kemp, C. D. (1966).
Class.: COP-GP-MI:osp:G

Gurland, J. 1958. 1151
A generalized class of contagious distributions. *Biometrics*, **14**, 229–49.
Review: PA **33** (1959), Ab. No. 7278; ZBL **81** (1959), 139.
Users: Cohen, A. C. (1959c), Gurland, J. (1959), Cohen, A. C. (1960c), Katti, S. K. (1960d), Shumway, R., and Gurland, J. (1960c), Bardwell, G. E. (1961), Khatri, C. G., and Patel, I. R. (1961), Martin, D. C., and Katti, S. (1962b), Pielou, E. C. (1962), Yassky, D. (1962), Subrahmaniam, K. (1964), Blischke, W. R. (1965), Katti, S. K., and Sly, L. E. (1965), Kemp, C. D., and Kemp, A. W. (1965), Martin, D. C., and Katti, S. K. (1965), Sprott, D. A. (1965b), Katti, S. K. (1966).
Class.: GP-COB:mb-gf-a:G-BM

Gurland, J. 1959. 1152
Some applications of the negative binomial and other contagious distributions. *Amer. J. Publ. Health*, **49**, 1388–99.
Review: BA **35** (1960), Ab. No. 12200; IJA **2** (1961), 256.
Users: Shumway, R., and Gurland, J. (1960c), Bartko, J. J. (1962), Gurland, J. (1965), Shenton, L. R., and Myers, R. (1965).

Gurland, J. 1965. 1153
A method of estimation for some generalized Poisson distributions. *Classical and Contagious Discrete Distributions*, Ed. G, P. Patil, Statistical Publishing Society, Calcutta and Pergamon Press, pp. 141–58.
Class.: NB-N-GP:pe:G

Gurland, J., and Edwards, C. B.
See Edwards, C. B., and Gurland, J. (1960).

Gurland, J., and Edwards, C. B.
See Edwards, C. B., and Gurland, J. (1961).

Gurland, J., and Hinz, Paul. 1966. 1154
Some relatively simple procedures for estimating parameters in the negative binomial and Neyman type A distributions (Abstract). *Biometrics*, **22**, 421.

Gurland, J., and Katti, S. K.
See Katti, S. K., and Gurland, J. (1958).

Gurland, J., and Katti, S. K.
See Katti, S. K., and Gurland, J. (1960).

Gurland, J., and Katti, S. K.
See Katti, S. K., and Gurland, J. (1961).

Gurland, J., and Katti, S. K.
See Katti, S. K., and Gurland, J. (1962a).

Gurland, J., and Katti, S. K.
See Katti, S. K., and Gurland, J. (1962b).

Gurland, J., and Shumway, R.
See Shumway, R., and Gurland, J. (1960a).

Gurland, J., and Shumway, R.
See Shumway, R., and Gurland, J. (1960b).

Gurland, J., and Shumway, R.
See Shumway, R., and Gurland, J. (1960c).

Gusak, D. V., and Koroljuk, V. S.
See Koroljuk, V. S., and Gusak, D. V. (1962).

Gusimano, G. 1958. 1155
On conformity of statistical data to a probability distribution (Italian). *Ann. Fac. Econ. Com. Palermo*, **12**, 209–19.
Review: IJA **1** (1959), 66.

Guthrie, D., and Birnbaum, A.
See Birnbaum, A., and Guthrie, D. (1956).

Guttman, I. 1958. 1156
A note on a series solution of a problem in estimation. *Biometrika*, **45**, 565–567.
Users: Edwards, A. W. F. (1960a), Narayana, T. V., and Sathe, Y, S. (1961), Patil, G. P. (1962a), Patil, G. P. (1963a), Crow, E. L., and Bardwell, G. E. (1965), Tweedie, M. C. K. (1965).
Class.: NB-P:pe:G

Guttman, I., and Csorgo, M.
See Csorgo, M., and Guttman, I. (1962).

Guttman, I., and Fraser, D. A. S.
See Fraser, D. A. S., and Guttman, I. (1952).

Gutzman, Wayne, and Benson, Robert.
See Benson, Robert, and Gutzman, Wayne (1961).

Gyrisco, George C., and Forsythe, H. Y., Jr.
See Forsythe, H. Y. Jr., and Gyrisco, George C. (1961).

H

Haáz, I. B. 1956. 1157
Une généralisation du théorème de Simmons. *Acta Sci. Math. (Szeged)*, **17**, 41–44.
Review: ZBL **70** (1957), 362.
Class.: B:osp:G

Haberman, Sol. 1964. 1158
Procedures for generating means and variances of statistics which measure 'variation' in data expressed in the form of outcome of Bernoulli trials. Unclassified Memo. TG-563, PAM-75, Appl. Physics Lab., Planning Analysis, John Hopkins Univ., pp. 72.
Class.: B:mi:G

Hadden, F. A. 1955. 1159
Machine testing for deviation of data from a Poisson distribution. *Trans. Amer. Inst. Elec. Engrs. I*, **74**, 155–57.
Review: SA **58**A (1955), Ab.No. 7620.

Hadley, G., and Whitin, T. M. 1961. 1160
Useful properties of the Poisson distribution. *Operations Res.*, **9**, 408–10.
Review: ZBL **101** (1963), 358.
Users: Benson, F., and Gregory, G. (1961)
Class.: P:m-osp:G

Hadwiger, H. 1945. 1161
Über Verteilungsgesetze vom Poissonschen Typus. *Mitt. Verein. Schweiz. Verisch.-Math.*, **45**, 257–77.
Review: MR **7** (1946), 310; ZBL **60** (1957), 286.

Haenschke, D. G. 1963. 1162
Analysis of delay in mathematical switching models for data systems. *Bell System Tech. J.*, **42**, 709–36.

Hagstroem, K. G. 1956. 1163
Variables fondamentales du hasard. *Giorn. 1st. Ital. Attuari.*, **19**, 84–91.
Review: MR **19** (1958), 466,

Haight, Frank A. 1957. 1164
Queuing with balking. *Biometrika*, **44**, 360–69.
Review: ZBL **85** (1961), 347.
User: Finch, P. D. (1959a).
Class.: B-NB-P-MI:pc:G

Haight, F. A. 1958. 1165
Two queues in parallel. *Biometrika*, **45**, 401–10.
Review: IJA **1** (1959), 128; ZBL **86** (1961), 338.
User: Wilkins, C. A. (1960).
Class.: P-G:pc:G

Haight, Frank A. 1959a. 1166
The generalized Poisson distribution. *Ann. Inst. Statist. Math. Tokyo*, **11**, 101–05.
Review: MR **22** (1961), 47; IJA **1** (1960), 596; ZBL **99** (1963), 145.
Users: Oliver, R. M. (1961), Haight, F. (1965b).
Class.: OPR:m-gf:G

Haight, Frank A. 1959b. 1167
Overflow at a traffic light. *Biometrika*, **46**, 420–24.
Review: ZBL **92** (1962), 342.
Class.: MI:pc:E

Haight, Frank A. 1960. 1168
Queuing with balking II. *Biometrika*, **47**, 285–96.
Review: ZBL **97** (1962), 130; IJA **2** (1961), 753.
Class.: MI:pc:G

Haight, Frank A. 1961a. 1169
A distribution analogous to the Borel-Tanner. *Biometrika*, **48**, 167–73.
Review: IJA **4** (1963), 404.
Class.: MI:pc-tc: G

Haight, Frank A. 1961b. 1170
Index to the distributions of mathematical statistics. *J. Res. Nat. Bur. Standards Sect. B*, **65**, 23–60.
Review: ZBL **96** (1962), 120; IJA **2** (1961), 776.
Users: Patil, G. P. (1965), Dubey, S. D. (1965a).

Haight, Frank A. 1964a. 1171
Accident proneness—the history of an idea. *Automobilismo*, **12**, 534–46.

Haight, Frank A. 1964. 1172
Special discrete distributions. *Internat. Encyclopedia of the Social Sci.*

Haight, Frank A. 1964. 1173
Annotated bibliography of scientific research in road traffic and safety. *Operations Res.*, **12**, 976–1039.

Haight, Frank A. 1965a. 1174
On the effect of removing persons with N or more accidents from an accident prone population. *Biometrika*, **52**, 298–300.
Review: STMA **6** (1965), 1218.
Class.: MI:mb:G-A

Haight, Frank A. 1965b. 1175
Counting distributions for renewal processes. *Biometrika*, **52**, 395–403.
Review: STMA **7** (1966), 396.
Class.: COP-MI:pc:G

Haight, F. A. 1965c. 1176
The discrete busy period distribution for various single server queues (Polish and Russian summaries). *Zastos. Mat.*, **8**, 37–46.
Review: MR **31** (1966), 146.

Haight, Frank A., and Breuer, M. 1960. 1177
The Borel-Tanner distribution. *Biometrika*, **47**, 143–50.
Review: ZBL **117** (1965), 140; IJA **2** (1961), 195.
Class.: BT:m-tc:G

Haight, Frank A., and Fisher, G. R. 1966. 1178
Maximum liklihood estimation for the truncated Poisson. *Biometrics*, **22**, 620–23.
Class.: TCP:pe:G

Haight, Frank A., and Jacobson, Allan S. 1962. 1179
Some mathematical aspects of the parking problem. *Highway Res. Board Proceedings*, **41**, 363–74.

Haight, Frank A., and Reichenback, Hans. 1965. 1180
Fisher's distribution, with tables for fitting it to discrete data by three different methods. Res. Rep. 38, Inst. Transportation and Traffic Eng., Univ. California (Los Angeles, Calif.), pp. 24.
Users: Patil, G. P., and Bildikar, S. (1966a).
Class.: LS:pe-tc:G

Haight, Frank A., and Whittlesey, J. R. B.
See Whittlesey, J. R. B., and Haight, Frank A. (1961-2).

Haight, Frank A., Whisler, Bertram F., and Mosher, Walter W. Jr. 1961. 1181
New statistical method for describing highway distribution of cars. *Highway Res. Board Proceedings*, **40**, 557–64.

Hairston, N. G. 1959. 1182
Species abundance and community organization. *Ecology*, **40**, 404–16.
Users: Hairston, N. G. (1959), Clark, P. J., Eckstrom, P. T., and Linden, L. C. (1964), Berthet, P., and Gerard, G. (1965).
Class.: LS:mb:BM

Hairston, N. G., and Byers, G. W. 1954. 1183
The soil arthropods of a field in southern Michigan. A study in community ecology. *Contri. Lab. Vert. Biol. Univ. Mich.*, **64**, 1–37.
Users: Clark, P. J., Eckstrom, P. T., and Linden, L. C. (1964).

Hájek, J. 1958. 1184
Some contributions to the theory of probability sampling. *Bull. Inst. Internat. Statist.*, **36** (3), 127–34.
Review: MR **22** (1961), 2182.

Hájek, J. 1962. 1185
Cost minimisation in multiparameter estimation (Czech). *Apl. Mat.*, **7**, 405–25.
Review: STMA **6** (1965), 227.
Class.: M:pe:G

Hald, A. 1952. 1186
Statistical tables and formulae. John Wiley and Sons, New York and Chapman and Hall, London, 97 pp.
Review: ZBL **48** (1953), 365.

Hald, A. 1960. 1187
The compound hypergeometric distribution and a system of single sampling inspection plans based on prior distributions and costs. *Technometrics*, **2**, 275–340.
Review: MR **22** (1961), 1031; ZBL **97** (1962), 137; IJA **2** (1961), 705.
Users: Van der Waerden, B. L. (1960), Samuel, E. (1963), Hald, A. (1963b), Hald, A. (1964), Blischke, W. R. (1965), Hald, A. (1965), Lindley, D. V., and Barnett, B. (1965), Hald, A. (1966), Dubey, S. D (1966a).

Hald, A. 1962. 1188
Some limit theorems for the Dodge-Romig LTPD single sampling inspections plans. *Technometrics*, **4**, 497–513.
Review: STMA 5 (1964), 155; MR 32 (1966), 317; ZBL 109 (1964-65), 121.
Users: Hald, A., and Kousgaard, E. (1963), Hald, A. (1963b), Hald, A. (1964), Hald, A. (1965).

Hald, A. 1963a. 1189
Efficiency of sampling inspection plans for attributes (French summary). *Bull. Inst. Internat. Statist.*, **40**, 681–98.
Review: MR **30** (1965), 510.
Class.: B:sqc:E

Hald, A. 1963b. 1190
Efficiency of sampling inspection plans for attributes. *Proc. 34th. Session Internat. Statist. Inst.*, pp. 681–97.
Class.: B:sqc:E

Hald, A. 1964. 1191
Single sampling inspection plans with specified acceptance probability and minimum average costs. Inst. Math. Statist., Univ. Copenhagen, pp. 38.
Class.: B:sqc-tc:E

Hald, A. 1964. 1192
Bayesian single sampling attribute plans for discrete prior distributions. Tech. Rep., Univ. Copenhagen pp. 88.

Hald, A. 1965a. 1193
Bayesian single sampling attribute plans for discrete prior distributions. *Mat.-Fys. Skr. Danske Vid. Selsk.* 3, *II*, 88 pp.
Review: STMA **7** (1966), 562; MR **32** (1966), 810.
Class.: B:sqc:E

Hald, A. 1965b. 1194
Single sampling inspection plans with specified acceptance probability and minimum costs. *Skand. Aktuarietidskr.*, **48**.
Class.: B:sqc-tc:E

Hald, A. 1965c. 1195
On the theory of single sampling inspection by attributes based on two quality levels. Tech. Rep., Univ. Copenhagen, pp. 36.
Class.: B:a-sqc:E

Hald, A. 1965d. 1196
Bayesian single sampling attribute plans for discrete prior distributions (Abstract). *Ann. Math. Statist.*, **36**, 1082.

Hald, A. 1966a. 1197
Attribute sampling plans based on prior distributions and costs. EOQC Conference, Stockholm, 13 pp.
Class: B:sqc:E

Hald, A. 1966b. 1198
The determination of single sampling attribute plans with given producer's and consumer's risk. Res. Rep., Inst. Math Statist. Univ. Copenhagen, 18 pp.
Class.: H-B-P:sqc:E

Hald, A. 1966. 1199
Asymptotic properties of Bayesian single sampling plans. Res. Rep. Inst. Math. Statist. Univ. Copenhagen, pp. 16.
Class.: B:sqc:E

Hald, A., and Kousgaard, E. 1963. 1200
Some limit theorems for the Dodge-Romig AOQL single sampling inspection plans. *Sankhyā*, **25**A, 255–68.
Review: STMA **6** (1965), 226.
Users: Hald, A. (1964), Hald, A. (1965).
Class.: B:sqc:E

Hald, A., and Kousgaard, E. 1966. 1201
A table for solving the equation $B(c,n,p) = p$ *for* $c = 0(1)100$ *and 15 values of p.* Res. Rep. Inst. Math. Statist. Univ. Copenhagen. 77 pp.
Class.: B:sqc-tc:E

Hald, A., and Thyregod, P. 1965. 1202
The composite operating characteristic under normal and tightened sampling inspection by attributes (French summary). *35th Session Internat. Statist. Inst., Beograd., No.* 40, pp. 17.
Class.: B:sqc:E

Haldane, J. B. S. 1937. 1203
The exact value of the moments of the distribution of χ^2 used as a test of goodness of fit, when expectations are small. *Biometrika*, **29**, 133–43.
Review: ZBL **16** (1937), 412.
Users: Haldane, J. B. S. (1939), David, F. N. (1950), Rao, C. R., and Chakravarti, I., (1956), Chakravarti, I. M., and Rao, C. R. (1959), Wise, M. E. (1963), Tweedie, M. C. K. (1965).
Class.: M:gf:G

Haldane, J. B. S. 1939. 1204
The cumulants and moments of the binomial distribution, and the cumulants of χ^2 for a $(n\times 2)$-fold table. *Biometrika*, **31**, 392–96.
Review: ZBL **23** (1941), 339.
Users: Curtiss, J. H. (1941), Patil, G. P. (1959), Patil, G. P. (1960c), Wise, M. E. (1963).
Class.: B:m:G

Haldane, J. B. S. 1941. 1205
The fitting of binomial distributions. *Ann. Eugenics*, **11**, 179–81.
Review: MR **4** (1943), 26.
Users: Fisher, R. A. (1941a), Wise, M. E. (1946), Anscombe, F. J. (1950a), Shenton, L. R. (1950), Sichel, H. S. (1951), Bliss, C. I., and Fisher, R. A. (1952), Hunter, G. C., and Quenouille, M. H. (1952), Binet, F. E. (1953), David, F. N. (1955a), Sampford, M. R. (1955), Roy, J., and Mitra, S. K. (1957), Shenton, L. R. (1958), Cohen, A. C. (1959d), Patil, G. P. (1959), Edwards, A. W. F. (1960), Bartko, J. J. (1961b), Gonin, H. T. (1961), Patil, G. P. (1961b), Martin, D. C., and Katti, S. K. (1962b), Martin, D. C., and Katti, S. K. (1962c), Patil, G. P. (1962c), Patil, G. P. (1962d), Patil, G. P. (1962e), Berthet, P., and Gerard, G. (1965), Cohen, A. Clifford, Jr. (1965), Katti, S. K., and Sly, L. E. (1965), Martin D., and Katti, S. (1965).
Class.: B-NB:pe-c:G

Haldane, J. B. S. 1945a. 1206
On a method of estimating frequencies. *Biometrika*, **33**, 222–25.
Review.: BA **20** (1946), Ab. No. 14726.
Users: McCarthy, P. H. (1947), Finney, D. J. (1947a), Plackett, R. L. (1948), Finney, D. J. (1949), Anscombe, J. B. S. (1950a), Sandelius, M. (1950), Bailey, N. T. J. (1951), Deevey, E. S. Jr., and Potzger, J. E. (1951), Sandelius, M. (1951a), Sandelius, M. (1951b), Sandelius, M. (1951c), Cox, D. R. (1952), Craig, C. C. (1953c) Goodman, L. A. (1953), Bennett, B. M. (1957a), Lindley, D. V. (1957), Armitage, P. (1958), Guttman, I. (1958), deGroot, M. H. (1959), Patil, G. P. (1959), Bol'shev, L. N. (1960), Edwards, A. W. F. (1960a), Nadler, J. (1960), Bartko, J. J. (1961b), Patil, G. P. (1961b), Patil, G. P. (1962d), Patil, G. P. (1962e), Nelson, A. C. Jr., Williams, J. S., and Fletcher, N. T. (1963), Patil, G. P. (1963a), Pathak, P. K. (1964), Knight, W. (1965), Tweedie, M. C. K. (1965).
Class.: NB-NM:pe:G

Haldane, J. B. S. 1945b. 1207
A labour-saving method of sampling. *Nature*, **155**, 49.
Users: Tweedie, M. C. K. (1945), Barnard, G. A. (1946), Finney, D. J. (1947a), Tweedie, M. C. K. (1947), Tiago de Olivira, J. (1952b), Craig, C. C. (1953c), de Groot, M. H. (1959).
Class.: NB:pe:G

Haldane, J. B. S. 1947–49. 1208
A test for homogeneity of records of familial abnormalities. *Ann. Eugenics*, **14**, 339–41.
Review: ZBL **34** (1950), 81.
Class.: TCB:h;BM

Haldane, J. B. S. 1955a. 1209
The calculation of mortality rates from ringing data. *Acta XI Congr. Int. Orn.*, 1954, 454–58.

Haldane, J. B. S. 1955b. 1210
A problem in the significance of small numbers. *Biometrics*, **42**, 266–67.
Review: ZBL **65** (1965), 116.

Haldane, J. B. S. 1955–56a. 1211
The estimation and significance of the logarithm of a ratio of frequencies. *Ann. Human Genetics*, **20**, 309-11.
User: Haldane, J. B. S. (1956b).

Haldane, J. B. S. 1956b. 1212
Almost unbiased estimates of functions of frequencies. *Sankhyā*, **17**, 201–08.
User: Tweedie, M. C. K. (1965).
Class.: B:pe:G

Hale, B. W., Cottam, G., and Curtis, J. T.
See Cottam, G., Curtis, J. T., and Hale, B. W. (1953).

Halevi, Hai. 1965. 1213
An alternative approach to the method of correct matching. *Psychometrika*, **30**, 197–205.
Class.: B-M:tp:BM

Hall, W. J. 1956. 1214
Some hypergeometric series distributions occurring in birth-and-death processes of equilibrium (Abstract). *Ann. Math. Statist.*, **27**, 221.
Users: Bardwell, G. E., and Crow, E. L. (1964).

Hall, W. J. 1962. 1215
On median unbiased estimation from discrete data. (Abstract). *Ann. Math. Statist.*, **33**, 1491.

Halperin, M., and Burrows, G. L. 1960. 1216
The effect of sequential batching for acceptance-rejection sampling upon sample assurance of total product quality. *Technometrics*, **2**, 19–26.
Review: MR **22** (1961), 187.

Halperin, M., and Burrows, G. L. 1961. 1217
An asymptotic distribution for an occupancy problem with statistical applications. *Technometrics*, **3**, 79–89.
Class.: MHR:a-sqc:G-E

Ham, W. T. Jr., and Turner, M. E.
See Turner, M. E., and Ham, W. R. Jr. (1960).

Hamaker, H. C. 1953. 1218
'Average confidence' limits for binomial probabilities. *Rev. Inst. Internat. Statist.*, **21**, 17–27.
Review: MR **15** (1954), 331; ZBL **51** (1954), 109.
Class.: B:ie:G

Hamaker, H. C. 1958. 1219
On hemacytometer counts. *Biometrics*, **14**, 558–59.
Review: BA **33** (1959), Ab. No. 24384.
Class.: H-B-P:cm:BM

Hamaker, H. C. 1960a. 1220
Attribute sampling in operation. *Bull. Inst. Internat. Statist.*, **37** (2), 263–81. Discussion: *Bull. Int. Statist. Inst.*, **37** (I), 134–38.
Review: IJA **2** (1961), 355.

Hamaker, H. C. 1960b. 1221
Qualitative control of the sample (French). *Rev. Statist. Appl.*, **8**, 5–40.
Review: IJA **2** (1961), 357.

Hamburg, Morris. 1962. 1222
Bayesian decision theory and statistical quality control. *Industrial Quality Control*, **19** (6), 10–14.

Hammersley, J. M. 1951. 1223
The sums of products of the natural members. *Proc. London Math. Soc.*, **1**, 435–52.
User: Hammersley, J. M. (1952).

Hammersley, J. M. 1952. 1224
Tauberian theory for the asymptotic forms of statistical frequency functions. *Proc. Cambridge Philos. Soc.*, **48**, 592–99.

Hancock, J. L. Rothschild, Lord, and Campbell, R. C..
See Campbell, R. C., Hancock, J. L., and Rothschild, Lord (1953).

Hannan, J. 1960. 1225
Consistency of maximum likelihood estimation of discrete distributions. *Contributions to Probability and Statistics*, Ed. Olkin, Stanford Univ. Press, Stanford, Calif., pp. 249–57.
Review: MR **22** (1961), 1958.

Hannan, J., and Harkness, W. 1960. 1226
Normal approximation to the distribution of two independent binomials, condition on fixed sum (Abstract). *Ann. Math. Statist.*, **31**, 525.

Hannan, J. and Harkness, W. 1963. 1227
Normal approximation to the distribution of two independent binomials, conditional on fixed sum. *Ann. Math. Statist.*, **34**, 1593–95.
Review: MR **28** (1964), 683; STMA **6** (1965), 81.
Users: Govindarajulu, Z. (1965), Harkness, W. L. (1965).
Class.: B:a:G

Hans, O. 1957. 1228
A note on negative binomial distribution (Czech; Russian and English summaries). *Apl. Mat.*, **2**, 222–26.
Review: MR **19** (1958), 586; ZBL **103** (1964), 372.

Hanson, W. R. 1957. 1229
Density of wood rat houses in Arizona chaparral. *Ecology*, **38**, 650.
Class.: P:gf:BM

Harcourt, D. G. 1960. 1230
Distribution of the mature stages of the diamondback moth (*Plutella maculipennis*) (Curt.) (Lepidoptera: Plutellidae) on cabbage. *Canad. Entomol.*, **92**, 517–21.
Users: Harcourt, D. G. (1961), Kuno, E. (1963), Harcourt, D. G. (1965), Kobayashi, S. (1965).
Class.: NB:gf-anovat:BM

Harcourt, D. G. 1961. 1231
Spatial pattern of the imported cabbage worm, *Pieris rapae* (*L*), on cultivated Cruciferae. *Canad. Entomologist.*, **93**, 945–52.
Users: Ito, Nakamura, Kondo, Miyashita, and Nakamura (1962), Kuno, E. (1963), Kobayashi, S. (1965).

Harcourt, D. G. 1963. 1232
Population dynamics of *Leptinotarsa decemlineata* (Say) in eastern Ontario. I. Spatial pattern and transformation of field counts. *Canad. Entomol.*, **95**, 813–20.
Users: Harcourt, (1965), Kobayashi, S. (1965).

Harcourt, D. G. 1965. 1233
Spatial pattern of the cabbage looper, *Trichoplusia ni*, on crucifers. *Ann. Entomol. Soc. Amer.*, **58**, 89–94.

Harcourt, D. G. 1966. 1234
Sequential sampling for use in control of the cabbage looper on cauliflower. *J. Econ. Entomol.*, **59**, 1190–92.
Class.: NB:tp:BM

Harding, J. P. 1949. 1235
The use of probability paper for the graphical analysis of polymodal frequency distributions. *J. Marine Biol. Assoc.*, **28**, 141–53.

Harkness, W. L. 1965. 1236
Properties of the extended hypergeometric distribution. *Ann. Math. Statist.*, **36**, 938–45.
Review: ZBL **31** (1966), 1135.
Class.: OHR:m-a-pe-ie:G

Harkness, W. L., and Bolger, E. M.
See Bolger, E. M., and Harkness, W. L. (1965).

Harkness, W. L., and Bolger, E. M.
See Bolger, E. M., and Harkness, W. L. (1966).

Harkness, W., and Hannan, J.
See Hannan, J., and Harkness, W. (1960).

Harkness, W., and Hannan, J.
See Hannan, J., and Harkness, W. (1963).

Harper, J. L. 1957. 1237
Ranunculus. *J. Ecol.*, **45**, 289–342.

Harris, B. 1956. 1238
Confidence intervals for the number of cells in a multinormal population with equal cell probabilities (Abstract). *Ann. Math. Statist.*, **27**, 867.

Harris, B. 1960a. 1239
Probability distributions related to random mappings. *Ann. Math. Statist.*, **31**, 1045–62.
Review: IJA **3** (1962), 141.
User: Riordan, J. (1962).
Class.: MI:mi:G

Harris, B. 1963. 1240
Some results on estimating the number of classes in a discrete uniform population (Abstract). *Ann. Math. Statist.*, **34**, 1620.

Harris, Bernard, and Cobb, E. B.
See Cobb, E. B., and Harris, Bernard (1966).

Harris, Eugene K. 1958. 1241
On the probability of survival of bacteria in sea water. *Biometrics*, **14**, 195–206.
Users: Dubey, S. D. (1966a).
Class.: COB:pe-gf:BM

Harris, E. K. 1960b. 1242
Analysis of experiments measuring threshold taste. *Biometrics*, **16**, 245–60.
Review: IJA **2** (1961), 30.

Harris, Lee B. 1952. 1243
On a limiting case for the distribution of exceedances with an application to life-testing. *Ann. Math. Statistics*, **23**, 295–98.
Review: ZBL **46** (1953), 358.

Harris, T. E. 1948. 1244
Some further results on the Bernoulli process (Abstract) *Ann. Math. Statist.*, **19**, 116.
User: Kendall, D. G. (1949).

Harrison, J. L. 1945. 1245
Stored products and the insects infesting them as examples of logarithmic series. *Ann. Eugenics*, **12**, 280–82.
Review: BA 20 (1946), Ab.No. 10.
Users: Williams, C. B. (1947b), Patil, G. P. (1959), Patil, G. P. (1962b), Nelson, W. C., and David, H. A. (1964).

Hart, Hornell. 1926. 1246
The reliability of a percentage. *J. Amer. Statist. Assoc.*, **21**, 40–46.
Class.: B:ie:G

Hart, J. L. 1943. 1247
Tagging experiments on British Columbia pilchards. *J. Fisheries Res. Board Canada*, **6**, 164–82.

Hartigan, J. A. 1966. 1248
Estimation by ranking parameters. *J. Roy Statist. Soc. Ser. B*, **28**, 32–44.
Class.: B:ie:G

Hartley, H. O. 1958. 1249
Maximum likelihood estimation from incomplete data. *Biometrics*, **14**, 174–94.
Users: Cohen, A. C. (1959d), Cohen, A. C. (1959e), Irwin, J. O. (1959), Cohen, A. C. (1960b), Cohen, A. C. (1961c), Wilkinson, G. N. (1961), Hughes, E. J. (1962), Doss, S. A. D. C. (1963), Cohen, A. Clifford Jr. (1965), Tester, John F., and Sinif, Donald B. (1965), Cohen, A. C. Jr. (1966).
Class.: MI-TCP-TCNB:pe-gf:G-BM

Hartley, H. O., and Fitch, E. R. 1951. 1250
A chart for the incomplete beta-function and the cumulative binomial distribution. *Biometrika*, **38**, 423-26.
Review: MR **14** (1953), 63.
Class.: B:tc:G

Hartley, H. O. and Pearson, E. S. 1950. 1251
Tables of the χ^2 integral and of the cumulative Poisson distribution. *Biometrika*, **37**, 313–25.
Review: MR **12** (1951), 344.
User: Tiku, M. (1964).
Class.: P:tc:G

Harville, David A. 1966. 1252
On bias in variance component estimation (Abstract). *Ann. Math. Statist.*, **37**, 1418–19.

Hasofer, A. M. 1965. 1253
Some perturbation results for the single-server queue with Poisson input. *J. Appl. Prob..*, **2**, 462–66.

Hawkes, A. G. 1965. 1254
Time-dependent solution of a priority queue with bulk arrival. *Operations Res.*, **13**, 586–95.
Review: MR **32** (1966), 307.
Class.: MI:pc-G

Hayakawa, R., and Ishii, G.
See Ishii, G., and Hayakawa, R. (1960).

Hayman, B. I., and Lowe, A. D. 1961. 1255
The transformation of counts of the cabbage aphid (Brevicoryne Brassicae (L.)). *New Zealand J. Sci.*, **4**, 271–78.
Class.: MI-NB:anovat:BM

Hayman, G. E., and Leone, F. C. 1965. 1256
Analysis of categorical data. *Biometrika*, **52**, 654–60.
Class.: M:gf:G

Hayman G. E., Chu, J. T., Topp, C. W., and Leone, F. C.
See Leone, F. C., Hayman, G. E., Chu, J. T., and Topp, C. W. (1960).

Hayne, D. W. 1949. 1257
Calculation of size of home range. *J. Mammalogy*, 30, 1–18.
User: Godfrey, G. K. (1954).

Hazen, S. W. Jr., and Becker, R. M.
See Becker, R. M., and Hazen, S. W. Jr. (1961).

Healy, M. J. R. 1964. 1258
A property of the multinomial distribution and the determination of appropriate scores. *Biometrika*, **51**, 265–66.
Review: STMA **5** (1964), 813; ZBL **124** (1966), 111.

Healy, W. C. Jr., and Birnbaum, A.
See Birnbaum, A., and Healy, W. C. Jr. (1960).

Heathcote, C. R. 1961. 1259
Preemptive priority queueing. *Biometrika*, **48**, 57–63.
Review: IJA **4** (1963), 524.
Class.: MI:pc:G

Hendricks, W. A. 1964. 1260
Estimation of the probability that an observation will fall into a specified class. *J. Amer. Statist. Assoc.*, **59**, 225–32.
Review: MR **28** (1964), 684; STMA **7** (1966), 79.
Class.: M:pe:G

Hendrickson, G. O., Kozicky, E. L., and Jessen, R. J.
See Kozicky, E. L., Jessen, R. J., and Hendrickson, G. O. (1956).

Henson, W. R. 1954. 1261
A sampling system for poplar insects. *Canad. J. Zool.*, **32**, 421–33.
Users: Waters, W. E., and Henson, W. R. (1959), Henson, W. R. (1959).

Henson, W. R. 1959. 1262
Some effects of secondary dispersive processes on distribution. *Amer. Naturalist.*, **93**, 315–20.
Class.: NB:gf:BM

Henson, W. R., and Waters, William E.
See Waters, William E., and Henson, W. R. (1959).

Heppes, A. 1956. 1263
On the determination of probability distributions of more dimensions by their projections (Russian summary). *Acta. Math. Acad. Sci. Hungar.*, **7**, 403–10.
Review: MR **19** (1958), 70.
Class.: MI:mi:G

Herback, Leon H. 1948. 1264
Bounds for some functions used in sequentially testing the mean of a Poisson distribution. *Ann. Math. Statist.*, **19**, 400–05.
Review: MR **10** (1949), 201
User: Chapman, D. G. (1952b).
Class.: P:tp:G

Herdan, G. 1957. 1265
The mathematical relation between the number of diseases and the number of patients in a community. *J. Roy. Statist. Soc. Ser. A*, **120**, 320–30.
Class.: LS:mi:BM

Herdan, G. 1958. 1266
The relation between the dictionary distribution-and the occurrence distribution of word length and its importance for the study of quantitative linguistics. *Biometrika*, **45**, 222–28.
User: Rider, P. R. (1965).
Class.: MI:mi:L

Herdan, G. 1960. 1267
Type-token mathematics. Mouton and Co.

Herdan, G. 1961. 1268
A critical examination of Simon's model of certain distribution functions in linguistics. *Appl. Statist.*, **10**, 65–76.
Review: MR **10** (1964), 201; ZBL **103** (1964), 126.
Class.: MI–COP:mb:L

Herdan, G. 1962. 1269
The calculus of linguistic observations. Mouton and Co.

Herdan, G. 1964. 1270
Quantitative linguistics. Butterworths.

Herrera, L., Sutcliffe, M. I., and Mainland, D.
See Mainland, D., Herrera, L., and Sutcliffe, M. I. (1956).

Hermann, Horst. 1965. 1271
Variationsabstand zwischen der Verteilung einer Summe unabhängiger nichtnegativer ganzzahliger Zufallsgrössen und Poissonschen Verteilungen. *Math. Nachr.*, **29**, 265–89.

Herrmann, Ursula. 1965. 1272
A generalization of a theorem of Prochorow and LeCam for random vectors (German). *Math. Nachr.*, **29**, 17–24.
Review: STMA **7** (1966), 47.
Class.: MI-P:mi:G

Herzel, A. 1963a. 1273
The central moments of the binomial and Poisson distributions (Italian). *Bibliotèca Metron Serie C Rome*, **2**, 145–62.
Review: STMA **6** (1965), 55.
Class.: B-P:m:G

Herzel, A. 1963b. 1274
On order statistics and on the mean deviation about the median value in samples taken with or without replacement from discrete and finite populations (Italian; English, French and German summaries). *Bibliotèca Metron Serie C Rome*, **2**, 163–98.
Review: STMA **5** (1964), 829.
User: Herzel, A. (1963c).
Class.: MI:os-mi:G

Herzel, A. 1963c. 1275
The mean value and the variance of the simple index of dissimilarity in the Bernoullian and exhaustive universe of samples (Italian; English, French, and German summaries). *Bibliotèca Metron Serie C Rome*, **2**, 199–224.
Review: STMA **5** (1964), 830.
Class.: B:mi:G

Herzog, F. 1947. 1276
Upper bound for terms of the binomial expansion. *Amer. Math. Monthly*, **54**, 485–87.
Review: ZBL **29** (1948), 152.
Class.: B:mi:G

Hetz, W., and Klinger, H. 1958. 1277
Untersuchungen zur Frage der Verteilung von Objekten auf Plätze. *Metrika*, **1**, 3–20.
Review: ZBL **82** (1960), 129.

Heuer, Gerald A. 1959. 1278
Estimation in a certain probability problem. *Amer. Math. Monthly*, **66** 704–06.
Review: MR **21** (1960), 1405; ZBL **85** (1961), 353.
Class.: MI:pe:G

Hida, Takeyuki. 1953. 1279
On some asymptotic properties of Poisson process. *Nagoya Math. J.*, **6**, 29–36.
Review: ZBL **53** (1961), 98.

Higgins, George M., Berkson, Joseph, and Dews, Peter B.
See Dews, Peter, B., Higgins, George M., and Berkson, Joseph (1954).

Hills, M. 1966. 1280
Allocation rules and their error rates. *J. Roy. Statist. Soc. Ser. B*, **28**, 1–31.
Class.: MI-M:mi:G

Hinz, Paul, and Gurland, John.
See Gurland, John, and Hinz, Paul (1966).

Hirata, S. 1933. 1281
The movement of people at the ticket office of Shinzyuku station. *Kagaku* (*Science*), **3**, 274–75.

Hobby, Charles, and Pyke, Ronald. 1963. 1282
A combinatorial theorem related to comparisons of empirical distribution functions. *Wahrscheinlichkeitstheorie und Verw*, **2**, 85–89.
Review: MR **28** (1964), 331.

Hodges, J. L. Jr. 1955. 1283
A bivariate sign test. *Ann. Math. Statist.*, **26**, 523–27.
Class.: MI:ctp:G

Hodges, J. L. Jr., and Blackwell, David.
See Blackwell, David, and Hodges, J. L. Jr. (1959).

Hodges, J. L. Jr., and Le Cam, L. 1960. 1284
The Poisson approximation to the Poisson binomial distribution. *Ann. Math. Statist.*, **31**, 737–40.
Review: MR **22** (1961), 1464; ZBL **100** (1963), 143; IJA **2** (1961), 240.
Users: Makable, H. (1962), Ellner, H. (1963), Govindarajulu, Z. (1965).
Class.: GBDP-P:a:G

Hodges, J. L. Jr., and Lehmann, E. L. 1950. 1285
Some problem in minimax point estimation. *Ann. Math. Statist.*, **21**, 182–97.
Review: ZBL **38** (1951), 98.
Users: Basu, D. (1952), Hodges, J. L., and Lehmann, E. (1952), Steinhaus, H. (1957), Trybula, S. (1958a), Goodman, L. A. (1961), Trybula, S. (1962), Wasan, M. T. (1964), Wasan, M. T. (1965).
Class.: B-H:pe-ctp:G

Hodges, J. L. Jr., and Lehmann, E. L. 1952. 1286
The use of previous experience in reaching statistical decisions. *Ann. Math. Statist.*, **23**, 396–407.
Class.: MI-B:tp:G

Hodges, J. L. Jr., and Lehmann, E. L. 1964. 1287
Basic concepts of probability and statistics. Holden-Day, San Francisco, 375 pp.
Review: MR **32** (1966), 538.

Hodgson, Vincent. 1965. 1288
Non-preemptive priorities in machine interference (Abstract). *Ann. Math. Statist.*, **36**, 1600.

Hoeffding, W. 1955. 1289
On the distribution of the number of successes in independent trials (Abstract). *Ann. Math. Statist.*, **26**, 538.

Hoeffding, W. 1956. 1290
On the distribution of the number of successes in independent trials. *Ann. Math. Statist.*, **27**, 713–21.
Review: MR **18** (1957), 240.
Users: Cohen, A. C. (1959b), David, H. A. (1960), Taylor, R. J., and David, H. A. (1962), Darroch, J. N. (1964), Samuels, S. M. (1964), Samuels, S. (1965).
Review: ZBL **73** (1960), 139.
Class.: GBDP:osp:G

Hoeffding, W. 1963. 1291
Large deviations in multinomial distributions (Abstract). *Ann. Math. Statist.*, **34**, 1620.
User: Hoeffding, W. (1965).

Hoeffding, W. 1964. 1292
Asymptotically optimal tests for multinomial distributions. Inst. Statist. Mimeograph Ser. 396, Dept. Statist., Univ. North Carolina.
Class.: M:tp:G

Hoeffding, W. 1965. 1293
Asymptotically optimal tests for multinomial distributions. *Ann. Math. Statist.*, **36**, 369–408.
Review: MR **30** (1965), 667.
Class.: M:tp-a:G

Hoffding, W. 1941. 1294
Mabstabinvariante Korrelantionsmasse für diskontinuierliche Verteilungen. *Arch. Math. Wirtsch.-u. Sozialforschg.*, **7**, 49–70.
Review: ZBL **25** (1942), 201.

Hoel, Paul. 1398. 1295
On the chi-square distribution for small samples. *Ann. Math. Statist.*, **9**, 158–165.
Review: ZBL **19** (1939), 357.
User: Watson, G. S. (1959).
Class.: M:gf:G

Hoel, Paul G. 1943. 1296
On indices of dispersion. *Ann. Math. Statist.*, **14**, 155–62.
Users: Bateman, G. I. (1950), Thomas, M. (1951), Kathirgamatamby, N. (1953), Okamoto, M. (1955), Bennett, B. M. (1956), Bennett, B. M. (1959), Bennett, B. M., and Hsu, P. (1961), Bennett, B. M. (1962a), Nicholson, W. L. (1963), Selby, B. (1965).

Hoel, Paul G. 1945. 1297
Testing the homogeneity of Poisson frequencies. *Ann. Math. Statist.*, **16**, 362–68.
Review: MR **7** (1946), 464; ZBL **60** (1957), 305.
User: Hoel, P. G. (1947).
Class.: P:ctp:G

Hoel, Paul G. 1947. 1298
Discriminating between binomial distributions. *Ann. Math. Statist.*, **18**, 556–64.
Review: MR **9** (1948), 295; ZBL **29** (1948), 308.
User: Binet, F. E. (1953).
Class.: B-P:tp:G

Hoetzl, George. 1963. 1299
Properties of the extended hypergeometric distribution. Master's Thesis, Pennsylvania State Univ.

Hofflander, A. E., Dickerson, O. D., and Katti, S. K.
See Dickerson, O. D., Katti, S. K., and Hofflander, A. E. (1961).

Hofmann, M. 1955. 1300
Über zusammengesetzte Poisson-Prozesse und ihre Anwendungen in der Unfallversicherung. *Mitt. Schweiz. Verich.-Math.*, **55**, 499–575.
Review: MR **17** (1956), 638; ZBL **66** (1957), 133.

Hogben, L., and Cross, K. W. 1952. 1301
Sampling from a discrete universe. *Acta Genet. Statist. Med.*, **3**, 305–42.

Holgate, P. 1964. 1302
Estimation for the bivariate Poisson distribution. *Biometrika*, **51**, 241–45.
Review: MR **30** (1965), 501; STMA **6** (1965), 106.

Holgate, P. 1964a. 1303
A modified geometric distribution arising in trapping studies. *Acta. Theriol.*, **9**, 353–56.
Users: Tanton, M. T. (1965), Holgate, D. (1966a).

Holgate, P. 1965. 1304
Test of randomness based on distance methods. *Biometrika*, **52**, 345–53.
Class.: B-P:mi:G

Holgate, P. 1966a. 1305
Contributions to the mathematics of animal trapping. *Biometrics*, **22**, 925–36.
Class.: G-TCP-TCNB-LS-MI:gf:BM

Holgate, P. 1966b. 1306
Bivariate generalizations of Neyman's type A distribution. *Biometrika*, **53**, 241-45.
Class.: N:mi-pe:G-BM

Holgate, P. 1966c. 1307
The use of distance methods in studying spatial pattern. *Res. Notes of the Second Conf. Internat. Advisory Group of Forest Statist. Royal College of Forestry Stockholm*, pp. 125–31.
Class.: P:h:BM

Holla, M. S., and Bhattacharya, S. K.
See Bhattacharya, S. K., and Holla, M. S. (1965).

Holme, N. A. 1950. 1308
Population-dispersion in Tellina tenuis Da Costa. *J. Marine Biol. Assoc.*, **29**, 267–80.

Holme, N. A. 1953. 1309
The biomass of the bottom fauna in the English Channel off Plymouth. *J. Marine Biol. Assoc.*, **32**, 1–49.

Holmes, R. W., and Widrig, T. M. 1956. 1310
The enumeration and collection of marine phytoplankton. *J. Conseil*, **22** (1), 21–32.
User: Kutkuhn, J. H. (1958).

Holt, S. B. 1958. 1311
Genetics, of dermal ridges: the relation between total ridge-count and the variability of counts from finger to finger. *Ann. Human Genetics*, **22**, 323–39.

Holt, S. J. 1955. 1312
On the foraging activity of the wood ant. *J. Animal Ecol.*, **24**, 1–34.
Class.: P:gf:BM

Holte, F. C. 1954. 1313
Noen egenskaper ved binomiale forfelingsfunksjon. (Some properties of the binomial distribution function.) Norwegian; English summary). *Nordisk. Mat. Tidskr.*, **2**, 113–15.
Review: MR **16** (1955), 376; ZBL **55** (1955), 365.

Homma, Tsuruchiyo, and Fujisawa, Takehisa. 1965. 1314
Some notes on the queues with multiple inputs. *Yokohama Math. J.*, **12**, 1–15.
Review: MR **32** (1966), 308.
Class.: MI:pc:G

Hopkins, B. 1955. 1315
The species–area relations of plant communities. *J. Ecol.*, **43**, 490–508.

Hopkins, C. E., and Cochran, W. G.
See Cochran, W. G., and Hopkins, C. E. (1960).

Hopkins, C. E., and Cochran, W. G.
See Cochran, W. G., and Hopkins, C. E. (1961).

Hopkins, D. E., New, W. D., Dudley, F. H., Bushland, R. C., Baumhover, A. H., Graham, A. J., and Bitter, B. A.
See Baumhover, A. H., Graham, A. J., Bitter, B. A., Hopkins, D. E., New, W. D., Dudley, F. H., and Bushland, R. C. (1955).

Hopkins, J. W. 1955. 1316
An instance of negative hypergeometric in practice. *Bull. Inst. Internat. Statist.*., **34** (4), 298–306.
Review: ZBL **64** (1965), 136.
Users: Sarkadi, K. (1957a), Sarkadi, K. (1960a).
Class.: NH:sqc:E

Horst, C., and Bennett, B. M.
See Bennett, B. M., and Horst, C. (1966).

Hsu, P., and Bennett, B. M.
See Bennett, B. M., and Hsu, P. (1960).

Hsu, P., and Bennett, B. M.
See Bennett, B. M., and Hsu, P. (1961).

Hsu, P., and Bennett, B. M.
See Bennett, B. M., and Hsu, P. (1962).

Huang, David S. 1956. 1317
Goodness of prediction concerning binomial variables. *Appl. Statist.*, **14**, 206–09.
Class.: B:rp:G

Hughes, Edwin Joseph. 1962. 1318
Maximum likelihood estimation of distribution parameters from incomplete data. Ph.D. Thesis, Iowa State Univ., pp. 87.

Hughes, R. D. 1955. 1319
The influence of the prevailing weather on the numbers of *Meromyza variegata* Meigen (Deptera, Chloropidae) caught with a sweepnet. *J. Animal Ecol.*, **24**, 324–35.
Class.: B:gf:BM

Hughes, R. D. 1956. 1320
British Ecological Society autumn meeting at London 19–20 September 1955. *J. Animal Ecol.*, **25**, 461–67.
Class.: B:mi:BM

Hunter, G. C., and Quenouille, M. H. 1952. 1321
A statistical examination of the worm egg count sampling technique for sheep. *J. Helminthology*, **26**, 157–70.
Class.: P-NB:gf-pe-rp-anovat:BM

Hunter, Larry, and Proschan, Frank. 1961. 1322
Replacement when constant failure rate precedes wearout. *Naval Res. Logist. Quart.*, **8**, 127–36.
Review: ZBL **123** (1966), 369.

Hurn, M., Berkson, J., and Magath, T. B.
See Berkson, J., Magath, T. B., and Hurn, M. (1935).

Hurn, M., Berkson, J., and Magath, T. B.
See Berkson, J., Magath, T. B., and Hurn, M. (1940).

Huron, R. 1955. 1323
Loi multinomiale et test du χ^2. *C. R. Acad. Sci. Paris*, **240**, 2047–48.
Review: MR **17** (1956), 56; ZBL **64** (1956), 387.
Users: Patil, G. P., and Bildikar, S. (1966a).
Class.: M:gf:G

Huron, Roger. 1956. 1324
Sur une transformation linéaire des fréquences observées dans les grands échantillons non exhaustifs, de taille fixée, et extraits au hasard d'une urne à k catégories. *C. R. Acad. Sci. Paris*, **242**, 1951–53.
Review: ZBL **71** (1958), 129.

Huron, R. 1959. 1325
Étude de la corrélation entre certaines variables aléatoires liées a la loi multinomiale. *C. R. Acad. Sci. Paris*, **249**, 2268–69.
Review: MR **22** (1961), 182.
Users: Patil, G. P., and Bildikar, S. (1966a)
Class.: OMR:osp:G

Huron, R., and Meric, J. 1954. 1326
Sur une application du schéma d'urnes de Poisson. *Ann. Fac. Sci. Univ. Toulouse*, **17** (4), 265–72.
Review: ZBL **55** (1955), 121.

Hurst, D. C., and Quesenberry, C. P.
See Quesenberry, C. P., and Hurst, D. C. (1964).

Huruya, A., and Kitagawa, T.
See Kitagawa, T., and Huruya, S. (1941).
Huruya, S., and Kitagawa, T.
See Kitagawa, T., and Huruya, S. (1941).
Huruya, S., Yazima, T., and Kitagawa, T.
See Kitagawa, T., Huryua, S., and Yazima, T. (1942).
Hurwitz, H. Jr., and Kac, M. 1944. 1327
Statistical analysis of certain types of random functions. *Ann. Math. Statist.*, **15**, 173–81.
Class.: P:pc:G
Hutchinson, D. W., and Blyth, Colin R.
See Blyth, Colin R., and Hutchinson, D. W. (1960).
Hutchinson, D. W., and Blyth, Colin R.
See Blyth, Colin R., and Hutchinson, D. W. (1961).
Huyett, J., Sobel, M., and Gupta, S. S.
See Gupta, S. S., Huyett, J., and Sobel, M. (1957).
Huyett, J., and Sobel, M.
See Sobel, M., and Huyett, J. (1957).
Users: Beachhofer, R. E., Elmaghraby, S., and Morse, N., (1959).
Huzurbazar, V. S. 1950. 1328
Probability distributions and orthogonal parameters. *Proc. Cambridge Philos. Soc.*, **46**, 281–84.
Class.: NB-MI:pe:G
Huzurbazar, V. S. 1965. 1329
Some invariants of some discrete distributions admitting sufficient statistics for parameters. *Classical and Contagious Discrete Distributions*, Ed. G. P. Patil, Statistical Publishing Society, Calcutta and Pergamon Press, pp. 231–40. Reprinted in *Sankhyā*, Ser. A., **28**, (1966).
Class.: B-P-M:osp:G

I

Ibarra, E. L., Wallwork, J. A., and Rodriguez, J. G. 1965. 1330
Ecological studies of mites found in sheep and cattle pastures. I. Distribution patterns of Oribatid mites. *Ann. Entomol. Soc. Amer.*, **58**, 153–59.
Ijiri, Y., and Simon, H. A. 1964. 1331
Business firm growth and size. *Amer. Econ. Rev.*, **54**, 77–89.
Class.: MI:mb:S
Imbrie, J. 1955. 1332
Biofacies analysis. *Geol. Soc. Amer. Spec. Paper*, **62**, 449–64.
Irwin, J. O. 1937. 1333
The frequency distribution of the difference between two independent variates following the same Poisson distribution. *J. Roy. Statist. Soc. Ser. A*, **100**, 415–16.
Users: Skellam, J. G. (1946), Skellam, J. G. (1952), Fisz, M. (1955a), Johnson, N. L. (1959), Ramasubban, T. A. (1960), Bennett, B. M. (1962a), Irwin, J. O. (1963), Shanfield, F. (1964), Kamat, A. R. (1965).
Class.: P:a-m:G
Irwin, J. O. 1941. 1334
Comments on the paper " Chambers, E. C., and Yule, G. U. Theory and Observation in the Investigation of Accident Causation." *J. Roy. Statist. Soc. Supple.*, **7**, 89–109.
Users: Haight, F. (1965a), Haight, F. (1965b), Mellinger, G., Sylwester, D., Gaffey, W., and Manheimer, D. (1965).
Irwin, J. O. 1943. 1335
A table of the variance of $\sqrt{x}$ when x has a Poisson distribution. *J. Roy. Statist. Soc. Ser. A*, **106**, 143–44.
Review: MR **5** (1944), 209.
Class.: P:anovat-tc:G
Irwin, J. O. 1953. 1336
On " transition probabilities " corresponding to any accident distribution. *J. Roy. Statist. Soc. Ser. B*, **15**, 87–89.
Review: ZBL **50** (1954), 353.
Class.: MI:mi:G
Irwin, J. O. 1954. 1337
A distribution arising in the study of infectious diseases. *Biometrika*, **41**, 266–68.
Users: Sarkadi, K. (1957), Kemp, C. D., and Kemp, A. W. (1965b).
Class.: NH:mb:BM
Irwin, J. O. 1955. 1338
A unified derivation of some well-known frequency distributions of interest in biometry and statistics. *J. Roy. Statist. Soc. Ser. A*, **118**, 389–404.
Review: MR **17** (1956), 380.
Users: Barton, D., and David, F. (1956), Barton, D. E. (1958), Haight, F. A., and Fisher, G. R. (1966).
Class.: MI:mb:G
Irwin, J. O. 1959. 1339
On the estimation of the mean of a Poisson distribution from a sample with the zero class missing. *Biometrics*, **15**, 324–26.
Review: BA **34** (1959), Ab.No. 243; PA **34** (1960), Ab. No. 5009.
Users: Cohen, A. C. (1959e), Cohen, A. C. (1960b), Cohen, A. C. Jr. (1961a), Irwin, J. O. (1963).
Class.: TCP:pe:G
Irwin, J. O. 1963. 1340
The place of mathematics in medical and biological statistics. *J. Roy. Statist. Soc. Ser. A*, **126**, 1–44.
Users: Kemp, C. D., and Kemp, A. W. (1965).
Class.: MI:mi:BM
Irwin, J. O. 1965. 1341
Inverse factorial series as frequency-distributions. *Classical and Contagious Discrete Distributions*, Ed. G. P. Patil, Statistical Publishing Society, Calcutta and Pergamon Press, pp. 159–74.
Class.: IFS:m-gf:G-BM
Isabel, M. M., and Camargo, M. C.
See Camargo, M. C., and Isabel, M. M. (1956).
Ishii, Goro, and Hayakawa, Reiko. 1960. 1342
On the compound binomial distribution. *Ann. Inst. Statist. Math. Tokyo*, **12**, 69–80.
Review: MR **23A** (1962), 565; ZBL **104** (1964), 134; IJA **2** (1961), 469.
Users: Bhattacharya, S., and Holla, M. (1965).

Ishii, Goro, and Yamasaki, Mitsuru. 1960-61. 1343
A note on the testing of homogeneity of k binomial experiments based on the range. *Ann. Inst. Statist. Math. Tokyo*, **12**, 273–78.
Review: MR **25** (1963), 1088; IJA **3** (1962), 198.

Isii, K. 1957. 1344
Some investigations of the relation between distribution functions and their moments. *Ann. Inst. Statist. Math. Tokyo*, **9**, 1–12.

Isii, K. 1958. 1345
Note on a characterization of unimodal distributions. *Ann. Inst. Statist. Math. Tokyo*, **9**, 173–84.

Isii, K. 1964. 1346
On a limit theorem for a stochastic process related to quantum biophysics of vision. *Ann. Inst. Statist. Math. Tokyo*, **15**, 167–75.
Review: STMA **7** (1966), 261.
Class.: P:pc:P

Itô, Y. 1962. 1347
Distribution of the over wintering arrow-head scale, *Prontaspis yanonensis Kuwana*, on the satsuma orange leaves (Japanese; English summary). *Jap. J. Appl. Entomol. Zool.*, **6**, 183–89.
Users: Kuno, E. (1963), Kobayashi, S. (1965).

Itô, Y., Gotoh, A., and Miyashita, K. 1960. 1348
On the spatial distribution of Pieris rapae crucivora population (Japanese; English summary). *Jap. J. Appl. Entomol. Zool.*, **4**, 141–45.
User: Kobayashi, S. (1966).

Itô, Y., Nakamura, Masako, Kindo, Masaki, Miyashita, Kazuyashi, and Nakamura, Kazuo. 1962. 1349
Population dynamics of the chesnut gall-wasp *Dryocosmus kuriphilus* Yasumatsu. (Hymenoptera: Cynipidae) II. Distribution of individuals in bud of chestnut tree. *Res. Popul. Ecol.*, **4**, 35–46.
Class.: P-TCP-NB:gf:BM

Ivcenko, G. I., and Medvedev, Yu. I. 1965. 1350
Some multidimensional theorems on a classical problem of permutations. *Theor. Probability Appl.*, **10**, 144–49.
Class.: OMR-P:a:G

Ivcenko, G. I., and Medvedev, Yu. I. 1965a. 1351
Some multidimensional theorems concerning one classical problem on permutation (Russian; English summary). *Teor. Verojatnost. i Primenen.*, **10**, 156–62.

Ivcenko, G. I., and Medvedev, Yu. I. 1966. 1352
Asymptotic behaviour of a number of groups of particles in a classical problem of permutation (Russian; English summary). *Teor. Verojatnost. i Primenen.*, **11**, 701–08.
User: Kobayashi, S. (1965).
Class.: MI:a:G

Ives, W. G. H. 1955. 1353
Estimation of egg populations of the larch sawfly *Pristiphora erichsonii* (Htg). *Canad. J. Zool.*, **33**, 370–88.
Users: Waters, W. E., and Henson, W. R. (1959).

Iwao, S. 1956. 1354
The relation between the distribution pattern and the population density of the large 28-spotted lady beetle, *Epilachna 28 maculata Motschulsky*, in egg-plant field (Japanese; English summary). *Jap. J. Ecol.*, **5**, 130-35.

Iwaszkiewicz, K., and Neyman, J. 1931. 1355
Counting virulent bacteria and particles of virus. *Acta Biologiae Experimentalis*, **6**, 101-42.

Iyer, P. V. K. 1950a. 1356
The theory of probability distributions of points on a lattice. *Ann. Math. Statist.*, **21**, 198–217.
Users: Iyer, P. V. K., and Rao, A. S. (1953), Iyer, P. V. K. (1963).
Class.: MI:mb-m:G

Iyer, P. V. K. 1950. 1357
Difference equations of moment-generating functions for some probability distributions. *Nature*, **165**, 370.
Review: ZBL **36** (1951), 85.
Class.: MI:m:G

Iyer, P. V. K. 1951. 1358
The use of difference equation in solving distribution problems. *Bull. Inst. Internat. Statist.*, **33** (2), 97–104.
User: Iyer, P. V. K. (1955).
Class.: MI:mb:G

Iyer, P. V. K. 1954. 1359
Some distributions arising in matching problems. *J. Indian Soc. Agric. Statist.*, **6**, 5–29.
Users: Iyer, P. V. K. (1955), Barton, D. E. (1958).
Class.: MI:m-tc-ctp:G

Iyer, P. V. K. 1957. 1360
Some new methods for testing randomness of a binomial sequence and its applications in two sample problems. *Defence Sci. J.*, **7**, 9–26.
User: Iyer, P. V. K. (1963).

Iyer, P. V. K. 1958. 1361
A theorem on factorial moments and its applications. *Ann. Math. Statist.*, **29**, 254–61.
Review: MR **20** (1959), 62; ZBL **89** (1961), 155.
Users: Edwards, C. B. (1962), Iyer, P. V. K. (1963), Fuchs, C. E., and David, H. (1965a).
Class.: B-OBR:m:G

Iyer, P. V. K. 1958b. 1362
A simplified method for evaluating the product cumulants of some distributions arising from a sequence of observations. *J. Indian Soc. Agri. Statist.*, **10**, 33–40.
Review: MR **22** (1961), 51; ZBL **111** (1964-65), 341.

Iyer, P. V. K. 1963. 1363
Some methods of analysis of a sequence of observations and their applications. *50th Indian Sci. Congress*, Delhi.

Iyer, P. V. K., and Bhattacharyya, M. N. 1955. 1364
On some statistics comparing two binomial sequences. *J. Indian Soc. Agric. Statist.*, **7**, 187–217.
Review: MR **19** (1958), 329.
Class.: B:ctp-a-tc:G

Iyer, P. V. K., and Rao, A. S. P. 1953. 1365
Theory of the probability distribution of runs in a sequence of observations. *J. Indian Soc. Agric. Statist.*, **5**, 29–77.
User: Iyer, P. V. K. (1963).

Iyer, P. V. K., and Shakuntala, N. S. 1959. 1366
Cumulants of some distributions arising from a two-state Markoff chain. *Proc. Cambridge Philos. Soc.*, **55** 273–76.
Review: ZBL **92** (1962), 347.
User: Iyer, P. V. K. (1963).

J

Jacobsen, A. C., Marino, R. T., and Davis, S. A.
See Davis, S. A., Jacobsen, A. C., and Marino, R. T. (1965).

Jacobsen, Robert Leland. 1966. 1367
Probability mixtures. Master's Thesis, Cornell Univ.

Jacobsen, Allan S., and Haight, Frank A.
See Haight, Frank A., and Jacobson, Allan S. (1962).

James, W. H. 1963–64. 1368
Estimates of fecundability. *Population Studies*, **17**, 57–65. *Class.*: B:pe:BM

Janossy, L. 1955. 1369
Statistical problems of an electron multiplier (Russian). *Z. Eksper. Teoret. Fiz.* **28** (6), 679–94.
Review: SA **59**A (1956), Ab.No. 5601.
Notes: English translation in *Soviet Physics JETP* (*New York*), **1** (3), 520–31.

Janossy, L., Renyi, A., and Aczel, J. 1950. 1370
On composed Poisson distributions (Russian summary). *Acta Math. Acad. Sci. Hungar.*, **1**, 209–24.
Review: MR **13** (1952), 663; ZBL **41** (1952), 249; ZBL **54** (1956), 58; MR **13** (1952), 958.
Users: Rényi, A. (1951a), Aczel, J. (1952), Prékopa, A. (1957a).
Notes: Hungarian version: Osszetett Poisson-eloszlasokrol. I. *Magy. Tud. Akad. III Mat. Fiz. Oszt. Közl.*, **1**, 315–28.

Jarratt, P., and Chambers, M. L.
See Chambers, M. L., and Jarratt, P. (1964).

Jauho, P. 1954. 1371
The number of artificial nuclear reactions as a random variable. *Ann. Acad. Sci. Fenn. Ser. A I*, **165**, pp. 12.

Jeffreys, Harold. 1936. 1372
Further significance tests. *Proc. Cambridge Philos. Soc.*, **32**, 416–45.
Class.: B-MI:tp:G

Jenkins, R. M. 1955. 1373
The effect of gizzard shad on the fish population of a small Oklahoma lake. *Trans. Amer. Fisheries Soc.*, **85**, 58–74.

Jennett, W. J., and Welch, B. L. 1939. 1374
The control of proportion defective as judged by a single quality characteristics varying on a continuous scale. *J. Roy. Statist. Soc. Suppl.*, **6**, 80–88.
Review: ZBL **21** (1940), 148.

Jensen, Paul. 1959. 1375
Fit of certain distribution functions to counts of two species of cryptozoa. *Ecology*, **40**, 447–53.
Class.: NB-N-MI:gf:BM

Jessen, R. J., Hendrickson, G. O., and Kozicky, E. L.
See Kozicky, E. L., Jessen, R. J., and Hendrickson, G. O. (1956).

Jewell, W. S. 1960. 1376
The properties of recurrent-event processes. *Operations Res.*, **8**, 446–72.
Users: Oliver, R. M. (1961), Whittlesey, John R. B. (1963), Haight, F. (1965b)
Class.: P-MI:pc:G

Johnson, A. L. S., and Foster, R. E.
See Foster, R. E., and Johnson, A. L. S. (1963).

Johnson, E. A., and Brown, B. W. 1961. 1377
The Spearman estimator for serial dilution assays. *Biometrics*, **17**, 79–88.
Review: IJA **2** (1961), 689.
Class.: P:pe:BM

Johnson, N. L. 1951. 1378
Estimators of the probability of the zero class in Poisson and certain related populations. *Ann. Math. Statist.*, **22**, 94–101.
Review: MR **12** (1951), 622; ZBL **54** (1956), 62.
Class.: P-OPR:pe:G

Johnson, N. L. 1957a. 1379
Uniqueness of a result in the theory of accident proneness. *Biometrika*, **44**, 530–31.
User: Subrahmaniam, K. (1966a).
Class.: NB:rp:G-A

Johnson, N. L. 1957b. 1380
A note on the mean deviation of the binomial distribution. *Biometrika*, **44**, 532–33.
Review: MR **19** (1958), 1204.
Users: Crow, E. L. (1958b), Ramasubban, T. A. (1958), Katti, S. K. (1960a), Kamat, A. R. (1965).
Class.: B:m:G
Notes: Correction; *Biometrika*, **45**, 587.

Johnson, N. L. 1959. 1381
On an extension of the connexion between Poisson and χ^2 distributions. *Biometrika*, **46**, 352–63.
Review: IJA **1** (1960), 375; ZBL **101** (1963), 358.
Users: Haynam, G. E., and Leone, F. C. (1965).
Class.: P-NB:osp:G

Johnson, N. L. 1960. 1382
An approximation to the multinomial distribution: some properties and applications. *Biometrika*, **47**, 93–102.
Review: IJA **2** (1961), 32; ZBL **94** (1962), 336.
Users: Johnson, N. L., and Young, D. H. (1960), Govindarajulu, Z. (1965).
Class.: M:m-pe-tp-a:G

Johnson, N. L. 1965. 1383
Quota fulfilment in finite populations. *Classical and Contagious Discrete Distributions*, Ed. G. P. Patil, Statistical Publishing Society, Calcutta and Pergamon Press, pp. 419–26.
Class.: H:mi:G

Johnson, N. L., and David, F. N.
See David, F. N., and Johnson, N. L. (1950).

Johnson, N. L., and David, F. N.
See David, F. N., and Johnson, N. L. (1952).

Johnson, N. L., and David, F. N.
See David, F. N., and Johnson, N. L. (1953).

Johnson, N. L., and David, F. N.
See David, F. N., and Johnson, N. L. (1956).

Johnson, N. L., and Leone, Fred C. 1964a. 1384
Statistics and experimental design in engineering and the physical sciences. Vol. I. John Wiley and Sons, New York, 523 pp.
Review: MR **30** (1965), 499.

Johnson, Norman L., and Leone, Fred C. 1964b. 1385
Statistics and experimental design in engineering and the physical sciences. Vol. II. John Wiley and Sons, New York, 399 pp.
Review: MR **30** (1965), 499.

Johnson, N. L., and Young, D. H. 1960. 1386
Some application of two approximations to the multinomial distribution. *Biometrika*, **47**, 463–69.
Review: STMA **6** (1965), 150; ZBL **97** (1962), 347.
User: Young, D. H. (1962).
Class.: M:a-tp:G

Johnson, Paul B. 1966. 1387
The washing of socks. *Math. Mag.*, **39**, 77–83.
Class.: P:mi:G

Johnston, R. F. 1956a. 1388
Population structure in salt marsh song sparrows. Part I. Environment and annual cycle. *Condor*, **58**, 24–44.
User: Johnston, R. F. (1956b).
Class.: P:gf:BM

Johnston, R. F. 1956b. 1389
Population structure in salt marsh song sparrows. Part II. Density, age structure, and maintenance. *Condor*, **58**, 254–72.
Class.: P:gf:BM

Jolicœur, Pierre, and Brunel, Pierre. 1966. 1390
Application du diagramme hexagonal a l'étude de la sélection de ses proies par la morue. *Vie et Milieu*, **17**, 419–33.

Jones, Edward R., and Webb, Wilse B.
See Webb, Wilse B., and Jones, Edward R. (1953).

Jones, E. W. 1937. 1391
Practical field methods of sampling soil for wireworms. *J. Agri. Res.*, **54**, 123–34.
Users: Wadley, F. M. (1950), Bliss, C. I., and Owen, A. R. G. (1958).

Jones, E. W. 1945. 1392
The regeneration of the douglas fir, Pseudotsuga taxifolia Britt., in the new forest. *J. Ecol.*, **33**, 44–56.

Jones, E. W. 1955. 1393
Ecological studies on the rain forest of southern Nigeria. IV. The plateau forest of the Okumu Forest Reserve. *J. Ecol.*, **43**, 595–605.

Jones, H. L. 1959. 1394
How many of a group of random numbers will be usable in selecting a particular sample? *J. Amer. Statist. Assoc.*, **54**, 102–22.
Class.: MI-B-P:mb-m-a:G

Jones, P. C. T., Mollison, J. E., and 1395
Quenouille, M. H. 1948.
A technique for the quantitative estimation of soil micro-organisms, and a statistical note. *J. Gen. Microbiol.*, **2**, 54–69.
Users: Quenouille, M. H. (1949), Bliss, C. I., and Fisher, R. A. (1952), Hunter, G. C., and Quenouille, M. H. (1952), Gurland, J. (1957), Gurland, J. (1959), Nelson, W. C., and David, H. A. (1964), Patil, G. P., and Wani, J. K. (1965).
Class.: LS-P-NB:gf-anovat:BM

Jordan, Ch. 1927. 1396
Sur un cas généralisé de la probabilité des épreuves répétées. *Acta Sci. Math. (Szeged)*, **3**, 193–210.
Users: Geiringer, H. (1938), Sarkadi, K. (1957).

Jordan, Ch. 1927a. 1397
Sur un cas généralisé de la probabilité des éprevues répétées. *C. R. Acad. Sci. Paris*, **184**, 315–17.
User: Sarkadi, K. (1957).

Jordan, Ch. 1939. 1398
Problèmes de la probabilité des épreuves répétées dans le cas général. *Bull. Soc. Math. France*, **67**, 223–42.
Review: MR **1** (1940), 340; ZBL **23** (1941), 57.

Jorgenson, Dale W. 1961. 1399
Multiple regression analysis of a Poisson process. *J. Amer. Statist. Assoc.*, **56**, 235–45.
Review: IJA **4** (1963), 437.

Jorgenson, Dale W. 1966. 1400
Rational distributed lag functions. *Econometrica*, **32**, 135–49.
Class.: MI:a-pe:G

Josephs, K. J., and Doust, J. F.
See Doust, J. F., and Josephs, K. J. (1941).

Josephson, N. S., and Russell, A. M.
See Russell, A. M., and Josephson, N. S. (1965).

Jowett, G. H. 1963. 1401
The relationship between the binomial and F Distributions. *The Statistician*, **13**, 55–67.
Review: STMA **6** (1965), 58.
Class.: B-P:osp:G

Joyce, T., and Parker-Rhodes, A. F.
See Parker-Rhodes, A. F., and Joyce, T. (1956).

Jung, Jan. 1963. 1402
A theorem on compound Poisson processes with time dependent change variables. *Skand. Akuarietidskr.*, **46**, 95–98.
Review: STMA **6** (1965), 290.
Class.: COP:pc-m:G-S

Junge, C. O., and Chapman, D. G.
See Chapman, D. G., and Junge, C. O. (1956).

K

Kaarsemaker, L., and Wijngaarden, A. 1952. 1403
Tables for use in rank correlation. Rep. R73, Comput. Dept., Math. Centre, Amsterdam, pp. 17.
Review: MR **5** (1952), 1616.

Kabir, A. B. M. Lutful, and Ali, Mir M.
See Ali, Mir M., and Kabir, A. B. M. Lutful (1965).

Kac, M., and Hurwitz, H. Jr.
See Hurwitz, H. Jr., and Kac, M. (1944).

Kadakia, Pravin L. 1965. 1404
Confidence interval limits for various percentiles in non-parametric case (Summary). *J. Amer. Statist. Assoc.*, **60**, 664.

Kalaba, R. E. 1953. 1405
A simple derivation of the Poisson distribution. Paper P-414, RAND Corp.

Kalbfleish, J. G., and Sprott, D. A.
See Sprott, D. A., and Kalbfleish, J. G. (1965).

Kale, B. K. 1961. 1406
A characterisation of the moments of a random variable assuming only a finite number of values. *J. Indian Soc. Agri. Statist.*, **12**, 175–81.
Review: IJA **3** (1962), 150.
Class.: MI:m:G

Kaller, C. L. 1956. 1407
The negative binomial distribution. Master's thesis, Univ. Saskatchewan, Saskatoon, Canada.

Kalmus, H. 1946-47. 1408
The incidence of Placenta Praevia and Antepartum Haemorrhage according to maternal age and parity (with a note on the mathematical treatment by Cedric A. B. Smith). *Ann. Eugenics*, **13**, 283–90.

Kamat, A. R. 1964. 1409
A generalization of a property of the mean deviation for a class of discrete distributions (Abstract). *Ann. Math. Statist.*, **35**, 1406.

Kamat, A. R. 1965. 1410
Incomplete and absolute moments of some discrete distributions. *Classical and Contagious Discrete Distributions*, Ed. G. P. Patil, Statistical Publishing Society, Calcutta and Pergamon Press, pp. 45–64.
Class.: PS-B-P-NB-LS-H-IH:m:G

Kamat, A. R. 1966. 1411
A generalization of Johnson's property of the mean deviation for a class of discrete distributions. *Biometrika*, **53**, 285–87.
Class.: MI:osp:G

Kamat, A. R., Wani, J. K., and Patil, G. P.
See Patil, G. P., Kamat, A. R., and Wani, J. K. (1964).

Kambo, N. Singh, and Kotz, Samuel. 1966. 1412
On exponential bounds for binomial probabilities. *Ann. Inst. Statist. Math.*, **18**, 277–87.
Class.: B:mi:G

Kambo, N. Singh, and Kotz, Samuel. 1966a. 1413
On exponential bounds for binomial probabilities. Res. Rep. Dept. Industrial Engg., Univ. Toronto. 13 pp.
Class.: B:mi:G

Kander, Z., and Zacks, S. 1966. 1414
Test procedures for possible changes in parameters of statistical distributions occurring at unknown time points. *Ann. Math. Statist.*, **37**, 1196–1210.
Class.: B:tp:G

Kanellos, S. G. 1952. 1415
On a conditional distribution. *Bull. Soc. Math. Grèce*, **26**, 24–28.
Review: MR **14** (1953), 566.
Class.: MI:m-tp:G

Kano, Seigo. 1961. 1416
Random controlled stochastic processes generated by Poisson processes. *Sci. Rep. Kagoshima Univ.*, **10**, 1–9.

Kano, Seigo. 1961a. 1417
Random prediction and random filtering based on a Poisson process. *Sci. Rep. Kagoshima Univ.* **10**, 31–35.

Kaplansky, Irving. 1945. 1418
The asymptotic distribution of runs of consecutive elements. *Ann. Math. Statist.*, **16**, 200–03.
Users: Katz, L. (1952), Mendelsohn, N. S. (1956).
Class.: MI-P:a:G

Karlin, Samuel, and MacGregor, James. 1965. 1419
Ehrenfest urn models. *J. Appl. Prob.*, **2**, 352–76.

Karlin, Samuel and Proschan, Frank. 1960. 1420
Polya type distributions of convolutions. *Ann. Math. Statist.*, **31**, 721–36.
Class.: MI:mi:G

Kastenbaum, M. A. 1958. 1421
Estimation of relative frequencies of four sperm types in Drosophila Melanogaster. *Biometrics*, **14**, 223–28.
User: Batschelet, E. (1960).

Kathirgamatamby, N. 1953. 1422
Note on the Poisson index of dispersion. *Biometrika*, **40**, 225–28.
Users: David, F. N. (1955a), Darwin, J. H. (1957).
Class.: P:id:G

Katti, S. K. 1960a. 1423
Moments of the absolute difference and absolute deviation of discrete distributions. *Ann. Math. Statist.*, **31**, 78–84.
Review: MR **22** (1961), 1954; ZBL **92** (1962), 362; IJA **3**, (1962), 335.
User: Kamat, A. R. (1965).
Class.: MI-B-NB-P-H:ctp-m:G

Katti, S. K. 1960b. 1424
Approximations to Neyman type A and negative binomial distributions in practical problems (Abstract). *Ann. Math. Statist.*, **31**, 817.

Katti, S. K. 1960c. 1425
Distribution of dental caries in children. *J. Dental Res.*, **39**, 501–05.
Class.: MI:mb-gf:BM

Katti, S. K. 1960d. 1426
Some aspects of statistical inference for contagious distributions. Ph.D. Thesis, Iowa State Univ.

Katti, S. K. 1962a. 1427
Approximation to hypergeometrics by the binomial and to the binomial by the Poisson (Abstract). *Ann. Math. Statist.*, **33**, 819.

Katti, S. K. 1962b. 1428
Some properties of $g_1\{g_2(z)\}$. Res. Rep., School of Aviation and Medicine, Texas, and Florida State Univ.
Class.: MI:m-osp-tc:G

Katti, S. K. 1963. 1429
Use of behavioristic models in analysing special (pertaining to space) data—Part II (Abstract). *Ann. Math. Statist.*, **34**, 1126.

Katti, S. K. 1965. 1430
The log-O-Poisson distribution (Abstract). *Ann. Math. Statist.*, **36**, 1601.

Katti, S. K. 1965a. 1431
Infinite divisibility of discrete distributions. FSU Statist. Rep. M105, Dept. Statist., Florida State Univ., pp. 20.
Class.: LS-MI:mi:G

Katti, S. K. 1965b. 1432
The log-O-Poisson distribution. Res. Rep., Florida State Univ.

Katti, S. K. 1966. 1433
Interrelations among generalized distributions and their components. *Biometrics*, **22**, 44–52.
Class.: MI:m-osp:G

Katti, S. K., and Gurland, J. 1958. 1434
On the fitting of some contagious distributions (Abstract). *Ann. Math. Statist.*, **29**, 616.

Katti, S. K., and Gurland, J. 1960a. 1435
Comparison of estimators for some generalized Poisson distributions (Abstract). *Ann. Math. Statist.*, **31**, 526.

Katti, S. K., and Gurland, J. 1960b. 1436
Some methods of estimation for the Poisson binomial distribution. MRC Tech. Summary Rep. 212, Univ. Wisconsin, pp. 14.
Class.: GP-COB:pe-gf-tc:G

Katti, S. K., and Gurland, J. 1961. 1437
The Poisson Pascal distribution. *Biometrics*, **17**, 527–38.
Review: ZBL **113** (1965), 351; IJA **3** (1962), 520.
Users: Katti, S. K., and Gurland, J. (1962a), Katti, S. K., and Gurland, J. (1962b), Blischke, W. R. (1965), Crow, E. L., and Bardwell, G. E. (1965), Govindarajulu, Z. (1965), Gurland, J. (1965), Katti, S. K., and Sly, L. E. (1965), Sprott, D. A. (1965b), Katti, S. K. (1966), Dubey, S. K. (1966a).
Class.: GP-CONB-MI:a-pe-gf-m:G

Katti, S. K., and Gurland, J. 1962a. 1438
Some methods of estimation for the Poisson binomial distribution. *Biometrics*, **18**, 42–51.
Review: MR **25** (1963), 317; STMA **5** (1964), 71.
Users: Blischke, W. R. (1965), Gurland, J. (1965), Kemp, C. D., and Kemp, A. W. (1965), Sprott, D. A. (1965b).
Class.: GP-COB:pe-gf:G

Katti, S. K., and Gurland, J. 1962b. 1439
Efficiency of certain methods of estimation for the negative binomial and the Neyman type A distributions. *Biometrika*, **49**, 215–26.
Review: ZBL **114** (1965), 355; IJA **3** (1962), 554.
Users: Bardwell, G. E., and Crow, E. L. (1964), Blischke, W. R. (1965), Shenton, L. R., and Myers, R. (1965).
Class.: NB-N:pe:G

Katti, S. K., and Martin, Donald C.
See Martin, Donald C., and Katti, S. K. (1961).

Katti, S. K., and Martin, D. C.
See Martin, D. C., and Katti, S. K. (1962).

Katti, S. K., and Martin, D. C.
See Martin, D. C., and Katti, S. K. (1965).

Katti, S. K., and Rao, A. Vijay. 1966. 1440
The Log-zero-Poisson distribution. Typescript pp. 15, Florida State University.

Katti, S. K., and Singh, J. 1966. 1441
Expressions for moments in terms of cumulants and for factorial moments of generalized distributions in terms of those of its components. Rep., Florida State Univ.' pp. 16.
Class.: MI-N:m:G

Katti, S. K., and Singh, J. 1965. 1442
Expressions for moments in terms of cumulants, Res. Rep., Florida State Univ.
Class.: MI-N:m:G

Katti, S. K., and Singh, J.
See Singh, J., and Katti, S. K. (1965).

Katti, S. K., and Sly, L. E. 1965. 1443
Analysis of contagious data through behaviouristic models. *Classical and Contagious Discrete Distributions*, Ed. G. P. Patil, Statistical Publishing Society, Calcutta and Pergamon Press, pp. 303–19.
Class.: GP-N-OPR:mb-gf-cm:BM

Katti, S. K., Hofflander, A. E., and Dickerson, O. D.
See Dickerson, O. D., Katti, S. K., and Hofflander, A. E. (1961).

Katz, Leo. 1945. 1444
Characteristics of frequency functions defined by first order difference equations. Ph.D. Thesis, Univ. Michigan.

Katz, Leo. 1946. 1445
On the class of functions defined by the difference equation $(x+1)f(x+1) = (a+bx)f(x)$ (Abstract). *Ann. Math. Statist.*, **17**, 501.
Users: Patil, G. P. (1963c), Bardwell, G. E., and Crow, E. L. (1964).

Katz, L. 1952. 1446
The distribution of the number of isolates in a social group. *Ann. Math. Statist.*, **23**, 271–76.
Review: ZBL **47** (1953), 126.
User: Govindarajulu, Z. (1965).
Class.: MI:mb-a:S

Katz, Leo. 1953. 1447
Confidence intervals for the number showing a certain characteristic in a population when sampling is without replacement. *J. Amer. Statist. Assoc.*, **48**, 256–61.
Review: ZBL **50** (1954), 363.
User: Wise, M. E. (1954).
Class.: H:ie:G-E

Katz, Leo. 1965. 1448
Unified treatment of a broad class of discrete probability distributions. *Classical and Contagious Discrete Distributions*, Ed. G. P. Patil, Statistical Publishing Society, Calcutta and Pergamon Press, pp. 175–82.
Class.: MI-B-P-NB:m-pe-gf:G-P

Katz, Leo, and Wilson, T. R. 1956. 1449
The variance of the number of mutual choices in sociometry. *Psychometrica*, **21**, 299–304.
Class.: MI:mb-osp:S

Katz, Morris W. 1961. 1450
Admissible and minimax estimates of parameters in truncated spaces. *Ann. Math. Statist.*, **32**, 136–42.
User: Katz, M. W. (1963).
Class.: MI-B-P:pe:G

Katz, Morris W. 1963. 1451
Estimating ordered probabilities. *Ann. Math. Statist.*, **34**, 967–72.
Review: STMA **5** (1964), 604.
Class.: B:pe:G

Kaufman, A., and Cruon, R. 1961. 1452
Temps d'attente dans une file avec arrivées suivant la loi de Poisson par " grappes ". *Chiffres*, **3**, 135–42.

Kawamura, Kazutomo. 1964. 1453
Asymptotic behavior of sequential design with costs of experiments. *Kodai Math. Sem. Rep.*, **16**, 169–82.
Review: MR **30** (1965), 315.
Class.: B:ctp:G

Kawamura, M., Asai, A., and Murakami, M. 1954. 1454
The estimation of Poisson parameter from a truncated distribution and a censored sample (Japanese). *J. College Arts Sci. Chiba Univ.*, **1**, 141–48.
Class.: TCP:pe-tc:G

Kawamura, Tomoo. 1964–65. 1455
Transient behavior of Poisson queue. *J. Operations Res. Soc. Japan*, **7**, 76–92.
Review: MR **31** (1966), 957.
Class.: MI:pc-a:G

Keats, J. A. 1963. 1456
Some generalizations of a theoretical distribution of mental test scores. Tech. Rep. RB-63-25, Educational Testing Service, Princeton, N. J., pp. 26.
Class.: OHR:mb-gf:S

Keats, J. A. 1964. 1457
Some generalizations of a theoretical distribution of mental test scores. *Psychometrika*, **29**, 215–31.
Class.: OHR:mb-gf:S

Keats, J. A., and Lord, F. M. 1962. 1458
A theoretical distribution for mental test scores. *Psychometrika*, **27**, 59–72.
Review: ZBL **100** (1963), 346.
Users: Lord, F. M. (1962), Keats, J. A. (1964), Lord, F. M. (1964a).
Class.: NH-MHR:mb-ie:S

Keeping, E. S. 1956. 1459
Statistical decisions. *Amer. Math. Monthly*, **63**, 147–59.
Class.: B:pe-tp:G

Keilin, J. E. 1961. 1460
The exact distribution of the differences of two binomial variables for small, equal size samples. Master's Thesis, George Washington Univ., Washington, D. C.

Keilson, J., and Kooharian, A. 1960. 1461
On time dependent queuing processes. *Ann. Math. Statist.*, **31**, 104–12.

Kelleher, T. M., and Kojima, Ken-Ichi.
See Kohima, Ken-Ichi, and Kelleher, T. M. (1962).

Kellerer, Hans G. 1964. 1462
Linearkombinationen zufälliger Grössen and ihre gemeinsame Verteilung. *Math. Z.*, **84**, 403–414.
Review: MR **29** (1965), 778.

Kemp, A. W., and Kemp, C. D.
See Kemp, C. D., and Kemp, A. W. (1956).

Kemp, A. W., and Kemp, C. D.
See Kemp, C. D., and Kemp, A. W. (1965).

Kemp, A. W., and Kemp, C. D. 1966. 1463
An alternative derivation of the Hermite distributions. *Biometrika*, **53**, 627–28.
Class.: GP:osp:G

Kemp, C. D. 1965. 1464
Fitting the " short " distribution for accidents (Abstract). *Biometrics*, **21**, 776.

Kemp, C. D., and Kemp, A. W. 1956a. 1465
The analysis of point quadrat data. *Austral. J. Bot.*, **4**, 167–74.
Users: Kemp, C. D., and Kemp, A. W. (1956b), Bliss, C. I., and Fleischer, D. A. (1964).
Class.: OHR:anovat-pe-gf:G-BM

Kemp, C. D., and Kemp, A. W. 1956b. 1466
Generalized hypergeometric distributions. *J. Roy. Statist. Soc. Ser. B*, **18**, 202–11.
Review: MR **18** (1957), 769.
Users: Kemp, C. D., and Kemp, A. W. (1965a), Sarkadi, K. (1957), Edwards, A. W. F. (1958), Govindarajulu, Z. (1965), Knight, W. (1965).
Class.: N-NH-IH-OHR:mb-m-pe:G

Kemp, C. D., and Kemp, A. W. 1965. 1467
Some properties of the Hermite distribution. *Biometrika*, **52**, 381–94.
Review: STMA **7** (1966), 264.
Class.: GP-MI:m-pe-osp-a-gf:G-BM

Kemp, C. D., and Kemp, A. W.
See Kemp, A. W., and Kemp, C. D. (1966).

Kempthorne, Oscar. 1966. 1468
Some aspects of experimental inference. *J. Amer. Statist. Assoc.*, **61**, 11–34.
Class.: MI:mi:G

Kendall, D. G. 1948. 1469
On some modes of population growth leading to R. A. Fisher's logarithmic series distribution. *Biometrika*, **35**, 6–15.
Users: Anscombe, F. J. (1949a), Kendall, D. G. (1949), Skellam, J. G., (1949), Oakland, G. B. (1950), Ramakrishnan, A. (1951), Skellam, J. G. (1952), Brian, M. V. (1953), David, F. N., and Moore, P. G. (1954), Robinson, P. (1954), David, F. N. (1955a), Simon, H. A. (1955), Patil, G. P. (1959), Pielou, D. P. (1960a), Cassie, R. M. (1962), Patil, G. P. (1962b), Tsao, C. M. (1962), Clark, P. J., Eckstrom, P. T., and Linden, L. C. (1964), Nelson, W. C., and David, H. A. (1964), Williamson, E., and Bretherton, M. (1964), Patil, G. P., and Shorrock, R. W. (1966).
Class.: LS:pc-mb:G

Kendall, David G. 1949. 1470
Stochastic processes and population growth. *J. Roy. Statist. Soc. Ser. B*, **11**, 230–64.
Review: ZBL **38** (1951), 88.
Users: Anscombe, F. J. (1950a), Bartlett, M. S. (1957), Ramakrishnan, A., and Srinivasan, S. K. (1958), Gurland, J. (1959), Bartko, J. J. (1961b), Blackwell, D., and Kendall, D. (1966).

Kendall, David G. 1953a. 1471
Stochastic processes and the growth of bacterial colonies. *Symposia Soc. Experimental Biology*, **7**, 55–65.
Review: ZBL **53** (1961), 416.

Kendall, D. G. 1953. 1472
Stochastic processes occurring in the theory of queues and their analysis by the method of the imbedded

Markov chain. *Ann. Math. Statist.*, **24**, 338–54.
Users: Benes, V. E. (1957), Haight, Frank A. (1957), Finch, P. D. (1959b), Haight, F. (1959a), Burrows, C. (1960), Conolly, B. W. (1960), Finch, P. D. (1960), Finch, P. D. (1960b), Prabhu, N. U. (1960), Finch, P. D. (1961-62).

Kendall, D. G. 1960. 1473
Birth-and-death processes, and the theory of carcinogenesis. *Biometrika*, **47**, 13–21.
Review: IJA **2** (1961), 201.

Kendall, D. G. 1960a. 1474
The bibliography of operational research. *Operational Res. Quart.*, **11**, 31–36.

Kendall, D. G. 1963. 1475
Information theory and the limit theorem for Markov chains and processes with a countable infinity of states. *Ann. Inst. Statist. Math.*, **15**, 137–43.
Review: MR **30** (1965), 307
Class.: MI:mi:G

Kendall, D. G., and Blackwell, David.
See Blackwell, David, and Kendall, David (1964).

Kendall, D. G., and Stuart, A. 1958. 1476
Advanced theory of statistics. Vol. I. Charles Griffin and Co. London.

Kendall, M. G. 1952. 1477
Moment-statistics in samples from a finite population. *Biometrika*, **39**, 14–15.
Review: ZBL **46** (1953), 358.

Kendall, M. G. 1961. 1478
Natural law in the social sciences. *J. Roy. Statist. Soc. Ser. A*, **124**, 1–16.
Users: Irwin, J. O. (1963), Chatfield, C., Ehrenverg, A. S. C., and Goodhardt, G. J. (1965), Irwin, J. O. (1965), Chatfield, C., Ehrenberg, A. S. C., and Goodhardt, G. J. (1966).

Kendall, Maurice G., and Doig, Alison G. 1965. 1479
Bibliography of statistical literature, Vol. 2, 1940-49. Oliver and Boyd, Edinburgh and London, 190 pp.

Kennedy, W. A. 1954. 1480
Tagging returns, age studies and fluctuations in abundance of Lake Winnipeg whitefish, 1931–1951. *J. Fisheries Res. Board Canada*, **11** (3), 284–309.

Ker, J. W. 1954. 1481
Distribution series arising in quadrat sampling of reproduction. *J. Forestry*, **52**, 838–41.
Users: Shive, C. J., and Beazley, R. (1957), Smith, J. H. G., and Ker, J. W. (1957).

Ker, J. W., and Smith, J. H. G.
See Smith, J. H. G., and Ker, J. .W. (1957)

Kerrich, J. C., and Arbous, A. G.
See Arbous, A. G., and Kerrich, J. C. (1951).

Kerridge, D. 1964. 1482
Probabilistic solution of the simple birth process. *Biometrika*, **51**, 258–59.
Review: STMA **6** (1965), 298; MR **31** (1966), 148.
Class.: B:pc:G

Kershaw, K. A. 1958. 1483
An investigation of the structure of a grassland community. *J. Ecol.*, **46**, 507–25.

Kerstan, J. 1964. 1484
Generalisation of a theorem of Prochorov and Lecam (German). *Zeit. Wahrscheinlichkeitsch.*, **2**, 173–79.
Review: STMA **6** (1965), 32.

Kerstan, J., and Matthes, K. 1964. 1485
Stationare zufallige Punktfolgen II. *Jber. Deutsch. Math.-Verein*, **66**, 106–18.

Kesten, Harry, and Morse, Norman. 1959. 1486
A property of the multinomial distribution. *Ann. Math. Statist.*, **30**, 120–27.
Review: MR **21** (1960), 837; IJA **1** (1959), 191.
Users: Bechhofer, R. E., Elmaghraby, S., and Morse, N. (1959), Patil, G. P., and Bildikar, S. (1966a).
Class.: B:tp:G

Kettman, G. 1959. 1487
Comparison of the principal acceptance sampling systems with a new German system (French). *Rev. Statist. Appl.*, **7**, 5–25.
Review: IJA **2** (1961), 358.

Keuls, M., Over, H. J., and Dewit, C. T. 1962. 1488
The distance method for estimating densities. *Statistica Neerlandica*, **17**, 71–91.
Review: STMA **5** (1964), 359.
Class.: P:pe-h:G-BM

Khratri, C. G. 1959. 1489
On certain properties of power-series distributions. *Biometrika*, **46**, 486–90.
Review: ZBL **90** (1961), 356; IJA **1** (1960), 576.
Users: Khatri, C. G. (1960), Patil, G. P. (1962a), Patil, G. P. (1962c), Patil, G. P. (1962d), Bardwell, G. E., and Crow, E. L. (1964), Nelson, W. C., and David, H. A. (1964), Shanfield, F. (1964), Govindarajulu, Z. (1965), Patil, G. P. (1965), Tweedie, M. C. K., (1965), Patil, G. P., and Bildikar, S. (1966a), Mathai, A. M. (1966).
Class.: PS:m:G

Khatri, C. G. 1961-62. 1490
On the distributions obtained by varying the number of trials in a binomial distribution. *Ann. Inst. Statist. Math. Tokyo*, **13**, 47–51.
Review: MR **24** B (1962), 186; IJA **3** (1962), 521
Class.: COB:mb-pe-gf:G-BM

Khatri, C. G. 1962a. 1491
A method for estimating approximately the parameters of a certain class of distributions from grouped observations. *Ann. Inst. Statist. Math. Tokyo*, **14**, 57–62.
Review: MR **27** (1964), 178.

Khatri, C. G. 1962b. 1492
A fitting procedure for a generalized binomial distribution. *Ann. Inst. Statist. Math. Tokyo*, **14**, 133–41.
Review: MR **30** (1965), 131; ZBL **114** (1965), 347; STMA **5** (1964), 33.
Class.: GB:pe-gf:G-BM

Khatri, C. G. 1962c. 1493
Distributions of order statistics for discrete case. *Ann. Inst. Statist. Math. Tokyo*, **14**, 167–71.
Review: MR **27** (1964), 181; STMA **6** (1965), 480; ZBL **114** (1965), 113.
User: Gupta, S. S. (1965).
Class.: MI:os:G

Khatri, C. G. 1962d. 1494
A simplified method of fitting the doubly or singly truncated negative-binomial distribution. *J. Maharaja Sayajirao Univ. Baroda*, **11**, 35–38.
Class.: TCNB:pe-gf:G

Khatri, C. G. and Patel, I. R. 1961. 1495
Three classes of univariate discrete distributions. *Biometrics*, **17**, 567–75.
Review: IJA **3** (1962), 336 ; ZBL **111** (1964-65), 159.
Users: Khatri, C. G. (1962b), Blischke, W. D. (1965), Kemp, C. D., and Kemp, A. W. (1965), Sprott, D. A. (1965b).
Class.: MI:mb-m-pe-gf:G-BM

Khazanie, R. G., and McKean, H. E. 1966. 1496
A Mendelian Markov process with binomial transition probabilities. *Biometrika*, **53**, 37–48.
Class.: B:pc:BM

Khinchin, A. Ya. 1929. 1497
Über einen neuen Grenzwertsatz der Wahrscheinlichkeitsrechnung. *Math. Ann.*, **101**, 745–52.

Khinchin, A. Ya. 1937. 1498
Zur theorie der unbeschränkt teilbaren Verteilungsgesetze. *Mat. Sb.*, **44**, 79–120.

Khinchin, A. Ya. 1956a. 1499
On Poisson sequences of chance events. *Theor. Probability Appl.*, **1**, 291–97.
Class.: P:pc:G

Khinchin, A. Ya. 1956b. 1500
Über Poissonache Folgen zufälliger Ereignisse (Russian; German summary). *Teor. Verojatnost. i Primenen.*, **1**, 320–27.
Review: ZBL **73** (1960), 128.

Khinchin, A. Ya. 1962. 1501
Erlang's formulae in queueing theory (Russian). *Teor. Verojatnost. i Primenen.*, **7**, 330–35.

Khinchin, A. Ya. 1963. 1502
Erlang's formulas in the theory of mass service. *Theor. Probability Appl.*, **7**, 320–25.

Kiefer, J., and Wolfowitz, J. 1956. 1503
Sequential tests of hypotheses about the mean occurrence time of a continuous parameter Poisson process. *Naval. Res. Logist. Quart.*, 3, 205–219.

Kiefer, J. and Wolfowitz, J. 1959. 1504
Asymptotic minimax character of the sample distribution function for vector chance variables. *Ann. Math. Statist.*, **30** 463–89.
Class.: M:pe:G

Kincaid, W. M. 1962. 1505
The combination of tests based on discrete distributions. *J. Amer. Statist. Assoc.*, **57**,10–19.
Review: ZBL **114** (1965), 354; STMA **6** (1965), 551.
Class.: MI-B:tp-c:G

Kingman, J. F. C. 1961. 1506
Two similar queues in parallel. *Ann. Math. Statist.*, **32**, 1314–23.
Review: STMA **5** (1964), 486.
Class.: MI:pc:G

Kingman, J. F. C. 1962. 1507
On queues in heavy traffic. *J. Ròy. Statist. Soc. Ser. B*, **24**, 383–92.
Review: IJA **4** (1963), 531.
Class.: MI:pc:G

Kingman, J. F. C. 1963. 1508
Poisson counts for random sequences of events. *Ann. Math. Statist.*, **34**, 1217–32.
Review: STMA **6** (1965), 300.
Class.: P:pc:G

Kingman, J. F. C. 1964. 1509
On doubly stochastic Poisson processes. *Proc. Cambridge Philos. Soc.*, **60**, 923–30.

Kingston, Charles E. 1965a. 1510
Applications of probability theory to criminalistics. *J. Amer. Statist. Assoc.*, **60**, 70–80.
User: Kingston, C. R. (1965b).
Class.: B-P:mb:S

Kingston, Charles E. 1965b. 1511
Applications of probability theory in criminalistics II. *J. Amer. Statist. Assoc.*, **60**, 1028–34.
Class.: B:mi:S

Kinney, E. C. 1956. 1512
The otter trawl as a fish sampling device in western Lake Erie. *Trans. Amer. Fisheries Soc.*, **86**, 58–60.

Kinney, J. R. 1962. 1513
A transient discrete time queue with finite storage. *Ann. Math. Statist.*, **33**, 130–36.
Review: STMA **5** (1964), 747.
Class.: MI:pc:G

Kirkham, W. J. 1935. 1514
Moments about the arithmetic mean of a binomial frequency distribution. *Ann. Math. Statist.*, **6**, 96–101.
Users: Larguier, E. H. (1936), Curtiss, J. H. (1941), Patil, G. P. (1959).
Class.: B:m:G

Kitagawa, T. 1941a. 1515
The limit theorems of the stochastic contagious processes. *Mem. Fac. Sci. Kyūshū Univ. Ser. A*, **1**, 167–94.
Review: MR **2** (1941), 230; ZBL **24** (1941), 160.
Users: Kitagawa, T. (1941b), Feller, W. (1943), Tsao, C. M. (1962), Kitagawa, T. (1965).
Class.: MI:pc:G

Kitagawa, T. 1941b. 1516
The weakly contagious discrete stochastic process. *Mem. Fac. Sci. Kyūshū Univ. Ser. A*, **2**, 37–65.
Users: Kitagawa, T., Huruya, S., and Yazima, T. (1942), Kitagawa, T. (1965).
Class.: MI:pc:G

Kitagawa, T. 1952. 1517
Tables of Poisson Distribution. Baifukan, Tokyo, pp. 156.
Review: MR **15** (1954), 724.

Kitagawa, T. 1965. 1518
The weakly contagious discrete stochastic process. *Classical and Contagious Discrete Distributions*, Ed. G. P. Patil, Statistical Publishing Society, Calcutta and Pergamon Press, pp. 18–32.
Class.: MI:pc:G

Kitagawa, Tosio. 1965a. 1519
The weakly contagious discrete stochastic process. *Sankhyā Ser. A*, **27**.

Kitagawa, T. and Huruya, S. 1941. 1520
The application of the limit theorem of the contagious stochastic process to the contagious diseases. *Mem. Fac. Sci. Kyūshū Univ. Ser. A*, **1**, 195–207.
Review: ZBL **24** (1941), 160.
Users: Kitagawa, T., Huruya, S., and Yazima, T. (1942), Feller, W. (1943), Tsao, C. M. (1962).
Class.: MI:pe:BM

Kitagawa, T., Huruya, S., and Yazima, T. 1942. 1521
The probabilistic analysis of the time-series of rare events. I. *Mem. Fac. Sci. Kyūshū Univ. Ser. A*, **2**, 151–204.
Class.: MI:h:G-P-BM

Kleczkowski, A. 1949. 1522
The transformation of local lesion counts for statistical analysis. *Ann. Appl. Biol.*, **36**, 139–52.
Users: Bliss, C. I., and Owen, A. R. G. (1958), Hayman, B. I., and Lowe, A. D. (1961), Cassie, R. M. (1962).

Klerk Grobben, G., and Prins, H. J. 1954. 1523
A test for comparing two small unknown probabilities, using samples of equal size, and its power (Dutch; English summary). *Statistica Neerlandica*, **8**, 7–20.
Review: MR **16** (1955), 499.

Klimov, V. N. 1957. 1524
On a local limit theorem for lattice distributions. *Theor. Probability Appl.*, **2**, 260–65.

Klinger, H., and Hetz, W.
See Hetz, W., and Klinger, H. (1958).

Knight, F. B. 1963. 1525
The distribution of twig-boring insects in the crowns of Aspen. *Proc. North Central Branch, E.S.A.*, **18**, 65–67.

Knight, William. 1965. 1526
A method of sequential estimation applicable to the hypergeometric, binomial, Poisson and exponential distributions. *Ann. Math. Statist.*, **36**, 1494–1503.
Review: MR **31** (1966), 1140.
Class.: H-B-P:se:G

Knox, E. G. 1964. 1527
The detection of space-time interactions. *Appl. Statist.*, **13**, 25–29.
Review: STMA **7** (1966), 358.
Class.: P:a:G

Knox, E. G. 1964a. 1528
Epidemiology of childhood leukemia in Northumberland and Durham. *Brit. J. Prev. Soc. Med.*, **18**, 17–24.

Kobayashi, S. 1957. 1529
Studies on the distribution structure of the common cabbage butterfly, *Pieris rapae* in cabbage farm. *Sci. Rep. Tohoku Univ. 4th Ser.*, **23** (1), 1–6.
Users: Kobayashi, S. (1960), Kobayashi, S. (1960a) and Kobayashi, S. (1966).
Class.: P-GP:gf:BM

Robayashi, Shirô. 1960. 1530
Studies on the distribution pattern of the eggs of the common cabbage butterfly, *Pieris rapae crucivora* in a cabbage farm and the factors affecting its concentrating trend. *Jap. J. Ecol.*, **10** (4), 154–60.
Class.: P-COP-GP:gf:BM

Kobayashi, Shirô. 1960a. 1531
Studies on the progressive change in the distribution pattern of the common cabbage butterfly, *Pieris rapae crucivora* Boisduval in a cabbage farm and its mechanisms. (Japanese; English summary). *Jap. J. Ecol.*, **10**, 233–38.
Class.: P-GP:gf:BM

Kobayashi, Shirô. 1965. 1532
Influence of parental density on the distribution pattern of eggs in the common cabbage butterfly, *Pieris rapae crucivora*. *Res. Population Ecol.*, **7**, 109–17.
User: Kobayashi, S. (1966).
Class.: NB-P:gf:BM

Kobayashi, Shirô. 1966. 1533
Process generating the distribution pattern of eggs of the common cabbage butterfly, *Pieris rapae crucivora*. *Res. Population Ecol.*, **8**, 51–61.
Class.: P-NB-GP:gf:BM

Kojima, Ken-Ichi, and Kelleher, T. M. 1962. 1534
Survival of mutant genes. *Amer. Naturalist*, **96**, 329–46.
Class.: NB-P:mb:BM

Kolčin, V. F. 1966. 1535
The speed of convergence to limiting distributions in a classical problem with balls. *Teor. Verojatnost. i Primenen.*, **11**, 144–56.
Class.: MI-B:a:G

Kolmogorov, A. N. 1950. 1536
Unbiased estimates (Russian). *Izv. Akad. Nauk SSSR Ser. Mat.*, **14**, 303–26.
User: Sirazhdinov, S. H. (1956).
Notes: Translation; *Amer. Math. Soc. Translation*, No. 98, pp. 28 (1953).

Kolomogorov, A. N. 1951. 1537
Generalization of Poisson's formula to the case of a sample from a finite set (Russian). *Uspehi, Mat. Nauk* **6, 3** (43), 133–34.
Review: MR **13** (1952), 258; ZBL **44** (1952), 137.
User: Bolshev, L. N. (1964a).

Komarek, E. V., Wood, J. E., and Davis, D. E.
See Wood, J. E., David, D. E., and Komarek, E. V. (1958)

Kondo, Masaki, Miyashita, Kazuyashi, Nakamura, Kazuo, Itô, Y., and Nakamura, Masako.
See Itô, Y., Nakamura, Masako, Kondo, Masaki, Miyashita, Kazuyashi, and Nakamura, Kazuo (1962).

Konig, Dieter and Matthes, Klaus. 1963. 1538
Verallgemeinerungen der Erlangschen Formeln I. *Math. Nachr.*, **26**, 45–56.

Kono, K. and Uranisi, H. 1961. 1539
Estimate of the number of defectives when inspectors may make miss checks. *Bull. Math. Statist.*, **10**, 57–68.
Review: ZBL **103** (1964), 366.

Kono, T. 1952. 1540
Time-dispersion curve. Experimental studies on the dispersion of insects, 2 (Japanese; English summary). *Res. Population Ecol.*, **1**, 109–18.

Kono, T. 1953. 1541
Basic unit of population observed in the distribution of the rice-stem borer, Chilo simplex, in a paddy field. Pattern of the spatial distribution of insects. 4th report (Japanese). *Res. Population Ecol.*, **2**, 95–105.

Kono, T., and Sugino, T. 1958. 1542
On the estimation of the density of rice stems infected by the rice stem borer (Japanese; English summary). *Jap. J. Appl. Entomol. Zool.*, **2**, 184–88.
Class.: P-NB:pe:BM

Kono, Tatsuro, Utida, Synnro, Yosida, Tosiharu, and Watanabe, Syozi. 1952. 1543
Pattern of spatial distribution of the rice-stem borer Chilo simplex, in a paddy field (Japanese; English summary). *Res. Population Ecol.*, **1**, 65–82.
Class.: P-NB-GP:gf:BM

Kono, T., Watanabe, S., Yosida, T., and Utida, S.
See Uitda, S., Kono, T., Watanabe, S., and Yosida, T. (1952).

Kooharian, A., and Keilson, J.
See Keilson, J., and Kooharian, A. (1960).

Koopman, B. O. 1950a. 1544
Necessary and sufficient conditions for Poisson's distribution. *Proc. Amer. Math. Soc.*., **1**, 813–23.
Review: MR **12** (1951), 424; ZBL (1952), 249.
Users: Koopman, B. O. (1951), Walsh, J. E. (1955b), Walsh, J. E. (1955c), Fitzpatrick, R. (1958), Eisenberg, H. B., Geoghagen, R. M., and Walsh, J. E. (1963), Govindarajulu, Z. (1965).
Class.: P:osp:G

Koopman, B. O. 1950b. 1545
A generalization of Poisson's distribution for Markoff chains. *Proc. Nat. Acad. Sci. U.S.A.*, **36**, 202–07.
Review: ZBL **37** (1951), 85.
Users: Koopman, B. O. (1951a), McCord, J. R. (1963).
Class.: OPR:pc:G

Koopman, B. O. 1951. 1546
A law of small numbers in Markoff chains. *Trans. Amer. Math. Soc.*, **70**, 277–90.
User: McCord, J. R. (1963).
Class.: OPR:pc:G

Koopmans, L. H., and Clark, C. R.
See Clark, C. R., and Koopmans, L. H. (1959).

Kordonskiĭ, H. B. 1958. 1547
The distribution of the number X of defective items in a lot (Russian). Proc. All-Union Conf. Theory Prov. and Math. Statist. (Erevan) (Russian). pp. 172–86. Izdat. Akad. Nauk. Armjan. SSR, Erevan, 1960.
Review: MR **32** (1966), 317

Kordonskii, KH. B. 1958. 1548
On a distribution of the number of defective items in manufactured lots. *Theor. Probability Appl.*, **3**, 329–33.
User: Kordonskii, Kh. B. (1961).
Class.: MI:sqc:G

Kordonskii, Kh. B. 1958a. 1549
On a distribution for the number of spoiled articles in shipments (Russian; English summary). *Teor. Verojatnost. i Primenen*, **3**, 354–58.

Kordonskii, Kh. B. 1961. 1550
The distribution of the number of defective items in manufactured lots. *Theor. Probability Appl.*, **6**, 314–20.
Class.: MI:sqc:G

Kordonskii, Kh. B. 1961a. 1551
The distribution of the number of defective items in lots (Russian; English summary). *Teor. Verojatnost. i Primenen.*, **6**, 342–49.
Review: MR **24**A (1962), 216.
Class.: MI:sqc:G

Koroljuk. V. S. 1956. 1552
Asymptotic expansions for distributions of maximum deviations in the scheme of Bernoulli (Russian). *Dokl. Akad. Nauk SSSR*, **108**, 183–86.
Review: MR **18** (1957), 241; ZBL **71** (1958), 347.
Users: Koroljuk, V. S. (1961), Govindarajulu, Z. (1965).

Koroljuk, V. S. 1959. 1553
Asymptotic analysis of maximum deviation distributions in the Bernoulli scheme (Russian; English summary). *Teor. Verojatnost. i Primenen.*, **4**, 369–79.
Review: MR **22** (1961), 1720; ZBL **101** (1963), 130.

Koroljuk, V. S. 1961. 1554
Asymptotic analysis of the distribution of the maximum deviation in the Bernoulli scheme. *Theor. Probability Appl.*, **4**, 339–66.
Review: ZBL **101** (1963), 130.

Koroljuk, V. S., and Gusak, D. V. 1962. 1555
On the asymptotic behaviour of distributions of maximal deviations in the Poisson process (Russian). *Ukrain. Mat. Z.*, **14**, 138–44.

Kosambi, D. D. 1949. 1556
Characteristic properties of series distributions. *Proc. Nat. Inst. Sci. India*, **15**, 109–13.
Review: MR **11** (1950), 42.
Users: Shanfield, F. (1964), Patil, G. P. (1965), Mathai, A. M. (1966).
Class.: PS:m-pe-osp:G

Kosambi, D. D. 1959. 1557
An application of stochastic convergence. *J. Indian Soc. Agri. Statist.*, **11**, 58–72.
Review: IJA **2** (1961), 608.
Class.: MI:mi:G

Kotz, Samuel. 1964. 1558
Russian-English dictionary of statistical terms and expressions and Russian reader in statistics. Univ. N. Carolina Monogr. Ser. Probab. Statist. No. 1., Univ. N. Carolina Press, Chapel Hill, N.C. 115 pp.
Review: MR **30** (1966), 587.

Kotz, Samuel, and Kambo, N. Singh.
See Kambo, N. Singh, and Kotz, Samuel, (1966).

Kousgaard, E., and Hald, A.
See Hald, A., and Kousgaard, E. (1963).

Kousgaard, E., and Hald, A.
See Hald, A., and Kousgaard, E. (1966).

Kovalenko, I. N. 1959a. 1559
On one class of optimal resolving functions for a binomial family of distributions. *Teor. Verojatnost. i Primenen.*, **4** 101–05.
Review: MR **12** (1960), 77; ZBL **102** (1963), 354.
Class.: B:tp:G

Kovalenko, I. N. 1959b. 1560
On a class of optimal decision functions for a binomial family of distributions. *Theor. Probability Appl.*, **4**, 95–99.
Review: MR **22** (1961), 2178.
Class.: B:tp:G

Kozelka, R. M. 1956. 1561
Approximate upper percentage points for extreme values in multinomial sampling. *Ann. Math. Statist.*, **27**, 507–512.
Review: MR **17** (1956), 1222.
Users: Johnson, N. L., and Young, D. H. (1960), Young, D. H. (1962), Patil, G. P., and Bildikar, S. (1966a).
Class.: M:os-a:G

Kozicky, E. L., Jessen, R. J., and Kendrickson, G. O. 1956. 1562
Estimation of fall population in Iowa. *J. Wildlife Managem.*, **20**, 97–104.

Kozniewska, I. 1954. 1563
The first absolute central moment for Polya's distributions (Polish; Russian and English summaries). *Zastos. Mat.*, **1**, 206–11.
Review: MR **16** (1955), 602; ZBL **55** (1955), 121.
User: Sarkadi, K. (1957a).

Krishnamoorthy, A. S. 1951. 1564
Multivariate binomial and Poisson distributions. *Sankhyā*, **11**, 117–24.
Review: MR **13** (1952), 478; ZBL **44** (1952), 135.
Users: Patil, G. P., and Bildikar, S. (1966).
Class.: MB:MP:osp:G

Krooth, R. S. 1952-53. 1565
The fertility of the parents of abnormals. *Ann. Eugenics*, **17**, 79–89.

Krumbein, W. C. 1954. 1566
Applications of statistical methods to sedimentary rocks. *J. Amer. Statist. Assoc.*, **49**, 51–66.
Class.: B-P:gf:P

Krysicki, Vladimir. 1961. 1567
Sur une formule asymptotique. *C. R. Acad. Sci. Paris*, **253**, 369–71.
Review: MR **24A** (1962), 321.
Class.: B:a:G

Krysicki, W. 1955. 1568
On the combined problem of Bayes and Bernoulli (Polish; Russian and English summary). *Zastos. Mat.*, **2**, 172–78.
Review: ZBL **66** (1957), 111.

Krysicki, Wlodzimierz. 1957. 1569
Remarques sur la loi de Poisson. *Bull. Soc. Sci. Lettres Lódź*, **8**, 1–24.
Review: ZBL **87** (1961), 136.

Krysicki, W., and Olekiewicz, M. 1963. 1570
On the generalized joint problem of Bayes and Bernouilli (Polish). *Zastos. Mat.*, **7**, 77–103.
Review: STMA **5** (1964), 801.
Class.: COB:osp-a:G

Kubik, L. 1959. 1571
The limiting distributions of cumulative sums of independent two-valued random variables. *Studia Math.*, **18**, 295–309.
Review: ZBL **113** (1965), 341.
Class.: MI:a:G

Kubik, L. 1960. 1572
On the limiting distributions of sums of r-valued random variables (Polish; Russian and English summaries). *Prace Mat.*, **4**, 111–18.
Review: ZBL **104** (1964), 370.

Kudô, A., and Furukawa, N. 1958. 1573
A model in probit analysis. *Bull. Math. Statist.*, **8**, 1–7.
Review: IJA **2** (1961), 97.
Class.: P:mi:G

Kudô, Akio, and Yao, Jing-Shing. 1964. 1574
Some considerations on the multiple inverse sampling method. *Bull. Math. Statist.*, **11** (1–2), 63–77.
Review: MR **30** (1965), 1011.
Class.: NB-NM-MI:tp-a:G

Kudô, A., Okuma, A. Yamato, H., and Yanagawa, T. 1965. 1575
On sequential multinomial estimation (Abstract). *Ann. Math. Statist.*, **36**, 732.

Kuenzler, E. J. 1958. 1576
Niche relations of three species of Lycosid spiders. *Ecology*, **39** (3), 494–500.
Class.: P:gf:BM

Kulkarni, N. V., and Bhat, B. R.
See Bhat, B. R., and Kulkarni, N. V. (1966).

Kullback, Solomon. 1935. 1577
On the Bernoulli distribution. *Bull. Amer. Math. Soc.*, **41**, 857–64.
User: Riordan, J. (1937).

Kullback, Solomon. 1937. 1578
On certain distributions derived from the multinomial distribution. *Ann. Math. Statist.*, **8**, 127–44.
Users: Patil, G. P., and Bildikar (1966a).
Class.: OMR:mb-gf:G

Kuller, Robert G. 1964. 1579
Coin tossing, probability, and Weierstrass approximation theorem. *Math. Mag.*, **37**, 262–65.

Kuno, E. 1963. 1580
A comparative analysis on the distribution of nymphal populations of some leaf and planthoppers on rice plant (Japanese summary). *Res. Population Ecol.*, **5**, 31–43.
Users: Shiyomi, M., and Nakamura, K. (1964), Kobayashi, S. (1965), Kobayashi, S. (1966).
Class.: P-NB:id-gf:BM

Kuno, E., Yamamoto, S., Satomi, H., Ōuti, Y., and Okada, T. 1963. 1581
On the assessment of insect populations in a large area of paddy field based on the negative binomial distribution (Japanese). *Proc. Assoc. Pl. Prot. Kyushu*, **9**, 33–36.
Class.: NB:h-anovat:BM

Kupper, Josef. 1962. 1582
Modelle mit Wahrscheinlichkeitsansteckung (French, English and Italian summaries). *Mitt. Verein. Schweiz. Versich.-Math.*, **62**, 183–94.

Kutkuhn, J. H. 1958. 1583
Notes on the precision of numerical and volumetric plankton estimates from small-sample concentrates. *Limnology and Oceanography*, **3** (1), 69–83.
User: Bamforth, S. S. (1958).

L

Ladegast, K. 1953. 1584
Einige Abschätzungen für endliche ustetige Verteilungen. *Mitteilungsbl. Math. Statist.*, **5**, 75–86.
Review: MR **14** (1953), 770.

Ladouceur, J. C., Narayana, T. V., and Mohanty, S. G.
See Narayana, T. V., Mohanty, S. G., and Ladouceur, J. C. (1960).

Lagler, K. F., and Cooper, G. P.
See Cooper, G. P., and Lagler, K. F. (1956).

Laha, R. G., and Lukacs, E.
See Lukacs, E., and Laha, R. G. (1964).

Lancaster, H. O. 1949a. 1585
The derivation and partition of χ^2 in certain discrete distributions. *Biometrika*, **36**, 117–29.
Review: ZBL **33** (1950), 74.
Users: Cochran, W. G. (1954), Lancaster, H. O. (1954), Corstein, L. C. A. (1957), Bodmer, W. F. (1959).
Class.: M-B:mi:G

Lancaster, H. O. 1949b. 1586
The combination of probabilities arising from data in discrete distributions. *Biometrika*, **36**, 370–82.
Review: ZBL **34** (1950), 229.
Users: David, F. N., and Johnson, N. L. (1950), Pearson, E. S. (1950), Lancaster, H. O. (1952), Kincaid, W. M. (1962), Ellner, H. (1963).
Class.: B:tp:G

Lancaster, H. O. 1950a. 1587
The theory of amoebic surveys. *J. Hyg. Camb.*, **48**, 257–76.

Lancaster, H. O. 1950b. 1588
Statistical control in haematology. *J. Hyg. Camb.*, **48**, 402–17.
Users: Lancaster, H. O. (1952), Lancaster, H. O. (1954), Turner, M. E., and Eadie, G. S. (1957).

Lancaster, H. O. 1950. 1589
The exact partition of χ^2 and its application to the problem of the pooling of small expectation. *Biometrika*, **37**, 267–70.
User: Corstein, L. C. A. (1957).

Lancaster, H. O. 1950–51. 1590
The sex ratios in sibships with special reference to Geissler's data. *Ann. Eugenics*, **15**, 153–58.
Users: Robertson, A. (1951), Edwards, A. W. F. (1958).

Lancaster, H. O. 1952. 1591
Statistical control of counting experiments. *Biometrika*, **39**, 419–22.
Users: Skellam, J. G. (1952), Kathirgamatamby, H. (1953), Turner, M. E., and Eadie, G. S. (1957), Bennett, B. M. (1959), Lancaster, H. O. (1961), Ellner, H. (1963), Nicholson, W. L. (1963).
Class.: MI:mi:G

Lancaster, H. O. 1953–54. 1592
A reconciliation of χ^2 considered from metrical and enumerative aspects. *Sankhyā*, **13**, 1–10.
Review: ZBL **52** (1955), 152.
Users: Corstein, L. C. A. (1957), Watson, G. S. (1959), Richter, W. (1964).
Class.: M-B:gf:G

Lancaster, H. O. 1961. 1593
Significance tests in discrete distributions. *J. Amer. Statist. Assoc.*, **56**, 223–34; correction: **57**, 919.
Review: ZBL **104** (1964), 132; IJA **4** (1963), 66.
User: Ellner, H. (1963).
Class.: B-H-P-M:tp-a:G

Lancaster, H. O. 1962. 1594
Significance tests in discrete distributions. *J Amer. Statist. Assoc.*, **57**, 919.
Review: IJA **4** (1963), 459.

Lancaster, H. O. 1966. 1595
Forerunners of the Pearson χ^2. *Austral. J. Statist.*, **8**, 117–26.
Class.: M:gf:G

Lancaster, H. O., and Brown, T. A. I. 1965. 1596
Sizes of the χ^2 test in the symmetrical multinomials. *Austral. J. Statist.*, **7**, 40–44.
Review: STMA **7** (1966), 334.
Class.: M-P:gf:G

Langenhop, C. E., and Mathews, J. C. 1964. 1597
Invariance of probabilities in finite sample spaces under stochastic operations. *Amer. Math. Monthly*, **71**, 841–49.
Review: MR **30** (1965), 994.
Class.: MI:mi:G

Laplace, Pierre S. 1812. 1598
Theorie analytique des probabilities.
Notes: Reprinted in *Euvres completes de Laplace, Treatise 7*, Gauthier Villars, Paris, 1886.

Larguier, Everett H. 1936. 1599
On a method for evaluating the moments of a Bernoulli distribution. *Ann. Math. Statist.*, **7**, 191–95.
Class.: B:m:G

Larsen, Harold D. 1939. 1600
Moments about the arithmetic mean of a hypergeometric frequency distribution. *Ann. Math. Statist.*, **10**, 198–201.
Review: ZBL **22** (1940), 244.
Class.: H:m:G

Larsh, J. E., Gilchrist, H., and Greenberg, B. G.
See Greenberg, B. G., Larsh, J. E., and Gilchrist, H. (1952).

Latif, A., and Blench, T.
See Blench, T., and Latif, A. (1950).

Laubscher, Nico F. 1960. 1601
On the stabilization of the Poisson variance. *Trabajos Estadist.*, **11**, 199–207.
Review: MR **25** (1963), 137; ZBL **99** (1963), 145; STMA **5** (1964), 569.
Class.: P:anovat:G

Laubscher, Nico F. 1961. 1602
On stabilizing the binomial and negative binomial variances. *J. Amer. Statist. Assoc.*, **56**, 143–50.

Review: IJA **4** (1963), 605; ZBL **109** (1964-65), 133.
Class.: B-NB:anovat:G

Laurent, A. G. 1965. 1603
Probability distributions, factorial moments, empty cell test. *Classical and Contagious Discrete Distributions*, Ed. G. P. Patil, Statistical Publishing Society, Calcutta and Pergamon Press, pp. 437–42
Class.: MI:m:G

Lea, D. E., Thoday, J. M., and Catcheside, D. G.
See Catcheside, D. G., Lea, D. E., and Thoday, J. M. (1945–46).

Leadbetter, M. R., and Watson, G. S.
See Watson, G. S., and Leadbetter, M. R. (1963).

Leander, E. K., and Finney, D. J. 1956. 1604
An extension of the use of the χ^2 test. *Appl. Statist.*, **5**, 132–36.
Review: MR **18** (1957), 78.
Class.: P-B:tp:E

Lecam, Lucien. 1960. 1605
An approximation theorem for the Poisson binomial distribution. *Pacific J. Math.*, **10**, 1181–97.
Review: ZBL **118** (1965–66), 336.
Users: Hodges, J. L., and LeCam, L. (1960), Makable, H. (1962), Govindarajulu, Z. (1965).
Class.: GBOP-B-P:a-mi:G

Lecam, Lucien, and Hodges, J. L., Jr.
See Hodges, J. L., Jr., and LeCam, Lucien (1960).

Lechner, J. A. 1962. 1606
Optimum decision procedures for a Poisson process parameter. *Ann. Math. Statist.*, **33**, 1384–1402.
Review: ZBL **114** (1965), 344.
Users: Lindley, D. V., and Barnett, B. (1965).

Lee, T. H., and Zellner, A.
See Zellner, A., and Lee, T. H. (1963).

Leech, J. W., and Davies, R. O.
See Davies, R. O., and Leech, J. W. (1954).

Lehman, E, H., Jr., and Mendenhall, W.
See Mendenhall, W., and Lehman, E. H., Jr. (1960).

Lehmann, E. L. 1952. 1607
Testing multiparameter hypotheses. *Ann. Math. Statist.*, **23**, 541-52.
Class.: M:tp:G

Lehmann, E. L. 1966. 1608
Some concepts of dependence. *Ann. Math. Statist.*, **37**, 1137–53.

Lehmann, E. L., and Hodges, J. L., Jr.
See Hodges, J. L., Jr., and Lehmann, E. L. (1950).

Lehmann, E. L., and Hodges, J. L., Jr.
See Hodges, J. L., Jr., and Lehmann, E. L. (1952).

Lehmann, E. L., and Hodges, J. L., Jr.
See Hodges, J. L., Jr., and Lehmann, E. L. (1964).

Lehmann, E. L., and Stein, C. 1949. 1609
On the theory of some non-parametric hypotheses. *Ann. Math. Statist.*, **20**, 28–45.

Lejeune, Jerome. 1958. 1610
Sur une solution "A priori" de la methode "A posterior" de Haldane. *Biometrics*, **14**, 513–20.
Review: IJA **1** (1959), 52; ZBL **85** (1961), 141.

Lellouch, J., and Wembersie, A. 1966. 1611
Estimation par la methode du maximum de vraisemblance des courbes de survie de microorganismes irradies (English summary). *Biometrics*, **22**, 673–83.
Class.: P:pe:BM

Leonard, W. H., and Clark, Andrew.
See Clark, Andrew, and Leonard, W. H. (1939).

Leone, F. C., and Hayman, G. E.
See Hayman, G. E., and Leone, F. C. (1965).

Leone, Fred C., and Johnson, Norman L.
See Johnson, Norman L., and Leone, Fred C. (1964).

Leone, F. C., Hayman, G. E., Chu, J. T., and Topp, C. W. 1960. 1612
Percentiles of the binomial distribution. Statist. Lab. Rep. 1030, Case Inst. Tech., pp. 36.
Class.: B:tc:G

Leslie, P. H., and Davis, D. H. S. 1939. 1613
An attempt to determine the absolute number of rats on a given area. *J. Animal Ecol.*, **8** (1), 94–113.
Users: Moran, P. A. P. (1951), Chapman, D. G. (1954).

Leslie, P. H., and Gower, J. C. 1958. 1614
The properties of a stochastic model for two competing species. *Biometrika*, **45**, 316–30.
Class.: MI:pc-gf:BM

Leslie, P. H., and Orians, G. H.
See Orians, G. H., and Leslie, P. H. (1958).

Leslie, R. T., and Binet, F. E.
See Binet, F. E., and Leslie, R. T. (1960).

Leslie, R. T., Weiner, S., Anderson, R. L., and Binet, F. E.
See Binet, F. E., Leslie, R. T., Weiner, S., and Anderson, R. L. (1956).

Letac, Gerard. 1964. 1615
Une propriété de fluctuation des processus de Poisson composés croissants. *C. R. Acad. Sci. Paris*, **258**, 1700–03.

Leti, G. 1961. 1616
The distributions of the extremes and of the range of samples from a discrete and finite population (Italian). *Metron*, **21**, 201–55.
Review: STMA **5** (1964), 346.
Class.: B-MI:mi:G

Levine, Jack. 1958. 1617
A binomial identity related to rhyming sequences. *Math. Mag.*, **32**, 71–74.
Review: MR **20** (2) (1959), 949.
Class.: B:mi:G

Lévy, P. 1937a. 1618
Sur les exponentielles de polynomes. *Ann. Sci. Ecole Norm. Sup.*, **73**, 231–92.
Users: Kemp, C. D., and Kemp, A. W. (1965).

Lévy, P. 1937b. 1619
L'arithmétique des lois de probabilité et les produits finis des lois de Poisson. *C. R. Acad. Sci. Paris*, **204**, 944–46.
Review: ZBL **16** (1937), 127.
Class.: P:mi:G

Lévy, P. 1937c. 1620
Nouvelle contribution à l'arithmétique des produits de lois de Poisson. *C. R. Acad. Sci. Paris*, **205**, 535–37.

Lévy, P. 1937. 1621
Complement à un theorème sur la loi de Gauss. *Bull. Sci. Math.*, **61**, 115–28.
Review: ZBL **16** (1937), 364.

Lewis, Peter A. W. 1964. 1622
A branching Poisson process model for the analysis of computer failure patterns. *J. Roy. Statist. Soc. Ser. B*, **26**, 398–456.

Lewis, Peter A. W. 1965. 1623
Some results on tests for Poisson processes. *Biometrika*, **52**, 67–77.
Review: STMA **6** (1965), 1225.

Lewontin, R. C., and Prout, Timothy. 1956. 1624
Estimation of the number of different classes in a population. *Biometrics*, **12**, 211–23.
User: Pathak, P. K. (1964).

Li, C. C. 1957. 1625
Repeated linear regression and variance components of a population with binomial frequencies. *Biometrics*, **13**, 225–33.
Class.: B:anovat:G-BM

Lighte, William H. 1947. 1626
A method and tables for obtaining standard errors of differences between proportions whenN_1 is equal to N_2. *J. Appl. Psychol.*, **31**, 449–56.
Review: PA **22** (1948), Ab. No. 1963.
Class.: B:ctp-tc:G

Lidstone, G. J. 1942. 1627
Notes on the Poisson frequency distribution. *J. Inst. Actuar.*, **71**, 284–91.
Review: MR **4** (1943), 20.

Lieberman, Gerald J., and Owen, Donald B. 1961. 1628
Tables of the hypergeometric probability distribution. Stanford Studies in Math. and Statist., III. Stanford Univ. Press, pp. 726.
Review: ZBL **93** (1962), 324; MR **22** (1961), 1463.

Lieblein, Julius. 1949. 1629
Partial sums of the negative binomial in terms of the incomplete β-function (Abstract). *Ann. Math. Statist.*, **20**, 623.
Users: Tsao, C. M. (1962), Patil, G. P. (1963c).

Lienau, C. C. 1941. 1630
Discrete bivariate distribution in certain problems of statistical order. *Amer. J. Hyg. Sect. A*, **33** (3), 65–85.
Review: BA **15** (1941), Ab. No. 14214.
Users: Patil, G. P., and Bildikar, S. (1966a).

Likeš, J. 1962. 1631
The efficiency of some estimators for the variance of a two-valued variable (German). *Metrika*, **5**, 184–93.
Review: STMA **5** (1964), 75.
Class.: B:pe:G

Limber, D. N. 1953. 1632
Analysis of counts of the extragalactic nebulae in terms of a fluctuating density field. *Astrophys. J.*, **117**, 134–44.
Users: Neyman, J., and Scott, E. (1954).
Class.: MI:mi:P

Limber, D. N. 1954. 1633
Analysis of counts of the extragalactic nebulae in terms of a fluctuating density field. *Astrophys. J.*, **119**, 655–81.
Class.: MI:mi:P

Limber, D. N. 1957. 1634
Analysis of counts of the extragalactic nebulae in terms of a fluctuating density field. *Astrophys. J.*, **125**, 9–14.
Class.: MI:mi:P

Linden, L. C., Clark, P. J., and Eckstrom, P. T.
See Clark, P. J., Eckstrom, P. T., and Linden, L. C. (1964).

Linder, Arthur. 1935. 1635
Wahrscheinlichkeitsansteckung und Differenzenglei-chungen. *Metron*, **12**, (4), 71–89.

Lindley, D. V. 1957. 1636
Binomial sampling schemes and the concept of information. *Biometrika*, **44**, 179–86.
Review: ZBL **84** (1960), 358; MR **19** (1958), 30.
Users: Edwards, A. W. F. (1960a), Novick, M. R., and Grizzle, J. E. (1965), Tweedie, M. C. K. (1965).
Class.: B:mi:G

Lindley, D. V., and Barnett, B. N. 1965. 1637
Sequential sampling: two decision problems with linear losses for binomial and normal random variables. *Biometrika*, **52**, 507–32.
Review: STMA **7** (1966), 335.
Users: Aitchison, J., and Sculthorpe, Diane (1965), Pratt, John W. (1966).

Linhart, H. 1959. 1638
Techniques for discriminant analysis with discrete variables. *Metrika*, **2**, 138–49.
Review: IJA **1** (1959), 248; ZBL **90** (1961), 361; MR **21** (1960), 1130.

Linnik, Yu. V. 1957a. 1639
On the composition of Gaussian and Poissonian probability laws (Russian). *Dokl. Akad. Nauk SSSR* **114**, 21–24.
Review: MR **19** (1958), 889; ZBL **84** (1960), 140.

Linnik, Yu. V. 1957b. 1640
Some theorems on the factorization of infinitely divisible laws. *Dokl. Akad. Nauk SSSR*, **116**, 549–51.

Linnik, Yu. V. 1957c. 1641
On the factorization of infinitely divisible laws (Russian). *Dokl. Akad. Nauk SSSR*, **116**, 735–37.

Linnik, Yu. V. 1957d. 1642
On the decomposition of the convolution of Gaussian and Poissonian laws. *Theor. Probability Appl.*, **2**, 31–57.
Class.: P:mi:G

Linnik, Yu. V. 1957e. 1643
On factorizing the composition of a Gaussian and a Poissonian law (Russian; English summary). *Teor. Verojatnost. i Primenen.*, **2**, 34–59.
Review: MR **19** (1958), 777; ZBL **88** (1961), 114.

Linnik, Yu. V. 1958. 1644
General theorems on the factorization of infinitely divisible laws. *Teor. Verojatnost. i Primenen.*, **3**, 3–40.
Review: ZBL **94** (1962), 325.
User: Ramachandran, B. (1961a).

Linnik, Yu. V. 1959a. 1645
General theorems of the factorization of infinitely divisible laws. *Teor. Verojatnost., i Primenen.*, **4**, 55–85.
Review: IJA **1** (1960), 577; ZBL **94** (1962), 325.
User: Ramachandran, B. (1961a).

Linnik, Yu. V. 1959b. 1646
General theorems on the factorization of infinitely divisible laws. III. Sufficient conditions (countable bounded Poisson spectrum; unbounded spectrum; "stability"). *Teor. Verojatnost. i Primenen.*, **4**, 150–71.
User: Ramachandran, B. (1961a).

Linnik, Yu. V. 1959c. 1647
Five lectures on some topics in number theory and probability theory (Russian; Hungarian and English summaries). *Magyar Tud. Akad. Mat. Kutató Int. Közl.*, **4**, 225–58.
Review: MR **28** (1964), 764.

Lipson, H. I., and Schrek, R.
See Schrek, R., and Lipson, H. I. (1941).

Littauer, S. R., and Peach, P.
See Peach, Paul, and Littauer, S. B. (1946).

Little, J. D. C. 1961. 1648
Approximate expected delays for several maneuvers by a driver in Poisson traffic. *Operations Res.*, **9**, 39–52.
User: Oliver, R. (1962).

Littleford, R. A., Newcombe, C. L., and Shepherd, B. B. 1940. 1649
An experimental study of certain quantitative plankton methods. *Ecology*, **12** (3), 309–22.
Users: Robertson, O. H. (1947), Kutkuhn, J. H. (1958).
Class.: P:h:BM

Locher, Milan P. 1963. 1650
On the combinatorial derivation of the coefficients in the product density method (German summary). *Z. Angew. Math. Phys.*, **14**, 368–72.
Review: MR **14** (1964), 9.

Locke, L. G. 1964. 1651
Bayesian statistics. *Industrial Quality Control*, **20** (10), 18–22.

Lockhart, E. E. 1951. 1652
Binomial systems and organoleptic analysis. *Food Tech.*, **5**, 428–31.
Users: Baker, G. A., Amerine, M. A., and Roessler, E. B. (1954).

Loève, Michel. 1950. 1653
On sets of probability laws and their limit elements. *Univ. California Publ. Statist.*, **1**, 53–88.
Users: Ahmed, M. S. (1961), Edwards, C. B. (1962), Fuchs, C. E., and David, H. (1965a).
Class.: MI-P:a:G

Lokki, O. 1961. 1654
An application of the common chi-square test. *Soc. Sci. Fenn., Comm. Physico-Math.*, **26**, VIII 1–10.
Review: IJA **3** (1962), 366.
Class.: MI:gf:G

Longuet-Higgins, M. S. 1952. 1655
On the statistical distributionof the heights of sea waves. *J. Marine Res.*, **11** (3), 245–66.

Lord, F. M. 1960. 1656
The negative hypergeometric distribution with practical application to mental test scores. Naval Res. Contract Nonr-2752(00), Educational Test. Ser., Princeton, N. J. pp. 19.
Class.: NH:m-osp-c-mb:S

Lord, F. M. 1962. 1657
Estimating true measurements from fallible measurements (binomial case)—expansion in a series of Beta distributions. Naval Res. Contract Nonr-2752(00), Educational Test. Ser., Princeton, N.J., pp. 26
Class.: B-NH:pe-gf:S

Lord, F. M. 1964. 1658
A strong true-score theory, with applications. Res. Bull. RB-64-19, Educational Test. Ser., Princeton, N. J., pp. 58.
Class.: COB:mb-gf-mi:S

Lord, F. M. 1964a. 1659
True score theory—the four-parameter beta model with binomial errors. Naval Res. Contract Nonr-2752(00), Educational Test. Ser. Princeton, N. J.
Class.: COB:mb-gf:S

Lord, Frederic M. 1965. 1660
A strong true-score theory, with applications. *Psychometrica*, **30**, 239–70.
Class.: COB:mb-pe:BM

Lord, F. M., and Keats, J. A.
See Keats, J. A., and Lord, F. M. (1962).

Lowe, A. D., and Hayman, B. I.
See Hayman, B. I., and Lowe, A. D. (1961).

Luce, R. D. 1966. 1661
A model for detection in temporally unstructured experiments with a Poisson distribution of signal presentations. *J. Math. Psychol.*, **3**, 48–64.

Luders, R. 1934. 1662
Die statistik der seltenen ereignisse. *Biometrika*, **26**, 108–28.
Review: BA **9** (1935), Ab. No. 15361.
Users: Cernuschi, F., and Castagnetto (1946), Anscombe, F. J. (1950a), Barton, D. E. (1957), Barton, D. E., and David, F. N. (1959a), Bartko, J. J. (1961b).

Ludwig, Otto. 1957. 1663
Die Pascalsche Fragestellung für Merkmalsiterationen. *Mitteilungsbl. Math. Statist.*, **9**, 1–26; 81–101.
Review: ZBL **81** (1959), 137.

Ludwig, W. 1937. 1664
Fehlerrechnung bei biologischen Messungen. *Naturwissenschaften*, **25**, 459–60.

Lukacs, E. 1954. 1665
Sur une caracterisation de la distribution de Poisson. *C. R. Acad. Sci. Paris*, **239**, 1114–16.
Review: MR **16** (1955), 377; ZBL **58** (1958), 342.
User: Lukacs, E. (1960).
Class.: P:osp:G

Lukacs, E. 1960. 1666
On the characterization of a family of populations which includes the Poisson population (Abstract). *Ann. Math. Statist.*, **31**, 239.

Lukacs, E. 1960a. 1667
Characteristic functions. Charles Griffin and Co., London and Hafner Publishing Co., New York, 216 pp.

Lukacs, E. 1960–61. 1668
On the characterization of a family of populations which includes Poisson populations. *Ann. Univ. Sci. Budapest. Eötvös Sect. Math.*, **3–4**, 159–75.
User: Lukacs, E. (1965).
Class.: P-MI:osp:G

Lukacs, E. 1963. 1669
Applications of characteristic functions in statistics. *Sankhyā*, **25A**, 175–88.
Review: MR **30** (1965), 311; STMA **6** (1965), 34.
Class.: B-NB:osp:G

Lukacs, E. 1965. 1670
Characterization problems for discrete distributions. *Classical and Contagious Discrete Distribution*, Ed. G. P. Patil, Statistical Publishing Society, Calcutta and Pergamon Press, pp. 65–74. Reprinted in *Sankhyā A*, **27** (1966).
Class.: MI-B-P-NB-G:osp:G

Lukacs, E., and Laha, R. G. 1964. 1671
Applications of characteristic functions. Hafner Publishing Co., New York, 202 pp.
Review: MR **30** (1965), 499.

Lundberg, F. 1930. 1672
Über die Wahrscheinlichkeitsfunktion einer Risikenmasse. *Skand, Aktuarietidskr.*, **13**, 1–

Lundberg, Ove. 1940. 1673
On random processes and their application to sickness and accident statistics. Ph.D. Thesis, Univ. of Stockholm, Uppsala.
Users: Feller, W. (1943), Govindarajulu, Z. (1965).

Lundkvist, K. 1936. 1674
Calculation of the grade of service in automatic telephone systems. *Ericsson Technic*, **4**, 75–81.

Lyons, L. A. 1962. 1675
The effect of aggregation on egg and larval survival in Neodiprion swainei Midd (Hymenoptera: Diprionidae). *Canad. Entomologist*, **94** (1), 49–58.

Lyons, L. A. 1964. 1676
The spatial distribution of two pine saw lies and methods of sampling for the study of population dynamics. *Canad. Entomologist*, **96**, 1373–1457.

M

MacArthur, R. H. 1957. 1677
On the relative abundance of bird species. *Proc. Nat. Acad. Sci. U.S.A.*, **43**, 293–95.
Users: Hairston, N. G. (1959), MacArthur, R. H. (1960), Clark, P. J., Eckstrom, P. T., and Linden, L. C. (1964).
Class.: MI:mb BM

MacArthur, R. H. 1960. 1678
On the relative abundance of species. *Amer. Naturalist*, **94**, 25–36.
Users: Clark, P. J., Eckstrom, P. T., and Linden, L. C. (1964).

Maceda, E. C. 1948. 1679
On the compound and generalized Poisson distributions. *Ann. Math. Statist.*, **19**, 414–16.
Review: MR **10** (1949), 552; ZBL **32** (1950), 37.
Users: Janossy, L., Renyi, A., and Aczel, J. (1950), Crow, E. L. (1958a), Teicher, H. (1960a), Tsao, C. M. (1962), Blischke, W. R. (1965).
Class.: GP-COP:osp:G

MacFadyen, A. 1952. 1680
The small arthropods of a Molinia fen at Cothill. *J. Animal Ecol.*, **21** (1), 87–117.
Class.: P:h:BM

MacFadyen, A. 1953. 1681
Notes on methods for the extraction of small soil arthropods. *J. Animal Ecol.*, **22** (1), 65–77.
Class.: GP:mi:BM

Mack, C. 1950. 1682
The expected number of aggregates in a random distribution of *n* points. *Proc. Cambridge Philos. Soc.*, **46**, 285–92.
Review: ZBL **36** (1951), 209.
User: Naus, J. I. (1965).
Class.: MI:mi:G

Mack, C. 1953. 1683
The effect of overlapping in bacterial counts of incubated colonies. *Biometrika*, **40**, 220–22.
Review: ZBL **50** (1954), 365.
Class.: MI:mi:BM

MacKay, J. H. 1959. 1684
Asymptotically efficient tests based on the sums of observations. *Ann. Math. Statist.*, **30**, 806–13.
Class.: B:tp:G

MacKenzie, W. A., Fisher, R. A., and Thornton, H. G.
See Fisher, R. A., Thornton, H. G., and MacKenzie, W. A. (1922).

MacKinnon, William J. 1959. 1685
Compact table of twelve probability levels of the symmetric binomial cumulative distribution for sample sizes to 1,000. *J. Amer. Statist. Assoc.*, **54**, 164–72.
Review: ZBL **85** (1961), 138; IJA **2** (1961), 425.
Class.: B:tp-tc:G
Correction: *J. Amer. Statist. Assoc.* (1959), **54**, 811.

MacLeod, P., Anderson, E. O., Bliss, C. I., and Morgan, M. E.
See Morgan, M. E., MacLeod, P., Anderson, E. O., and Bliss, C. I. (1951).

Maclulich, D, A. 1951. 1686
A new technique of animal census, with examples. *J. Mammology*, **32** (3), 318–28.

Maclulich, D. A. 1957. 1687
The place of chance in population processes. *J. Wildlife Managem.*, **21** (3), 293–99.

MacNaughton-Smith, P. 1965. 1688
Some statistical and other numerical techniques for classifying individuals. *Res. Rep., Home Office Res. Unit*, London.
Class.: M:mi:G

MacStewart, W. 1941. 1689
A note on the power of the sign test. *Ann. Math. Statist.*, **12**, 236–39.
Review: ZBL **25** (1942), 200.
Class.: B:ctp-tp:G

Madow, William G. 1941. 1690
On some new results in the sampling of discrete random variables (Abstract). *Ann. Math. Statist.*, **12**, 123.

Madow, W. G. 1948. 1691
On the limiting distributions of estimates based on samples from finite universes. *Ann. Math. Statist.*, **19**, 535–45.
Review: ZBL **37** (1951), 86.
User: Morgenstern, D. (1961).

Maéda, Hiroshi. 1963. 1692
Distribution pattern of groundfishes hooked along a row of setline in the shallower part of the continental slope in the Bering Sea 2. Bathymetric difference in relative abundance in relation to distribution pattern (Japanese summary). *Res. Population Ecol.*, **5**, 74–86.
Class.: P-GP:gf:BM

Magath, Thomas B., Hurn, M., and Berkson, J.
See Berkson, J., Magath, T. B., and Hurn, M. (1935).

Magath, T. B., Hurn, M., and Berkson, J.
See Berkson, J., Magath, T. B., and Hurn, M. (1940).

Magistad, J. G. 1961. 1693
Some discrete distributions associated with life testing. *Proc. 7th Nat. Symp. Rel. Qual. Contr.* pp. 1–11.
Review: STMA **5** (1964), 571.
Class.: B-NB-MI:pc:G-E

Maguire, B. A., Pearson, E. S., and Wynn, A. H. A. 1952. 1694
The time intervals between industrial accidents. *Biometrika*, **39**, 168–80.
Users: Birnbaum, A. (1953), Cox, D. R. (1953), Birnbaum, A. (1954b), Moore, P. G. (1954b), Taylor, W. F. (1956), Thomissen, F. X. (1956), Fitzpatrick, R. (1958), Haight, F. (1965b).

Maguire, B. A., Pearson, E. S., and Wynn, A. H. A. 1953. 1695
Further notes on the analysis of accident data. *Biometrika*, **40**, 214–16.

Mahamunulu, D. M. 1962. 1696
Summation of an infinite series. *Math. Student*, **30**, 207–08.
Class.: TCNB:mi:G

Mahamunulu, D. M. 1964. 1697
Some remarks on the regression in the multivariate Poisson distribution (Abstract). *Ann. Math. Statist.*, **35**, 1850.

Mainland, D., Herrera, L., and Sutcliffe, M. I. 1956. 1698
Statistical tables for use with binomial samples—contingency tests, confidence limits, and sample size estimates. Dept. Medical Statist., New York Univ. College of Medicine, pp. 83.
Class.: B:tp-ie-pe-tc:G

Makabe, H. 1955. 1699
A normal approximation to binomial distribution. *Rep. Statist. Appl. Res. Un. Japan Sci. Engrs.*, **4**, 47–53.
Review: MR **17** (1956), 756.
Users: Makabe, H., and Morimura, H. (1956), Govindarajulu, Z. (1965).

Makabe, H. 1962. 1700
On the approximations to some limiting distributions with some applications. *Kōdai. Math. Sem. Rep.*, **14**, 123–33.
Class.: P-GBDP:a-sqc:G-E

Makabe, H. 1964. 1701
On approximations to some limiting distributions with applications to the theory of sampling inspections by attributes. *Kōdai. Math. Sem. Rep.*, **16**, 1–17.
Review: MR **28** (1964), 1064.
User: Govindarajulu, Z. (1965).
Class.: B-P-NB-GBDP:a-sqc:G-E

Makabe, H., and Morimura, H. 1955. 1702
A normal approximation to Poisson distribution. *Rep. Statist. Appl. Res. Un. Japan Sci. Engrs.*, **4**, 37–46.
Review: MR **17** (1956), 756.
Users: Makabe, H., and Morimura, H. (1956), Makabe, H. (1962), Govindarajulu, Z. (1965).

Makabe, H., and Morimura, H. 1956. 1703
On the approximation to some limiting distributions. *Kōdai Math. Sem. Rep.*, **8**, 31–40.
Review: MR **18** (1957), 423.
Users: Makabe, H. (1962), Govindarajulu, Z. (1965).
Class.: B-P-MI:a:G

Malmquist, S. 1947. 1704
A statistical problem connected with the counting of radioactive particles. *Ann. Math. Statist.*, **18**, 255–64.
Users: Sandelius, M. (1950), Nadler, J. (1960).
Class.: P:pc:P

Malý, V. 1960. 1705
Sequential problems with multiple decisions and sequential estimation. I: General theory and application to the binomial distribution (German). *Biom. Zeit.*, **2**, 45–64.
Review: IJA **2** (1960), 80; ZBL **101** (1963), 354.
Class.: B:tp:G

Malý, V. 1961. 1706
Sequential problems with different decisions and sequential estimation. II: Special case of application of the sequential analysis to the binomial distribution (German). *Biom. Zeit.*, **3**, 149–77.
Review: IJA **3** (1962), 368; ZBL **101** (1963), 354.
Class.: B-P:tp-se:G

Malý, V. 1961a. 1707
Sequenzprobleme mit mehereren Entscheidungen und Sequenzschätzung. III: Anwendung auf die Poisson-Verteilung. *Biom. Zeit.*, **3**, 157–65.
Review: ZBL **101** (1963), 354.

Malý, V. 1961b. 1708
Sequenzprobleme mit mehereren Entscheidungen und Sequenzschätzung. IV: Anwendung auf die Normalverteilung. *Biom. Zeit.*, **3**, 166–77.
Review: ZBL **101** (1963), 354.

Malý, V., and Grimm, H.
See Grimm, H., and Malý, V. (1961).

Malý, V., and Grimm, H.
See Grimm, H., and Malý, V. (1962).

Manheimer, D. I., Mellinger, G. D., Slywester, D. L., and Gaffey, W. R.
See Mellinger, G. D., Sylwester, D. L., Gaffey, W. R., and Manheimer, D. I. (1965).

Mantel, Nathan. 1966a. 1709
F-ratio probabilities from binomial tables. *Biometrics*, **22**, 404–07.
Class.: B:c:G

Mantel, Nathan. 1966b. 1710
Light bulb statistics. *J. Amer. Statist. Assoc.*, **61**, 633–39.
Class.: B:pc:G

Mantel, N., and Patwary, K. M. 1961. 1711
Interval estimation of single parametric functions. *Bull. Inst. Internat. Statist.*, **38** (4), 227–40.
Review: IJA **4** (1963), 256.
Class.: B:pe:G

Mantel, Nathan, Ederer, Fred, and Myers, Max H.
See Ederer, Fred, Myers, Max H., and Mantel, Nathan (1964).

Marczewski, E. 1953. 1712
Remarks on the Poisson stochastic process. II. *Studia Math.*, **13**, 130–36.
Review: ZBL **50** (1954), 139.

Marczewski, E., Ryll-Nardzewski, C., and Florek, K.
See Florek, K., Marczewski, E., and Ryll-Nardzewski, C. (1953).

Mardia, K. V. 1961. 1713
An important integral and applications in partial fractions. *Math. Student*, **29**, 15–20.
Class.: NB:osp:G

Margalef, R. 1957. 1714
La toeria de la informacion en ecologia. *Mem. Real Acad. Ci. Art. Barcelona*, **32**, 373–449.
Users: Hairston, N. G. (1959), Clark, P. J., Eckstrom, P. T., and Linden, L. C. (1964).

Margolin, Barry H., and Winokur, Herbert S., Jr. 1715
1966.
Exact moments of the order statistics of the geometric distribution and their application to a stage-dependent binomial sampling scheme (Abstract). *Ann. Math. Statist.*, **37**, 765.

Margolis, J., and Bishop, Y. M. M.
See Bishop, Y. M. M., and Margolis, L. (1955).

Marino, R. T., Davis, S. A., and Jacobsen, S. C.
See Davis, S. A., Jacobsen, S. C., and Marino, R. T.

Maritz, J. S. 1950. 1716
On the validity of inferences drawn from the fitting of Poisson and negative binomial distributions to observed accident data. *Psychol. Bull.*, **47**, 434–43.
Users: Blum, M. L., and Mintz, A. (1951), Sichel, H. S. (1951), Webb, M., and Jones, E. R. (1953), Arbous, A. G., and Sickel, H. S. (1954a), Arbous, and Sichel, H. S. (1954b), Fitzpatrick, R. (1958), Edwards, C. B., and Gurland, J. (1960), Edwards, C. B., and Gurland, J. (1961), Haight, F. (1965a), Gonin, H. T. (1966).
Class.: MP-NM:gf:A

Maritz, J. S. 1952. 1717
Note on a certain family of discrete distributions. *Biometrika*, **39**, 196–198.
Review: MR **13** (1952), 956; ZBL **47** (1953), 373.
Users: Khatri, C. G. (1960), Khatri, C. G., and Patel, I. R. (1961), Kemp, C. D., and Kemp, A. W. (1965).
Class.: MP:cm:G

Maritz, J. S. 1966. 1718
Smooth empirical Bayes estimation for one-parameter discrete distributions. *Biometrika*, **53**, 417–29.
Class.: MI-G-B-P:pe:G

Marlow, W. H. 1965. 1719
Factorial distributions. *Ann. Math. Statist.*, **36**, 1066–68.
Review: MR **31** (1966), 495.
Class.: MI:mi:G

Marsaglia, G. 1961. 1720
Generating exponential random variables. *Ann. Math. Statist.*, **32**, 899–900.
Review: IJA **3** (1962), 522.
User: Sibuya, M. (1961).
Class.: MI:mb:G

Marsden, E., and Barratt, R. 1911. 1721
The probability distribution of the time intervals of α particles with application to the number of particles emitted by uranium. *Proc. Phys. Soc.*, **23**, 367–73.

Marshall, Albert W., and Olkin, Ingram. 1964. 1722
Inclusion theorems for eigen-values from probability inequalities. *Numer. Math.*, **7**, 98–102.
Review: MR **29** (1965), 549.

Marshall, J. 1936. 1723
The distribution and sampling of insect populations in the field with special reference to the American boll-worm, *Heliothus obsoleta*, Fabre. *Ann. Appl. Biol.*, **23**, 133–52.
Users: Beall, G., and Rescia, R. R. (1953), Jensen, P. (1959).

Marshall, S. M., and Barnes, H.
See Barnes, H., and Marshall, S. M. (1951).

Martin, D. C. 1964. 1724
Validity of certain contagious distributions. Tech. Rep. Dept. Statist., Florida State Univ.
Class.: N-P-MI:a-gf-mb:BM

Martin, D. C., and Katti, S. K. 1961. 1725
Maximum likelihood estimates for certain contagious distributions using high speed computers. *Ann. Math. Statist.*, **32**, 1349.

Martin, D. C., and Katti, S. K. 1962a. 1726
Approximations to the Neyman type A distribution for practical problems. *Biometrics*, **18**, 354–64.
Review: MR **26** (1963), 1077; ZBL **113** (1965), 355; STMA **6** (1965), 62.
Users: Blischke, W. R. (1965), Govindarajulu, Z. (1965), Katti, S. K., and Sly, L. E. (1965), Martin D. C., and Katti, S. K. (1965).
Class.: N:a:G

Martin, D. C., and Katti, S. K. 1962b. 1727
Fitting of certain contagious distributions to some available data by the maximum likelihood method. Res. Rep. School of Aerospace Medicine and Florida State Univ., pp. 16.
Class.: N-NB-P-COB:pe-gf-c:G

Martin, D. C., and Katti, S. K. 1962c. 1728
Fitting routines for contagious distributions. Res. Rep., School of Aerospace Medicine and Florida State Univ., pp. 24.
Class.: N-NB-P-COB:c:G

Martin, D. C., and Katti, S. K. 1965. 1729
Fitting of some contagious distributions to some available data by the maximum likelihood method. *Biometric*, **21**, 34–48.
Review: MR **31** (1966), 1139.
Users: Martin, D. C., and Katti, S. K. (1962c), Katti, S. K., and Sly, L. E. (1965).
Notes: Correction: *Biometrics*, **21**, 514.

Martin, L. 1961. 1730
Etude Biometrique de la natalite en belgigue sur la base du recensement de 1947. *Bull. Assoc. Licencies en Sci. Actuarielles, Univ. Libre de Bruxelles* (*Bruxelles*), **2**, 25–53.
Class.: COB-P-OBR:gf:S

Martin, L. 1962a. 1731
Les processus de Poisson et leurs applications en biologie. *Biometrie-Praximetrie*, **3**, 55–82.
Class.: P:pc-mb:BM

Martin, L. 1962b. 1732
Biomathematical model of adaptometric curves. *Biom. Zeit.*, **4**, 73–84.
Review: ZBL **113** (1965), 141.
Class.: P:mb:BM

Martin, L. 1963. 1733
Les processus de Poisson et leurs applications en biologie (suite). *Biometrie-Praximetrie*, **4**, 81–97.
Class.: P:pc-mb-osp:BM

Martin, Paul S., and Mosimann, James E. 1965. 1734
Geochronology of pluvial Lake Cochise, Southern Arizona, III. Pollen statistics and pleistocene metastability. *Amer. J. Sci.*, **263**, 313–58.
Class.: M-MNH:rp:BM

Martin, Sanelius. 1964. 1735
Positive unbiased estimators of a positive mean of a Poisson distribution (Abstract). *Ann. Math. Statist.*, **35**, 1407.

Martin-Löf, P. 1965. 1736
Probability theory on discrete semigroups. *Zeit Wahrscheinlichkeitsth.*, **4**, 78–102.
Review: STMA **7** (1966), 32.
Class.: MI-P:mi-osp:G

Martof, B. S. 1953. 1737
Territoriality in the green grog, Rana clamitans. *Ecology*, **34** (1), 165–74.
Class.: MI:h:BM

Marushin, M. N. 1965a. 1738
Some remarks on convergence to Poisson's limit law. *Teor. Verojatnost. i Primenen.*, **10**, 371–75.
Class.: P:a:G

Marushin, M. N. 1965b. 1739
Some remarks on convergence to a limiting Poisson distribution. *Theor. Probability Appl.*, **10**, 341–44.
Class.: P:a:G

Maruyama, G. 1955. 1740
On the Poisson distribution derived from independent random walks. *Natur. Sci. Rep. Ochanonizu Univ.*, **6**, 1–6.
Review: MR **18** (1957), 314.
Class.: P:pc-mb:G

Masaru, Ozawa, and Siotani, Minoru.
See Siotani, Minoru, and Masaru, Ozawa (1958).

Massa, J., Sirull, R., and Earles, D.
See Earles, D., Massa, J., and Sirull, R. (1963).

Masuyama, M. 1951. 1741
An improved binomial probability paper and its use with tables. *Rep. Statist. Appl. Res. Un. Japan. Sci Engrs.*, **1**, 15–22.
Review: MR **13** (1952), 961.
Notes: Correction. *Rep. Statist. Appl. Res. Un. Japan. Sci. Engrs.*, **1** (3), 32–33.

Masuyama, M. 1960. 1742
Table of n, $\log_e n$, $n \log_e n$ and n $(\log_e n)$ for $n=1$ through 500 with applications. *Rep. Statist. Appl. Res. Un. Japan. Sci. Engrs.*, **7**, 56–64.
Review: MR **22** (1961), 528.

Matérn, B. 1960. 1743
Spatial variation. Stochastic models and their application to some problems in forest surveys and other sampling investigations. *Meddelanden fran Statens skogsforskningsinstitut 49–5*, 144 pp.
Review: IJA **2** (1961), 359.

Mathai, A. M. 1966c. 1744
Some characterizations of the one-parameter family of probability distributions. *Canad. Math. Bull.*, **9**, 95–102.
Class.: PSD-B-P-NB-MPSD:osp:G

Mathai, A. M., and Saxena, R. K. 1966a. 1745
On a generalized hypergeometric distribution (Abstract). *Ann. Math. Statist.*, **37**, 1077.

Mathai, A. M., and Saxena, R. K. 1966b. 1746
A short table of the generalized hypergeometric distribution (Abstract). *Ann. Math. Statist.*, **37**, 1864.

Mathai, A. M., and Saxena, R. K. 1966. 1747
On a generalized hypergeometric distribution. *Metrika*, **11**, 127–32.

Mathen, K. K., and Chakravorty, P. N. 1950. 1748
A statistical study on multiple cases of disease in households. *Sankhyā*, **10**, 387–92.

Mather, K. 1949. 1749
The analysis of extinction time data in bioassay. *Biometrics*, **5**, 127–43.

Mathews, J. C., and Langenhop, C. E.
See Langenhop, C. E., and Mathews, J. C. (1964).

Matthai, A., Rao, C. R., and Mitra, S. K.
See Rao, C. R., Mitra, S. K., and Matthai, A. (1966).

Matthes, K., and Kerstan, J.
See Kerstan, J., and Matthes, K. (1964).

Matthes, K., and Konig, Dieter.
See Konig, Dieter, and Matthes, K. (1963).

Matula, Miloš. 1965. 1750
Zur Frage der Häufigkeitsverteilung der Worte. I. *Comment. Math. Univ. Carolinae*, **6**, 213–37.
Review: MR **32** (1966), 318.

Matusita, Kameo. 1954. 1751
On the estimation by the minimum distance method. *Ann. Inst. Statist. Math.*, **5**, 59–65.
Review: ZBL **55** (1955), 376.

Matusita, K. 1961. 1752
Interval estimation based on the notion of affinity. *Bull. Inst. Internat. Statist.*, **38** (4), 241–44.
Review: IJA **4** (1963), 258.
Class.: MI:mi:G

Matuszewski, T. I. 1962. 1753
Some properties of Pascal distribution for finite population. *J. Amer. Statist. Assoc.*, **57**, 172–74.
Review: STMA **6** (1965), 448.
User: Knight, W. (1965).
Class.: IH:mi:G

May, Kenneth. 1948. 1754
Probabilities of certain election results. *Amer. Math. Monthly*, **55**, 203–09.

Maynard, James M., and Chow, Bryant. 1966. 1755
A Pitman-type close estimator for the parameter of a Poisson distribution (Preliminary report). *Ann. Math. Statist.*, **37**, 1859–60.

McCall, C. H., Jr., Thomas, R. E., and Roberts, H. R.
See Roberts, H. R., McCall, C. H., Jr., and Thomas, R. E. (1958).

McCarthy, Philip J. 1947. 1756
The approximate solutions for means and variances in a certain class of box problems. *Ann. Math. Statist.*, **18**, 349–83.
Review: MR **11** (1950), 41; ZBL **32** (1950), 45.
User: Wise, E. (1950).
Class.: NB-MNB:m:G

McCord, James R. 1963. 1757
Generalized multidimensional Poisson distributions for finite Markov chains. *J. Math. Anal. Appl.*, **6**, 349–72.
Review: MR **27** (1964), 168; ZBL **117** (1965), 359.

McCulloch, Warren S. 1948. 1758
The statistical organization of nervous activity. *Biometrics*, **4**, 91–99.

McFadden, J. A. 1955. 1759
Urn models of correlation and a comparison with the multivariate normal integral. *Ann. Math. Statist.*, **26**, 478–89.
Review: ZBL **65** (1956), 112.

McFadden, J. A. 1965. 1760
The mixed Poisson process. *Sankhyā*, **27**A, 83–92.

McFeely, F. S. 1956. 1761
Decision procedures for the comparison of exponential and geometric populations. Ph.D. Thesis, Virginia Polytechnic Institute.

McGinnies, W. G. 1934. 1762
The relation between frequency index and abundance as applied to plant populations in a semi-arid region. *Ecology*, **15**, 263–82.
Users: Blackman, G. E. (1935), Fracker, S. B., and Brischke, H. (1944), Evans, F. C. (1952), Thomson, G. W. (1952).
Class.: MI:mi:BM

McGregor, James, and Karlin, Samuel.
See Karlin, Samuel, and McGregor, James (1965).

McGuire, J. U., Brindley, T. A., and Bancroft, T. A. 1763
1957.
The distribution of European corn borer larvae pyrausta nubilalis (hbn) in field corn. *Biometrics*, **13**, 65–78.
Review: BA **33** (1959), Ab. No. 16430.
Users: Neyman, J., and Scott, E. L. (1957), Sprott, D. A. (1958), Pruess, K. P., and Weaver, C. R. (1959), Pielou, E. C. (1960), Katti, S. K., and Gurland, J. (1961), Katti, S. K., and Gurland, J. (1962b), Khatri, C. G. (1962b), Martin, D. C., and Katti, S. K. (1962c), Staff, P. J. (1964), Blischke, W. R. (1965), Katti, S. K., and Sly, L. E. (1965), Kemp, C. D., and Kemp, A. W. (1965), Dubey (1966a), Katti, S. K. (1966).
Class:. N-P-GP-COB:gf:BM
Notes: Errata and Extensions: *Biometrics*, **14**, 432–34.

McIntosh, R. P., and Curtis, J. T.
See Curtis, J. T., and McIntosh, R. P. (1950).

McKean, H. E., and Khazanie, R. G.
See Khazanie, R. G., and McKean, H. E. (1966).

McKendrick, A. G. 1914. 1764
Studies on the theory of continuous probabilities with special reference to its bearing on natural phenomena of a progressive nature. *Proc. London Math. Soc.*, **13** (2), 401–16.
Users: Kendall, D. G. (1949), Anscombe, F. J. (1950a), Irwin, J. O. (1953), Bartko, J. J. (1961b), Irwin, J. O. (1963).

McKendrick, A. G. 1915a. 1765
A test for residual spidemicity. *Indian J. Medical Res.*, **2**, 882–87.

McKendrick, A. G. 1915b. 1766
The epidemiological significance of repeated infections and relapses. *Indian J. Medical Res.*, **3**, 266–70.

McKendrick, A. G. 1915c. 1767
The epidemiological significance of spenic enlargement in malaria. *Indian J. Medical Res.*, **3**, 271–74.

McKendrick, A. G. 1926. 1768
Applications of mathematics to medical problems. *Proc. Edinburgh Math. Soc.*, **44**, 98–130.
Users: Bartlett, M. S. (1957), Irwin, J. O. (1963), Kerridge, D. (1964), Kemp, C. D., and Kemp, A. W. (1965).

Medgyessy, P. 1954a. 1769
Diszkrét valószinuség-eloszlások keverékének felbontása összetevöire. (Decomposition of discrete compound probability distributions) (Hungarian; English and Russian summaries). *Magyar Tud. Akad. Alkalm. Mat. Int. Közl.*, **3**, 139–53.
Review: MR **17** (1956), 862; ZBL **68** (1957), 127.
User: Blischke, W. R. (1965).

Medgyessy, P. 1954b. 1770
Ujabb eredmények valószínüségeloszlásfüggvények keverékének összetevöire bontásaval kapcsolatban. (Some recent results concerning the decomposition of

compound probability distributions.) (Hungarian; English and Russian summaries). *Magyar Tud. Akad. Alkalm.*
Mat. Int. Közl., **3**, 155–69; 331–41.
User: Blischke, W. R. (1965).

Medvedev, Yu. I., and Ivchenko, G. I.
See Ivchenko, G. I., and Medvedev, Yu, I. (1965).

Medvedev, Yu. I., and Ivčenko, G. I.
See Ivčenko, G. I., and Medvedev, Yu. I. (1966).

Mellinger, G. D., Sylwester, D. L., Gaffey, W. R., 1771
and Manheimer, D. L. 1965.
A mathematical model with applications to a study of accident repeatedness among children. *J. Amer. Statist. Assoc.*, **60**, 46–1059.
Class.: NM:mb-gf:A

Mendelsohn, N. S. 1956. 1772
The asymptotic series for a certain class of permutation problem. *Canad. J. Math.*, **8**, 234–44.
Review: MR **17** (1956), 935; **71** (1958), 347.
Class.: MI-P:a:G

Mendenhall, W. 1958. 1773
A bibliography on life testing and related topics. *Biometrika*, **45**, 521–43.
Users: Mendenhall, W., and Lehman, E. H. (1960).

Mendenhall, W., and Lehman, E. H., Jr. 1960. 1774
An approximation to the negative moments of the positive binomial useful in life testing. *Technometrics*, **2**, 227–42.
Review: MR **22** (1961), 861; ZBL **105** (1964), 123; IJA **2** (1961), 637.
Users: Govindarajulu, Z. (1962c), Govindarajulu, Z. (1962d), Govindarajulu, Z. (1963), Tiku, M. (1964).
Class.: TCB:m-a:E

Mengido, R. M., and David, H. T.
See David, H. T., and Mengido, R. M. (1963).

Menon, M. V. 1966. 1775
Characterization theorems for some univeriate probability distributions. *J. Roy. Statist. Soc., Ser. B*, **28**, 143–45.
Class.: MI:mi:G

Meric, J. 1953. 1776
Test profressif de l'hypothètse que le paramètre d'une loi binomiale est voisin d'une valeur donnée. *C.R. Acad. Sci. Paris*, **237**, 1390–92.
Review: MR **15** (1954), 727; ZBL **53** (1961), 412.
Class.: B:tp:G

Meric, J. 1954a. 1777
Ajustement des contantes d'un test binomial de Wald permettant d'obtenir les expressions exactes de ses caractéristiques. *C. R. Acad. Sci. Paris*, **238**, 2142–43.
Review: MR **16** (1955), 272; ZBL **55** (1955), 131.
Class.: B:tp:G

Meric, J. 1954b. 1778
Étude de la formule de Walker donnant la fonction " O.C." du test binomial de Wald. *C. R. Acad. Sci. Paris*, **239**, 1117–19.
Review: MR **16** (1955), 383; ZBL **55** (1955), 130.
Class.: B:tp:G

Meric, J. 1957. 1779
Sur le calcul de la fonction " O.C." du test binomial de Wald, a partir de la relation de récurrence de Pólya. *C. R. Acad. Sci. Paris*, **245**, 1500–02.
Review: ZBL **79** (1959), 355.
Class.: B:tp:G

Meric, J. 1958. 1780
Sur une méthode matricielle pour le calcul de la fonction " O.C." du test binomial de Wald. *C. R. Acad. Sci. Paris*, **246**, 884–47.
Review: ZBL **79** (1959), 355.
Class.: B:tp-c:G

Meric, J. 1961. 1781
Caracteristiques d'un test binomial sequential avec echantillonnage par groupes. *C. R. Acad. Sci. Paris*, **253**, 2195–97.
Review: MR **24**A (1962), 214.
Class.: B:tp:G

Meric, J., and Huron, R.
See Huron, R., and Meric, J. (1954).

Merrell, Margaret. 1933. 1782
On certain relationships between β_1 and β_2 for the point binomial. *Ann. Math. Statist.*, **4**, 216–28.
Class.: B:osp:G

Merrington, M., Barton, D. E., and David, F. N.
See Barton, E. D., David, F. N., and Merrington, M. (1960).

Merrington, M., Barton, D. E., and David, F. N.
See Barton, D. E., David, F. N., and Merrington, M. (1965).

Mesălkin, L. D. 1960. 1783
Approximation of a multinomial distribution by infinitely divisible laws (Russian; English summary). *Teor. Verojatnost. i Primenen.*, **5**, 114–24.
Review: MR **24**À (1962), 448.
Users: Patil, G. P., and Bildikar, S. (1966a).
Class.: B-M:a:G

Mesălkin, L. D. 1960a. 1784
On the approximation of polynomial distributional by infinitely divisible laws. *Theor. Probability Appl.*, **5**, 106–14.
Review: ZBL **99** (1963), 132.
Class.: B-M:a:G

Mesălkin, L. D. 1960b. 1785
A lower estimate of the rate of approach of the distribution of sums of a set of infinitely divisible laws. *Soviet Math., Doklady*, **1**, 648–51.
Review: ZBL **101** (1963), 116.

Meyer, Paul L., and Birnbaum, Z. W.
See Birnbaum, Z. W., and Meyer, Paul L. (1951).

Meyer-König, W., and Zeller, K. 1959. 1786
Pascal-Verteilung und Approximation durch Bernsteinsche Potenzreihen. *Z. Angew. Math. Mech.*, **39**, 380.
Review: ZBL **95** (1962), 128.

Michael, E. L. 1920. 1787
Marine ecology and the coefficient of association: a plea in behalf of quantitative biology. *J. Ecol.*, **8** (1), 54–59.

Mielke, Paul W., and Siddiqui, M. M. 1965. 1788
A combinatorial test for independence of dichotomous responses. *J. Amer. Statist. Assoc.*, **60**, 437–41.
Class.: B:mi:G-BM

Mihoc, G. 1952. 1789
La loi des événements rares pour les chaînes de Markov (Romanian; Russian and French summaries). *Acad. R. P. Romine Bul. Sti. Mat. Fiz.*, **4**, 783–90.
User: McCord, J. R. (1963).

Mihoc, G. 1954. 1790
Extension de la loi de Poisson pour les chaînes de Markov, multiples et homogènes (Romanian; Russian and French summaries). *Acad. R. P. Romine Bul. Sti. Mat. Fiz.*, **6**, 5–15.
Review: ZBL **58** (1958), 122.

Mihoc, G. 1956. 1791
Über verschiedene Ausdehnungen des Poissonschen Gesetzes auf endliche konstante Markoffsche Ketten. *Ber. Tagung Wahrsch. Rechnung Math. Statist., Berlin*, 43–49.
Review: ZBL **73** (1960), 128.

Mihoc, G. 1963. 1792
Various generalizations of the Poisson distribution (Romanian; Russian and French summaries). *An. Univ. Bucureşti Ser. Şti. Natur. Mat.-Fiz.*, **12** (40), 9–14.
Review: MR **31** (1966), 498.
Class.: P:pc:G

Mihoc, G., and Onicescu, Octav.
See Onicescu, Octav, and Mihoc, G. (1937).

Mihoc, G., and Onicescu, Octav.
See Onicescu, Octav, and Mihoc, G. (1938).

Mikami, Misao. 1956. 1793
On a multi-level sampling inspection plan for continuous production. *Bull. Math. Statist.*, **7**, 1–10.
Review: ZBL **72** (1958), 365.

Mikhalevich, V. S. 1956. 1794
Sequential Bayes' solutions and optimal methods of statistical acceptance control. *Theor. Probability Appl.*, **1**, 395–421.
Users: Kordonskii, H. B. (1958), Lindley, D. W., and Barnett, B. (1965).
Class.: B-P:sqc:G

Mikhalevich, V. S. 1956a. 1795
Consecutive Bayes' solutions and optimal methods of statistical acceptance control (Russian; English summary). *Teor. Verojatnost. i Primenen.*, **1**, 437–65.
Review: MR **19** (1958), 694.
User: Kovalenko, I. N. (1959b).
Class.: B-P:sqc:G

Mikhalevich, V. S. 1958. 1796
Sequential selection between two solutions for a Poisson process. *Theor. Probability Appl.*, **3**, 430–34.

Miles, R. E. 1964. 1797
A wide class of distributions in geometrical probability (Abstract). *Ann. Math. Statist.*, **35**, 1407.

Miles, S. R. 1935. 1798
A rapid and easy method of testing the reliability of an average and a discussion of the normal and binomial methods. *J. Amer. Soc. Agron.*, **27** (1), 21–31.
Review: BA **9** (1935), Ab. No. 17712.

Miller, A. J. 1961. 1799
A queueing model for road traffic flow. *J. Roy. Statist. Soc. Ser. B*, **23**, 64–90.
Review: IJA **3** (1962), 276.
User: Daniels, H. E. (1961).
Class.: B-BT-MI:pc:G

Miller, A. J. 1962. 1800
Road traffic considered as a stochastic process. *Proc. Cambridge Philos. Soc.*, **58**, 312–25.

Miller, G. A., Newman, E. B., and Friedman, E. A. 1801
1958.
Length-frequency statistics for written English. *Information and Control*, **1**, 370–89.
Users: Simon, H. (1960), Mosteller, F., and Wallace, D. (1963).

Miller, K. W., and Adler, H. A.
See Adler, H. A., and Miller, K. W. (1946).

Miller, R. G., Jr. 1959. 1802
A contribution to the theory of bulk queues. *J. Roy. Statist. Soc. Ser. B*, **21**, 320–37.
Review: STMA **5** (1964), 975.
User: Conolly, P. W. (1960).
Class.: MI:pc:G

Miller, R. G., Jr. 1960. 1803
Priority queues. *Ann. Math. Statist.*, **31**, 86–103.
User: Welch, D. D. (1965).

Milne, A., and Clark, R. B.
See Clark, R. B., and Milne, A. (1955).

Mintz, A., and Blum, M. L. 1949. 1804
A re-examination of the accident process concept. *J. Appl. Psychology*, **33**, 195–211.

Mintz, Alexander, and Blum, Milton L.
See Blum, Milton, L., and Mintz, Alexander (1951).

Mirasol, Noel M. 1963. 1805
The output of an $M/G/\infty$ queuing system is Poisson. *Operations Res.*, **11**, 282–84.
Review: ZBL **114** (1965), 94.

Mirkovich, A. R., Youngs, J. W. T., and Geisler, M. A.
See Youngs, J. W. T., Geisler, M. A., and Mirkovich, A. R. (1954).

Mises, R. Von. 1942. 1806
On the correct use of Bayes' formula. *Ann. Math. Statist.*, **13**, 156–65.
Class.: B:ie:G

Mitra, S. K., and Roy, J.
See Roy, J., and Mitra, S. K. (1957).

Mitra, S. K., Roy, S. N., and Diamond, E. L.
See Diamond, E. L., Mitra, S. K., and Roy, S. N. (1960).

Mitra, S. K., Matthai, A., and Rao, C. R.
See Rao, C. R., and Mitra, S. K., and Matthai, A. (1966).

Miyashita, K., Itô, Y., and Gotoh, A.
See Itô, Y., Gotoh, A., and Miyashita, K. (1960).

Miyashita, Kazuyashi, Nakamura, Kazuo, Itô, Y., Nakamura, Masako, and Kondao, Masaki.
See Itô, Y., Nakamura, Masako, Kondo, Masaki, Miyashita, Kazuyashi, and Nakamura, Kazuo (1962).

Mode, C. J. 1962. 1807
Some multi-dimensional birth and death processes and their applications in population genetics. *Biometrics*, **18**, 543–67; Correction **19**, 667.
Review: IJA **4** (1963), 757.

Mohanty, S. G., Ladouceur, J. C., and Narayana, T. V.
See Narayana, T. V., Mohanty, S. G., and Ladouceur, J. C. (1960).

Molenaar, W. 1965. 1808
Some remarks on mixtures of distributions. Tech. Rep. S 337 (UP24), Math. Centrum (Amersterdam), pp. 15.

Molina, E. C. 1915. 1809
An interpolation formula for Poisson's exponential binomial limit. *Amer. Math. Monthly*, **22**, 223–24.
Class.: P:c:G

Molina, E. C. 1927. 1810
Application of the theory of probability to telephone trunking problems. *Bell. System. Tech J.*, **6**, 461–94.

Molina, E. C. 1929. 1811
Application to the binomial summation of a Laplacian method for the evaluation of definite integrals. *Bell System Tech., J.* **8**, 99–108.
Class.: B:c:G

Molina, E. C. 1942. 1812
Poisson's exponential binomial limit. Table I—Individual terms. Table II—Cumulated terms. Van Nostrand, N. J., pp. 45.
Review: PA **17** (1943), Ab. No. 2560.

Mollison, J. E., Quenouille, M. H., and Jones, P. C. T.
See Jones, P. C. T., Mollison, J. E., and Quenouille, M. H. (1948).

Mood, A. M. 1940. 1813
The distribution theory of runs. *Ann. Math. Statist.*, **11**, 367–92.
Review: ZBL **24** (1941), 53.
Users: Barton, D. E., and David, F. N. (1957), Edwards, C. B. (1962), Fuchs, C. E., and David, H. (1965a).
Class.: MI:mi:G

Moore, P. G. 1952. 1814
Estimation of the Poisson parameter from a truncated distribution. *Biometrika*, **39**, 247–51.
Review: MR **14** (1953), 391; ZBL **48** (1953), 121.
Users: Plackett, R. L. (1953), Cohen, A. C. (1954), Moore, P. G. (1954a), Rider, P. R. (1955b), Moore, P. G. (1956b), Roy, J., and Mitra, S. K. (1957), Brass, W. (1958a), Hartley, H. O. (1958), Tate, R. F., and Goen, R. L. (1958), Cohen, A. C. (1959e), Patil, G. P. (1959), Cohen, A. C. (1960b), Cohen, A. C. (1960d), Cohen, A. C. (1960e), Cohen, A. C. (1961a), Patil, G. P. (1961b), Hughes, E. J. (1962), Patil, G. P. (1962e), Doss, S. A. D. C. (1963).
Class.: TCP:pe:G

Moore, P. G. 1953. 1815
A test for non-randomness in plant populations. *Ann. Bot. Lond., N.S.*, **17**, 57–62.
Users: David, F. N., and Moore, P. G. (1945).

Moore, P. G. 1954a. 1816
A note on truncated Poisson distributions. *Biometrics*, **10**, 402–06.
Review: MR **16** (1955), 498; ZBL **58** (1958), 127.
Users: Moore, P. G. (1956b), Roy, J., and Mitra, S. K. (1957), Tate, R. F., and Goen, R. L. (1958), Patil, G. P. (1959), Hughes, E. J. (1962), Patil, G. P. (1962e), Staff, P. J. (1964).
Class.: TCP:pe:G

Moore, P. G. 1954b. 1817
Spacing in plant populations. *Ecology*, **35**, 222–27.
Users: Clark, P. J., and Evans, F. C. (1954a), Morisita, M. (1959), Waters, W. E., and Henson, W. R. (1959), Holgate, P. (1965).
Class.: P:mi-h:BM

Moore, P. G. 1956a. 1818
The geometric, logarithmic and discrete Pareto forms of series. *J. Inst. Actuar.*, **82**, 130–36.
Review: ZBL **73** (1960), 356.

Moore, P. G. 1956b. 1819
The transformation of a truncated Poisson distribution. *Skand. Aktuartidskr.*, **39**, 19–25.
Review: MR **19** (1958), 472; ZBL **73** (1960), 153.
Class.: TCP:anovat:G

Moore, P. G. 1957. 1820
Transformations to normality using fractional powers of the variable. *J. Amer. Statist. Assoc.*, **52**, 237–46.
Review: ZBL **84** (1960), 139.
Class.: P:anovat:G

Moore, P. G., and David, F. N.
See David, F. N., and Moore, P. G. (1954).

Moore, P. G., and David, F. N.
See David, F. N., and Moore, P. G. (1957).

Moran, P. A. P. 1951. 1821
A mathematical theory of animal trapping. *Biometrika*, **38**, 307–11.
Users: Craig, C. C. (1953a), Chapman, D. G. (1954), Sen, P. K. (1960), Pathak, P. K. (1964).
Class.: M:pe:BM

Moran, P. A. P. 1952. 1822
A characteristic property of the Poisson distribution. *Proc. Cambridge Philos. Soc.*, **48**, 206–07.
Review: ZBL **47** (1953), 373.
Class.: P:osp:G

Moran, P. A. P. 1955. 1823
A probability theory of dams and storage systems—modification of the release rule. *Austral. J. Appl. Sci.*, **6**, 117–30.
Users: Prabhu, N. U. (1958), Prabhu, N. U. (1959-60). Phatarfod, R. M. (1963).

Moran, P. A. P. 1958. 1824
The distribution of gene frequency in a bisexual diploid population. *Proc. Cambridge Philos. Soc.*, **54**, 468–74.
Class.: MI:pc:BM

Moran, P. A. P. 1961. 1825
The Engset distribution in telephone congestion theory. *Austral. J. Appl. Sci.*, **12**, 257–64.
Review: IJA **3** (1962), 279.
Class.: MI:pc:G

Moran, P. A. P., and Watterson, G. A. 1959. 1826
The genetic effects of family structure in natural populations. *Aust. J. Biol. Sci.*, **12**, 1–15.
Review: IJA **1** (1959), 137; IJA **1** (1959), 308.
Class.: MI:pc:BM

Morant, G. 1921. 1827
On random occurrences in space and time when followed by a closed interval. *Biometrika*, **13**, 309–37.

Morant, G. M., and Welch, B. L. 1939. 1828
A bibliography of the statistical and other writings of Karl Pearson. Cambridge Univ. Press VIII, 119 pp.
Review: ZBL **24** (1941), 424.

Morgan, M. E., MacLeod, P., Anderson, E. O., and Bliss, C. I. 1951. 1829
A sequential procedure for grading milk by microscopic counts. Bull. 276, Storrs Agri. Expt. Stn., Conn.
Users: Bliss, C. I., and Fisher, R. A. (1952).
Class.: MI:srp:BM

Morgan, R. W., and Welsh, D. J. A. 1965. 1830
A two-dimensional Poisson growth process. *J. Roy Statist. Soc. Ser. B*, **27**, 497–504.
Class.: P:pc:G

Morgenstern, D. 1961. 1831
The central limit theorem for the probability of exceedances (German). *Metrika*, **4**, 173–77.
Review: IJA **4** (1963), 18.
Class.: MI:a:G

Morgenstern, D. 1966. 1832
Eine diskrete Verteilung für ein Warteschlangen-problem. *Metrika*, **11**, 81–84.
Class.: OBR-BT:mi:G

Morgenthaler, G. W. 1960. 1833
On Bernoulli play with limited resources. *Amer. Math. Monthly*, **67**, 344–48.

Morignti, S., and Robbins, H. 1962. 1834
A Bayes test of $p \leqq 1/2$ versus $p > 1/2$. *Rep. Statist. Appl. Res. Un. Japan. Sci. Engrs.*, **9**, 39–60.
Users: Lindley, D. V., and Barnett, B. (1965).

Morimura, H. 1961. 1835
A note on sums of independent random variables. *Kōdai Math. Sem. Rep.*, **13**, 255–60.
Class.: MI:mi:G

Morimura, H. 1962. 1836
On the relation between the distributions of the queue size and the waiting time. *Kōdai Math. Sem. Rep.*, **14**, 6–19.
Review: STMA **5** (1964), 237.

Morimura, H., and Makabe, H.
See Makabe, H., and Morimura, H. (1955).

Morimura, H., and Makabe, H.
See Makabe, H., and Morimura, H. (1956).

Morisita, M. 1959. 1837
Measuring of the dispersion of individuals and analysis of the distributional patterns. *Mem. Fac. Sci. Kyūshū Univ. Ser. E*, **2**, 215–35.

Morista, M. 1962. 1838
I_δ-index, a measure of dispersion of individuals (Japanese summary). *Res. Population Ecol.*, **4**, 1–7.

Morisita, M. 1964. 1839
Application of I_δ-index to sampling techniques (Japanese summary). *Res. Population Ecol.*, **6**, 43–53.

Morlat, G. 1952. 1840
Sur une generalization de la loi de Poisson. *C. R. Acad. Sci. Paris*, **235**, 933–35.
Review: ZBL **48** (1953), 107.
Class.: OPR:mb:G

Morris, K. W. 1963. 1841
A note on direct and inverse binomial sampling. *Biometrika*, **50**, 544–45.
Review: MR **28** (1964), 1055; ZBL **115** (1965), 366; STMA **5** (1964), 574.
User: Mantel, N. (1966b).
Class.: B-NB:osp:G

Morris, R. F. 1954. 1842
A sequential sampling technique for spruce budworm egg surveys. *Canad. J. Zool.*, **32**, 302–13.
User: Waters, W. E. (1955), Bliss, C. I., and Fisher, A. R. G. (1958), Kuno, E. (1963), Kuno, Yamamoto, Satomi, Outi, and Okada (1963), Harcourt, D. G. (1960)

Morris, R. F. 1955. 1843
The development of sampling techniques for forest insect defoliators, with particular reference to the spruce budworm. *Canad. J. Zool.*, **33**, 225–294.
Users: Harcourt, D. G. (1960), Harcourt, D. G. (1961), Itô, Nakamura, Kondo, Miyashita, and Nakamura (1962), Kuno, E. (1963), Kuno, Yamamoto, Satomi, Outi, and Okada (1963), Embree, and Sisojevic (1965), Kobayashi, S. (1965).

Morse, Norman, and Kesten, Harry.
See Kesten, Harry, and Morse, Norman (1959).

Morse, Norman, Bechhofer, Robert E., and Elmaghraby, Salah.
See Bechhofer, Robert, E., Elmaghraby, Salah, and Morse, Norman (1959).

Morse, Phillip M., Simond, M., and Galliher, H. P.
See Galliher, H. P., Morse, Phillip M., and Simond, M. (1959).

Mortara, Giorgio. 1912. 1844
Sulle variazioni di frequenza di alcuni fenomeni demografici rari. *Ann. Statist.*, **4**, 5–81.
Users: Whitaker, L. (1914), Thorndike, F. (1926).

Morton, A. Q. 1965. 1845
The authorship of Greek prose. *J. Roy. Statist. Soc. Ser. A*, **128**, 169–233.
Class.: B-P-NB:gf:L

Mosher, W. W. Jr., Haight, F. A., and Whisler, B. F.
See Haight, F. A., Whisler, B. G., and Mosher, W. W. Jr. (1961).

Mosimann, J. E. 1962. 1846
On the compound multinomial distribution, the multivariate β-distribution, and correlations among proportions. *Biometrika*, **49**, 65–82.
Review: ZBL **105** (1964), 125; IJA **4** (1963), 33.
Users: Mosimann, J, E. (1963), Tallis, G. M. (1964), Blischke, W. R. (1965), Martin, P. S., and Mosimann, J. E. (1965), Patil, G. P., and Bildikar, S. (1966a).
Class.: COM:osp-pe:G-BM

Mosimann, J. E. 1963. 1847
On the compound negative multinomial distribution and correlations among inversely sampled pollen counts. *Biometrika*, **50**, 47–54.
Review: ZBL **118** (1965/66), 349; STMA **5** (1964), 907.
Users: Martin, P. S., and Mosimann, J. E. (1965).

Bhat, B. R., and Kulkarni, N. V. (1966b), Patil, G. P., and Bildikar, S. (1966a).
Class.: CONM:osp-pe:G-BM

Mosimann, J. E. 1965. 1848
Statistical methods for the pollen analyst: multinomial and negative multinomial techniques. *Handbook of Paleontological Techniques*, Ed. B. Kummel and D. Raup. W. H. Freeman Company, pp. 636–73.

Mosimann, J. E., and Martin, Paul S.
See Martin, Paul S., and Mosimann, J. E. (1965).

Mosimann, J. E., Gurian, Joan M., and Cornfield, J.
See Gurian, Joan M., Cornfield, J., and Mosimann, J. E. (1964).

Mosteller, F., and Tukey, J. W. 1949. 1849
The uses and usefulness of binomial probability paper. *J. Amer. Statist. Assoc.*, **44**, 174–212.
Users Blom, G. (1954), Raff, M. S. (1956).
Class.: B:mi:G-E

Mosteller, F., and Wallace, D. L. 1850
Disputed authorship. *Proc. Harvard Symp. Dig. Computers Appl.*

Mosteller, F., and Wallace, D. L. 1963. 1851
Inference in an authorship problem. *J. Amer. Statist. Assoc.*, **58**, 275–307.
Review: ZBL **124** (1966), 104; STMA **5** (1964), 660.
Users: Kendricks, W. (1964), Roberts, H. V. (1965).

Mosteller, Frederick, and Wallace, David L. 1964. 1852
Inference and disputed authorship: The federalist. Addison-Wesley Publishing Co., Reading, Mass. 287 pp.
Review: ZBL **122** (1966), 141.

Mosteller, F., and Youtz, C. 1961. 1853
Tables of the Freeman-Tukey transformations for the binomial and Poisson distributions. *Biometrika*, **48**, 433–40.
Review: STMA **5** (1964), 275.
User: Govindarajulu, Z. (1965).
Class.: B-P:anovat-tc:G

Mosteller, F., Savage, L. J., and Girshick, M. A.
See Girshick, M. A., Mosteller, F., and Savage, L. J. (1946).

Mote, V. L., and Anderson, R. L. 1965. 1854
An investigation of the effect of misclassification on the properties of χ^2-tests in the analysis of categorical data. *Biometrika*, **52**, 95–109.
Review: STMA **7** (1966), 105.
Class.: M:gf:G

Mott-Smith, J. C. 1960. 1855
Estimation of the binomial distribution. Res. Rep. Air Force Cambridge Lab., Bedford, Mass.
User: Mott-Smith, J. C. (1964).
Class.: B:pe:G

Mott-Smith, J. C. 1964. 1856
Two estimates of the binomial distribution. *Ann. Math. Statist.*, **35**, 809–16.
Review: MR **29** (1965), 138.
Class.: B:pe:G

Moustafa, M. D. 1958. 1857
Testing of hypotheses on a multivariate population; some of the variates being continuous and the rest categorical. *Egyptian Statist. J.*, **2**, 73–96.
Review: IJA **4** (1963), 653.
Class.: MI:tp:G

Moyal, J. E. 1949. 1858
The distribution of wars in time. *J. Roy. Statist. Soc. Ser. A.*, **112**, 446–49.

Muench, Hugo. 1936. 1859
The probability distribution of protection test results. *J. Amer. Statist. Assoc.*, **31**, 677–89.
Users: Muench, H. (1938), Chiang, C. L. (1951), Lord, F. (1960), Blischke, W. R. (1965).
Class.: B-NH:mb-gf:BM

Muench, H. 1938. 1860
Discrete frequency distributions arising from mixtures of several single probability values. *J. Amer. Statist. Assoc.*, **33**, 390–98.
Review: ZBL **19** (1939), 73.
Users: von Schilling, W. (1947), Cohen, A. C. (1961c), Blischke, W. R. (1964), Blischke, W. R. (1965).
Class.: COB-COP:mb-gf:BM

Münch, Guido. 1957. 1861
Stochastic processes of astronomical interest. *Proc. Symp. Appl. Math.*, **7**, 51–66.
Review: ZBL **93** (1962), 147.

Münzner, H. 1956. 1862
Zur Frage: Binomialverteilung oder Poissongesetz? *Bl.-Deutsch. Ges. Versicherungs-Math.*, **2**, 405–12.
Review: MR **17** (1956), 1217; ZBL **70** (1957), 155.

Münzner, H. 1963. 1863
Confidence limits for the frequency of defectives in samples of consecutive elements, when the defective items occur in series (German). *Metrika*, **6**, 95–99.
Review: IJA **4** (1963), 709.
Class.: MI:ie-sqc:G

Murakami, Masayasu. 1961. 1864
Censored sample from truncated Poisson distribution (Japanese; English summary). *J. College Arts Sci. Chiba Univ.*, **3**, 263–68.
Review: MR **26** (1963), 1076; ZBL **113** (1965), 344.

Murakami, M., Kawamura, M., and Asai, A.
See Kawamura, M., Asai, A., and Murakami, M. (1954).

Murteira, B. 1957–58. 1865
On type A regions for Polya distributions. *Univ. Lisboa Revista Fac. Ci. A.*, **6** (2), 327–30.

Murthy, V. K., and Rac Uppuluri, V. R. 1965. 1866
An extension of Ferguson's characterization of the geometric distribution (Abstract). *Ann. Math. Statist.*, **36**, 1603–04.

Murty, V. N. 1956. 1867
A note on Bhattacharyya bounds for the negative binomial distribution. *Ann. Math. Statist.*, **27**, 1182–83.
Review: MR **18** (1957), 772; ZBL **73** (1960), 138.
Class.: NB:pe:G

Myers, E., and Chapman, V. J. 1953. 1868
Statistical analysis applied to a vegetation type in New Zealand. *Ecology*, **34**, 175–85.
Class.: LS:mi:BM

Myers, Max H., Mantel, Nathan, and Ederer, Fred.
See Ederer, Fred, Myers, Max H., and Mantel, Nathan (1964).

Myers, R. 1963. 1869
Orthogonal statistics and some sampling properties of moment estimators for the negative binomial distribution. Ph.D. Thesis, Virginia Polytechnic Institute.
Users: Bowman, K. O., and Shenton, L. R. (1965a).

Myers, R., and Shenton, L. R.
See Shenton, L. R., and Myers, R. (1965).

Myhre, J. M., and Saunders, S. C. 1965. 1870
On confidence limits for the reliability of systems (Abstract). *Ann. Math. Statist.*, **36**, 1604.

N

Nabeya, S. 1950. 1871
On a relation between exponential law and Poisson's law. *Ann. Inst. Statist. Math. Tokyo*, **2**, 13–16.
Review: MR **12** (1951), 424.

Nadkarni, V. G. 1964. 1872
On some discrete models in branching processes. *J. Indian Soc. Agri. Statist.*, **16**, 72–82.
Review: MR **30** (1965), 999.

Nadler, Jack. 1960. 1873
Inverse binomial sampling plans when an exponential distribution is sampled with censoring. *Ann. Math. Statist.*, **31**, 1201–04.
Review: MR **22** (1961), 1961; IJA **3** (1962), 245.
Users: Bartholomew, D. J. (1965), Knight, W. (1965).
Class.: NB:mi:G

Nagel, K. 1966. 1874
On selecting a subset containing the best of several discrete populations (Abstract). *Ann. Math. Statist.*, **37**, 543.

Nagnur, B. N., and Bhat, B. R.
See Bhat, B. R., and Nagnur, B. N. (1965).

Nair, A. N. K. 1942. 1875
On the probability of obtaining k sets of consecutive successes in n trials. *Math. Student*, **10**, 83–84.
Review: MR **4** (1943), 248.
Class.: B:mi:G

Nakamura, K., and Shiyomi, M.
See Shiyomi, M., and Nakamura, K. (1964).

Nakamura, Masako, Kondo, Masaki, Miyashita, Kazuyachi, Nakamura, Kazuo, and Itô, Y.
See Itô, Y., Nakamura, Masako, Kondo, Masaki, Miyashita, Kazuyashi, and Nakamuro, Kazuo (1962).

Nakamura, Kazuo, Itô, Y., Nakamura, Masako, Kondo, Masaki, and Miyashita, Kazuyashi.
See Itô, Y., Nakamura, Masako, Kondo, Masaki, Miyashita, Kazuyashi, and Nakamuro, Kazuo (1962).

Nambiar, K. K. 1964. 1876
Representations of Distributions of Binary Random Variables. Ph.D. Thesis, Univ. Pennsylvania, Philadelphia.

Naor, P., and Ben-Israel, A.
See Ben-Israel A., and Naor, P. (1960).

Narayan, T. V. 1954. 1877
A problem in the theory of probability. *J. Indian Soc. Agri. Statist.*, **6**, 139–46.
Review: MR **18** (1953), 423.
Users: Narayana, T. V., and Sathe, Y. S. (1961).
Class.: B:pc:G

Narayana, T. V. 1962. 1878
An analogue of the multinomial theorem. *Canad. Math. Bull.*, **5**, 43–50.
Class.: M:mi:G

Narayana, T. V., and Brainerd, B.
See Brainerd, B., and Narayana, T. V. (1961).

Narayana, T. V., and Sathe, Y. S. 1961. 1879
Minimum variance unbiased estimation in coin tossing problems. *Sankhyā*, **23A**, 183–86.
Review: MR **24A** (1962), 704; ZBL **99** (1963), 356; IJA **3** (1962), 172.
Class.: B:pc-pe:G

Narayana, T. V., Chorneyko, I., and Sathe, Y. S. 1960. 1880
Sufficient partitions for a class of coin tossing problems. *Biom. Zeit.*, **2**, 267–75.
Review: ZBL **96** (1962), 127.

Narayana, T. V., Mohanty, S. G., and Ladouceur, J. C. 1960. 1881
A combinatorial problem and its applications to probability theory. II. *J. Indian Soc. Agric. Statist.*, **12**, 182–89.
Review: MR **24A** (1962), 107; IJA **4** (1963), 19.
Class.: B:pc:G

Nataf, A. 1962. 1882
Determination of probability distributions with given margins (French). *C. R. Acad. Sci., Paris*, **255**, 42–43.
Review: STMA **6** (1965), 36.
Class.: MI:mi:G

Nat. Bur. Standards. 1950. 1883
Tables of the binomial probability distribution. *Appl. Math. Ser.* 6.
Review: SA **53A** (1950), Ab.No. 7685; ZBL **37** (1951), 363.

Naus, J. I. 1965a. 1884
Clustering of random points in two dimensions. *Biometrika*, **52**, 263–66.
Review: STMA **6** (1965), 1226.
Class.: MI:mi:G

Naus, Joseph I. 1965b. 1885
The distribution of the size of the maximum cluster of points on a line. *J. Amer. Statist. Assoc.*, **60**, 532–38.
Class.: B-P-MI:mi:G

Navratil, J. 1958. 1886
Determination of the parameters of a compound distributions (Czech). *Pokroky Mat. Fyz. Astronom.*, **3**, 41–45.

Nawrotzki, Kurt. 1962. 1887
Ein Grenzwertsatz für homogene zufallige Punktfolgen (Verallgemeinerung eines Satzes von A. Rényi). *Math. Nachr.*, **24**, 201–17.
Review: ZBL **106** (1964), 333.

Naylor, A. F. 1964. 1888
Comparisons of regression constants fitted by maximum likelihood to four common transformations of binomial data. *Ann. Human Genetics*, **27**, 241–46.

Nazarov, I. M., Pressman, A. Ja., and Berljand, O.S.
See Berljand, O. S., Nazarov, I. M., and Pressman, A. Ja. (1962).

Neess, J., Emlen, J. T. Jr., and Young H.
See Young, H., Neess, J., and Emlen, J. T. Jr. (1952).

Nef, L. 1958. 1889
The distribution of acarina in the soil. *Progress in Soil Zool.*, **1**, 56–58. Ed. P. W. Murphy, Butterworths, London.
Class.: NB:gf:BM

Nef, L. 1959. 1890
Etude d'une population de Larves de *Retinia buoliana* (Schiff)(German and English summaries). *Zeit. Angew. Entomol.*, **44**, 167–86. (Translated in English by S. A. Emilson, East Lansing, Mich., Res. Center, Lake States Forest Experiment Station Forest Service, U.S. Dept. of Agriculture.)
Users: Nef, L. (1958), Berthet, P., and Gerard, G. (1965).
Class.: P-N-NB-GP:pe-gf:BM

Negi, G. S., and Chacko, V. J.
See Chacko, V. J., and Negi, G. S. (1966)

Nelson, A. C. Jr., Williams, J. S., and Fletcher, N. T. 1963. 1891
Estimation of the probability of defective failure from destructive tests. *Technometrics*, **5**, 459–68.
Review: STMA **7** (1966), 177; STMA **6** (1965), 521
Class.: B-NB-MI:pe:E

Nelson, L. S. 1965. 1892
Relationships between the common statistical distributions. *Annual Tech. Conf. Trans. Amer. Soc. Qual. Contr.* pp. 371–77.
Review: STMA **7** (1966), 288
Class.: B-NB-P:mi:G

Nelson, L., and Albert, G. E.
See Albert, G. E., and Nelson, L. (1953).

Nelson, W. C., and David, H. A. 1964. 1893
The logarithmic distribution. Tech. Rep. 58, Dept. Statist., Virginia Polytechnic Inst., pp. 98.
Users: Patil, G. P., and Bildikar, S. (1966a).

New, W. D., Dudley, F. H., Bushland, R. C., Baumhover, A. H., Graham, A. J., Bitter, B. A., and Hopkins, D. E.
See Baumhover, A. H., Graham, A. J., Bitter, B. A., Hopkins, D. E., New, W. D., Dudley, F. H., and Bushland, R. C. (1955).

Newbold, E. M. 1927. 1894
Practical applications of the statistics of repeated events, particularly to industrial accidents. *J. Roy. Statist. Soc. Ser. A*, **90**, 487.
Users: Feller, W. (1943), Anscombe, F. J. (1950a), Maritz, J. S. (1950), Arbous, A., and Kerrich, J. (1951), Adelstein, A. M. (1952), Good, I. J. (1953b), Webb, W., and Jones, E. R. (1953), Arbous, A. G., and Sichel, H. S. (1954b), Grainger, R. M., and Reid, D. B. (1954), Fitzpatrick, R. (1958), Gurland, J. (1958), Barton, D. E., and David, F. N. (1959a).
Notes: Appendix: *J. Roy Statist. Soc. Ser. A*, **90**, 518–35.

Newcombe, C. L., Shepherd, B. B., and Littleford, R. A.
See Littleford, R. A., Newcombe, C. L., and Shepherd, B. B. (1940).

Newell, D. J. 1963. 1895
Immediate admissions to hospital. *Proc. Third Int. Con. Oper. Res. Oslo*, **3**, 224–33.
Class.: P:gf:BM

Newell, D. J. 1965. 1896
Unusual frequency distributions. *Biometrics*, **21**, 159–68.
Review: MR **31** (1966), 958.

Newell, G. F. 1960. 1897
Queues for a fixed-cycle traffic light. *Ann. Math. Statist.*, **31**, 589–97.

Newman, E. B., Friedman, E. A., and Miller, G. A.
See Miller, G. A., Newman, E. B., and Friedman, E. A. (1958).

Neyman, J. 1933. 1898
" The law of small numbers " and its applications (Polish). *Wiadomo. Mat.*, **35**, 1–18.

Neyman, J. 1935. 1899
On the problem of confidence intervals. *Ann. Math. Statist.*, **6**, 111–16.
Users: Buehler, R. J. (1957), Steinhaus, H. (1957), Pachares, J. (1960).
Class.: MI-B:ie:G

Neyman, J. 1938. 1900
A historical note on Karl Pearson's deduction of the moments of the binomial. *Biometrika*, **30**, 11–15.
Review: ZBL **19** (1939), 35
Class.: B:m:G

Neyman, J. 1939. 1901
On a new class of " contagious " distributions, applicable in entomology and bacteriology. *Ann. Math. Statist.*, **10**, 35–57.
Users: Beall, G. (1940), Feller, W. (1943), Fracker, S. B., and Brischke, H. (1944), Wadley, F. M. (1945), Garwood, F. (1947), Archibald, E. E. A. (1948), Ashby, E. (1948), Shenton, L. R. (1949), Thomas, M. (1949), Archibald, E. E. A. (1950), Bateman, C. I. (1950), Curtis, J., and McIntosh, R. (1950), Anscombe, F. J. (1950a), Wadley, F. M. (1950), Arbous, A., and Kerrich, J. (1951), Barnes, H., and Marshall, S. M. (1951), Barnes, H., and Stanbury, F. (1951), Johnson, N. L. (1951), Morgan, M. E., MacLeod, P., Anderson, E. O., and Bliss, C. I. (1951), Consael, R. (1952b), Kono, Utida, Yosida, and Watanabe (1952), Skellam, J. G. (1952), Thomson, G. W. (1952), Beall, G., and Rescia, R. R. (1953), Bliss, C. I., and Fisher, R. A. (1953), MacFadyen, A. (1953), Moore, P. G. (1953), Clark, P. J., and Evans, F. C. (1954a), David, F. N., and Moore, P. G. (1954), Moore, P. G. (1954b), Robinson, P. (1954), Rutherford, R. S. G. (1954), Douglas, J. B. (1955), McGuire, J., Brindley, T. A., and Bancroft, T. A. (1957), Neyman, J., and Scott, E. L. (1957), Pielou, E. C. (1957), MacLulich, D. A. (1957), Bliss, C. I. (1958), Gurland, J. (1958), Skellam, J. G. (1958b), Cohen, A. C. (1959c), Hairston, N. G. (1959), Jensen,

P. (1959), Nef, L. (1959), Neyman, J., and Scott, E. L. (1959), Patil, G. P. (1959), Waters, W. E., and Hensen, W. R. (1959), Cohen, A. C. (1960c), Katti, S. K. (1960d), Pielou, D. P. (1960a), Pielou, E. C. (1960), Shumway, R., and Gurland, J. (1960b), Shumway, R., and Gurland, J. (1960c), Bardwell, G. E. (1961), Khatri, C. G., and Patel, I. R. (1961), Martin, L. (1961), Cassi, R. M. (1962), Itô, Nakamura, Kondo, Miyashita, and Nakamura (1962), Martin, D. C., and Katti, S. K. (1962a), Martin, D. C., and Katti, S. K. (1962b), Martin, D. C., and Katti, S. K. (1962c), Pielou, E. C. (1962), Tsao, C. M. (1962), Bardwell, G. E., and Crow, E. L. (1964), Shiyomi, M., and Nakamura, K. (1964), Subrahmaniam, K. (1964), Blischke, W. R. (1965), Crow, E. L., and Bardwell, G. E. (1965), Katti, S. K., and Sly, L. E. (1965), Kemp, C. D., and Kemp, A. W. (1965), Lukacs, E. (1965), Neyman, J. (1965), Sprott, D. A. (1965b), Butler, Margaret Sauls (1966), Holgate, P. (1966), Katti, S. K. (1966).
Class.: N-MI:mb-gf:G-BM

Neyman, J. 1949. 1902
On the problem of estimating the number of schools of fish. *Univ. California Publ. Statist.*, **1**, 21–36.
Review: ZBL40 (1951), 367.
Users: David, F. N. (1950), Chapman, D. G. (1954).
Class.: MI-P-B:pe:BM

Neyman, J. 1963. 1903
On finiteness of the process of clustering. *Sankhyā Ser. A*, **25**, 69–74.
Review: MR **30** (1965), 303.

Neyman, J. 1965. 1904
Inaugural address: Certain chance mechanisms involving discrete distributions. *Classical and Contagious Discrete Distributions*, Ed. G. P. Patil, Statistical Publishing Society, Calcutta and Pergamon Press, pp. 4–14. Reprinted in *Sankhyā A*, **27** (1966).
Class.: NM-MI:mb-osp-pc:A-G

Neyman, J. 1966. 1905
Behavioristic point of view on mathematical statistics. On Political Economy and Econometrics Essays in honour of Oskar Lange, PWN-Polish Scientific Publishers, Warszawa, 445–62.

Neyman, J., and Bates, G. E.
See Bates, G. E., and Neyman, J. (1952).

Neyman, J., and Iwaszkiewicz, K.
See Iwaszkiewicz, K., and Neyman, J. (1931).

Neyman, J., and Pearson, E. S. 1932-33. 1906
The testing of statistical hypotheses in relation to probabilities a priori. *Proc. Cambridge Philos. Soc.*, **29**, 492–510.

Newman, J., and Scott, E. L. 1952. 1907
A theory of the spatial distribution of galaxies. *Astrophys. J.*, **116**, 144–63.
Users: Neyman, J., and Scott, E. L. (1953), Neyman, J., and Scott, E. L. (1954), Neyman, J., and Scott, E. L. (1957), Neyman, J., and Scott, E. L. (1962), Neyman, J. (1963), Neyman, J. (1965).

Neyman, J., and Scott, E. L. 1953. 1908
Frequency of separation and of interlocking of clusters of galaxies. *Proc. Nat. Acad. Sci. U.S.A.*, **39**, 737–43.
Class.: MI:mb:P

Neyman, J., and Scott, E. L. 1954. 1909
Spatial distribution of galaxies: analysis of the theory of fluctuations. *Proc. Nat. Acad. Sci. U.S.A.*, **40**, 873–82.
Class.: MI:mi:P

Neyman, J., and Scott, E. L. 1957. 1910
On a mathematical theory of populations conceived as conglomerations of clusters. *Cold Spring Harbor Symp. on Quantitative Biol.*, **22**, 109–120.
Users: Neyman, J. (1965), Scott, E. L. (1965).

Neyman, J., and Scott, E. L. 1959. 1911
Stochastic models of population dynamics. *Science*, **130**, 303–08.

Neyman, J., and Scott, E. L. 1962. 1912
Contribution to the study of the abundance of multiple galaxies. *Studies in Mathematical Analysis and Related Topics*, Ed. Gilbarg, Solomon, etc., Stanford Univ. Press, pp. 262–69.
User: Scott, E. L. (1965).

Neyman, J., Scott, E. L., and Shane, C. D. 1956. 1913
Statistics of images of galaxies with particular reference to clustering. *Proc. 3rd Berkeley Symp. Math. Statist. Prob.*, **3**, 75–82.

Neyman, J., Puri, P. S., Buhler, W., Fein, H., and Goldsmith, D.
See Buhler, W., Fein, H., Goldsmith, D., Neyman, J., and Puri, P. S. (1965).

Nicholson, W. L. 1956. 1914
On the normal approximation to the hypergeometric distribution. *Ann. Math. Statist.*, **27**, 471–83.
Review: MR **19** (1958), 326
Users: Hannan, J., and Harkness, W. (1963), Govindarajulu, Z. (1965).
Class.: H-B:a:G

Nicholson, W. L. 1960. 1915
Occupancy probability tables based on the multinomial distribution for equally probable events. Res. and Develop. Rep. HW-57502 REV, Hanford Labs., General Electric pp. 44.
Users: Nicholson, W. L. (1961), Patil, G. P., and Bildikar, S. (1966a).
Class.: M-OMR:tc:G

Nicholson, W. L. 1961. 1916
Occupancy probability distribution critical points. *Biometrika*, **48**, 175–80.
Review: ZBL **109** (1964-65), 126; IJA **4** (1963), 387.
Users: Sarkadi, K., Schenell, Edit, and Vincze, I. (1962).
Class.: M-OMR:tp-tc:G

Nicholson, W. L. 1963. 1917
Asymptotic distribution theory for the noncentral Poisson index statistic. Res. Rep., Pacific Northwest Lab., Battelle Memorial Inst., Richland, Washington.
Class.: P:id:G

Nishida, T. 1962. 1918
On the multiple exponential channel queueing system with hyper-Poisson arrivals. *J. Operations Res. Soc. Japan*, **5**, 57–66.

Nisida, Tosio. 1952. 1919
On the inverse function of Poisson process. *Math. Japon.*, **2**, 135–42.
Review: ZBL **48** (1953), 112.

Nisida, T. 1953. 1920
On some probability distributions concerning Poisson process. *Math. Japon.*, **3**, 7–12.
Review: ZBL **53** (1961), 407.
Class.: MI:pc:G

Noack, A. 1950. 1921
A class of random variables with discrete distribution. *Ann. Math. Statist.*, **21**, 127–32.
Review: MR **11** (1950), 445; ZBL **36** (1951), 86.
Users: Roy, J., and Mitra, S. K. (1957), Khatri, C. G. (1959), Patil, G. P. (1959), Khatri, C. G. (1960), Bardwell, G. E. (1961), Williams, E. J. (1961b), Patil, G. P. (1962a), Patil, G. P. (1962c), Patil, G. P. (1962d), Tsao, C. M. (1962), Patil, G. P. (1963a), Bardwell, G. E., and Crow, E. L. (1964), Nelson, W. C., and David, H. A. (1964), Shanfield, F. (1964), Crow, E. L., and Bardwell, G. E. (1965), Govindarajulu, Z. (1965), Kamat, A. R. (1965), Tweedie, M. C. K. (1965), Mathai, A. M. (1966), Van Heerden, D. F. I., and Gonin, H. T. (1966).
Class.: PS-P-NB-LS:m:G

Noether, G. E. 1957. 1922
Two confidence intervals for the ratio of two probabilities and some measures of effectiveness. *J. Amer. Statist. Assoc.* **52**, 36–45.
Review: MR **19** (1958), 74; ZBL **84** (1960), 149.
Class.: B:ctp-ie:G

Noether, G. E. 1963. 1923
Note on the Kolmogorov statistic in the discrete case (German summary). *Metrika*, **7**, 115–16.
Review: MR **28** (1964), 338; ZBL **115** (1965), 367; STMA **5** (1964), 396.

Noether, G. E. 1964. 1924
An identity in probability. *Amer. Math. Monthly*, **71**, 903–04.
Class.: B:mi:G

Nolfi, P. 1943. 1925
Wahrscheinlichkeit unstetiger Vorgänge bei kontinuierlich wirkenden Ursachen. *Comment. Math. Helv.*, **15**, 36–44.
Review: MR **6** (1945), 5; ZBL **27** (1943), 113.

Nolfi, P. 1944. 1926
Zur bestimmung der rückschlusswahrscheinlichkeit einer geschlossenen gesamtheit. *Mitt. Verein. Schweiz. Versich.-Math.*, **44**, 217–20.
Review: MR **7** (1946), 311.

Nolfi, P. 1945. 1927
eiten. *Mitt. Verein. Schweiz. Versich.-Math.*, **45**, 311–21.
Zur mathematischen Darstellung wachsender Gesamth
Review: MR **7** (1946), 311

Norman, J. E. Jr., and David, H. A.
See David, H. A., and Norman, J. E. Jr. (1965).

Norris, Nilah. 1959. 1928
Nominal confidence limits for the expectation of a Poisson variable (Summary). *J. Amer. Statist. Assoc.*, **54**, 502.

Novick, Melvin R., and Grizzle, James E. 1965. 1929
A Bayesian approach to the analysis of data from clinical trials. *J. Amer. Statist. Assoc.*, **60**, 81–96.
Class.: B-P-M:mb:BM

Numata, M. 1950. 1930
Plant community as a stochastic population. *Biol. Sci. Tokyo*, **2**, 108–16.
Users: Goodall, D. W. (1952), Hairston, N. G. (1959), Morisita, M. (1959).

Nyunt, K. M., and Foster, F. G.
See Foster, F. G., and Nyunt, K. M. (1961).

O

Oakland, G. B. 1950. 1931
An application of sequential analysis to whitefish sampling. *Biometrics*, **6**, 59–67.
Users: Bliss, C. I., and Fisher, R. A. (1953), Waters, W. E. (1955), Bliss, C. I., and Owen, A. R. G. (1958), Grimm, H., and Maly, V. (1961).

O'Carroll, F. M. 1962. 1932
Fitting a negative binomial distribution to coarsely grouped data by maximum likelihood. *Appl. Statist.* **11**, 196–201.
Review: MR **27** (1964), 403; STMA **6** (1965), 1122.
Class.: NB:c:G

Ochoa, Merid J. 1962. 1933
Bounds of moments (Spanish). *Actas 2ª Reunion Mat. Espanoles (1961)*, pp. 53–55, *Seminario Matematico de Zarajoza.*

Oderfeld, J. 1951. 1934
On the dual aspect of sampling plans. *Colloq. Math.*, **2**, 89–97.
Review: ZBL **42** (1952), 382.
Users: Steinhaus, H., and Zubrzyski, S. (1957).

Ohlsen, Sally. 1963. 1935
Further models for phage reproduction in a bacterium. *Biometrics*, **19**, 441–49.
Review: ZBL **115** (1965), 373.

Okada, T., Kuno, E., Yamamoto, S., Satomi, H., and Outi, Y.
See Kuno, E., Yamamoto, S., Satomi, H., Outi, Y., and Okada, T. (1963).

Okamoto, M. 1955. 1936
Fit of a Poisson distribution by the index of dispersion. *Osaka Math. J.*, **7**, 7–13.
Review: MR **17** (1956), 53; ZBL **64** (1956), 384.
Class.: P:id:G

Okamoto, M. 1958. 1937
Some inequalities relating to the partial sum of binomial probabilities. *Ann. Inst. Statist. Math. Tokyo*, **10**, 29–35.
Review: MR **20** (1959), 1020; IJA **1** (1959), 17; ZBL **84** (1960), 140.
Users: Blischke, W. R. (1962), Kambo, N. Singh, and Kotz, S. (1966).
Class.: B:osp:G

Okamoto, M. 1959. 1938
A convergence theorem for discrete probability distributions. *Ann. Inst. Statist. Math. Tokyo*, **11**, 107–12.

Review: MR **22** (1961), 47; IJA **1** (1960), 553; ZBL **99** (1963), 130.

Okamoto, M. 1963. 1939
Chi-squared statistic based on the pooled frequencies of several observations. *Biometrika*, **50**, 524–28.
Review: STMA **6** (1965), 89; ZBL **117** (1965), 369.
Class.: M:gf:G-BM

Okuma, A., Yamato, H., Yanagawa, T., and Kudô, A.
See Kudô, A., Okuma, A., Yamato, H., Yanagawa, T. (1965).

Olbrich, E. 1965. 1940
Geometrical interpretation of the multinomial distribution and some conclusions (German). *Biom. Zeit.*, **7**, 96–101.
Review: STMA **7** (1966), 50.
Class.: M:ctp:G

Olds, E. G. 1960. 1941
Power characteristics of control charts. *Annual Conv. Trans., Amer. Soc. Qual.* pp. 195–208.
Review: IJA **2** (1961), 709.
Class.: P:sqc:E

Olekieweicz, M., and Krysicki, W.
See Krysicki, W., and Olekiewicz, M. (1963).

Oliver, R. M. 1961. 1942
A traffic counting distribution. *Operations Res.*, **9**, 802–10.
Class.: MI:pc-mb:E

Oliver, R. M. 1962. 1943
Distribution of gaps and blocks in a traffic system. *Operations Res.*, **10**, 197.
User: Weiss, G. H. (1963).

Oliver, R. M., and Bisbee, E. F. 1962. 1944
Queuing for gaps in high flow traffic streams. *Operations Res.*, **10**, 105–14.
Users: Oliver, R. (1962), Weiss, G. H. (1963).

Olkin, Ingram, and Marshall, Albert W.
See Marshall, Albert W., and Olkin, Ingram (1964).

Olkin, Ingram, and Sobel, Milton. 1965. 1945
Integral expressions for tail probabilities of the multinomial and negative multinomial distributions. *Biometrika*, **52**, 167–79.
Review: STMA **6** (1965), 1124.
Class.: M-NM:mi:G

Olkin, Ingram, and Tate, R. F.
See Tate, R. F., and Olkin, Ingram (1960).

Olkin, Ingram, and Tate, R. F. 1961. 1946
Multivariate correlation models with mixed discrete and continuous variables. *Ann. Math. Statist.*, **32**, 448–65.
Review: MR **27** (1964), 400; MR **30** (1965), 669; IJA **4** (1963), 95; ZBL **113** (1965), 351.
Class.: M:mb-osp-pe:G
Correction: *Ann. Math. Statist.*, **36**, 343–44.

Olmstead, P. S. 1940. 1947
Note on theoretical and observed distributions of repetitive occurrences. *Ann. Math. Statist.*, **11**, 363–66.
Review: ZBL **23** (1941), 339.
Class.: B:pe:G

Olmstead, P. S. 1946. 1948
Distribution of sample arrangements for runs up and down. *Ann. Math. Statist.*, **17**, 24–33.
Users: Barton, D. E., and David, F. N. (1959c), Govindarajulu, Z. (1965).
Class.: MI:mb-tc:G

Olson, E. C. 1957. 1949
Size-frequency distributions in samples of extinct organisms. *J. Geology*, **65**, 309–33.

O'Mathuna, D., and Serra, A.
See Serra, A., and O'Mathuna, D. (1966).

Onicescu, Octav, and Mihoc, G. 1937. 1950
Sur une généralisation de l'urne de Bernoulli. *Bull. Math. Phys. École Polytechn. Bucarest*, **8**, 61–77.
Review: ZBL **22** (1940), 369.

Onicescu, Octav, and Mihoc, G. 1938. 1951
Sur une généralisation de l'urne de Bernoulli. *Bull. Math. Phys. École Polytechn. Bucarest*, **9**, 57–75.
Review: ZBL **22** (1940), 369.

Orazov, G. 1965. 1952
A local theorem for a random number of random variables. (Russian). *Izv. Akad. Nauk Turkmen SSR Ser. Fix.-Tehn. Him. Geol. Nauk. No. 1*, 3–8.
Review: MR **31** (1966), 496.
Class.: MI:a:G

Orians, G. H., and Leslie, P. H. 1958. 1953
A capture-recapture analysis of a shearwater population. *J. Animal Ecol.*, **27**, 71–86.
Class.: GBDP:mi:BM

Orts, Aracil J. M. 1943. 1954
On the behaviour of certain probabilities. *Rev. Mat. Hisp.-Amer.*, **3** (4), 157–63.
Review: MR **5** (1944), 206.

Orts, José Ma. 1941. 1955
Die Legendreschen Polynome und das Schema der wiederholten Proben (Spanish). *Rev. Mat. Hisp.-Amer. IV*, **1**, 198–201.
Review: ZBL **26** (1942), 137.
Class.: B:mi:G

Osborne, M. F. M. 1962. 1956
Periodic structure in the Brownian motion of stock prices. *Operations Res.*, **10**, 345–79.

Ostrovskiĭ, I. V. 1963. 1957
Infinitely divisible laws with unbounded Poisson spectrum (Russian). *Dokl. Akad. Nauk SSSR*, **152**, 1301–04.

Ostrovskiĭ, I. V. 1965a. 1958
The multidimensional analogue of Yu. V. Linnik's theorem on decompositions of a convolution of Gaussian and Poisson laws. *Theor. Probability Appl.*, **10**, 673–77.
Class.: MP:mi:G

Ostrovskiĭ, I. V. 1965b. 1959
The multidimensional analog of Yu, V. Linnik's theorem of decompositions of a composition of Gaussian and Poisson Laws (Russian; English summary). *Teor. Verojatnost. i Primenen*, **10**, 742–45.

Ottaviani, G. 1938. 1960
Su una Fondamentale Proprietà della leggi di Gauss e di Poisson. *Gion. Inst. Ital. Attuairi*, **16**, 170–90.
Review: ZBL **19** (1939), 72.

Ottaviani, G. 1957. 1961
Su una porprietà di un sistema di due variabili casuali seguenti la legge di Poisson, deli eventi rari. *R. C. Accad. Lincei*, **23** (8), 230–32.
Review: ZBL **80** (1959), 340.

Ottestad, P. 1934. 1962
Letter to the editor. *J. Conseil*, **9** (2), 249–53.

Ottestad, P. 1937. 1963
On some discontinuous frequency functions and frequency distributions. *Skand. Aktuarietidskr.*, **20**, 75–86.
Review: ZBL **16** (1937), 313.
User: Tsao, C. M. (1962).
Class.: B-NB-P:m-gf:G

Ottestad, P. 1939. 1964
On the use of the factorial moments in the study of discontinuous frequency distributions. *Skand. Aktuarietidskr.*, **22**, 22–31.
Review: MR **1** (1940), 22; ZBL **21** (1940), 147.
Class.: B-P-H:gf:G

Ottestad, P. 1943. 1965
On Bernoullian, Lexis, Poisson and Poisson-Lexis series. *Skand. Aktuarietidskr.*, **26**, 15–67.
Review: MR **7** (1946), 211; ZBL **28** (1944), 170.
User: Tsao, C. M. (1962).
Class.: N-COB-P-GBDP:m:osp:G

Ottestad, P. 1944. 1966
On certain compound frequency distributions. *Skand. Aktuarietidskr.*, **27**, 32–42.
Review: MR **7** (1946), 211; ZBL **60** (1957), 296.
Users: Teicher, H. (1960a), Tsao, C. M. (1962).
Class.: COB-COP:m-rp:G

Ottestad, Per. 1948. 1967
On some discrete frequency functions. *Skand. Aktuarietidskr.*, **31**, 1–13.
Review: ZBL **31** (1949), 169.
User: Tsao, C. M. (1962).

Ottestad, Per. 1951. 1968
On the test of the hypothesis that the probability of an event is contained within given limits. *Skand. Aktuarietidskr.*, **34**, 197–201.
Review: ZBL **44** (1952), 146.

Ottestad, P. 1952. 1969
On the analysis of variance of percentage fractions. *Skand. Aktuarietidskr.*, **35**, 152–59.
Review: ZBL **48** (1953), 365.

Ottestad, P., and Faegri, K.
See Faegri, K., and Ottestad, P. (1949).

Outi, Y., Okada, T., Kuno, E., Yamamoto, S., and Satomi, H.
See Kuno, E., Yamamoto, S., Satomi, H., Outi, Y., and Okada, T. (1963).

Over, H. J., Dewit, C. T., and Keuls, M.
See Keuls, M., Over, H. J., and DeWit, C. T. (1962).

Overall, John E., and Williams, Clyde M. 1961. 1970
Models for medical diagnosis. *Behavioral Sci.*, **6**, 134–41.
Class.: MI:mi:BM

Owen, A. R. G., and Bliss, C. I.
See Bliss, C. I., and Owen, A. R. G. (1958).

Owen, D. B. 1961. 1971
Distribution free tolerance limits for an additional finite sample as obtained from the hypergeometric distribution. Reprint SCR-285, Sandia Corp.
Class.: H:sqc:G

Owen, D. B. 1962. 1972
Handbook of statistical tables. Addison-Wesley Publ. Co., Inc., Reading, Mass.—London.
Review: MR **28** (1964), 894.

Owen, D. B., and Lieberman, Gerald J.
See Lieberman, Gerald, J., and Owen, D. B. (1961).

Owen, D. B., Gilbert, E. J., Steck, G. P., and Young, D. A. 1959. 1973
A formula for determining sample size in hypergeometric sampling when zero defectives are observed in the sample. Tech. Memo. SCTM 178–59 (51), Sandia Corp.
Users: Clark, C. R., and Koopmans, L. H. (1959), Owen, D. B. (1961).
Class.: H:ie-sqc-osp:E

Ozawa, M., and Siotani, M.
See Siotani, M., and Ozawa, M. (1958).

P

Pachares, James. 1960. 1974
Tables of confidence limits for the binomial distribution. *J. Amer. Statist. Assoc.*, **55**, 521–33.
Review: PA **35** (1961), Ab. No. 1485; ZBL **92** (1962), 370; IJA **4** (1963), 371.
Class.: B:ie-tc:G

Pacioni, G. 1963. 1975
On the application of the theory of queues with one server to the approximate solution of problems relating to queues with several server (Italian). *Giorn. Ist. Ital. Attuari.*, **26**, 306–16.
Review: STMA **5** (1964), 977.
Class.: MI:pc:G

Page, E. S. 1959. 1976
The distribution of vacancies on a line. *J. Roy. Statist. Soc. Ser. B*, **21**, 364–74.
Review: IJA **1** (1960), 554.
Class.: MI:mb:G

Page, E. S., and Anscombe, F. J.
See Anscombe, F. J., and Page, E. S. (1954).

Palásti, Ilona. 1961. 1977
On the distribution of the number of trees which are isolated subgraphs of a chromatic random graph (Russian summary). *Magyar. Tud. Akad. Mat. Kutató Int. Közl.*, **6**, 405–09.
Review: IJA **3** (1962), 502.
Class.: MI-P:mi-a:G

Palm, C. 1936. 1978
Calcuexact de la perte dans les groupes de circuits échelonnes. *Ericsson Technics*, **4**, 41–71.

Palm, C. 1937a. 1979
Inhomogeneous telephone traffic in full-availability groups. *Ericsson Technics*, **5**, 3–36.

Palm, C. 1937b. 1980
Étude des délais d'áttente. *Ericsson Technics*, **5**, 39–56.

Palm, C. 1938. 1981
Analysis of the Erlang traffic formulae for busy-signal arrangements. *Ericsson Technics*, **6**, 39–58.
Users: Takacs, L. (1957c), Prékopa, A. (1958).

Palm, C. 1947. 1982
The distribution of repairmen in servicing automatic machines (Swedish). *Industritidningen Norden*, **75**, 75–80, 90–94, 119–23.
Users: Benson, F., and Cox, D. R. (1951).

Panse, V. G., and Sukhatme, P.V.
See Sukhatme, P. V., and Panse, V. G. (1943).

Papamichail, D., and Antoneas, G. 1964. 1983
Sur la distribution de la somme d'un nombre aléatoire de variables aléatoires independantes et isonômes. *Bull. Soc. Math. Grēce* (*NS*), **5** (1), 93–97.
Review: MR **30** (1965), 1000.
Class.: GP:mi:G

Parker, R. A. 1963. 1984
On the estimation of population size, mortality, and recruitment. *Biometrics*, **19**, 318–23.
Review: STMA **6** (1965), 115.
Class.: P:pe:BM

Parker-Rhodes, A. F., and Joyce, T. 1956. 1985
A theory of word frequency distribution. *Nature*, **178**, 1308.
Users: Good, I. J. (1957b), Rider, P. R. (1965).
Class.: MI:mb:L

Patel, I. R., and Khatri, C. G.
See Khatri, C. G., and Patel, I. R. (1961).

Patel, J. N. 1962. 1986
The relationships between discrete and continuous probability distributions. Master's Thesis, Virginia Polytechnic Inst.

Pathak, K. B. 1966a. 1987
A note on inflated power series distributions (Abstract). *Ann. Math. Statist.*, **37**, 553.

Pathak, K. B. 1966b. 1988
On the first conceptive delay (Abstract). *Ann. Math. Statist.*, **37**, 554.

Pathak, P. K. 1961. 1989
On the evaluation of moments of distinct units in a sample. *Sankhyā Ser. A*, **23**, 415–20.
Review: IJA **4** (1963), 498; ZBL **101** (1963), 121.
Class.: MI:m:G

Pathak, P. K. 1964. 1990
On estimating the size of a population and its inverse by capture mark method. *Sankhyā*, **26**, 75–80.
Review: STMA **7** (1966), 764.
Class.: MI:pe:BM

Patil, G. P. 1957. 1991
Problems of estimation in a class of discrete distributions. A.I.S.I. Thesis, Indian Statistical Inst.
Users: Patil, G. P. (1961b), Patil, G. P. (1962c), Patil, G. P. (1962e), Patil, G. P. (1963c), Patil, G. P. (1964b).

Patil, G. P. 1959. 1992
Contributions to estimation in a class of discrete distributions. Ph.D. Thesis, Univ. Michigan.
Users: Patil, G. P. (1960c), Patil, G. P. (1961b), Patil, G. P. (1962b), Patil, G. P. (1962d), Patil, G. P. (1962e), Shanfield, F. (1964), Kamat, A. R. (1965), Rao, C. R. (1965).

Patil, G. P. 1960a. 1993
Generalized power series distribution and certain characterization theorems (Abstract). *Ann. Math. Statist.*, **31**, 240.
Users: Bardwell, G. E. (1961), Govindarajulu, Z. (1965).

Patil, G. P. 1960b. 1994
On evaluation of negative binomial distribution function (Abstract). *Ann. Math. Statist.*, **31**, 527.
User: Bartko, J. J. (1961b).

Patil, G. P. 1960c. 1995
On the evaluation of the negative binomial distribution with examples. *Technometrics*, **2**, 501–05.
Review: MR **22** (1961), 1720; ZBL **96** (1962), 134; IJA **2** (1961), 636.
Users: Patil, G. P., and Cota, P. (1962), Patil, G. P. (1963c), Kerridge, D. (1964), Blischke, W. R. (1965).
Class.: NB:osp:G-E

Patil, G. P. 1961a. 1996
On some methods of estimation for the logarithmic series distribution (Abstract). *Ann. Math. Statist.*, **32**, 922.

Patil, G. P. 1961b. 1997
Asymptotic bias and variance of ratio estimates in generalized power series distributions and certain applications. *Sankhyā*, **23**A, 269–80.
Review: MR **25** (1963), 139; IJA **4** (1963), 261.
Users: Patil, G. P. (1962a), Patil, G. P. (1962b), Patil, G. P. (1962c), Patil, G. P. (1962d), Patil, G. P. (1962e), Patil, G. P. (1963a), Nelson, W. C., and David, H. A. (1964), Patil, G. P. (1964b), Shanifield, F. (1964), Staff, P. J. (1964).
Class.: PS-TCB-TCP:pe:G

Patil, G. P. 1962a. 1998
Certain properties of the generalized power series distribution. *Ann. Inst. Statist. Math. Tokyo*, **14**, 179–82.
Review: STMA **5** (1964), 35.
Users: Patil, G. P. (1963a), Nelson, W. C., and David, H. A. (1964), Shanfield, F. (1964), Patil, G. P. (1965).
Class.: PS:m-osp:G

Patil, G. P. 1962b. 1999
Some methods of estimation for the logarithmic series distribution. *Biometrics*, **18**, 68–75.
Review: MR **25** (1963), 139; STMA **6** (1965), 68; ZBL **114** (1965), 106.
Users: Birch, M. W. (1963), Patil, G. P. (1963a), Nelson, W. C., and David, H. A. (1964), Williamson, E., and Bretherton, M. (1964), Bliss, C. (1965), Chatfield, C., Ehrenberg, A. S. C., and Goodhardt, G. J. (1965), Chatfield, C., Ehrenberg, A. S. C., and Goodhardt, C. (1966), Patil, G. P., and Wani, J. K. (1965).
Class.: LS:pe:G

Patil, G. P. 1962c. 2000
On homogeneity and combined estimation for the generalized power series distribution and certain applications. *Biometrics*, **18**, 365–74.
Review: ZBL **114** (1965), 354; IJA **4** (1963), 442.
Users: Patil, G. P. (1962a), Patil, G. P. (1963a), Nelson, W. C., and David, H. A. (1964), Shanfield, F. (1964).
Class.: PS-TCB-NB:h-pe:G

Patil, G. P. 1962d. 2001
Maximum likelihood estimation for generalized power series distributions and its application to a truncated binomial distribution. *Biometrika*, **49**, 227–38.
Review: ZBL **114** (1965), 355; IJA **4** (1963), 59.
Users: Bardwell, G. E., and Crow, E. L. (1964), Nelson, W. C., and David, H. A. (1964), Shanfield, F. (1964), Staff, P. J. (1964), Crow, E. L., and Bardwell, G. E. (1965), Govindarajulu, Z. (1965), Patil, G. P., and Wani, J. K. (1965).
Class.: PS-TCB:pe-tc:G

Patil, G. P. 1962e. 2002
Estimation by two-moments method for generalized power series distribution and certain applications. *Sankhyā*, **24**B, 201–14.
Review: MR **32** (1966), 1128.
Users: Patil, G. P. (1962a), Patil, G. P. (1962d).
Class.: PS-TCB-TCP:pe:G

Patil, G. P. 1963a. 2003
Minimum variance unbiased estimation and certain problems of additive number theory. *Ann. Math. Statist.*, **34**, 1050–56.
Review: STMA **5** (1964), 610; ZBL **116** (1965), 372.
Users: Shanfield, F. (1964), Harkness, W. L. (1965), Patil, G. P. (1965), Tweedie, M. C. K. (1965), Patil, G. P. (1966), Patil, G. P., and Shorrock, R. (1966).
Class.: PS-P-B-NB-LS:pe:G

Patil, G. P. 1963b. 2004
A characterisation of the exponential-type distribution. *Biometrika*, **50**, 205–07.
Review: ZBL **114** (1965), 338.
Users: Patil, G. P. (1965), Bolger, E. M. (1966), Bolger, E. M., and Harkness, W. L. (1966).
Class.: P:osp:G

Patil, G. P. 1963c. 2005
On the equivalence of the binomial and inverse binomial acceptance sampling plans and an acknowledgement. *Technometrics*, **5**, 119–21.
Review: STMA **6** (1965), 452; ZBL **116** (1965), 116.
Class.: B-NB:sqc:E

Patil, G. P. 1964a. 2006
A conjecture. *Canad. Math. Bull.*, **7**, 305–06.

Patil, G, P. 1964b. 2007
On certain compound Poisson and compound binomial distributions. *Sankhyā*, **26**, 293–94.
Review: STMA **7** (1966), 478.
Users: Patil, G. P., and Bildikar, S. (1966a), Kemp, A. W., and Kemp, C. O. (1966).
Class.: COP-COB:mi:G

Patil, G. P. 1964c. 2008
On minimum variance unbiased estimation for the logarithmic series distribution. *Bull. Inst. Internat. Statist.*, **40**, 766.
Review: STMA **6** (1965), 884.
Class.: LS:pe:G

Patil, G. P. 1964d. 2009
Estimation for the generalized power series distribution with two parameters and its application to binomial distribution. *Contributions to Statistics* (70th Birthday Volume in honour of Professor P. C. Mahalanobis), Pergamon Press and Statistical Publishing Society, Calcutta. pp. 335–44.
Class.: PS-B:pe-gf:G

Patil, G. P. 1965a. 2010
Certain characteristic properties of multivariate discrete probability distributions akin to the Bates-Neyman model in the theory of accident proneness (Abstract). *Biometrics*, **21**, 765.
Users: Patil, G. P., and Bildikar, S. (1966a).

Patil, G. P. 1965b. 2011
Opening remarks. *Classical and Contagious Discrete Distributions*, Ed. G. P. Patil, Statistical Publishing Society, Calcutta and Pergamon Press, pp. 2–4.

Patil, G. P. 1965c. 2012
On multivariate generalized power series distribution and its application to the multinomial and negative multinomial. *Classical and Contagious Discrete Distributions*, Ed. G. P. Patil, Statistical Publishing Society, Calcutta and Pergamon Press, pp. 183–94. Reprinted in *Sankhyā A*, **28** (1966).
Class.: MPS-M-NM:m-osp-pe:G-TCP
Reprinted in *Sancbyā* A, 28 (1966).

Patil, G. P. 1965d. 2013
A proposed bibliography on discrete distributions, *Classical and Contagious Discrete Distributions*, Ed. G. P. Patil, Statistical Publishing Society, Calcutta and Pergamon Press, pp. 465-68.

Patil, G. P. 1965e. 2014
A selected bibliography of statistical literature on classical and contagious discrete distributions. *Classical and Contagious Discrete Distributions*, Ed. G. P. Patil, Statistical Publishing Society, Calcutta and Pergamon Press, pp. 469-552.

Patil, G. P. 1965f. 2015
Classical and Contagious Discrete Distributions Proc. of the International Symposium held at Montreal, 1963. Statistical Publishing Society, Calcutta and Pergamon Press, Ed. G. P. Patil.

Patil, G. P. 1965g. 2016
Certain characteristic properties of multivariate discrete probability distributions akin to the Bates-Neyman model in the theory of accident proneness. *Sankhyā. Ser. A*, **27**, 259–70.
Class.: MPS-M-NM-P:osp:G

Patil, G. P., and Bildikar, Sheela. 1965. 2017
Identifiability of countable mixtures of discrete probability distributions using methods of infinite matrices (Abstract). *Biometrics*, **21**, 765.

Patil, G. P., and Bildikar, Sheela. 1966a. 2018
Certain studies on the multivariate logarithmic series distribution. Tech. Rep., Aerospace Res. Lab., Wright-Patterson Air Force Base, pp. 48.

Patil, G. P., and Bildikar, Sheela. 1966b. 2019
Identifiability of countable mixtures of discrete distributions. Tech. Rep., Aerospace Res. Lab., Wright-Patterson Air Force Base, pp. 49.

Patil, G. P., and Bildikar, Sheela. 1966c. 2020
On minimum variance unbiased estimation for the logarithmic series distribution. *Sankhyā. Ser. A*, **28.**
Class.: LS:pe:G

Patil, G. P. and Bildikar, Sheela. 1966d. 2021
Identifiability of countable mixtures of discrete probability distributions using methods of infinite matrices. *Proc. Cambridge Philos. Soc.*, **62**, 485–94.
Class.: N-COM-NM-CONM-LS:mi:G

Patil, G. P., and Brown, J. L. Jr.
See Brown, J. L. Jr., and Patil, G. P. (1966).

Patil, G. P., and Cota, P. 1962. 2022
On certain strategies of signal detection using clipper cross-correlator. Tech. Rep. 128, 3674-1-T, Office Res. Admin., Univ. Michigan, pp. 37.

Patil, G. P., and Seshadri, V. 1964a. 2023
A characterization of a bivariate distribution by the marginal and the conditional distributions of the same component. *Ann. Inst. Statist. Math. Tokyo*, **15**, 215–21.
Review: MR **30** (1965), 811.
Class.: MI-P-B:osp:G

Patil, G. P., and Seshadri, V. 1964b. 2024
Characterization theorems for some univariate probability distributions. *J. Roy. Statist. Soc. Ser. B*, **26**, 286–92.
Review: MR **30** (1965), 311; STMA **6** (1965), 453; ZBL **123** (1966), 365.
Class.: MI-P-B-G-NB-NH:osp:G

Patil, G. P., and Shorrock, Richard. 1965a. 2025
On certain properties of the exponential families (Abstract). *Ann. Math. Statist.* **36**, 1594.

Patil, G. P., and Shorrock, Richard. 1965. 2026
On certain properties of the exponential-type families. *J. Roy. Statist. Soc. Ser. B*, **27**, 94–99.
Review: MR **32** (1966), 796.
Class.: MI-B-P:osp:G

Patil, G. P., and Shorrock, Richard. 1966. 2027
Stochastic and sampling processes for the logarithmic series distribution. Tech. Rep., Aerospace Res. Lab., Wright-Patterson Air Force Base, pp. 82.
Users: Patil, G. P., and Bildikar, S. (1966a).

Patil, G. P., and Wani, J. K. 1965. 2028
Maximum likelihood estimation for the complete and truncated logarithmic series distributions. *Classical and Contagious Discrete Distributions*, Ed. G. P. Patil, Statistical Publishing Society, Calcutta and Pergamon Press, pp. 398–409. Reprinted in *Sankhyā A*, **27** (1966).
Class.: LS:pe-tc:G

Patil, G. P., and Wani, J. K. 1966. 2029
Minimum variance unbiased estimation of the distribution function admitting a sufficient statistic. *Ann. Inst. Statist. Math. Tokyo*, **18**, 39–47.
Class.: PS:pe:G

Patil, G. P., and Wani, J. K. 1966a. 2030
On certain structural properties of the logarithmic series distribution and the first type Stirling distribution (Abstract). *Ann. Math. Statist.*, **37**, 543.

Patil, G. P., Kamat, A. R., and Wani, J. K. 1964. 2031
Certain studies on the structure and statistics of the logarithmic series distribution and related tables. Tech. Rep., Aerospace Res. Lab., Wright-Patterson Air Force Base. pp. 389.
Review: STMA **7** (1966), 479
Class.: LS:m-osp-pe-tc:G-BM

Patil, S. A. 1962. 2032
The maximum likelihood estimation of the parameter of the truncated censored Poisson distribution. *J. Indian Soc. Agric. Statist.*, **14**, 177–87.
Review: MR **29** (1965), 562.

Patnaik, P. B. 1954-55. 2033
A test of significance of a difference between two sample proportions when the proportions are small. *Sankhyā*, **14**, 187–202.
Review: MR **16** (1955), 727; ZBL **57** (1956), 118.
Class.: B-P:ctp:G

Patwary, K. M., and Mantel, N.
See Mantel, N., and Patwary, K. M. (1961).

Paulson, E. 1952. 2034
On the comparison of several experimental categories with a control. *Ann. Math. Statist.*, **23**, 239–46.
Review: MR **14** (1953), 299.
Users: Gupta, S. S., Huyett, M. J., and Sobel, M. (1957), Gupta, S. S., and Sobel, M. (1958), Ahmed, M. S. (1961), Gupta, S. S. (1965).
Class.: B:srp:G

Paulson, Edward. 1965. 2035
Sequential procedures for selecting the best one of several binomial populations (Abstract). *Ann. Math. Statist.*, **36**, 1322–23.

Peach, Paul, and Littauer, S. B. 1946. 2036
A note on sampling inspection. *Ann. Math. Statist.*, **17**, 81–84.
Users: Hald, A. (1960), Hald, A. (1964), Hald, A. (1965).

Pearce, C., and Finch, P. D.
See Finch, P. D., and Pearce, C. (1965).

Pearson, E. S. 1950. 2037
On questions raised by the combination of tests based on discontinuous distributions. *Biometrika*, **37**, 383–98.
Review: ZBL **40** (1951), 75.
Users: Crow, E. L. (1956), Blyth, C. R., and Hutchinson, D. W. O. (1960), Pfanzagl, E. J. (1960), Lancaster, H. O. (1961), Kincaid, W. M. (1962), Ellner, H. (1963).
Class.: MI:mi:G
Notes: Correction: *Biometrika*, **38**, 265.

Pearson, E. S. 1960. 2038
Editorial note of Bennett and Hsu's paper " On the power functions of the exact test for the 2×2 contingency table." *Biometrika*, **47**, 397–98.
Review: IJA **2** (1961), 507.
Class.: OHR:tp:G

Pearson, E. S. 1962. 2039
Frequency surfaces. Tech. Rep. 49 STRG, Princeton Univ.

Pearson, E. S., and Clopper, C. J.
See Clopper, C. J., and Pearson, E. S. (1934).

Pearson, E. S., and Hartley, H. O.
See Hartley, H. O., and Pearson, E. S. (1950)

Pearson, E. S., and Neyman, J.
See Neyman, J., and Pearson, E. S. (1932-33).

Pearson, E. S., Wynn, A. H. A., and Maguire, B. A.
See Maguire, B. A., Pearson, E. S., and Wynn, A. H. A. (1952).

Pearson, E. S., Wynn, A. H. A., and Maguire, B. A.
See Maguire, B. A., Pearson, E. S., and Wynn, A. H. A. (1953).

Pearson, K. 1895. 2040
Contributions to the mathematical theory of evolution. II. Skew variation in homogeneous material. *Philos. Trans.: Roy. Soc. London Ser. A*, **186**, 343–414.
Users: Pearson, K. (1899), Pearson, K. (1906-07), Neyman, J. (1938).

Pearson, K. 1899. 2041
On certain properties of the hypergeometrical series, and on the fitting of such series to observation polygons in the theory of chance. *Philos. Mag. Ser.* 5, **47**, 236–46.
Users: Pearson, K. (1905-06), Pearson, K. (1906-07), Riordan, J. (1937).

Pearson, K. 1900. 2042
On a criterion that a given system of deviations from the probable in the case of a correlated system of variables is such that it can be reasonably supposed to have arisen from random sampling. *Philos. Mag. Ser.*, **550**, 157-175.
Users: Haldane, J. B. S. (1937), Lancaster, H. O. (1949a), Rybarz, J. (1959), Steyn, H. S. (1959), Kullback, S., Kupperman, M., and Ku, H. H., (1962), Young, D. H. (1962), Quesenberry, C. P., and Hurst, (1964), Richter, W. (1964), Lancaster, H. O. (1966),

Pearson, K. 1905-06. 2043
Skew variation, a rejoinder. *Biometrika*, **4**, 169–212.
Users: Waters, W. E., and Henson, W. R. (1959).
Class.: B-H:a-gf:G-BM

Pearson, K. 1906-07. 2044
On the curves which are most suitable for describing the frequency of random samples of a population. *Biometrika*, **5**, 172–75.
User: Pearson, K. (1924).
Class.: B-H:a:G

Pearson, K. 1924a. 2045
On the moments of the hypergeometrical series. *Biometrika*, **16**, 157–62.
Users: Frisch, R. (1925), Ram, S. (1954), Ram, S. (1956).
Class.: H:m:G

Pearson, K. 1924b. 2046
On a certain double hypergeometrical series and its representation by continuous frequency surfaces. *Biometrika*, **16**, 172–88.
Class.: MH:a:G

Pearson, K. 1924c. 2047
Note on the relationship of the incomplete β-function to the sum of the first p terms of the binomial $(a+b)^n$. *Biometrika*, **16**, 202–03.
Users: Frisch, R. (1925), Scheffe, H. (1943-46), Katz, L. (1945), Wise, E. (1950), Govindarajulu, J. (1965).
Class.: B:mi:G

Pearson, K. 1928. 2048
On a method of ascertaining limits to the actual number of marked members in a population of given size from a sample. *Biometrika*, **20**, 149–74.
User: Davies, O. L. (1933).
Class.: H:mi:G

Pearson, K. 1934. 2049
Tables of the incomplete beta-function. Univ. Press; Cambridge.

Pearson, K., and Fieller, E. C. 1933. 2050
On the applications of the double bessel function to statistical problems. *Biometrika*, **25**, 160–61.
Users: Bartko, J. J. (1961b), Bartko, J. J. (1962).

Pearson, P. G. 1955. 2051
Population ecology of the spadefoot toad, *Scaphiopus h. holbrooki* (Harlan). *Ecol. Monogr.*, **25**, 233–67.
Class.: P:gf:BM

Penfound, W. T., and Rice, E. L.
See Rice, E. L., and Penfound, W. T. (1955).

Perera, A. G. A. D., and Foster, F. G.
See Foster, F. G., and Perera, A. G. A. D. (1965).

Peritz, Eric. 1966. 2052
On some models for segregation analysis. *Ann. Human Genetics*, **30**, 183–92.
Class.: B-M:mb:BM

Perkal, J. 1962. 2053
A design for a genetic test (Polish). *Zastos. Mat.*, **6**, 257–85.
Review: IJA **4** (1963), 460.
Class.: B-OBR:cm:BM

Perks, Wilfred, and Beard, R. E.
See Beard, R. E., and Perks, Wilfred (1949).

Pessin, Vivian. 1961. 2054
Some asymptotic properties of the negative binomial distribution (Abstract). *Ann. Math. Statist.*, **32**, 922.

Pessin, Vivian. 1962 2055
A new derivative with applications to asymptotic limits of discrete probability distributions. Ph.D. Thesis, Univ. Buffalo.
Users: Pessin, Vivian (1965).

Pessin, Vivian. 1965. 2056
Some discrete distribution limit theorems using a new derivative. *Classical and Contagious Discrete Distributions*, Ed. G. P. Patil, Statistical Publishing Society, Calcutta and Pergamon Press, pp. 109–22.
Class.: NB:a:G

Peterson, R. L. 1955. 2057
A graphic method for estimating the significance of differences between proportions or percentages. *Educ. Psychol. Measmt.*, **15**, 186–94.
Review: PA **30** (1956), Ab.No. 1912.

Peterson, R. O., Fitzpatrick, R., and Vasilas, J. N.
See Fitzpatrick, R., Vasilas, J. N., and Peterson, R. O. (1953).

Petro, S. 1953. 2058
A dose-response equation for the invasion of micro-organisms. *Biometrics*, **9**, 320–335.
Users: Armitage, P., and Spicer, C. C. (1956).

Petrov, V. V. 1957. 2059
A local theorem for lattice distributions (Russian). *Dokl. Akad. Nauk SSSR*, **115**, 49-52.
Review: ZBL **78** (1959), 315.

Petrov, V. V. 1962a. 2060
Local limit theorems for non-identical lattice distributions. *Theor. Probability Appl.*, **7**, 333–35.
Class.: MI:a:G

Petrov, V. V. 1962b. 2061
On a local limit theorem for non-identical lattice distributions (Russian; English summary). *Teor. Verojatnost. i Primenen*, **7**, 344–46.

Petz, B. 1957. 2062
Statisticka analiza nesreca (Statistical analysis of accidents). *Arh. hig. rada.*, **8**, 25–38.
Review: PA **33** (1959), Ab.No. 2532.

Pfanzagl, E. J. 1960. 2063
Tests and confidence intervals for exponential distributions and their application to some discrete distributions. *Metrika*, **3**, 1–25.
Review: IJA **1** (1960), 628;ZBL **89** (1941), 359.
Class.: B-P-NB:tp-ie:G

Pfanzagl, J., and Puntigam, F. 1961. 2064
Statements on the quotient of two Poisson-parameters and their applications to a problem of vaccination (German). *Biom. Zeit.*, **3**, 135–42.
Review: IJA **3** (1962), 523; ZBL **95** (1962), 341.
Class.: P:ctp-pe-ie:G-BM

Phatarfod, R. M. 1963. 2065
Application of methods in sequential analysis to dam theory. *Ann. Math. Statist.*, **34**, 1588-92.
Class.: COP-G:pc:G

Philip, U., and Evans, D. A.
See Evans, D. A., and Philip, U. (1964).

Philipson, Carl. 1965. 2066
A note on different models of stochastic processes dealt with in the collective theory of risk. *Skand. Aktuarietidskr.*, **39**, 26–37.

Philipson, Carl. 1957. 2067
On some distribution functions related to a specified class of stochastic processes. *Trans. 15th Inter. Congress Actuaries*, **2**, 1–13.

Philipson, Carl. 1960a. 2068
Note on the application of compound Poisson processes to sickness and accident statistics. *ASTIN Bull.*, **1**, 224–37.
User: Philipson, C. (1961c).

Philipson, Carl. 1960b. 2069
The theory of confluent hypergeometric functions and its application to compound Poisson processes. *Skand. Aktuarietidskr.*, **43**, 136-62.
Review: MR **24**A (1962), 700; ZBL **107** (1964), 352.

Philipson, Carol. 1961a. 2070
On a class of distribution functions as applied to different stochastic processes. *Skand. Aktuarietidskr.*, **44**, 20–54.
Review: MR **28** (1964), 895.

Philipson, Carl. 1961b. 2071
An extension of the models usually applied to the theory of risk. *Skand. Aktuarietidskr.*, **44**, 223–39.
Review: ZBL **115** (1965), 375.

Philipson, Carl. 1961c. 2072
Some estimation problems connected with compound Poisson processes. *Skand. Aktuarietidskr.*, **44**, 240–50.
Review: ZBL **115** (1965), 375.

Philipson, Carl. 1962a. 2073
A generalization of the stochastic processes commonly applied to the theory of causalty risk. Probability and intensity-functions of some compound Poisson processes. *Bull. Inst. Internat. Statist.*, **39**, 249–60.
Review: ZBL **112** (1965), 119; STMA **5** (1964), 495.

Philipson, Carl. 1963. 2074
A note on moments of a Poisson probability distribution. *Skand. Aktuarietidskr.*, **46**, 243–44.
Review: MR **30** (1965), 805; STMA **6** (1965), 1126.

Philipson, Carl. 1963a. 2075
Quelques processes applicables dans l'assurance et dans la biologie. *Assoc. Roy. Actuaries Belges, Bull.*, **61**, 51–65.

Philipson, Carl. 1963b. 2076
On Esscher transforms of distribution functions defining a compound Poisson process for large values of the parameter. *Skand. Aktuarietidskr.*, **46**, 226–36.
Review: MR **31** (1966), 310.

Philipson, Carl. 1963c. 2077
On the difference between the concepts " compound " and " composed " Poisson processes. *ASTIN Bull.*, **2**, 445–51.

Philipson, Carl. 1964. 2078
The transformed parameter of compound Poisson processes and the effect of an increase of that parameter. *Trans. 17th Inter. Congress Actuaries*, 627–49.

Philipson, Carl. 1965. 2079
A generalized model for the risk process and its application to a tentative evaluation of outstanding liabilities. *ASTIN Bull.*, **3**, 215–38.

Phipps, P. H., Boulter, E. A., and Westwood, J. C. N.
See Westwood, J. C. N., Phipps, P. H., and Boulter, E. A. (1957).

Pielou, D. P. 1960a. 2080
Contagious distribution in the European red mite, panonychus ulmi (Koch), and a method of grading population densities from a count of mite-free leaves. *Canad. J. Zool.*, **38**, 645–53.
Class.: P-NB-B-GP:gf-ie:BM

Pielou, E. C. 1957. 2081
The effect of quadrat size on the estimation of the parameters of Neyman's and Thomas' distributions. *J. Ecol.*, **45**, 31–47.
User: Morisita, M. (1959).
Class.: N-T:pe-gf:BM

Pielou, E. C. 1960. 2082
A single mechanism to account for regular, random and aggregated populations. *J. Ecol.*, **48**, 575–84.
Class.: MI–P:mb-cm-gf:BM

Pielou, E. C. 1962. 2083
Runs of one species with respect to another in transects through plant populations. *Biometrics*, **18**, 579–93.
Users: Pielou, E. C. (1963b), Pielou, E. C. (1965).
Class.: G-COG-GG:mb-gf:BM

Pielou, E. C. 1963a. 2084
The distribution of diseased trees with respect to healthy ones in a patchily infected forest. *Biometrics*, **19**, 450–459.
Review: STMA **5** (1964), 367.
Users: Potthoff, Richard F., and Whittinghill, Maurice (1966a).
Class.: GB:pe-gf:BM

Pielou, E. C. 1963b. 2085
Runs of healthy and diseased trees in transects through an infected forest. *Biometrics*, **19**, 603–14.
Review: STMA **6** (1965), 727.
User: Pielou, E. C. (1965).
Class.: G:mb-gf:BM

Pielou, E. C. 1965. 2086
The concept of segregation pattern in ecology: some discrete distributions applicable to the run lengths of plants in narrow transects. *Classical and Contagious Discrete Distributions*, Ed. G. P. Patil, Statistical Publishing Society, Calcutta and Pergamon Press, pp. 410–18.
Class.: CONB-G:mb:BM

Pielou, E. C. 1965a. 2087
The spread of disease in patchily-infected forest stands. *Forest Sci.*, **11**, 18–26.
Class.: G:gf:BM

Pielou, E. C., and Foster, R. E. 1962. 2088
A test to compare the incidence of disease in isolated and crowded trees. *Canad. J. Bot.*, **40**, 1176–79.
Users: Pielou, E. C. (1963a), Pielou, E. C. (1965).
Class.: B:h:BM

Pierce, D. A., Stewart, C., and Folks, J. L.
See Folks, J. L., Pierce, D. A., and Stewart, C. (1965).

Pierce, J. A. 1943. 2089
Correction formulas for moments of a grouped-distribution of discrete variates. *J. Amer. Statist. Assoc.*, **38**, 57–62.
Class.: MI:m:G

Pillai, K. C. S. 1954. 2090
On the distribution of Hotelling's generalized T test (Abstract). *Ann. Math. Statist.*, **25**, 412.

Pillai, K. S. 1943. 2091
A note on Poisson distribution. *Proc. Indian Acad. Sci. Sect. A*, **18**, 179–89.
Review: MR (1944), 128; SA **47A** (1944), Ab. No. 807.
Class.: P:pe:G

Pinel, Èmile. 1948. 2092
Essai d'interprétation cinématique des courbes en cloche de Gauss. *C. R. Acad. Sci., Paris*, **227**, 236–38.
Review: ZBL **34** (1950), 81.

Plackett, R. L. 1948. 2093
Boundaries of minimum size in binomial sampling. *Ann. Math. Statist.*, **19**, 575–80.
Review: MR **10** (1949), 313; ZBL **32** (1950), 39.
Users: Wasan, M. T. (1964), Wasan, M. T. (1965).
Class.: B:tp-se:G

Plackett, R. L. 1953. 2094
The truncated Poisson distribution. *Biometrics*, **9**, 485–88.
Review: MR **15** (1954), 543.
Users: Moore, P. G. (1954a), Finney, D. J., and Varley, G. C. (1955), Moore, P. G. (1956b), Roy, J., and Mitra, S. K. (1957), Brass, W. (1958a), Tate, R. F., and Goen, R. L. (1958), Cohen, A. C. (1959e), Patil, G. P. (1959), Cohen, A. C. (1960b), Cohen, A. C. (1960d), Cohen, A. C. (1960e), Cohen, A. C. (1961a), Patil, G. P. (1961b), Cacoullos, J. T. (1962), Patil, G. P. (1962b), Patil, G. P. (1962e), Staff, P. J. (1964), Subrahmaniam, K. (1965), Tweedie, M. C. K. (1965).
Class.: TCP:pe:G

Plackett, R. L. 1966. 2095
Current trends in statistical inference. *J. Roy. Statist. Soc. Ser. A*, **129**, 249–67.
Class.: B-MI:pe-ie:G

Pleszczynska, E. 1963. 2096
Some generation methods of realizing a Poisson process (Polish). *Algorytmy*, **1**, 31–42.

Pollaczek, Félix. 1957. 2097
Problèmes stochastiques posés par le phénomène de formation d'une queue d'attente à un guichet et par des phénomènes apparentes. *Mém. Sci. Math.*, **136**, 122p.
Review: ZBL **79** (1959), 351.

Pollsczek, F. 1958. 2098
The discrimination of different distribution functions relating to a group of telephone lines without a queueing device (French). *C. R. Acad. Sci., Paris*, **247**, 1826–29.
Review: STMA **6** (1965), 455.
Class.: MI:pc:G

Pollaczek-Geiringer, H. 1928a. 2099
Die Charlier'sche Entwicklung wilkürlicher Verteilungen. *Skand. Aktuarietidskr.*, **11**, 98–111.

Pollaczek-Geiringer, H. 1928b. 2100
Über die Poissonsche Verteilung und die Entwicklung willürlicher Verteilungen. *Z. Angew. Math. Mech.*, **8**, 292–309.
Users: Boas, R. P. Jr. (1949), Janossy, L., Rényi, A., and Aczel, J. (1950).

Pollack, H. O., and Gilbert, E. N.
See Gilbert, E. N., and Pollack, H. O. (1957).

Pólya, G. 1931. 2101
Sur quelques points de la theorie des probabilitiés. *Ann. Inst. H. Poincare*, **1**, 117–62.
Users: Linder, A. (1935), Beall, G. (1940), Kitagawa, T. (1940), Cole, L. C. (1946), Cole, L. C. (1946a), Ashby, E. (1948), Barnes, H., and Marshall, S. M. (1951), Johnson, N. L. (1951), Rutherford, R. S. G. (1954), McFadden, J. A. (1955), McGuire, J., Brindley, T. A., and Bancroft, T. A. (1957), Gurland, J. (1959), Jensen,

P. (1959), Waters, W. E., and Henson, W. R. (1959), Darwin, J. H. (1960), Lukacs, E. (1960), Lord, F. M. (1960), Martin, L. (1961), Cassie, R. M. (1962), Mosimann, J. E. (1962), Freedman, David A. (1965), Kitagawa, T. (1965).

Pólya, George. 1946. 2102
Sur une généralisation d'un problème élémentaire classique, importante dans l'inspection des produits industriels. *C. R. Acad. Sci. Paris*, **222**, 1422–24.
Review: ZBL **60** (1957), 291.

Pólya, G., and Eggenberger, F.
See Eggenberger, F., and Pólya, G. (1923).

Pollyak, Yu. G. 1963. 2103
On comparison of two probabilities (Russian). *Teor. Verojatnost. i Primenen.*, **8**, 195–96.
Review: STMA **6** (1965), 658.

Pompilj, Giusseppe. 1947. 2104
Sulla media geometrica e sopra un indice di mutabilità calcolati mediante un campione. *Mem. Soc. Ital. Sci. III S.*, **26**, 299–339.
Review: ZBL **37** (1951), 87.

Porcelli, P., and Anselone, P. M.
See Anselone, P. M., and Porcelli, P. (1960).

Post, R. F., and Schiff, L. I. 1950. 2105
Statistical limitations on the resolving time of a scintillation counter. *Phys. Rev.*, **80** (2), 1113.
Review: MR **12** (1951), 727.

Poti, S. J. 1955. 2106
Measures of over-all efficiency of sample multinomial tables. *Bull. Calcutta Statist. Assoc.*, **6**, 102–12.
Review: MR **17** (1956), 1101.
Users: Patil, G. P., and Bildikar, S. (1966a).
Class.: M:mi:G

Potter, R. G. Jr. 1960. 2107
Some relationships between short range and long range risks of unwanted pregnancy. *Millbank Memorial Fund Quart.*, **38**, 255–63.
Class.: OBR:ie:BM

Potter, R. G. Jr. 1961. 2108
Some physical correlates of fertility control in the United States. *Proc. Conf. Inter. Population Union.*

Potter, R. G. Jr., Sagi, Philip C., and Westoff, Charles F. 1962. 2109
Improvement of contraception during the course of marriage. *Population Studies*, **16**, 160–74.
Class.: MI:mi:BM

Potthoff, Richard, F., and Whittinghill, Maurice. 2110
1965.
Maximum-likelihood estimation of the proportion of nonpaternity. *Amer. J. Hum. Genetics*, **17**, 480–94.
Class.: B:pe:BM

Potthoff, Richard F. and Whittinghill, Maurice. 2111
1966a.
Testing for homogeneity. I. The binomial and multinomial distributions. *Biometrika*, **53**, 167–82.
Users: Potthoff, Richard F., and Whittinghill, Maurice (1966b).
Class.: B-M:h:G-BM

Potthoff, Richard F., and Whittinghill, Maurice. 1966b. 2112
Testing for homogeneity. II. The Poisson distribution. *Biometrika*, **53**, 183–90.
Class.: P:h:G

Potts, R. B. 1964. 2113
Note on the factorial moments of standard distributions. *Austral. J. Phys.*, **6**, 498–99.
Class.: B-P-PO-H:mi:G

Potzger, J. E., and Deevey, E. S., Jr.
See Deevey, E. S., Jr., and Potzger, J. E. (1951).

Prabhu, N. U. 1958. 2114
Some exact results for the finite dam. *Ann. Math. Statist.*, **29**, 1234–43.
User: Phatarfod, R. M., (1963).

Prabhu, N. U. 1959–60. 2115
Application of generating functions to a problem in finite dam theory. *J. Austral. Math. Soc.*, **1**, 116–20.

Prabhu, N. U. 1960. 2116
Some results for the queue with Poisson arrivals. *J. Roy. Statist. Soc. Ser. B*, **22**, 104–07.
Review: ZBL **91** (1962), 302.
Users: Finch, P. D. (1961-62), Prabhu, N. U. (1961), Shanbhag, D. N. (1963), Takacs, L. (1963a), Prabhu, N. U. (1964).

Prabhu, N. U. 1961. 2117
On the ruin problem of collective risk theory. *Ann. Math. Statist.*, **32**, 757–64.
Review: ZBE 103 (1964), 133.

Prabhu, N. U. 1964. 2118
A waiting time process in the queue. GI/M/1 *Acta. Math. Acad. Sci. Hungar.*, **15**, 363–71.
Class.: P-MI:pc:G

Prabhu, N. U., and Bhat, U. Narayan. 1963. 2119
Further results for the queue with Poisson arrivals. *Operations Res.*, **11**, 380–86.
Review: MR **31** (1966), 147.

Prabhu, N. U., and Gani, J.
See Gani, J., and Prabhu, N. U. (1959).

Prasad, A. 1956. 2120
A new discrete distribution. *Sankhyā*, **17**, 353–54.
Review: ZBL **86** (1961), 123.
User: Bardwell, G. E. (1961).
Class.: MI:osp:G

Pratt, John W. 1961. 2121
Length of confidence intervals. *J. Amer. Statist. Assoc.*, **56**, 549–67.
Class.: B:ie-tp:G

Pratt, John W. 1966. 2122
The outer needle of some Bayes sequential continuation regions. *Biometrika*, **53**, 455–67.
Class.: B-P:tp:G

Prékopa, A. 1953. 2123
On composed Poisson distributions. IV. Remarks on the theory of differential processes (Russian summary). *Acta. Math. Acad. Sci. Hungar.*, **3**, 317–25.
Review: MR **14** (1953), 993; ZBL **49** (1959), 218.
User: Prékopa, A. (1957a).
Class.: COP:pc-osp:G
Notes: Hungarian version: *Magy. Tud. Akad. III. Mat. Fiz. Oszt. Közl.*, **4**, 505–12; *Review*: MR **16** (1955), 723.

Prékopa, A. 1957a. 2124
On Poisson and compound Poisson stochastic set functions. *Studia Math.*, **16**, 142–55.
User: Prékopa, A. (1958).
Class.: P-COP:mi-osp:G

Prékopa, A. 1957b. 2125
On the compound Poisson distribution. *Acta Sci. Math.* (*Szeged*), **18**, 23–28.
Review: ZBL **81** (1959), 353.
Class.: COP:pc:G

Prékopa, A. 1958. 2126
On secondary processes generated by a random point distribution of Poisson type (Hungarian). *Ann. Univ. Sci. Budapest-Sect. Math.*, **1**, 153–70.
Review: IJA **1** (1950), 140.

Pressman, A. J., Berljand, O. S., and Nazarov, I. M.
See Berljand, O. S., Nazarov, I. M., and Pressman, A. J. (1962).

Preston, F. W. 1948a. 2127
The cowbird (*M. ater*) and the Cuckoo (*C. canorus*). *Ecology*, **29**, 115–16.
Class.: P:gf:BM

Preston, F. W. 1948b. 2128
The commonness, and rarity of species. *Ecology*, **29**, 254–83.
Users: Archibald, E. E. A. (1949a), Anscombe, F. J. (1950a), Curtis, J., and McIntosh, R. (1950), Brian, M. V. (1953), Good, I. J. (1953b), Williams, C. B. (1953), Lewontin, R. C., and Prout, T. (1956), Barton, D. E. (1957), Hairston, N. G. (1959), Cassie, R. M. (1962), Clark, P. J., Eckstrom, P. T., and Linden, L. C. (1964), Nelson, W. C., and David, H. A. (1964), Bliss, C. I. (1956), Patil, G. P., and Bildikar, S. (1966a), Patil, G. P., and Shorrock, R. W. (1966).
Class.: LS-P-MI:gf:BM

Preston, F. W. 1956. 2129
The migrant loons of western Pennsylvania. *Auk*, **73**, 235–51.
Class.: P:mi:O

Price, G. B. 1946. 2130
Distributions derived from the multinomial expansion. *Amer. Math. Monthly*, **53**, 59–74.
Review: MR **7** (1946), 309.
Users: Tsao, C. M. (1962), Laurent, A. G. (1965).
Class.: M-OMR:gf-tp-osp-mi:G

Prigge, R. 1937, 2131
Fehlerrechnung bei biologischen Messungen. *Naturwissenschaften*, **25**, 169–70.

Prins, H. J. 1954. 2132
Prüfmethoden und Anwendungen der Poisson-Verteilung. *Math. Centrum Amsterdam Rap. ZW, 1954-0005.*
Review: ZBL **55** (1955), 373.

Prins, H. J. 1960. 2133
Transforms for finding probabilities and variate values of a distribution function in tables of a related distribution function. *Statistica, Neerlandica*, **14**, 1–17.
Review: IJA **2** (1961), 36.
Class.: B-P-NB:mi-tc:G

Prins, H. J., and Doornbos, R.
See Doornbos, R., and Prins, H. J. (1958).

Prins, H. J., and Klerk-Grobben, G.
See Klerk-Grobben, G., and Prins, H. J. (1954).

Prins, H. J., and Van Klinken, J.
See Van Klinken, J., and Prins, H. J. (1954).

Prokhorov, Yu. V. 1953. 2134
Asymptotic behaviour of the binomial distribution. *Uspehi Mat. Nauk N.S.*, **8**, 135–42.
Review: MR **15** (1954), 138; ZBL (1954), 103.
Users: LeCam, L. (1960), Makabe, H. (1962), Govindarajulu, Z. (1965).
Notes: English translation: *Selected Translations in Math. Statist. and Prob.*, **1**, 87–95 (1961).

Prokhorov, Yu. V. 1954. 2135
On a local limit theorem for lattice distributions (Russian). *Dokl. Akad. Nauk SSSR*, **98**, 535–38.
Review: ZBL **58** (1958), 122.
Users: Klimo, V. N. (1957), Govindarajulu, Z. (1965), Gamkrelidze, N. G. (1966a).

Prokhorov, Yu. V. 1961. 2136
Asymptotic behaviour of the binomial distribution. *Math. Statist. and Prob.*, **1**, 87–95.
Review: MR **22** (1961), 1225; ZBL **112** (1965), 101.

Proschan, Frank, and Hunter, Larry.
See Hunter, Larry, and Proschan, Frank (1961).

Proschan, Frank, and Karlin, Samuel.
See Karlin, Samuel, and Proschan, Frank (1960).

Prout, Timothy, and Lewontin, R. C.
See Lewontin, R. C., and Prout, Timothy (1956).

Pruess, K. R., and Weaver, C. R. 1959. 2137
Sampling studies of the clover root borer. Res. Bull. 827, Ohio Agric. Expt. Stn., Ohio.
Users: Hayman, B. I., and Lowe, A. D. (1961).
Class.: NB-N-T:gf-anovat-tc:BM

Przyborowski, J., and Wilenski, H. 1935. 2138
Statistical principles of routine work in testing clover seed for dodder. *Biometrika*, **27**, 273–92.
Users: Garwood, F. (1936), Ricker, W. E. (1937), Rietz, H. L. (1938), Przyborowski, J., and Wilenski, H. (1939).
Class.: P:ie:tp-gf:BM

Przyborowski, J., and Wilenski, H. 1935a. 2139
Sur les erreurs de la première et de la second catégorie dans la vérification des hypothèses concernant la loi de Poisson. *C. R. Acad. Sci. Paris*, **200**, 1460–62.

Przyborowski, J., and Wilenski, H. 1939. 2140
Homogeneity of results in testing samples from Poisson series. *Biometrika*, **31**, 313–23.
Users: Lancaster, H. O. (1952), Patnaik, P. B. (1954), Pfanzagl, E. J. (1960).
Class.: P:ctp:BM

Puntigam, F., and Pfanzagl, J.
See Pfanzagl, J., and Puntigam, F. (1961).

Puri, P. S., Buhler, W., Fein, H., Goldsmith, D., and, Neyman, J.
See Buhler, W., Fein, H., Goldsmith, D., Neyman, J. and Puri, P. S. (1965).

Putnam, L. G., and Shklov, N. 1956. 2141
Observations on the distribution of grasshopper egg-pods, in western Canadian stubble fields. *Canad. Entomol.*, **88**, 110–17.
User: Bliss, C. I. (1958).
Class.: NB:gf:BM

Putter, Joseph. 1964. 2142
The χ^2 goodness-of-fit test for a class of cases of dependent observations. *Biometrika*, **51**, 250–52.
Review: STMA **6** (1965), 159.

PWR APPARATUS SYSTEM. 2143
Tables of binomial probability distribution to six decimal places. No. 1, 597–620.
Review: SA **56**A (1953), Ab. No. 4.
Class.: B:tc:G

Pyke, Ronald. 1959. 2144
The supremum and infimum of the Poisson process. *Ann. Math. Statist.*, **30**, 568–76.
Review: IJA **1** (1959), 311; ZBL **89** (1961), 136.
User: Gallot, S. (1966).

Pyke, Ronald, and Hobby, Charles.
See Hobby, Charles, and Pyke, Ronald (1963).

Q

Quay, W. B. 1953. 2145
Seasonal and sexual differences in the dorsal skin gland of the kangaroo rat (Dipodomys). *J. Mammalogy*, **34**, 1–14.

Quenouille, M. H. 1949. 2146
A relation between the logarithmic, Poisson, and negative binomial series. *Biometrics*, **5**, 162–64.
Review: MR **19** (1949), 722.
Users: Anscombe, F. J. (1950a), Skellum, J. G. (1952), Bliss, C. I., and Fisher, R. A. (1953), David, F. N., and Moore, P. G. (1954), Gurland, J. (1957), Patil, G. P. (1959), Waters, W. E., and Henson, W. R. (1959), Pielou, D. P. (1960a), Bartko, J. J. (1961b), Tsao, C. M. (1962), Williamson, E., and Bretherton, M. (1964), Chatfield, C., Ehrenberg, A. S. C., and Goodhardt, G. J. (1966).
Class.: LS-P-NB:osp:G

Quenouille, M. H., and Hunter, G. C.
See Hunter, G. C., and Quenouille, M. H. (1952).

Quenouille, M. H., Jones, P. C. T., and Mollison, J. E.
See Jones, P. C. T., Mollison, J. E., and Quenouille, M. H. (1948).

Quesenberry, Charles P. 1959. 2147
Asymptotic simultaneous confidence intervals for the probabilities of a multinomial distribution. Master's Thesis, Virginia Polytechnic Inst.
Users: Patil, G. P., and Bildikar, S. (1966a)

Quesenberry, Charles P. 1964. 2148
Controlling the proportion defective from classification data. *Technometrics*, **6**, 99–100.

Quesenberry, Charles P., and Hurst, D. C. 1964. 2149
Large sample simultaneous confidence intervals for multinomial proportions. *Technometrics*, **6**, 191–95.
Review: STMA **6** (1965), 885; MR **32** (1966), 311.
Class.: M:ie:G

R

Ractliffe, J. F. 1964. 2150
The significance of the difference between two Poisson variables: An experimental investigation. *Appl. Statist.*, **13**, 84–86.
Review: STMA **7** (1966), 339; MR **31** (1966) 959.
Class.: P:ctp:G

Råde, L. 1965. 2151
Some waiting time problems. *Metrika*, **9**, 222–27.
Class.: OBR-IH-OHR:mi:G

Raff, Morton S. 1951. 2152
The distribution of blocks in an uncongested stream of automobile traffic. *J. Amer. Statist. Assoc.*, **46**. 114–23.
User: Weiss, G. H. (1963).
Class.: P:pc-gf:E

Raff, M. S. 1956. 2153
On approximating the point binomial. *J. Amer. Statist. Assoc.*, **51**, 293–303.
Review: MR **18** (1957), 160; ZBL (1958), 129.
Users: MacKinnon, W. J. (1959), Bartko, J. J. (1966).
Class.: M-P:tc:G

Raikov, D. 1937. 2154
On the decomposition of Poisson laws. *C. R. (Doklady) Acad. Sci. URSS N.S.*, **14**, 9–11.
Users: Teicher, H. (1954), Linnik, Y. V. (1957), Ramachandran, B. (1961a).

Raikov, D. A. 1938. 2155
On the decomposition of Gauss and Poisson laws. *Izv. Akad. Nauk SSSR Ser. Mat.*, **2**, 91–124.
User: Linnik, Y. V. (1957).

Rajski, C. 1955. 2156
On the verification of hypotheses concerning two populations consisting of items marked by attributes (Polish ; English and Russian summaries). *Z. Mat.*, **2**, 179–89.
Review: ZBL **66** (1957), 122

Rajski, C. 1961. 2157
A metric space of discrete probability distributions. *Information and Control*, **4**, 371–77.
Review: ZBL **103** (1964), 358; IJA **4** (1963), 760.
Class.: MI:mi:G

Ram, S. 1954. 2158
A note on the calculation of moments of the two-dimensional hypergeometric distribution. *Ganita*, **5**, 97–101.
Review: ZBL **58** (1958), 343.
Users: Patil, G. P., and Bildikar, S. (1966a).
Class.: MH:m:G

Ram, S. 1955. 2159
Multidimensional hypergeometric distribution. *Sankhyā*, **15**, 391–98.
Review: MR **17** (1956), 753; ZBL **67** (1958), 114.
Users: Ram, S. (1956), Patil, G. P., and Bildikar, S. (1966a).
Class.: MH:m:G

Ram, S. 1956. 2160
On the calculation of moments of hypergeometric distribution. *Ganita*, **7**, 1–5.
Review: ZBL **73** (1960), 137
Class.: MH:m:G

Ramabhadran, V. K. 1954. 2161
A statistical study of the persistency of rain days during the monsoon season at Poona. *Indian J. Meteo. Geo.*, **5**, 48–55.
Users: Srinivasan, T. R. (1956), Siromoney, G. (1962), Tikkha, R. N. (1962), Patil, G. P., and Bildikar, S. (1966a).
Class.: LS:gf:P

Ramachandran, B. 1960. 2162
On the decomposition of certain characteristic functions (Abstract). *Ann. Math. Statist.*, **31**, 240.

Ramachandran, B. 1961a. 2163
On the decomposition of certain random variables. *Publ. Inst. Statist. Univ. Paris*, **10**, 1–7.
Class.: B-P:mi:G

Ramachandran, B. 1961b. 2164
On the decomposition of certain random variables. *Publ. Inst. Statist. Univ. Paris*, **10**, 267–73.
Review: MR **27** (1964), 166–67.
Class.: B-P:mi:G

Ramachandran, B. 1963. 2165
A stability theorem for the binomial law. *Sankhyā*, **25**A, 85–90.
Review: MR **30** (1965), 660; STMA **6** (1965), 70.
Class.: B:mi:G

Ramachandran, B. 1964. 2166
Application of a theorem of Mamay's to a " Denumerable decomposition " of the Poisson law. *Publ. Inst. Statist. Univ. Paris*, **13**, 13–19.
Class.: P:mi:G

Ramakrishnan, A. 1951. 2167
Some simple stochastic processes. *J. Roy. Statist. Soc. Ser. B*, **13**, 131–40.
Class.: NB-OPR-MI:pc:G

Ramakrishnan, A. and Srinivasan, S. K. 1958. 2168
On age distribution in population growth. *Bull. Math. Biophys.*, **20** (4), 289–303.
Review: BA **33** (1959), Ab. No. 16433.

Ramakrishnan, A., and Vasudevan, R. 1957. 2169
On the distribution of visible stars. *Astrophys. J.*, **126**, 573–78.
Class.: MI:mb:P

Ramasubban, T. A. 1958. 2170
The mean difference and the mean deviation of some discontinuous distributions. *Biometrika*, **45**, 549–56.
Review: ZBL **88** (1961), 352; IJA **1** (1959), 31.
Users: Ramasubban, T. A. (1959), Bardwell, G. E. (1960), Katti, S. K. (1960a), Ramasubban, T. A. (1960), Bardwell, G. E. (1961), Kamat, A. R. (1965).
Class.: B-P-LS-G-H-NB:mi:G

Ramasubban, T. A. 1959. 2171
The generalized mean differences of the binomial and Poisson distributions. *Biometrika*, **46**, 223–29.
Review: IJA **1** (1959), 195; ZBL **87** (1961), 338.
Users: Ramasubban, T. A. (1960), Kamat, A. R. (1965)
Class.: B-P:mi:G

Ramasubban, T. A. 1960. 2172
Some distributions arising in the study of generalized mean differences. *Biometrika*, **47**, 469–73.
Review: ZBL **97** (1962), 141.
Users: Shanfield, F. (1964, Kamat, A. R. (1965).
Class.: OBR-OPR-G:mi:G

Rao, A. G., and Griffiths, B.
See Griffiths, B., and Rao, A. G. (1965).

Rao, A. Vijay, and Katti, S. K.
See Katti, S. K., and Rao, A. Vijay (1966).

Rao, B. Raja. 1959. 2173
Properties of the invariant I_m(m-odd) for distributions admitting sufficient statistics. *Sankhyā*, **21**, 355–62.
Review: MR **22** (1961), 182; ZBL **92** (1962), 363; IJA **2** (1961), 262.
Class.: B-P:osp:G

Rao, C. R. 1957. 2174
Maximum likelihood estimation for multinomial distribution. *Sankyā*, **18**, 139–48.
Review: MR **21** (1960), 727; ZBL **86** (1961), 354.
Users: Rao, C. R. (1958), Rao, C. R. (1963), Patil, G. P., and Bildikar, S. (1966a).
Class.: M:pe:G

Rao, C. R. 1958. 2175
Maximum likelihood estimation for the multinomial distribution with infinite number of cells. *Sankhyā*, **20**, 211–18.
Review: MR **21** (1960), 1128; IJA **1** (1959), 56; ZBL **88** (1961), 125.
Users: Rao, C. R. (1963), Birch, M. W. (1964), Patil, G. P., and Bildikar, S. (1966a).
Class.: M:pe:G

Rao, C. R. 1961. 2176
A study of large sample test criteria through properties of efficient estimates. Part I: Tests for goodness of fit and contingency tables. *Sankhyā*, ***23A***, 25–40.
Class.: M:gf-tp:G

Rao, C. R. 1963. 2177
Criteria of estimation in large samples. *Sankhyā Ser. A*, **25**, 189–206.
Review: MR **30** (1965) 1005.
Class.: MI-M:pe:G

Rao, C. R. 1965. 2178
On discrete distributions arising out of methods of ascertainment. *Classical and Contagious Discrete Distributions*, Ed. G. P. Patil, Statistical Publishing Society, Calcutta and Pergamon Press, pp. 320–32. Reprinted in *Sankhyā A*. **27** (1966).
Class.: PS-COB-P-NB-LS:osp-pe:G

Rao, C. R., and Chakravarti, I. M. 1956. 2179
Some small sample tests of significance for a Poisson distribution. *Biometrics*, **12**, 264–82.
Review: MR **18** (1957), 425.
Users: Good, I. J. (1957a), Hetz, Klinger (1958), Chakravarti, I. M., and Rao, C. R. (1959), Watson, G. S. (1959), Rao, C. R. (1961), Tweedie, M. C. K. (1965).
Class.: P-TCP-B:h-gf-tc:G

Rao, C. R., and Chakravarti, I. M.
See Chakravarti, I. M., and Rao, C. R. (1959)

Rao, C. R., and Rubin, Herman. 1964. 2180
On a characterization of the Poisson distribution. *Sankhyā-A*, **26**, 295–98.
Review: MR **32** (1966), 210.

Rao, C. R., Mitra, S. K., and Matthai, A. 1966. 2181
Formulae and tables for statistical work. Statist. Publishing Soc. Calcutta. 234 pp.

Rao Uppuluri, V. R., and Murthy, V. K.
See Murthy, V. K., and Rao Uppuluri, V. R. (1965).

Rao, V. R. 1964. 2182
A characterization of the geometric distribution (Abstract). *Ann. Math. Statist.*, **35**, 1841.

Rapoport, A. 1951. 2183
The probability distribution of distinct hits on closely packed targets. *Bull. Math. Biophys.*, **13**, 133–38.
User: Sprott, D. A. (1957).
Class.: MI:mb:G

Rapoport, A. 1958. 2184
Nets with reciprocity bias. *Bull. Math. Biophys.*, **20**, 191–201.
Review: BA **33** (1959), Ab.No. 8643.

Rasch, G. 1964. 2185
On a class of discontinuous distributions related to the Jacobian theta functions. Tech. Rep., State Serum Inst., Copenhagen.
Class.: MI:mb-a:G

Rashevsky, N. 1955. 2186
Note on a combinatorial problem in topological biology. *Bull. Math. Biophys.*, **17**, 45–50.

Raup, D., and Kummel, B.
See Kummel, B., and Raup, D. (1965).

Ray, D., and Iyer, P. V. K.
See Iyer, P. V. K., and Ray, D. (1958).

Ray, S. N. 1963. 2187
Bayes sequential procedures for some binomial problems (Abstract). *Ann. Math. Statist..*, **34**, 684.

Ray, S. N., and Chernoff, Herman.
See Chernoff, Herman, and Ray, S. N. (1965).

Rayleigh, Lord. 1899. 2188
On James Bernoulli's theorem in probabilities. *Philos. Mag. Ser.* 5, **47**, 246–51.

Read, R. R. 1956. 2189
A probabilistic model describing drop count data for certain closed chamber experiments (Abstract). *Ann. Math. Statist.*, **27**, 862.

Redheffer, R. M. 1951. 2190
A note on the surprise index. *Ann. Math. Statist.*, **22**, 128–30.
Review: ZBL **42** (1952), 373.
Class.: MI-P-B:mi:G

Redheffer, R. M. 1953. 2191
A note on the Poisson law. *Math. Mag.*, **26**, 185–88.
Review: MR **14** (1953), 1098; ZBL **53** (1961), 267.
User: Meizler, D. (1965).
Class.: P:mi:G

Regan, Francis, and Copeland, Arthur H.
See Copeland, Arthur H., and Regan, Francis (1936).

Reichenbach, Hans, and Haight, F. A.
See Haight, Frank A., and Reichenbach, Hans (1965).

Reid, D. B. W., and Grainger, R. M.
See Grainger, R. M., and Reid, D. B. W. (1954).

Reid, D. B. W., Crawley, J. F., and Rhodes, A. J. 1949. 2192
A study of fowl pox virus titration on the chorioallantois by the pock counting technique. *J. Immunology*, **63**, 165–71.

Reid, R. W. 1957. 2193
The bark beetle complex associated with lodgepole pine slash in Alberta. Part IV. Distribution, population densities, and effects of several environmental factors. *Canad. Entomol.*, **89**, 437–47.
Users: Waters, W. E., and Henson, W. R. (1959).
Class.: P-NB-N:gf:BM

Reiersol, O. 1954a. 2194
Tests of linear hypotheses concerning binomial experiments. *Skand. Aktuartidskr.*, **37**, 38–59.
Review: MR **16** (1955), 605; ZBL **56** (1955), 373.
Users: Bhat, B. R., and Kulkarni, S. R. (1966a).

Reirsol, O. 1954b. 2195
Analysis of binomial experiments. Memo., Inst. Econ., Univ. Oslo, September 24, 1954, pp. 41.

Rényi, A. 1951a. 2196
On composed Poisson distributions. II (English; Russian summary). *Acta. Math. Acad. Sci. Hungar.*, **2**, 83-98.
Review: MR **13** (1952), 663; ZBL **44** (1952), 139.
Users: Aczel, J. (1952), Prékopa, A. (1952), Fisz, M. (1953a), Prékopa, A. (1957a).
Class.: COP:pc-osp:G.
Notes: Hungarian version (1951), *Magyar, Tud. Akad. Mat. Fiz. Oszt. Közl.*, **1**, 329–41.

Rényi, A. 1951b. 2197
Poisson-eloszlás problémaköréröl (On problems connected with Poisson distribution) (Hungarian). *Magyar. Tud. Akad. Mat. Fiz. Oszt. Közl.*, **1**, 202–12.
Review: MR **13** (1952), 958.

Rényi, A. 1951. 2198
On some problems concerning Poisson processes. *Publ. Math.*, **2**, 66–73.
Review: ZBL **54** (1956), 58.
User: Prékopa, A. (1958).

Rényi, A. 1952. 2199
On projections of probability distributions. *Acta. Math. Acad. Sci. Hungar.*, **3**, 131–42.
Review: MR **14** (1953), 771; ZBL **48** (1953), 108.
User: Heppes, A. (1956).
Class.: MI:mi:G
Notes: Hungarian version (1953): Valószinüségeloszlások vetületeiröl. *Magyar. Tud. Akad. Mat. Fiz. Oszt. Közl.*, **3**, 60–69.

Rényi, A. 1956. 2200
A characterization of Poisson processes (Hungarian). *Magyar. Tud. Akad. Mat. Kutató Int. Közl.*, **1**, 519–27.
Review: ZBL **103** (1964), 115.

Rényi, A. 1958. 2201
On the probabilistic generalization of the large sieve of Linnik (English; Hungarian and Russian summaries). *Magyar. Tud. Akad. Mat. Kutató. Int. Közl.*, **3**, 199–206.
Review: MR **22** (1961), 346; IJA **1** (1959), 253.

Rényi, A. 1959a. 2202
Some remarks on the theory of trees (Hungarian and Russian summaries.) *Magyar. Tud. Akad. Mat. Kutató. Int. Közl.*, **4**, 73–85.
Review: IJA **1** (1959), 196.

Rényi, A. 1959b. 2203
Summation methods and probability theory (Hungarian and Russian summaries.) *Magyar. Tud. Akad. Mat. Kutató, Int. Közl.*, **4**, 389–99.
Review: IJA **1** (1960), 556.
User: Riordan, J. (1962).

Rényi, A. 1960a. 2204
On the central limit theorem for the sum of a random number of independent random variables. *Acta. Math. Acad. Sci. Hungar.*, **11**, 97–102.
Review: IJA **1** (1960), 557.
User: Teicher, Henry (1965).
Class.: MI:a:G

Rényi, A. 1960b. 2205
Some fundamental problems of information theory. *Magyar. Tud. Akad. III Oszt. Közl.*, **10**, 251–82.
Review: IJA **2** (1961), 577.
Users Rényi, A. (1964a).
Class.: MI:mi:G

Rényi, A. 1961a. 2206
On measures of entropy and information. *Proc. 4th Berkeley Symp. Math. Statist. Prob..*, **1**, 547–61.
Review: IJA **3** (1962), 663.
Class.: MI:mi:G

Rényi, A. 1961b. 2207
On different measures of information. Second Hungarian Mathematical Congress.
Review: IJA **2** (1961), 580.
Class.: MI:mi:G

Rényi, A. 1964. 2208
On an extremal property of the Poisson process. *Ann. Inst. Statist. Math. Tokyo*, **16**, 129–33.

Rènyi, Alfred. 1964a. 2209
On the amount of information concerning an unknown parameter in a sequence of observations (Russian summary). *Magyar. Tud. Akad. Mat. Kutato Int. Közl.*, **9**, 617–25.

Rényi, A., and Erdos, P.
See Erdös, P., and Rényi, A. (1959).

Rényi, Alfred, and Takács, Lajos. 1952. 2210
Sur les processus d'événements dérivés par un processus de Poisson et sur leurs applications techniques et physiques (Hungarian; Russian and French summaries.) *Publ. Inst. Math. Appl. Acad. Sci. Hongrie*, **1**, 139–46.
Review: ZBL **49** (1959), 217.
User: Prékopa, A. (1958).

Rényi, A., Aczel, J., and Janossy, L.
See Janossy, L., Rényi, A., and Aczel, J. (1950).

Rescia, R. R., and Beall, G.
See Beall, G., and Rescia, R. R. (1953).

Restle, Frank. 1961. 2211
Statistical methods for a theory of learning. *Psychometrika*, **26**, 291–306.

Restrepo, Rodrigo, A. 1965. 2212
A queue with simultaneous arrivals and Erlang service distribution. *Operations Res.*, **13**, 375–81.
Review: MR **31** (1966), 1134.
Class.: MI:pc:G

Reza, F. M. 1959. 2213
An introduction to probability theory discrete schemes. Tech. Note RADC-TN-59-129, Rome Air Develop. Center,Air Res. and Develop. Command,U.S.A.F.,N.Y.

Rhodes, A. J., Reid, D. B. W., and Crawley, J. F.
See Reid, D. B. W., Crawley, J. F., and Rhodes, A. J. (1949).

Rhoades, B. E. 1961. 2214
Some special series. *Math. Mag.*, **34**, 165–67.
Class.: MI:mi:G

Riauba, B. 1958. 2215
Asymptotic laws of a trinomial distribution (Lithuanian; Russian summary). *Vilniaus Valst. Univ. Mokslo Darbai Mat. Fiz.*, **8**, 17–22.
Review: MR **32** (1966), 528.
Class.: M:a:G

Rice, E. L., and Penfound, W. T. 1955. 2216
An evaluation of the variable-radius and paired-tree methods in the blackjack-post oak forest. *Ecology*, **36**, 315–20.
Class.: P:gf:BM

Richardson, L. F. 1944. 2217
The distribution of wars in time. *J. Roy. Statist. Soc. Ser. A*, **107**, 242–50.

Richter, Wolfgang. 1964a. 2218
Multidimensional limit theorems for large deviations and their application to the χ^2 distribution. *Theor. Probability Appl.*, **9**, *Class.*:

Richter, Wolfgang. 1964b. 2219
Mehrdimensionale Grenzwertsatze für grosse Abweichungen und ihre Anwendung auf die Verteilung von χ^2 (Russian summary). *Teor. Verojatnost. i Primenen.*, **9**, 31–42.
Class.: MI:a-gf:G

Ricker, William E. 1937. 2220
The concept of confidence or fiducial limits applied to the Poisson frequency distribution. *J. Amer. Statist. Assoc.*, **32**, 349–56.
Review: ZBL **16** (1937), 313.
Users: Rietz, H. L. (1938), Robertson, O. H. (1947), MacLulich, D. A. (1951), Young, H., Neess, J., and Emlen, J. T. (1952), Birnbaum, A. (1954b), Walsh, J. E. (1954), Geiss, A. D. (1955), Kozicky, E. L., Jessen, R. J., and Hendrickson, G. O. (1956), Kutkuhn, J. H. (1958).
Class.: P:ie:G

Ricker, W. E. 1938. 2221
On adequate quantitative sampling of the pelagic net plankton of a lake. *J. Fisheries Res. Board Canada*, **4**, 19–32.

Riddell, W. J. B. 1944. 2222
The relation between the number of speakers and the number of contributions to the transactions of the opthalmological society of the United Kingdom between 1887 and 1890. *Ann. Eugenics*, **12**, 274–79.
Users: Nelson, W. C., and David, H. A. (1964).

Rider, P. R. 1950. 2223
The distribution of ranges from a discrete rectangular population. *Proc. Int. Congr. Math. Camb., Mass.*, **1**, 583.
Class.: MI:mi:G

Rider, Paul R. 1951. 2224
The distribution of the range in samples from a discrete rectangular population. *J. Amer. Statist. Assoc.*, **46**, 375–78.
Review: ZBL **43** (1952), 136.
Users: Goodman, L. A. (1952b), Abdel-Aty, S. H. (1954).

Rider, P. R. 1952. 2225
Truncated Poisson distributions (Abstract). *Ann. Math. Statist.*, **23**, 638.

Rider, P. R. 1953. 2226
Truncated Poisson distributions. *J. Amer. Statist. Assoc.*, **48**, 826–30.
Review: MR **15** (1954), 544; ZBL **52** (1955), 155.
Users: Cohen, A. C. (1954), Moore, P. G. (1954a), Finney, D. J., and Varley, G. C. (1955), Rider, P. R. (1955b), Moore, P. G. (1956b), Roy, J. and Mitra, S. K. (1957), Tate, R. F., and Goen, R. L. (1958), Cohen, Cohen, A. C. (1959e), Patil, G. P. (1959), Cohen, A. C. (1960b), Cohen, A. C. (1960d), Cohen, A. C. (1960e), Bardwell, G. E. (1961), Cohen, A. C. (1961a), Patil, G. P. (1961b), Cacoullos, T. (1962), Hughes, E. J. (1962), Patil, G. P. (1962e).
Class.: TCP:pe:G

Rider, P. R. 1955a. 2227
Truncated binomial and negative binomial distributions (Abstract). *Ann. Math. Statist.*, **26**, 774.

Rider, P. R. 1955b. 2228
Truncated binomial and negative binomial distributtions. *J. Amer. Statist. Assoc..*, **50**, 877–83.
Review: MR **17** (1956), 169; ZBL **65** (1956), 127.
Users: Clark, F. E. (1957), Patil, G. P. (1959), Patil, G. P. (1961b), Shad, S. M. (1961), Hughes, E. J. (1962), Patil, G. P. (1962e), Rider, P. R. (1962b).
Class.: TCB-TCNB:pe:G
Notes: Corrigenda: *J. Amer. Statist. Assoc.*, **50**, 1332.

Rider, P. R. 1961. 2229
Estimating the parameters of mixed Poisson, binomial and Weibull distributions. *Bull. Inst. Internat. Statist.*, **38**, 1–8.
User: Blischke, W. R. (1965).

Rider, P. R. 1962. 2230
Estimating the parameters of mixed Poisson, binomial and Weibull distributions. *Bull. Inst. Internat. Statist.*, **39**, 225–32.
Review: STMA **5** (1964), 369.
User: Blischke, W. R. (1964).

Rider, P. R. 1962a. 2231
The negative binomial distribution and the incomplete beta function. *Amer. Math. Monthly*, **69**, 302–24.
User: Patil, G. P. (1963c).
Class.: NB:mi:G

Rider, P. R. 1962b. 2232
Expected values and standard deviations of the reciprocal of a variable from a decapitated negative binomial distribution. *J. Amer. Statist. Assoc.*, **57**, 439–45.
Review: STMA **6** (1965), 457; ZBL **102** (1963), 146.
Users: Govindarajulu, Z. (1962b), Govindarajulu, Z. (1962e).
Class.: TCNB:m-tc:G

Rider, P. R. 1964. 2233
Distribution of product and quotient of maximum values in samples from a power function population. *J. Amer. Statist. Assoc.*, **59**, 877–80.

Rider, P. R. 1965. 2234
The zeta distribution. *Classical and Contagious Discrete Distributions*, Ed. G. P. Patil, Statistical Publishing Society, Calcutta and Pergamon Press, pp. 443–444.
Class.: MI:mi:G

Riebesell, Paul. 1941. 2235
Die mathematischen Grundlagen der Sachversicherung. *Ber.* 12. *Internat. Kongr. Versich.-Math. Luzern* 1940, **4**, 27–34.
Review: ZBL **28** (1944), 178.

Rietz, H. L. 1938. 2236
On a recent advance in statistical inference. *Amer. Math. Monthly*, **45**, 149–58.
Class.: B:ie:G

Riordan, John. 1937. 2237
Moment recurrence relations for binomial, Poisson and hypergeometric frequency distributions. *Ann. Math. Statist.*, **8**, 103–11.
Class.: B-P-H:m:G

Riordan, John. 1962. 2238
Enumeration of linear graphs for mappings of finite sets. *Ann. Math. Statist.*, **33**, 178–85.
User: Gould, H. W. (1963).
Class.: B:mi:G

Risser, Rene. 1945. 2239
Sur le mode de tirages contagieux. *C. R. Acad. Sci. Paris*, **220**, 210–12.
Review: MR **7** (1946), 128.
Class.: PO:a:G

Risser, R. 1949. 2240
Note relative aux tirages contagieux. *Bull. Assoc. Actuaries Belgès*, **55**, 25–51.
Review: ZBL **41** (1952), 250.

Risser, R. 1951. 2241
Note relative aux tirages contagieux. *Bull. Trimestr. Inst. Actuaires Françs.*, **62**, 235–58.
Review: ZBL **44** (1952), 335.

Risser, R. 1957. 2242
Essai sur les surfaces de probabilité. *Bull. Inst. Internat. Statist.*, **35** (2), 105–30.
Review: ZBL **87** (1961), 337.

Rizzi, A. 1964. 2243
On an inversion problem relating to two samples. *Metron*, **23**, 123–36.
Review: STMA **7** (1966), 107.
Class.: MI:mi:G

Robbins, H. 1948. 2244
Mixture of distributions. *Ann. Math. Statist.*, **19**, 360–69.
Review: ZBL **37** (1951), 363.
Users: Teicher, H. (1960a), Daniels, H. E. (1961), Blischke, W. R. (1962), Blischke, W. R. (1965), Jacobsen, Robert Leland (1966), Patil, G. P., and Bildikar, S. (1966a).
Class.: MI:mi:G

Robbins, H. 1961. 2245
Recurrent games and the Petersburg paradox. *Ann. Math. Statist.*, **32**, 187–94.
Review: IJA **3** (1962), 318.
Class.: MI:mi:G

Robbins, H. 1964. 2246
The empirical Bayes approach to statistical decision problems. *Ann. Math. Statist.*, **35**, 1–20.
User: Van Ryzin, J. (1966).
Class.: P:pe:G

Robbins, H., and Morignti, S.
See Morignti, S., and Robbins, H. (1962).

Roberts, H. R., McCall, C. H. Jr., and Thomas, R. E. 1958. 2247
Some statistical considerations for small sample evaluation in triangular taste tests. *Food Res.*, **23**, 388–95.
Review: PA **33** (1959), Ab.No. 5128.
Class.: M-B:tp:O

Roberts, Harry V. 1965. 2248
Probabilistic prediction. *J. Amer. Statist. Assoc.*, **60**, 50–62.
Class.: MI:mi:G

Robertson, A. 1951. 2249
The analysis of heterogeneity in the binomial distribution. *Ann. Eugenics*, **16**, 1–15.
Review: MR **13** (1952), 260; PA **26** (1952), Ab.No. 2520; ZBL **44** (1952), 342.
Users: Edwards, A. W. F. (1958), Potthoff, Richard F., and Whittinghill, Maurice (1966a).
Class.: COB:pe-h:BM

Robertson, O. H. 1947. 2250
An ecological study of two high mountain trout lakes in the Wind River Range, Wyoming. *Ecology*, **28**, 87–112.
Class.: P:pe:BM

Robertson, W. H. 1960a. 2251
Tables of the binomial distribution function for small values of p. Monograph SCR-143, Sandia Corp.

Robertson, W. H. 1960b. 2252
Programming Fisher's exact method of comparing two percentages. *Technometrics*, **2**, 103–07.
Review: MR **22** (1961), 183; PA **35** (1961), Ab.No. 1489; IJA **2** (1961), 694.
Class.: B:ctp-c:G

Robinson, H. F., and Comstock, R. E.
See Comstock, R. E., and Robinson, H. F. (1952).

Robinson, P. 1954. 2253
The distribution of plant populations. *Ann. Bot. Lond., N.S.*, **18**, 35–45.
Users: David, F. N., and Moore, P. G. (1954), David, F. N. (1955a), Robinson, P. (1955), Kemp, C. D., and Kemp, A. W. (1956a).

Robinson, P. 1955. 2254
The estimation of ground cover by the point quadrat method. *Ann. Bot. Lond., N.S.*, **19**, 59–66.

Robson, D. S. 1955. 2255
Admissible and minimax integer-valued estimators of an integer-valued parameter. Ph.D. Thesis, Cornell Univ.

Robson, D. S. 1958. 2256
Admissible and minimax integer-valued estimators of an integer-valued parameter. *Ann. Math. Statist.*, **29**, 801–12.
Class.: MI-B:pe:G

Robson, D. S. 1960. 2257
An unbiased sampling and estimation procedure for creel censuses of fishermen. *Biometrics*, **16**, 261–77.
User: Robson, D. S. (1961).

Robson, D. S. 1961. 2258
On the statistical theory of a roving creel census of fishermen. *Biometrics*, **17**, 415–37.

Robson, D. S., and Chapman, D. G.
See Chapman, D. G., and Robson, D. S. (1960).

Roby, N. 1959. 2259
Sur certains processus remarquables generalisant les processus de Poisson. *C. R. Acad. Sci. Paris*, **248**, 2945–47.

Roby, N., and Fuchs, A.
See Fuchs, A., and Roby, N. (1960).

Rodgers, Eric. 1940. 2260
Probable error for Poisson distributions. *Phys. Rev.*, **57**, 735–37.
Review: MR **1** (1940), 246; ZBL **23** (1941), 146.
Users: Gumbel, E. J. (1941), Govindarajulu, Z. (1965).

Rodriguez, J. G., Ibarra, E. L., and Wallwork, J. A.
See Ibarra, E. L., Wallwork, J. A., and Rodriguez, J. G. (1965).

Roessler, E. B., Baker, G. A., and Amerine, M. A. 1954.
See Baker, G. A., Amerine, M. A., and Roessler, E. B. (1954).

Roessler, E. B., Baker, G. A., and Amerine, M. A. 1956. 2261
One-tailed and two-tailed tests in organoleptic comparisons. *Food. Res.*, **21**, 117–21.
Users: Baker, G. A., Amerine, M. A., Roessler, E. B., and Filipello, F. (1960).
Class.: B:tp-tc:O

Roessler, E. B., Filipello, F., Baker, G. A., and Amerine, M. A.
See Baker, G. A., Amerine, M. A., Roessler, E. B., and Filipello, F. (1960).

Rogers, A. 1964. 2262
A stochastic analysis of intraurban retail spatial structure. Ph.D. Thesis, Univ. North Carolina.

Rogers, A. 1965. 2263
A stochastic analysis of the spatial clustering of retail establishments. *J. Amer. Statist. Assoc.*, **60**, 1094–1103.
Class.: P-NB-N-T:gf:S

Roma, Maria Sofia, and Ghizzetti, Aldo.
See Ghizzetti, Aldo, and Roma, Maria Sofia (1964).

Romani, J. 1956. 2264
Distribución de la suma algebraica de variables de Poisson (Spanish; English summary). *Trabajos Estadist.*, **7**, 175–81.
Review: MR **18** (1957), 521; ZBL **74** (1962), 343.
User: Shanfield, F. (1964).

Romanovsky, V. 1923. 2265
Note on the moments of a binomial $(p+q)^n$ about its mean. *Biometrika*, **15**, 410–12.
Users: Riordan, J. (1937), Haldane, J. B. S. (1939), Curtiss, J. H. (1941), Patil, G. P. (1959), Bardwell, G. E. (1961).
Class.: B:m:G

Romanovsky, V. 1925. 2266
On the moments of the hypergeometrical series. *Biometrika*, **17**, 57–60.
Users: Ram, S. (1954), Ram, S. (1956).
Class.: H:m:G

Romanovskii, V. I. 1952a. 2267
A comparison of hypergeometric, Bernoulli, and Poisson probabilities (Russian). *Dokl. Akad. Nauk. Uzbek. SSSR.*

Romanovskii, V. I. 1952b. 2268
On the dual theorem of the hypergeometric distribution (Russian). *Dokl. Akad. Nauk. Uzbek. SSR.*

Romanovskii, V. I. 1953. 2269
Duality theorems for the hypergeometric distribution (Russian). *Trudy Inst. Mat. Mekh., Akad. Nauk. Uzbek. SSR*, **11**, 22–28.

Romig, Harry B. 1953. 2270
50-100 *binomial tables*. John Wiley and Sons, New York.
Review: BA **29** (1955), Ab.No. 222; ZBL **50** (1954), 352.

Romig, Harry G., and Dodge, Harold F.
See Dodge, Harold F., and Romig, Harry G. (1959).

Rosenblatt, Alfred. 1940a. 2271
Sur les théorèmes des petits nombres de Poisson, de Bortkiewicz et G. Pólya. Application aux phénomènes rares. I. Propagation des maladies contagieuses: peste bubonique au Brésil. *Actas Acad. Ci. Lima*, **3**, 160–67.
Review: MR **4** (1943), 28.

Rosenblatt, Alfred. 1940b. 2272
Sur le concept de contagion de M. G. Pólya dans le calcul del probabilités. Divers schèmes. Application à la peste bubonique au Pérou. *Actas Acad. Ci. Lima*, **3**, 186–204.
Review: MR **3** (1942), 2.
User: Feller, W. (1943).

Rosenblatt, M., and Blum, J. R.
See Blum, J. R., and Rosenblatt, M. (1959).

Rothschild, Lord, Campbell, R. C., and Hancock, J. L.
See Campbell, R. C., Hancock, J. L., and Rothschild, Lord (1953).

Rott, N. 1946. 2273
Ueber Wahrscheinlichkeitsprobleme der Garfestigkeitsprüfung. *Schweiz. Arch. Angew. Wiss. Tech.*, **12**, 93–95.
Review: MR **7** (1946), 457; ZBL **60** (1957), 286.

Roy, J., and Mitra, Sujit Kumar. 1957. 2274
Unbiased minimum variance estimation in a class of discrete distributions. *Sankhyā*, **18**, 371–78.
Review: MR **19** (1958), 1096.
Users: Patil, G. P. (1961b), Patil, G. P. (1962a) Patil, G. P. (1962c), Nelson, W. C., and David, H. A (1964), Shanfield, D. F. (1964), Crow, E. L., and Bardwell, G. E. (1965), Tweedie, M. C. K. (1965).
Class.: PS-NB-TCP:pe:G

Roy, S. N. 1961. 2275
On the planning and interpretation of multifactor multiresponse experiments. *Bull. Inst. Internat. Statist.*, **38** (4), 59–72.
Review: IJA **4** (1963), 324.
Class.: M:anovat:G

Roy, S. N., Diamond, E. L., and Mitra, S. K.
See Diamond, E. L., Mitra, S. K., and Roy, S. N. (1960).

Rozanov, Ju, A, (Yu. A.) 1957. 2276
On a local limit theorem for lattice distributions (Russian; English summary). *Teor. Verojatnost. i Primemen.*, **2**, 275–81.
Review: ZBL **83** (1960), 141.

Ruark, A., and Devol, L. 1936. 2277
A general theory of fluctuations in radioactive disintegration. *Phys. Rev.*, **49**, 355–67.

Rubin, H., and Rao, C. R.
See Rao, C. R., and Rubin, H. (1964).

Rubin, H., Sitgreaves, R., and Girshick, M. A.
See Girshick, M. A., Rubin, H., and Sitgreaves, R. (1955).

Rudemo, M., 1964. 2278
Dimenson and entropy for a class of stochastic processes. *Publ. Math. Inst. Hungar. Acad. Sci.*, **9**, 73–88.

Runnenburg, J. T., and Van Eeden, Constance.
See Van Eeden, Constance, and Runnenburg, J. T. (1960).

Russell, A. M., and Josephson, N. S. 1965. 2279
Measurement of area by counting. *J. Appl. Prob.*, **2**, 339–51.

Rust, H., Van der Waerden, B. L., and Egg, K.
See Egg, K., Rust, H., and Van der Waerden, B. L. (1965).

Rutherford, C. E. 1920. 2280
Radiations from radioactive substances. Univ. Press, Cambridge.

Rutherford, R. S. G. 1954. 2281
On a contagious distribution. *Ann. Math. Statist.*, **25**, 703–13.
Review: MR **19** (1958), 585; ZBL **56** (1955), 359.
Class.: OBR:mb-osp-a-gf:G

Rutledge, George, and Douglass, R. D. 1936. 2282
Integral functions associated with certain binomial coefficient sums. *Amer. Math. Monthly*, **43**, 27–32.
Class: B:mi:G

Rybarz, J. 1959. 2283
Ein einfacher beweis für das dem χ^2-Verfahren zugrundeliegende Theorem. *Metrika*, **2**, 89–93.
Review: ZBL **86** (1961), 126.

Ryll-Nardzewski, C. 1953. 2284
On the non-homogeneous Poisson process (I). *Studia Math.*, **14**, 124–28.
Review: ZBL **53** (1961), 98.
User: Prékopa, A. (1957a).

Ryll-Nardzewski, C. 1954. 2285
Remarks on the Poisson stochastic process (III). *Studia Math.*, **14**, 314–18.
Review: ZBL **64** (1965), 132.
User: Prékopa, A. (1957a).

Ryll-Nardzewski, C. 1955. 2286
On the non-homogeneous Poisson processes. *Colloq. Math.*, **3** 192–93.

Ryll-Nardzewski, C., Florek, K., and Marczewski, E.
See Florek, K., Marczewski, E., and Ryll-Nardzewski, C. (1953).

S

Sagi, Philip C., Westoff, Charles F., and Potter, Robert G. Jr.
See Potter, Robert G. Jr., Sagi, Philip C., and Westoff, Charles F. (1962).

Saito, K. 1956. 2287
Maximum-likelihood estimate of proportion using supplementary information. *Bull. Math. Statist.*, **7** (1-2), 11–17.
Review: MR **19** (1958), 472; ZBL **73** (1960), 148.
Class.: B:pe:G

Sakoda, J. M., and Cohen, B. H. 1957. 2288
Exact possibilities for contingency taboles using binomial coefficients. *Psychometrika*, **22**, 83–86.
Review: MR **19** (1958), 73.

Salaevskii, O. V. 1959. 2289
Stability in Raĭkov's theorem. *Vestnik Leningrad. Univ.*, **14** (7), 41–49.
Review: MR **21** (1960), 828.

Saleh, A. K. Md. E., and Subrahmaniam, K.
See Subrahamaniam, K., and Saleh, A. K. Md. E. (1965).

Saleme, Ernesto, and Cernuschi, Felix.
See Cernuschi, Felix, and Saleme, Ernesto (1944).

Salmon, S. C. 1930. 2290
The point binomial formula for evaluating agronomic experiments. *J. Amer. Soc. Agron.*, **22**, 77–81.
Review: BA **8** (1934), Ab.No. 12924.

Salvemini, T. 1958. 2291
Distribution de l'étendue d'échantillons obtenus par des tirages en bloc d'un ensemble de nombres equidistribues. *Bull. Inst. Internat. Statist.*, **36** (3), 71–78.
Review: MR **21** (1960), 1730; ZBL **89** (1961), 110.

Salvemini, T. 1958a. 2292
Sui campioni di una massa discreta equidistribuità: il campo di variazione. *Statistica*, **18**.

Salvemini, T. 1958b. 2293
Distribuzione della differenza media dei campioni ricavati da una massa discreta equidistribuità. *Atti XVII Riun. Roma Soc. Ital. Statist.*

Salvi, Filippo. 1949. 2294
Estensione di alcuni teoremi classici del calcolo delle probabilità. *Rend. Mat. Sue Appl., Univ. Roma* 1*st Naz. Alta Mat., V. S.*, **8**, 282–308.
Review: ZBL **39** (1951), 344.

Sampford, M. R. 1955. 2295
The truncated negative binomial distribution. *Biometrika*, **42**, 58–69.
Review: ZBL **65** (1956), 127.
Users: Finney, E. J., and Varley, G. C. (1955), Bliss, C. I. (1958), Brass, W. (1958a), Harris, E. K. (1958), Hartley, H. O. (1958), Cohen, A. C. (1959d), Patil, G. P. (1959), Bartko, J. J. (1961b), Cohen, A. C. (1961c), Hughes, E. J. (1962), Ito, Nakamura, Kondo, Miyashita, and Nakamura (1962), Khatri, C. G. (1962d), Chatfield, C., Ehrenberg, A. S. C., and Goodhardt, G. J. (1965), Tanton (1965), Cohen, A. Clifford, Jr., (1965), Chatfield, C., Ehrenberg, A. S. C. and Goodhardt, G. J. (1966), Cohen, A. C., Jr., (1966), Holgate, P. (1966a).
Class.: TCNB:m-pe-gf:G

Samuel, E. 1963. 2296
An empirical Bayes approach to the testing of certain parametric hypotheses. *Ann. Math. Statist.*, **34**, 1370-1385.
Review: STMA **6** (1965), 164.
Users: Robbins, H. (1964), Samuel, E. (1965), Van Ryzin, J. (1966).
Class.: P-G-NB-B:tp:G

Samuel, E. 1965. 2297
Sequential compound estimators. *Ann. Math. Statist.*, **36**, 879–89.
Class.: B-P-G-MI:se:G

Samuel, E. 1966. 2298
Estimators with prescribed bound on the variance for the parameters in the binomial and Poisson distributions, based on two-stage sampling. *J. Amer. Statist. Assoc.*, **61**, 220–27.
Class.: B-P:pe:G

Samuels, Stephen M. 1964. 2299
The model number of successes in independent trials (Abstract). *Ann. Math. Statist.*, **35**, 1841.

Samuels, Stephen M. 1965. 2300
On the number of successes in independent trials. *Ann. Math. Statist.*, **36**, 1272–78.
Review: MR **31** (1966), 737.
Class.: GBDP:osp:G

Samuels, S. M., and Anderson, T. W. 1966. 2301
Some inequalities among binomial and Poisson probabilities (Abstract). *Ann. Math. Statist.*, **37**, 544.

Sandelius, D. M. 1950. 2302
A truncated sequential procedure for interval estimation, with applications to the Poisson and negative binomial distributions (Preliminary report) (Abstract). *Ann. Math. Statist.*, **21**, 314.

Sandelius, D. M. 1953. 2303
Some unbiased estimates for a type of two-phase sampling. *Kungl. Lantbrukshögskolans Annaler*, **19**, 113–26.

Sandelius, D. M. 1962. 2304
A simple randomization procedure. *J. Roy. Statist. Soc. Ser. B*, **24**, 472–81.
Class.: MI:mi:G

Sandelius, Martin. 1950. 2305
An inverse sampling procedure for bacterial plate counts. *Biometrics*, **6**, 291–92.
Users: Sandelius, M. (1951a), Chapman, D. G. (1952b), Sandelius, M. (1952b), Nadler, J. (1960), Knight, W. (1965).
Class.: P:pe-ie:BM

Sandelius, Martin. 1951a. 2306
Inverse sampling applied to bacterial plate counts. I. Unrestricted and truncated sampling in the Poisson case. *Kungl. Lantbrukshögskolans Annaler*, **18**, 86–94.
Review: ZBL **43** (1952), 349.
User: Sandelius, M. (1951c).
Class.: P:pe-ie:BM

Sandelius, Martin. 1951b. 2307
Unbiased estimation based on inverse hypergeometric sampling. *Kungl. Lantbrukshögsholans Annaler*, **18**, 123–27.
Review: BA **27** (1953), Ab.No. 8573; ZBL **43** (1952) 348.
User: Bennett, B. M. (1957a).
Class.: H-IH:pe:G

Sandelius, Martin. 1951c. 2308
Truncated inverse binomial sampling. *Skand. Aktuarietidskr.*, **34**, 41–44.
Review: MR **14** (1953), 665; ZBL **43** (1952), 137.
Class.: B-NB:pe:G

Sandelius, Martin. 1952a. 2309
Confidence interval for the smallest proportion of a binomial population. *J. Roy. Statist. Soc. Ser. B*, **14**, 115–16.
Review: MR **14** (1953), 488; ZBL **47** (1953), 133.
Class.: B:ie:G

Sandelius, Martin. 1952b. 2310
Inverse sampling applied to bacterial plate counts. II. Cases when technical errors cannot be neglected. *Kungl. Lantbrukshögskolans Annaler*, **19**, 197–204.
Review: ZBL **50** (1954), 364.
Class.: P:pe-ie:BM

Sandelius, Martin. 1961. 2311
On an optimal search procedure. *Amer. Math. Monthly*, **68**, 133–34.
Class.: MI:mi:G

Sandiford, Peter J. 1960. 2312
A new binomial approximation for use in sampling from finite populations. *J. Amer. Statist. Assoc.*, **55**, 718–22.
Review: IJA **3** (1962), 12; ZBL **114** (1965), 100.
User: Govindarajulu, Z. (1965).
Class.: H-B:a:G

Sankaranarayanan, G. 1958. 2313
Some asympotic properties of Poisson processes. *Tôhoku Math. J.*, **10**, 60–68.
Review: ZBL **89** (1961), 341.

Santacroce, G. 1959. 2314
On binomial distributions with negative exponent (Italian). *Giorn. Economisti*, **18**, 359–72.
Review: IJA **4** (1963), 40

Santacroce, G. 1962. 2315
On the determination of the parameters of Makeham's survival law through the method of binomial moments (Italian). *Giorn. Economisti*, **21**, 222–28.
Review: STMA **6** (1965), 11.
Class.: B:m-pe:S

Sapogov, N. A. 1951. 2316
The stability problem for a theorem of Cramer. *Izv. Akad. Nauk SSSR Ser. Mat.*, **15**, 205–18.
Users: Ramachandran, B. (1961a), Ramachandran, B. (1963).

Sarkadi, K. 1953a. 2317
A selejtarány Bayes-féle valószinüsegi határaira vonatkozó dualitási elvröl (On the "duality law" concerning the Bayes' probability limit of the fraction defective) (Hungarian; English and Russian summaries). *Magyar Tud. Akad. Alkalm. Mat. Int. Közl.*, **2**, 275–86.
Review: ZBL **59** (1965)66), 129.
Users: Steinhous, H., and Zubrzycki, S. (1957), Sarkadi, K. (1960a).

Sarkadi, K. 1953b. 2318
A selejtarány a priori bétaeloszlásarol (On the *a priori* beta distribution of fraction defective) (Hungarian; English and Russian summaries). *Magyar Tud. Akad. Alkalm. Mat. Int. Közl.*, **2**, 287–98.
Review: ZBL **59** (1965-66), 129
Users: Sarkadi, K. (1957a), Sarkadi, K. (1960a).

Sarkadi, K. 1957a. 2319
On the distribution of the number of exceedances. *Ann. Math. Statist.*, **28**, 1021–23.
Review: ZBL **80** (1959), 357.
Users: Sarkadi, K. (1957b), Sarkadi, K. (1960).
Class.: MI:osp:G

Sarkadi, K. 1957b. 2320
Generalized hypergeometric distributions (Hungarian and Russian summaries). *Magyar Tud. Akad. Mat. Kutató Int. Közl*, **2**, 59–69.
Review: MR **20** (1959), 1018; ZBL **107** (1964), 142.
Users: Sarkadi, K. (1960); Sarkadi, K. (1960a).

Sarkadi, K. 1960. 2321
On the median of the distribution of exceedances. *Ann. Math. Statist.*, **31**, 225–26.
Review: IJA **2** (1961), 643.
Class.: MI:mi:G

Sarkadi, K. 1960a. 2322
A rule of dualism in mathematical statistics. *Acta. Math. Acad. Sci. Hungar.*, **11**, 83–92.
Review: IJA **2** (1961), 539; ZBL **96** (1962), 343.
Class.: B-P-H-MI:mi:G

Sarkadi, K., Schnell, Edit, and Vincze, I. 1962. 2323
On the position of the sample mean among the ordered sample elements. *Magyar. Tud. Akad. Mat. Kutató Int. Közl.*, **7**, 239–63.
Review: STMA **6** (1965), 92.
Class.: MI:os-a:G

Sarmanov, O. V. 1956. 2324
Necessary and sufficient conditions of existence of a discrete limit law for Markov chains with two states (Russian). *Dokl. Akad. Nauk SSSR*, **110**, 735–38.

Sarmanov, O. V., and Vistelius, A. B.
See Vistelius, A. B., and Sarmanov, O. V. (1947).

Sarndal, Carl-Erik. 1964. 2325
A unified derivation of some nonparametric distributions. *J. Amer. Statist. Assoc.*, **59**, 1042–53.
Review: MR **31** (1966), 151.
User: Sarndal, Carl-Erik (1965).

Sarndal, Carl-Erik. 1965. 2326
Derivation of a class of frequency distributions via Bayes's theorem. *J. Roy. Statist. Soc. Ser. B*, **27**, 290-300.
Review: STMA **7** (1966), 324.
Users: Patil, G. P., and Bildikar, S. (1966a).
Class.: MHR-NB-P-M:mb-osp:G

Sastry, M. P. 1962. 2327
On some stochastic models. Ph.D. Thesis, Andhra Univ. India.

Sathe, Y. S., and Narayana, T. V.
See Narayana, T. V., and Sathe, Y. S. (1961).

Sathe, Y. S., Narayana, T. V., and Chorneyko, I.
See Narayana, T. V., Chronoeyko, I., and Sathe, Y. S. (1960).

Satomi, H., Outi, Y., Okada, T., Kuno, E., and Yamamoto, S.
See Kuno, E., Yamamoto, S., Satomi, H. Outi, Y., and Okada, T. (1963).

Satterthwaite, F. E. 1942a. 2328
Generalized Poisson distribution. *Ann. Math. Statist.*, **13**, 410–17.
Review: MR **4** (1943), 163.
Users: Feller, W. (1943), Janossy, L., Renyi, A., and Aczel, J. (1950), Gurland, J. (1957), Tsao, C. M. (1962), Blischke, W. R. (1965).
Class.: GP-COP:m-osp-gf:G-A.

Satterthwaite, F. E. 1942b. 2329
Generalized Poisson distribution (Abstract). *Ann. Math. Statist.*, **13**, 451.

Satterthwaite, F. E. 1956. 2330
Comparison of two fractions defective. *Industrial Quality Control*, **13** (5), 17–18.
Class.: B:sqc:E

Satterthwaite, F. E. 1957. 2331
Binomial and Poisson confidence limits. *Industrial Quality Control*, **13** (11), 56–59.
Class.: B-P:ie:G

Saunders, S. C., and Myhre, J. M.
See Myhre, J. M., and Saunders, S. C. (1965).

Savage, I. R. 1962. 2332
Contributions to the theory of rank order statistics: applications of lattice theory. Tech. Rep. 15, Dept. Statist., Univ. Minnesota, pp. 20.
Class.: MI:os:G

Savage, I. R. 1964. 2333
Contributions to the theory of rank order statistics: Applications of lattice theory. *Rev. Inst. Internat. Statist.*, **32**, 52–64.
Review: MR **32** (1955), 100.
Class.: MI:os:G

Savage, I. R. 2334
Surveillance problems: Poisson models with noise. *Naval Res. Logist. Quart.* 1–13.
Class.: P:pc:G

Savage, I. R., and Grab, E. L.
See Grab, E. L., and Savage, L. R. (1954).

Savage, L. J. 1947. 2335
A uniqueness theorem for unbiased sequential binomial estimation. *Ann. Math. Statist..*, **18**, 295–97.
Review: MR **9** (1948), 152; ZBL **32** (1950), 43.
Users: deGroot, M. H. (1959), Wasan, M. T. (1965).
Class.: B:se:G

Savkevitch, V. 1940. 2336
Sur le schéma des urnes à composition variable. *C. R. (Doklady) Acad. Sci. URSS N.S.*, **28**, 8–12.
Review: MR **2** (1941), 229.
Users: Freedman, David A. (1965).

Sawkins, D. T. 1947. 2337
A new method of approximating the binomial and hypergeometric probabilities. *J. Proc. Roy. Soc. New South Wales*, **81**, 38–47.
Review: MR **9** (1948), 193.
Class.: B-H:a:G

Saxena, R. K., and Mathai, A. M.
See Mathai, A. M., and Saxena, R. I. (1966).

Sazonov, V. V. 1966. 2338
On multidimensional concentration functions (Russian; English summary). *Teor. Verojatnost. i Primenen.*, **11**, 683–90.

Schaefer, M. B., and Bishop, Y. M. M. 1958. 2339
Particulate iron in offshore waters of the Paname Bight and in the Gulf of Panama. *Limnology and Oceanographt*, **3**, 137–49.

Schafer, R. E. 1964. 2340
Bayesian operating characteristic curves for reliability and quality sampling plans. *Industrial Quality Control*, **21**, 118–22.

Schafer, W. 1937. 2341
Fehlerrechnung bei biologischen Messungen. *Naturwissenschaften*, **25**, 218–19.

Schafer, W. 1953. 2342
Bayes-Funktion ohne Hypothese. *Mitteilungsbl. Math. Statist.*, **5**, 70–74.
Review: ZBL **50** (1954), 148.

Schäffer, K. A. 1957. 2343
Der likelihood-Anpassungstest. *Mitteilungsbl. Math. Statist.*, **9**, 27–54.
Review: ZBL **78** (1959), 333.

Scheffe, Henry. 1943-46. 2344
Note on the use of the tables of percentage points of the incomplete beta function to calculate small sample confidence intervals for a binomial p. *Biometrika*, **33**, 181.

Review: ZBL **60** (1957), 295.
User: Sandelius, M. (1951c).
Class.: B:ie-c:G

Schiff, L. I., and Post, R. F.
See Post, R. F., and Schiff, L. I. (1950).

Schilling, Walter. 1947. 2345
A frequency distribution represented as the sum of two Poisson distributions. *J. Amer. Statist. Assoc.*, **42**, 407–24.
Users: Dandekar, V. M. (1955), Blischke, W. R. (1965).
Class.: P:pe-gf:BM-P

Schmetterer, L. 1952. 2346
Über ein Beispiel aus der Statistik. *Z. Angew. Math. Mech.*, **32**, 281–84.
Review: MR **14** (1953), 391.

Schmid, Paul. 1958. 2347
On the Kolmogorov and Smirnov limit theorems for discontinuous distribution functions. *Ann. Math. Statist.*, **29**, 1011–27.
Review: ZBL **94** (1962), 132.
Class.: MI:a:G

Schmidt, E. 1933. 2348
Über die Charlier-Jordansche Entwicklung einer willkürlichen Funktion nach der Poissonschen Funktion und ihren Ableitungen. *A. Angew. Math. Mech.*, **13**, 139–42.
User: Bogs, R. P. Jr. (1949).

Schmutz, Paul. 1964. 2349
Des espérances mathématiques de X^{-1} et X^{-2}. *Publ. Inst. Statist. Univ. Paris*, **13**, 153–68.
Review: MR **30** (1965), 129.
Class.: MI:m-a:G

Schnabel, Zoe E. 1938. 2350
The estimation of the total fish population of a lake. *Amer. Math. Monthly*, **45**, 348–52.
Users: Bailey, N. T. J. (1951), Bailey, N. T. J. (1952), Chapman, D. G. (1952b), Wohlschlag, D. E. (1952), Young, H., Neess, J., and Emlen, J. T. (1952), Chapman, D. G. (1954), Wohlschlag, D. E. (1954), Geiss, A. D. (1955), Jenkins, R. M. (1955).
Class.: B-P:pe:O

Schneeweiss, Hans. 1965. 2351
Das Aggregationsproblem. *Statistische Hefte*, **6**.

Schneiderman, M., and Brecher, G. 1950. 2352
The relative frequency of sparse cell elements—an application of reticulocyte blood counts. *Biometrics*, **6**, 390–94.
Class.: P:pe:BM

Schoderbek, J. J. 1962. 2353
Some weapon system survival probability models. II. Random time between firings. *Operations Res.*, **10**, 168–79.

Schoenberg, I. J. 1951. 2354
On Pólya frequency functions. I. The totally positive functions and their Laplace transforms. *J. Analyse Math.*, **1**, 331–74.
Users: Karlin, S., and Proschan, F. (1960), Efron, B. (1965).
Class.: MI:mi:G

Schrek, R., and Lipson, H. I. 1941. 2355
Logarithmic frequency distributions. *Hum. Biol.*, **13** 1–22.
Review: PA **16** (1942), Ab.No. 865; BA **15** (1941), Ab.No. 10050.

Schuhl, A., and Gerlough, D. L.
See Gerlough, D. L., and Schuhl, A. (1955).

Schultz, Vincent. 1961. 2356
An annotated bibliography on the uses of statistics in ecology—a search of 31 *periodicals*. AEC Report No. Tid-3908. Office Tech. Information, U.S. Atomic Energy Comm., pp. 315.

Schultz, V., and Byrd, M. A. 1957. 2357
An analysis of covariance of cottontail rabbit population data. *J. Wildlife Managem.*, **21**, 315–19.

Schützenberger, M. F. 1949. 2358
Résultats d'une enquête sur la distribution du sexe dans les familles nombreuses. *La Semaine des Hôpitaux* **25**, 2579–82.

Schützenberger, M. F. 1950. 2359
Nouvelles recherches sur la distribution du sexe à la naissance. *La Semaine des Hôpitaux*, **26**, 4458–65.

Schwartz, Mischa. 1956. 2360
A coincidence procedure for signal detection. *IRE Transactions* **IT-2**, 135–39.
Class.: B:mi:E

Schwarz, A. 1955. 2361
Zur Theorie seltener Ereignisse; alte Zufallsverteilungen in neuem Licht. *Rev. Suisse Econ. Polit. Statist.*, **92**, 175–82.

Schwarz, Gideon. 1962. 2362
Asymptotic shapes of Bayes sequential testing regions. *Ann. Math. Statist.*, **33**, 224–36.
Users: Lindley, D. V., and Barnett, B. (1965).
Class.: B-P-G:tp:G

Scott, A. D. 1951. 2363
Bibliography of applications of mathematical statistics to economics. *J. Roy. Statist. Soc. Ser. A*, **114**, 372–93.
Notes: Also; *J. Roy. Statist. Soc. Ser. A*, **116**, 177–85.

Scott, A. D. 1953. 2364
Bibliography of applications of mathematical statistics to economics. *J. Roy. Statist. Soc. Ser. A*, **116**, 177–85.
Notes: Also; *J. Roy. Statist. Soc. Ser. A*, **114**, 373–93.

Scott, E. L. 1965. 2365
Subclustering. *Classical and Contagious Discrete Distributions*, Ed. G. P. Patil, Statistical Publishing Society, Calcutta and Pergamon Press, pp. 33–44.
Class.: MI:mb:P

Scott, E. L., and Neyman, J.
See Neyman, J., and Scott, E. L. (1952).

Scott, E. L., and Neyman, J.
See Neyman, J., and Scott, E. L. (1953).

Scott, E. L., and Neyman, J.
See Neyman, J., and Scott, E. L. (1954).

Scott, E. L., and Neyman, J.
See Neyman, J., and Scott, E. L. (1957).

Scott, E. L., and Neyman, J.
See Neyman, J., and Scott, E. L. (1959).

Scott, E. L., and Neyman, J.
See Neyman, J., and Scott, E. L. (1960).

Scott, E. L., and Neyman, J.
See Neyman, J., and Scott, E. L. (1962).

Scott, E. L., Shane, C. D., and Neyman, J.
See Neyman, J., Scott, E. L., and Shane, C. D. (1956).

Scrase, F. 1935. 2366
The sampling errors of the Aitken nucleus counter. *Quart. J. Roy. Meteoro. Soc.*, **61**, 367–79.
Users: Moore, P. G. (1954a), Duker, S. (1955), Moore, P. G. (1956b), Glasser, G. J. (1962b).

Sculthorpe, Diane, and Aitchison, J.
See Aitchison, J., and Sculthorpe, Diane (1965).

Seal, H. L. 1947. 2367
A probability distribution of deaths at age x when policies are counted instead of lives. *Skand. Aktuarietidskr.*, **30**, 18–43.
Review: MR **9** (1948), 96.
Users: Seal, H. L. (1949), Seal, H. L. (1953), Rider, P. R. (1965).
Class.: OMR-P-MI:mb-gf:S

Seal, H. L.. 1949. 2368
Mortality data and the binomial probability law. *Skand. Aktuarietidskr.*, **32**, 188–216.
Review: MR **11** (1950), 449; ZBL **41** (1952), 469.
Class.: B:gf:S

Seal, H. L. 1952. 2369
The maximum likelihood fitting of the discrete Pareto law. *J. Inst. Actuar.*, **78**, 115–21.
Review: ZBL **46** (1953), 372.
Class.: MI:pe-gf:S

Seal, H. L. 1953. 2370
The maximum likelihood fitting of the discrete Pareto law. *Skand. Aktuarietidskr.*, **36**, 115–21.
User: Rider, P. R. (1965).
Class.: MI:pe-gf:S

Seegrist, D. W. 1964. 2371
Some results of fitting various distributions by means of an IBM 1620 (Abstract). *Biometrics*, **20**, 906.

Seguinot, J., and Bizard, G.
See Bizard, G., and Seguinot, J. (1954).

Selby, B. 1965. 2372
The index of dispersion as a test statistic. *Biometrika*, **52**, 627–29.
Review: STMA **7** (1966), 340.
Class.: P-COB:id-cm:G

Sen, P. K. 1960. 2373
On the estimation of the population size by capture-recapture method. *Bull. Calcutta Statist. Assoc.*, **9**, 93–110.
Review: MR **22** (1961), 859.

Serra, A., and O'Mathuna, D. 1966. 2374
A theoretical approach to the study of genetic parameters of histocompatibility in man. *Ann. Human Genetics*, **30**, 97–118.
Class.: M-B:mb:BM

Seshadri, V. 1964a. 2375
A characterization of the logarithmic and geometric distributions (Abstract). *Ann. Math. Statist.*, **35**, 1841.

Seshadri, V., and Patil, G. P.
See Patil, G. P., and Seshadri, V. (1964).

Sette, O. E., and Ahlstrom, E. H. 1947. 2376
Estimations of abundance of the eggs of the Pacific pilchard (Sardinops caerulea) off southern California during 1940 and 1941. *J. Marine Res.*, **7**, 511–42.

Sevast'yanov, B. A. 1957. 2377
An ergodic theorem for Markov processes and its application to telephone systems with refusals (Russian). *Teor. Verojatnost. i Primenen.*, **2**, 106–16.

Sevast'yanov, B. A. 1961. 2378
Confidence intervals for the mean probability in Poisson schemes. *Theor. Probability Appl.*, **6**, 220–22.
Review: ZBL **101** (1963), 128.
Class.: GDBP:ie:G

Sevast'yanov, B. A. 1966. 2379
Limit theorems in a scheme of disposal of particles in cells (Russian; English summary). *Teor. Verojatnost. i Primenen.*, **11**, 696–700.

Sevast'yanov, B. A., and Chistyakov, V. P. 1964. 2380
Asymptotic normality in the classical ball problem. *Theor. Probability Appl.*, **9**, 198–211.
User: Chistyakov, V. P. (1964b).
Class.: MI-P:a:G

Shah, S. M. 1961. 2381
The asymptotic variances of method of moments estimates of the parameters of the truncated binomial and negative binomial distributions. *J. Amer. Statist. Assoc.*, **56**, 990–94.
Review: STMA **6** (1965), 460; ZBL **101** (1963), 130.
User: Shah, S. M. (1966b).
Class.: TCB-TCNB:pe:G

Shah, S. M. 1966a. 2382
A note on Craig's paper on the minimum of binomial variates. *Biometrika*, **53**, 614–15.
Class.: B:os:G

Shah, S. M. 1966b. 2383
On estimating the parameter of a doubly truncated binomial distribution. *J. Amer. Statist. Assoc.*, **61**, 259–63.
Class.: TCB:pe:G

Shah, B. K., and Venkataraman, V. K. 1963. 2384
A note on modified Poisson distribution. *Metron*, **22** (3-4), 27–35.
Review: STMA **6** (1965), 840.
Class.: OPR:m-pe:G

Shah, Gamanalal, P. 1965. 2385
Bounds for lattice distributions having monotone hazard rate with applications. Res. Rep. ORC 65-12 Operations Res. Center College Eng. Univ. California Berkeley.
Class.: MI:mi:G-BM

Shakuntala, N. S., and Iyer, P. V. K. 1959.
See Iyer, P. V. K., and Shakuntala, N. C. (1959).

Shanbhag, D. N. 1963. 2386
On queues with Poisson service time. *Austral. J. Statist.*, **5**, 57–61. Corregenga: **6**, 1088.
Review: ZBL **116** (1965), 106.
Class.: P:pc:G

Shanbhag, D. N. 1965. 2387
A note on queueing systems with Erlangian service time distributions. *Ann. Math. Statist.*, **36**, 1574–78.
Review: MR **131** (1966), 957.
Class.: MI:pc:G

Shane, C. D., Neyman, J., and Scott, E. L.
See Neyman, J., Scott, E. L., and Shane, C. D. (1956).

Shanfield, Florence. 1964. 2388
Positive and negative integral valued random variables. Master's Thesis, McGill Univ.

Shanks, R. E. 1953. 2389
Forest composition and species association in the beechmaple forest region of western Ohio. *Ecology*, **34**, 455–66.
Class.: MI:h:BM

Shannon, S. 1942. 2390
Comparative aspects of the point binomial polygon and its associated normal curve of error. *Record Amer. Inst. Actuar.*, **31**, 208–26.
Review: MR **4** (1943), 279.
Class.: B:a:G

Shapiro, J. M. 1955. 2391
Error estimates for certain probability limit theorems. *Ann. Math. Statist.*, **26**, 617–30.
User: Govindarajulu, Z. (1965).

Shapiro, J. M. 1958. 2392
Sums of powers of independent random variables. *Ann. Math. Statist.*, **29**, 515–22.
User: Govindarajulu, Z. (1965).
Class.: P:pc-osp:G

Shapiro, J. M. 1960. 2393
Sums of small powers of independent random variables. *Ann. Math. Statist.*, **31**, 222–24.

Shaw, H. W. 1936. 2394
Poisson probability summation. *Junior Inst. Eng. J.*, **46**, 479–98.
Review: SA **39B** (1936), Ab.No. 2819.

Shcheglova, M. V., Bol'shev, L. N., and Gladkov, B. V.
See Bol'shev, L. N., Gladkov, B. V., and Shcheglova, M. V. (1961).

Shenton, L. R. 1949. 2395
On the efficiency of the method of moments and Neyman's type A distribution. *Biometrika*, **36**, 450–54.
Users: Anscombe, F. J. (1950a), Goodall, D. W. (1952), Thomson, G. W. (1952), Beall, G., and Rescia, R. R. (1953), Evans, D. E. (1953), Douglas, J. B. (1955), Sprott, D. A. (1958), Katti, S. K. (1960d), Katti, S. K., and Gurland, J. (1960b), Shumway, R., and Gurland, J. (1960a), Shumway, R., and Gurland, J. (1960b), Shumway, R., and Gurland, J. (1960c), Katti, S. K., and Gurland, J. (1961), Katti, S. K., and Gurland, J. (1962a), Katti, S. K., and Gurland, J. (1962b), Tsao, C. M. (1962), Blischke, W. R. (1965).
Class.: N:pe:G

Shenton, L. R. 1950. 2396
Maximum likelihood and the efficiency of the method of moments. *Biometrika*, **37**, 111–16.
Review: ZBL **36** (1951), 212.
Users: Shenton, L. R. (1958), Gonin, H. T. (1961).
Class.: NH:pe:G

Shenton, L. R. 1958. 2397
Moment estimators and maximum likelihood. *Biometrika*, **45**, 411–20. Corrections: *Biometrika*, **46**, 502, *Biometrika*, **48**, 474.
Review: IJA **1** (1959), 58; MR **20** (1959), 1107.
Users: Gonin, H. T. (1961), Shenton, L. R., and Wallington, P. A. (1962), Shenton, L. R., and Myers, R. (1965), Van Heerden, D. F. I., and Gonin, H. T. (1966).
Class.: NB:pe:G

Shenton, L. R. 1963. 2398
A note on bounds for the asymptotic sampling variance of the maximum likelihood estimator of a parameter in the negative binomial distribution. *Ann. Inst. Statist. Math. Tokyo*, **15**, 145–51.
Review: MR **30** (1965), 503.
Class.: NB:pe:G

Shenton, L. R. and Bowman, K. 1963. 2399
Higher moments of a maximum-likelihood estimate. *J. Roy. Statist. Soc. Ser. B*, **25**, 305–17.
Users: Bowman, K. O., and Shenton, L. R. (1965a), Patil, G. P., and Wani, J. K. (1965), Shenton, L. R., and Myers, R. (1965).
Class.: MI:pe:G

Shenton, L. R. and Bowman, K. O. 1965. 2400
Biases and covariances of maximum likelihood estimators. Report K-1633 Oak Ridge National Lab. Oak Ridge, Tennessee.

Shenton, L. R., and Bowman, K. O.
See Bowman, K. O., and Shenton, L. R. (1965).

Shenton, L. R., and Bowman, K. O.
See Bowman, K. O., and Shenton, L. R. (1966).

Shenton, L. R., and Myers, R. 1965a. 2401
Comments on estimation for the negative binomial distributions. *Classical and Contagious Discrete Distributions*, Ed. G. P. Patil, Statistical Publishing Society, Calcutta and Pergamon Press, pp. 241–62.
Class.: NB:pe:G

Shenton, L. R., and Myers, R. 1965b. 2402
Orthogonal statistics. *Classical and Contagious Discrete Distributions*, Ed. G. P. Patil, Statistical Publishing Society, Calcutta and Pergamon Press, pp, 445–58.
Class.: MI-NB:osp:G

Shenton, L. R., and Skellan, J. G.
See Skellam, J. G., and Shenton, L. R. (1957).

Shenton, L. R., and Wallington, P. A. 1962. 2403
The bias of moment estimators with an application to the negative binomial distribution. *Biometrika*, **49**, 193–204.
Review: IJA **3** (1962), 557.
Users: Shenton, L. R. (1963), Blischke, W. R. (1965), Bowman, K. O., and Shenton, L. R. (1965a), Shenton, L. R., and Myers, R. (1965), Bowman, K. O., and Shenton, L. R. (1966).
Class.: NB:pe:G

Shepherd, B. B., Littleford, R. A., and Newcombe, C. L.
See Littleford, R. A., Newcombe, C. L., and Shepherd, B. B. (1940).

Sheps, Mindel C. 1963-64. 2404
Effects on family size and sex ratios of preferences regarding the sex of children. *Population Studies*, **17**, 66–72.

Sherman, R. E. 1965. 2405
Design and evaluation of a repetitive group sampling plan. *Technometrics*, **7**, 11–21.
Review: STMA **7** (1966), 374.
Class.: B:sqc:G

Shimizu, Ryoichi, Sibuya, Masaaki, and Yoshimura, Isao.
See Sibuya, Masaaki, Yoshimura, Isao, and Shimizu, Ryoichi (1964).

Shiue, C-J., and Beazley, R. 1957. 2406
Classification of the spatial distribution of trees using the area sampling method. *Forest Sci.*, **3**, 22–31.
Class.: MI:tp:BM

Shiyomi, M., and Nakamura, K. 1964. 2407
Experimental studies on the distribution of the aphid counts (Japanese summary). *Res. Population Ecol.*, **6**, 79–87.
Class.: P-NB:gf:BM

Shklov, N., and Putnam, L. G.
See Putnam, L. G., and Shklov, N. (1956).

Shook, B. L. 1930. 2408
Synopsis of elementary mathematical statistics. *Ann. Math. Statist.*, **1**, 224–59.
Class.: B:mi:G

Shorrock, Richard W. 1965. 2409
Stochastic and sampling processes for the logarithmic series distribution. Master's Thesis, McGill Univ.

Shorrock, Richard W., and Patil, G. P.
See Patil, G. P., and Shorrock, Richard W. (1965).

Shorrock, Richard W., and Patil, G. P.
See Patil, G. P., and Shorrock, Richard W. (1966).

Shortley, G. 1965. 2410
A stochastic model for distribution of biological response times. *Biometrics*, **21**, 562–82.

Shumway, Robert, and Gurland, John. 1960a. 2411
Fitting the Poisson binomial distribution. *Biometrics*, **16**, 522–33.
Review: MR **23**A (1962), 122; IJA **2** (1961), 493; ZBL **97** (1962), 348.
Users: Katti, S. K. (1960d), Katti, S. K., and Gurland, J. (1960b), Shumway, R., and Gurland, J. (1960c), Katti, S. K., and Gurland, J. (1962a), Staff, P. J. (1964), Blischke, W. R. (1965), Gurland, J. (1965), Kemp, C. D., and Kemp, A. W. (1965), Sprott, D. A. (1965b).
Class.: GP-COB:c:G

Shumway, Robert, and Gurland, John. 1960b. 2412
A fitting procedure for some generalized Poisson distributions. *Skand. Aktuarietidskr.*, **43**, 87–108.
Review: MR **24**A (1962), 325; IJA **3** (1962), 13; ZBL **98** (1963), 116.
Class.: GP-COB-CONB:pe-c:G

Shumway, Robert, and Gurland, John. 1960c. 2413
A fitting procedure for some generalized Poisson distributions. MRC Tech. Summary Rep. 205, Univ. Wisconsin, pp. 33.
Class.: GP-COB-CONB:pe-c:G

Sibuya, Masaaki. 1960. 2414
Cutting out procedures for material with Poisson defects. *Ann. Inst. Statist. Math. Tokyo*, **12**, 151–59.
Review: MR **22** (1961), 1965; IJA **3** (1962), 14.

Sibuya, Masaaki. 1961. 2415
On exponential and other random variable generators. *Ann. Inst. Statist. Math. Tokyo*, **13**, 231–37.
Class.: P-G:9sp:G

Sibuya, Masaaki. 1963. 2416
Randomized unbiased estimation of restricted parameters. *Ann. Inst. Statist. Math. Tokyo*, **15**, 61–66.
Class.: B-H:pe:G

Sibuya, Masaaki, Yoshimura, Isao, and Shimizu, Ryoichi. 1964. 2417
Negative multinomial distribution. *Ann. Inst. Statist. Math. Tokyo*, **16**, 409–26.
Review: MR **30** (1965), 312.
Users: Patil, G. P. and Bildikar, S. (1966a).
Class.: NM:MI:mb-m-pe-osp:G

Sichel, H. S. 1951. 2418
The estimation of the parameters of a negative binomial distribution with special reference to psychological data. *Psychometrika*, **16**, 107–27.
Review: MR **13** (1952), 53.
Users: Bliss, C. I., and Fisher, R. A. (1953), Arbous, A. G., and Sichel, H. S. (1954a), Arbous, A. G., and Sichel, H. S. (1954b), Patil, G. P. (1959), Edwards, C. B., and Gurland, J. (1960), Patil, G. P. (1960c), Edwards, C. B., and Gurland, J. (1961), Berthet, P., and Gerard, G. (1965), Shenton, L. R., and Myers, R. (1965).

Sichel, H. S., and Arbous, A. G.
See Arbous, A. G., and Sichel, H. S. (1954).

Siddiqui, M. M., and Mielke, Paul W.
See Mielke, Paul W., and Siddiqui, M. M. (1965).

Silva, Giovanni. 1941. 2419
Una generalizzazione del problema delle concordanze. *Atti Ist. Veneto Sci. Etc.*, **100** (2), 689–709.
Review: ZBL **28** (1944), 168.

Simon, Herbert A. 1955. 2420
On a class of skew distributions functions. *Biometrika*, **42**, 425–39.
Review: ZBL **66** (1957), 112.
Users: Herdan, G. (1957), Herdan, G. (1958), Simon, H. A., and Bonini, C. P. (1958), Simon, H. (1960), Herdan, G. (1961), Simon, H. A. (1961a), Simon, H. A. (1961b), Simon, H., and Van Wormer, T. (1963), Ijiri, Y., and Simon, H. A. (1964), Rider, P. R. (1965).
Class.: MI:mb-gf:L-BM-S

Simon, Herbert A. 1960. 2421
Some further notes on a class of skew distribution functions. *Information and Control*, **3**, 80–88.
Review: IJA **4** (1963), 609.
Users: Simon, H. A. (1961a), Simon, H. A., and Van Wormer, T. (1963).
Class.: MI:mb:L

Simon, Herbert A. 1961a. 2422
Reply to 'final note' by Benoit Mandelbrot. *Information and Control*, **4**, 217–23.
Review: ZBL **103** (1964), 126.
User: Simon, H. A. (1961b).
Class.: MI:mb:L

Simon, Herbert A. 1961b. 2423
Reply to Dr. Mandelbrot's Post Scriptum. *Information and Control*, **4**, 305–08.
Review: ZBL **103** (1964), 126.
Class.: MI:mb:L

Simon, Herbert A., and Bonini, C. P. 1958. 2424
The size distribution of business firms. *Amer. Econ. Rev.*, **48**, 607–17.
Users: Ijiri, Y., and Simon, H. A. (1964).
Class.: MI:mb-gf:S

Simon, Herbert A., and Ijiri, Y.
See Ijiri, Y., and Simon, H. A. (1964).

Simon, Herbert A., and Van Wormer, T. A. 1963. 2425
Some Monte Carlo estimates of the Yule distribution. *Behavioral Sci.*, **8**, 203–10.
Users: Ijiri, Y., and Simon, H. A. (1964).
Class.: MI:gf:L

Simond, M., and Galliher, H. P., and Morse, Philip M.
See Galliher, H. P., Morse, Philip M., and Simond, M. (1959).

Simpson, E. H. 1949. 2426
Measurement of diversity. *Nature*, **163**, 688.
Review: ZBL **32** (1950), 39.
Users: Anscombe, F. S. (1950a), Morisita, M. (1959), Kuno, E. (1963).
Class.: MI-NB-M:mi:G-BM

Singal, M. K., and Dhruvarajan, P. S.
See Dhruvarajan, P. S., and Singal, M. K. (1957).

Singh, B. N., and Chalam, G. V. 1937. 2427
A quantitative analysis of the weed flora on arable land. *J. Ecol.*, **25**, 213–21.
Users: Singh, B. N., and Chalam, G. (1938), Singh, B. N., and Das, K. (1939), Blackman, G. E. (1942), Archibald, E. E. A. (1948), Ashby, E. (1948), Bliss, C. I., and Fisher, R. A. (1953).

Singh, B. N., and Das, K. 1938. 2428
Distribution of weed species on arable land. *J. Ecol.*, **26**, 455–66.
Users: Singh, B. N., and Das, K. (1939), Archibald, E. E. A. (1948), Ashby, E. (1948).

Singh, B. N., and Das, K. 1939. 2429
Percentage frequency and quadrat size in analytical studies of weed flora. *J. Ecol.*, 66–77.
Users: Blackman, G. E. (1942), Ashby, E. (1948).

Singh, J., and Katti, S. K. 1965. 2430
Some properties of compound distributions (Abstract). *Biometrics*, **21**, 765.

Singh, J., and Katti, S. K.
See Katti, S. K., and Singh, J. (1965).

Singh, N. 1960. 2431
Estimation of parameters of a mixture of two or more Poissonian populations from a censored sample. *J. Indian Soc. Agric. Statist.*, **12**, 88–94.
Review: MR **23A** (1962), 253.
Class.: GP:pe:G

Singh, Rajinder. 1963. 2432
Existence of bounded length confidence intervals. *Ann. Matb. Statist.*, **34**, 1474–85.
Class.: P:se:G

Singh, Rajinder. 1964. 2433
Estimating log p (Abstract). *Math. Student*, **32** Appendix 2, 32–33.

Singh, S. N. 1962. 2434
A note on inflated Poisson distribution. (Abstract). *Ann. Math. Statist.*, **33**, 1210.

Singh, S. N. 1962–63. 2435
Inflated Poisson distribution. *J. Sci. Res. Banaras Hindu Univ.*, **13**, 317–26.
Review: MR **29** (1965), 561.
User: Cohen, A. C., Jr. (1966).

Singh, S. N. 1963. 2436
A note on inflated Poisson distribution. *J. Indian Statist. Assoc.*, **1**, 140–44.
Review: MR **28** (1964), 1065; STMA **5** (1964), 372.
User: Cohen, A. C., Jr. (1966).
Class.:

Singh, S. N. 1964. 2437
On the time of first birth. *Sankhyā Ser. B*, **26**, 95–102.

Singh, S. N. 1966. 2438
Testing the homogeneity of truncated Poisson distributions (Abstract). *Ann. Math. Statist.*, **37**, 775.

Sinha, P. 1953. 2439
Distribution of total number of runs in samples from Poisson populations. *Bull. Calcutta Statist. Assoc.*, **4** 171–72.
Review: MR **14** (1953), 1102.

Sinif, Donald B., and Tester, John R.
See Tester, John R., and Sinif, Donald B. (1965)

Siniscalco, M., Smith, C. A. B., and Ceppellini, R.
See Ceppellini, R., Siniscalco, M., and Smith, C. A. B. (1955).

Siotani, M. 1956. 2440
Order statistics for discrete case with a numerical application to the binomial distribution. *Ann. Inst. Statist. Math. Tokyo*, **8**, 95–104.
Review: MR **19** (1958), 331; ZBL **73** (1960), 139.
Users: Khatri, C. G. (1962), Gupta, S. S. (1965).

Siotani, M., and Oazwa, M. 1958. 2441
Tables for testing the homogeneity of k independent binomial experiments on certain event based on the range. *Ann. Inst. Statist. Math. Tokyo*, **10**, 47–63.
Review: IJA **1** (1959), 157.
Class.: B:h:G

Sirazhdinov, S. Kh. 1956a. 2442
On estimators with minimum bias for a binomial distribution. *Theor. Probability Appl.*, **1**, 150–56.
Class.: B:pe:G

Sirazhdinov, S. Kh. 1956b. 2443
Concerning estimations with minimum bias for a binomial distribution. (Russian; English summary). *Teor. Verojatnost. i. Primenen.*, **1**, 168–74.
Review: MR **19** (1958), 783; ZBL **73** (1960), 152.

Siromoney, Gift. 1962. 2444
Entropy of logarithmic series distributions. *Sankhyā*, **24**A, 419–20.
Review: STMA **6** (1965), 462; ZBL **108** (1964), 325.
Users: Nelson, W. C., and David, H. A. (1964), Patil, G. P., and Wani, J. K. (1965).
Class.: LS:osp:G

Siromoney, Gift. 1964. 2445
The general Dirichlet's series distribution. *J. Indian Statist. Assoc.*, **2**, 69–74.
Review: MR **30** (1965), 491; STMA **7** (1966), 294.
Class.: MI-PS:mi:G-P

Sirull, R., Earles, D., and Massa, J.
See Earles, D., Massa, J., and Sirull, R. (1963).

Sisojevic, P., and Embree, D. G.
See Embree, D. G., and Sisojevic, P. (1965).

Siskind, V. 1965. 2446
A solution of the general stochastic epidemic. *Biometrika*, **52**, 613–16.
Review: STMA **7** (1966), 416.
Class.: MI:pc:BM

Sitgreaves, R., and Girshick, M. A., and Rubin, H.
See Girshick, M. A., Rubin, H., and Sitgreaves, R. (1955).

Sittig, J., de Jong, A. J., and de Wolff, P.
See de Wolff, P., Sittig, J., and de Jong, A. J. (1950).

Skellam, J. G. 1946. 2447
The frequency distribution of the difference between two Poisson variates belonging to different populations. *J. Roy. Statist. Soc. Ser. A*, **109**, 296.
Users: Skellan, J. G. (1952), Johnson, N. L. (1959), Irwin, J. O. (1963), Shanfield, F. (1964).
Class.: OPR:mi:G

Skellam, J. G. 1948. 2448
A probability distribution derived from the binomial distribution by regarding the probability of success as variable between the sets of trials. *J. Roy. Statist. Soc. Ser. B*, **10**, 257–61.
Review: MR **10** (1949), 463; ZBL **32** (1950), 419.
Users: Skellam, J. G. (1949), Shenton, L. R. (1950), Binet, F. E. (1953), Kemp, C. D., and Kemp, A. W. (1956a), Kemp, C. D., and Kemp, A. W. (1956b), Sarkadi, K. (1957), Sarkadi, K. (1957a), Edwards, A. W. F. (1958), Patil, G. P. (1959), Darwin, J. H. (1960), Lord, F. M. (1960), Sarkadi, K. (1960a), Mosimann, J. E. (1962), Pielou, E. C. (1962), Bosch, A. J. (1963), Tallis, G. M. (1964), Blischke, W. R. (1965), Potthoff, Richard F., and Whittinghill, Maurice (1966a), Holgate, P. (1966a).
Class.: NH:mb-pe-gf:G-E

Skellam, J. G. 1949. 2449
The probability distribution of gene-differences in relation to selection, mutation, and random extinction. *Proc. Cambridge Philos. Soc.*, **45**, 364–67.
Review: ZBL **33** (1950), 294.
User: Skellam, J. G. (1952).

Skellam, J. G. 1949a. 2450
The distribution of the moment statistics of samples drawn without replacement from a finite population. *J. Roy. Statist. Soc., Ser. B*, **11**, 291–96.
Review: ZBL **37** (1951), 86.
User: Wishart, John (1952).

Skellam, J. G. 1952. 2451
Studies in statistical ecology. *Biometrika*, **39**, 346–62.
Review: ZBL **47** (1953), 386.
Users: Chapman, D. G. (1954), Clark, P. J., and Evans, F. C. (1954a), Clark, P. J., and Evans, F. C. (1954b), Thompson, H. R. (1954), Gurland, J. (1957), McGuire, J., Brindley, T. A., and Bancroft, T. A. (1957), Neyman, J., and Scott, E. L. (1957), Pielou, E. C. (1957), Bliss, C. I. (1958), Gurland, J. (1958), Skellam, J. G. (1958b), Gurland, J. (1959), Jensen, P. (1959), Waters, W. (1959), Waters, W. E., and Henson, W. R. (1959), Katti, S. K. (1960d), Pielou, E. C. (1960), Pielou, D. P. (1960a), Shumway, R., and Gurland, J. (1960c), Bardwell, G. E. (1961), Harcourt, D.G.(1961), Khatri, C. G., and Patel, I. R. (1961), Ito, Nakamura, Kondo, Miyashita, and Nakamura, (1962), Pielou, E. C. (1962), Shanfield, F. (1964), Gurland, J. (1965), Holgate, P. (1965), Katti, S. K., and Sly, L. E. (1965), Kemp, C. D., and Kemp, A. W. (1965), Kobayashi, S. (1966).
Class.: COP-MI:mb-gf:BM

Skellam, J. G. 1954. 2452
Estimation of animal populations by extraction processes considered from the mathematical standpoint. *Progress Soil. Zool.*, **1**, 26–36. Ed. P. W. Murphy, Butterworths, London.
User: Holgate, P. (1966a).

Skellam, J. G. 1958a. 2453
The mathematical foundations underlying the use of line transects in animal ecology. *Biometrics*, **14**, 385–400.
Review: ZBL **83** (1960), 156.

Skellam, J. G. 1958b. 2454
On the derivation and applicability of Neyman's type A distribution. *Biometrika*, **45**, 32–36.
Review: ZBL **93** (1962), 325.
User: Holgate, P. (1965).
Class.: N:mb:G

Skellam, J. G., and Shenton, L. R. 1957. 2455
Distributions associated with random walk and recurrent events. *J. Roy. Statist. Soc. Ser. B*, **19**, 64–111.
Review: ZBL **87** (1961), 132.
User: Haight, F. (1965b).
Class.: P-B-MI:pc:G

Skibinsky, M., and Cote, L. 1963. 2456
On the admissibility of some standard estimates in the presence of prior information. *Ann. Math. Statist.*, **34**, 439–548.
Review: STMA **5** (1964), 617.
Class.: B:pe-c:G

Sloan, R. E. 1955. 2457
Paleoecology of the Pennsylvanian Marine shales of Palo Pinto County, Texas. *J. Geology*, **63**, 412–27.

Sly, L. E., and Katti, S. K.
See Katti, S. K., and Sly, L. E. (1964).

Smith, C. A. B.
See Kalmus, H. (1946-47).

Smith, C. A. B. 1957. 2458
Counting methods in genetical statistics. *Ann. Human Genetics*, **21**, 254–76.
Class.: B:pe-h:BM

Smith, C. A. B., Ceppellini, R., and Siniscalco, M.
See Ceppellini, R., Siniscalco, M., and Smith, C. A. B. (1955).

Smith, E. S. 1953. 2459
Binomial, normal and Poisson probabilities. Bel Air, Maryland. Publ. by the author, pp. 71.
Review: MR **14** (1953), 887.
Notes: Addenda; 1954.

Smith, J. H. G., and Ker, J. W. 1957. 2460
Some distributions encountered in sampling forest stands. *Forest Sci.*, **3**, 137–44.
Class.: P-NB:mi:BM

Smith, K. 1961. 2461
On the 'best' values of the constants in frequency distributions. *Biometrika*, **11**, 262–76.

Smith, W. L. 1957. 2462
On renewal theory, counter problems, and quasi-Poisson processes. *Proc. Cambridge Philos. Soc.*, **53**, 175–93.

Smith, W. L. 1958. 2463
Renewal theory and its ramifications. *J. Roy. Statist. Soc. Ser. B*, **20**, 243–302.
Users: Page, E. S. (1959), Conolly, B. W. (1960), Jewell, W. S. (1960), Weiss, G. H. (1963), Morgan, R. W., and Welsh, D. J. A. (1965).

Smith, W. L., and Cox, D. R.
See Cox, D. R., and Smith, W. L. (1957).

Smyly, W. J. P. 1952. 2464
The entomostraca of the weeks of a mooreland pond. *J. Animal Ecol.*, **21**, 1–11.
Class.: P:ie:BM

Sneyers, R. 1964. 2465
On the precision of measuring radioactivity in a body (French). *Rev. Bèlge Statist. Rech. Opérat.*, **5**, 19–31.
Review: STMA **6** (1965), 888.
Class.: P:pe:P

Sobel, Milton. 1960. 2466
Group testing to classify efficiently all units in a binomial sample. *Information and Decision Processes*, McGraw-Hill, New York, pp. 127–61.
Review: MR **24**A (1962), 461.

Sobel, M. 1966. 2467
Binomial and hypergeometric group-testing (Abstract). *Ann. Math. Statist.*, **37**, 1865.

Sobel, M., and Bechhofer, R. E.
See Bechhofer, R. E., and Sobel, M. (1956).

Sobel, Milton, and Cacoullos, T.
See Cacoullos, T., and Sobel, Milton (1966).

Sobel, M., and Croll, P. A. 1959. 2468
Group testing to eliminate efficiently all defectives in a binomial sample. *Bell System Tech. J.*, **38**, 1179–1252.

Sobel, Milton, and Groll, Phyllis A. 1966. 2469
Binomial group-testing with an unknown proportion of defectives. *Technometrics*, **8**.

Sobel, M., and Gupta, S. S.
See Gupta, S. S., and Sobel, M. (1958).

Sobel, M., and Gupta, S. S.
See Gupta, S. S., and Sobel, M. (1960).

Sobel, M., and Huyett, M. J. 1957. 2470
Selecting the best one of several binomial populations. *Bell System Tech. J.*, **36**, 537–76.
Review: SA **60**A (1957), Ab.No. 4967.
Users: Gupta, S. S., Huyett, M. J., and Sobel, M. (1957), Gupta, S. S., and Sobel, M. (1960), Taylor, R. J., and David, H. A. (1962), Gupta, S. S. (1965).

Sobel, M., and Olkin, Ingram.
See Olkin, Ingram, and Sobel, M. (1965).

Sobel, M., Gupta, S. S., and Huyett, J.
See Gupta, S. S., Huyett, J., and Sobel, M. (1957).

Soest, J. L. Van. 1960. 2471
Order and disorder. *Statistica Neerlandica*, **14**, 249–58.
Review: IJA **2** (1961), 583.
Class.: MI:mi:G

Solari, Mary E. 1963. 2572
The distribution of the Chi Square test of fit statistic. *The Statistician*, **13**, 263–67.
Review: STMA **6** (1965), 95.

Solomon, M. J. 1954. 2473
Optimum operation of a complex activity under conditions of uncertainty. *J. Operations Res. Soc. Amer.*, **2**, 419–32.

Solow, Robert M. 1960. 2474
On a family of lag distributions. *Econometrica*, **28**, 393–406.
Review: ZBL **97** (1962), 347.

Somerville, P. N. 1957. 2475
Optimum sampling in binomial populations. *J. Amer. Statist. Assoc.*, **52**, 494–502.
Review: MR **19** (1958), 991; ZBL **84** (1960), 152.

Soper, H. E. 1914–15. 2476
Tables of Poisson's exponential binomial limit. *Biometrika*, **10**, 23–35.
User: Utida, S. (1943a).
Class.: P:tc:G

Špaček, Antonín. 1950. 2477
Sampling plans for per cent defective, which minimize the maximum of a given risk function. *Časopis Mat. Fys.*, **74** (4), 307–09.
Review: ZBL **39** (1951), 356.

Spicer, C. C., and Armitage, P.
See Armitage, P., and Spicer, C. C. (1956).

Springer, Melvin D., and Thompson, William E. 1965. 2478
Confidence limits for the product of *N* binomial parameters (Abstract). *Ann. Math. Statist.*, **36**, 1323.

Springer, M. D., and Thompson, W. E. 1966. 2479
Bayesian confidence limits for the product of *N* binomial parameters. *Biometrika*, **53**, 611–13.
Class.: B:ie:G

Sprott, D. A. 1957. 2480
Probability distributions associated with distinct hits on targets. *Bull. Math. Biophys.*, **19**, (3), 163–70.
Review: MR **19** (1958), 587; BA **32** (1958), Ab.No. 11163.
Users: Barton, D. E., and David, F. N. (1959d), Blyth, C. R., and Curme, G. L. (1960).
Class.: MI:mb-m-osp:G

Sprott, D. A. 1958. 2481
The method of maximum likelihood applied to the Poisson binomial distribution. *Biometrics*, **14**, 97–106.
Review: MR **20** (1959), 814; PA **33** (1959), Ab.No. 7318; ZBL **81** (1959), 139.
Users: Katti, S. K. (1960d), Katti, S. K., and Gurland, J. (1960b), Shumway, R., and Gurland, J. (1960a), Shumway, R. and Gurland, J. (1960b), Shumway, R., and Gurland, J. (1960c), Khatri, C. G. (1961), Khatri, C. G., and Patel, I. R. (1961), Katti, S. K., and Gurland, J. (1962a), Martin, D. C., and Katti, S. K. (1962b), Martin, D. C., and Katti, S. K. (1962c), Blischke, W. R. (1965), Katti, S. K., and Sly, L. E. (1965), Kemp, C. D., and Kemp, A. W. (1965), Martin, D. C., and Katti, S. K. (1965), Sprott, D. A. (1965b).
Class.: COB:pe-gf:G

Sprott, D. A. 1965a. 2482
Some comments on the question of identifiability of parameters raised by Rao. *Classical and Contagious Discrete Distributions*, Ed. G. P. Patil, Statistical Publishing Society, Calcutta and Pergamon Press, pp. 333–36. Reprinted in *Sankhyá A*, **27** (1966).
Class.: MI-B-P-LS-NB:osp:G

Sprott, D. A. 1965b. 2483
A class of contagious distributions and maximum likelihood estimation. *Classical and Contagious Discrete Distributions*, Ed. G. P. Patil, Statistical Publishing Society, Calcutta and Pergamon Press, pp. 337–51. Reprinted in *Sankbyá A*, **27** (1966).
User: Sprott, D. A. (1965a).
Class.: MI:osp-pe:G

Sprott, D. A., and Kalbfleish, J. G. 1965. 2484
Use of the likelihood function in inference. *Psychol. Bull.*, **64**, 15–22.
Class.: B:pe-tp:G-BM

Sproule, R. N. 1962. 2485
The Poisson distribution. Master's Thesis, McGill Univ.

Sprowls, R. C. 1950. 2486
Statistical decisions by the method of minimum risk: an application. *J. Amer. Statist. Assoc.*, **45**, 238–248.
Review: ZBL **38** (1951), 99.
Class.: B:tp:G

Srinath, Mandyam D., Yeh, Andrew T., and Fu, Yumin.
See Fy, Yumin, Srinath, Mandyam D., and Yeh, Andrew T. (1965).

Srinivasan, S. K., and Ramakrishnan, Alladi.
See Ramakrishnan, Alladi, and Srinivasan, S. K. (1958) .

Srinivasan, T. R. 1956. 2487
Spells of abnormally cold and hot days at Poona. *Indian J. Meteorol. Geol.*, **7**, 43–48.
User: Tikka, R. N. (1962).
Class.: G-LS:gf:P

Srivastva, J. N. 1957. 2488
The number of red hot lesions on the mid-rib of sugarcane leaves. *J. Indian Soc. Agri. Statist.*, **9**, 52–60.

Śródka, T. 1963-64. 2489
A recursive formula for the ordinary moments in the Polya distribution (Polish; Russian and English summaries). *Prace Mat.*, **8**, 217–20.
Review: MR **30** (1965), 995.
Class.: PO-B-H-P:m:G

Staff, P. J. 1964. 2490
The displaced Poisson distribution. *Austral. J. Statist.*, **6**, 12–20.
Review: MR **32** (1966), 1125; STMA **6** (1965), 74.
Class.: OPR:pc-mb-m-pe-osp-gf:G-BM

Stam, A. J. 1960. 2491
Some mathematical remarks on information theory (Dutch). *Statistica Neerlandica*, **14**, 259–65.
Review: IJA **2** (1961), 585.
Class.: MI:mi:G

Stanbury, F. A., and Barnes, H.
See Barnes, H., and Stanbury, F. A. (1951).

Stancu, D. D. 1964. 2492
On the moments of some discrete random variables (Romanian; Russian and English summaries). *Studia Univ. Babeş-Bolyai Ser. Math.-Phys.*, **9** (2), 35–48.
Review: MR **31** (1966), 142.
Class.: B-P-MI:m:G

Stange, K. 1964. 2493
Die Auflösung für den einfachen exponentiellen Bedienungskanal (mit beliebig vielen Warteplätzen), der für $t = 0$ leer ist. *Unternehmensforschung*, **8**, 1–24.
Review: ZBL **113** (1965), 340.

Stange, K. 1965. 2494
A generalisation of the graphic method for hypothesis testing by binomial probability paper of Mosteller and Tukey to three dimensions (German). *Qualitätskontrolle*, **10**, 45–52.
Review: STMA **7** (1966), 341.
Class.: B:tp:G

Stark, R. W. 1952. 2495
Sequential sampling of the lodgepole needle miner. *For. Chron.*, **28**, 57–60.
Users: Waters, W. E., and Henson, W. R. (1959).

Steck, G. P. 1964. 2496
Approximations for the binomial and hypergeometric distributions (Abstract). *Technometrics*, **6**, 124.

Steck, G. P., Young, D. A., Owen, D. B., and Gilbert, E. J.
See Owen, D. B., Gilbert, E. J., Steck, G. P., and Young, D. A. (1959).

Steffensen, J. F. 1923. 2497
Factorial moments and discontinuous frequency-functions. *Skand. Aktuarietidskr.*, **6**, 73–89.
User: Kullback, S. (1937).

Stehn, J. R., and Erickson, R. O.
See Erickson, R. O., and Stehn, J. R. (1945).

Stein, C., and Lehmann, E. L.
See Lehmann, E. L., and Stein, C. (1949).

Steinhaus, H. 1957. 2498
The problem of estimation. *Ann. Math. Statist.*, **28**, 633–48.
Review: ZBL **88** (1961), 355.

Steinhaus, H., and Urbanik, K. 1959. 2499
Poissonsche Folgen. *Math. Z.*, **72**, 127–45.
Review: IJA **2** (1961), 213; ZBL **92** (1962), 338.
Class.: P:pc:G

Steinhaus, H., and Zubrzycki, S. 1957. 2500
On the comparison of two production processes and the rule of dualism. *Colloq. Math.*, **5**, 103–15.
Review: MR **19** (1958), 1205; ZBL **82** (1960), 131.
Class.: P-B:ctp:E

Steinhaus, H., and Zubrzycki, S. 1958. 2501
On comparing two production processes and on the principle of dualism (Polish; Russian and English summaries). *Zastos. Mat.*, **3**, 229–57.
Review: ZBL **101** (1963), 353.

Stephen, F. F. 1945. 2502
The expected value and variance of the reciprocal and other negative powers of a positive Bernoullian variate. *Ann. Math. Statist.*, **16**, 50–61.
Review: MR **6** (1945), 232.
Users: Chapman, D. G. (1951), David, F. N., and Johnson, N. L. (1953), Grab, E. L., and Savage, R. I. (1954), Okamoto, M. (1955), David F. N., and Johnson, N. L. (1956), Patil, G. P. (1959), Mendenhall, W., and Lehman, E. H. (1960), Sen, P. K. (1960), Bennett, B. M. (1962a), Govindarajulu, Z. (1962c), Govindarajulu, Z. (1962e), Rider, P. R. (1962b), Govindarajulu, Z. (1963), Tiku, M. (1964).
Class.: TCB-OHR:m:G

Stephens, K. S., and Dodge, H. F.
See Dodge, H. F., and Stephens, K. S. (1965).

Sterne, T. E. 1954. 2503
Some remarks on confidence or fiducial limits. *Biometrika*, **41**, 275–78.
Users: Crow, E. L., (1956), Stevens, W. L. (1957), Pratt, J. W. (1961).
Class.: B:ie:G

Stevens, W. L. 1937. 2504
Significance of grouping. *Ann. Eugenics*, **8**, 57–69.
Users: Tukey, J. W. (1949a), David, F. N. (1950), Thomas, M. (1951), Craig, C. C. (1954a), Irwin, J. O. (1955), Thompson, K. R. (1955), Rao, C. R., and Chakravarti, I. (1956), Barton, D. E. (1957), Sprott, D. A. (1957), Barton, D. E., and David, F. N. (1959), Barton, D. E., and David, E. N. (1959a), Chakravarti, I., and Rao, C. R. (1959), Jones, H. L. (1959), Barton, D. E., David, F. N., and Merrington, M. (1960), Nicholson, W. L. (1960), Good, I. J. (1961c), Nicholson, W. L. (1961), Locher, M. P. (1963).

Stevens, W. L. 1950. 2505
Fiducial limits of the parameter of a discontinuous distribution. *Biometrika*, **37**, 117–29.
Review: MR **12** (1951), 37; ZBL **37** (1951), 367.
Users: Pearson, E. S. (1950), Lancaster, H. O. (1952), Sandelius, M. (1952a), Abdel-Aty, S. H. (1954), David, F. N. (1955a), Buehler, R. J. (1957), Stevens, W. L. (1957), Clunies-Ross, C. W. (1958), Crow, E. L. (1959), Blyth, C. R., and Hutchinson, D. W. (1960), Cacoullos, T. (1962), Thatcher, A. R. (1964).
Class.: B:ie:G

Stevens, W. L. 1957. 2506
Shorter intervals for the parameter of the binomial and Poisson distribution. *Biometrika*, **44**, 436–40.
Review: MR **19** (1958), 780; ZBL **82** (1960), 348.
User: Crow, E. L. (1959).
Class.: P:ie:G

Stewart, C., Folks, J. L., and Pierce, D. A.
See Folks, J. L., Pierce, D. A., and Stewart, C. (1965).

Steyn, H. S. 1951. 2507
On discrete multivariate probability functions. *Nederl. Akad. Wetensch. Proc. Ser. A*, **54**, (Indag. Math. 13), 23–30.
Review: MR **12** (1951), 722; ZBL **43** (1952), 339.
Users: Wiid, A. J. B. (1957-58), Patil, G. P., and Bildikar, S. (1966a).
Class.: MH-MIH:m-rp:G

Steyn, H. S. 1955. 2508
On discrete multivariate probability functions of hypergeometric type. *Nederl. Akad. Wetensch. Proc. Ser. A*, **58**, 588–95.
Review: MR **17** (1956), 634; ZBL **68** (1957), 127.
Users: Steyn, H. S. (1959), Patil, G. P., and Bildikar, (1966a).
Class.: MH-MIH:m-a-osp:G

Steyn, H. S. 1956. 2509
On the univariable series $F(t) = F(a;\ b_1, b_2,.., b_k;\ c;\ t, t^2,.., t^k)$ and its applications in probability theory. *Nederl. Akad. Wetensch. Proc. Ser. A*, **59**, (Indag. Math. 18), 190–97.
Review: MR **17** (1956), 981.
Class.: MH-MIH:m-a:G

Steyn, H. S. 1957. 2510
On regression properties of discrete systems of probability functions. *Nederl. Akad. Wetensch. Proc. Ser. A*, **60**, 119–27.
Review: ZBL **84** (1960), 140.
Users: Patil, G. P., and Bildikar, S. (1966a).
Class.: MI-M-NM-MH-MIH:rp:G

Steyn, H. S. 1959. 2511
On χ^2-tests for contingency tables of negative multinomial types (Dutch summary). *Statistica Neerlandica*, **13**, 433–44.
Review: MR **24A** (1962), 456; IJA **1** (1960), 419.
Users: Bartko, J. J. (1961b), Bennett, B. M. (1962).
Class.: NM:tp:G

Steyn, H. S. 1963. 2512
On approximations for the distributions obtained from multiple events. *Nederl. Akad. Wetensch. Proc. Ser. A*, **66**, 85–96.
Users: Patil, G. P., and Bildikar, S. (1966a).
Class.: N-NM-MH-MIH:a:G

Steyn, H. S., and Wiid, A. J. B.
See Wiid, A. J. B., and Steyn, H. S. (1956).

Steyn, H. S., and Wiid, A. J. B. 1958. 2513
On eightfold probability functions. *Nederl. Akad. Wetensch. Proc. Ser. A*, **61**, 129–38.
Review: ZBL **84** (1960), 358.
Users: Patil, G. P., and Bildikar, S. (1966a).
Class.: MH-MB-MNB:m-a-rp:G

Störmer, H. 1960. 2514
A queueing problem in telephone exchanges (German). *A. Angew. Math. Mech.*, **40**, 236–46.
Review: IJA **2** (1961), 584.
Class.: MI:pc:G

Strackee, J., and Van der Gon Denier, J. J. 1962. 2515
The frequency distribution of the difference between two Poisson variates. *Statistica Neerlandica*, **16**, 17–23.
Review: IJA **4** (1963), 41.
Class.: P:mi-tc-a:G

Stratton, H. H., Jr., and Tucker, H. G. 1964. 2516
Limit distributions of a branching stochastic process. *Ann. Math. Statist.*, **35**, 557–65.

Strauch, H. 1961. 2517
Continuous sampling inspection from the belt (German). *Qualitätskontr. Operat. Res.*, **6**, 21–25.
Review: IJA **2** (1961), 715.
Class.: B:sqc:E

Stuart, A. 1957. 2518
A singularity in the estimation of binomial variance. *Biometrika*, **44**, 262–64.
Review: ZBL **78** (1959), 335.
Class.: B:pe:G

Stuart, A., and Kendall, D. G.
See Kendall, D. G., and Stuart, A. (1958).

" Student". 1907. 2519
On the error of counting with a haemocytometer. *Biometrika*, **5**, 351–55.
Users: Soper, H. E. (1914-15), Whitaker, L. (1914), Thorndike, F. (1926), Luders, R. (1934), Matuszewsky, T., Supinska, J., and Neyman, J. (1936), Neyman, J. (1939), Katz, L. (1945), Cole, L. C. (1946), Cole, L. C. (1946a), Chamberlain, A. C., and Turner, F. (1952), Skellam, J. G. (1952), Bliss, C. I., and Fisher, R. A. (1953), Rutherford, R. S. G. (1954), Turner, M. E., and Eadie, G. S. (1957), Jensen, P. (1959), Griffiths, J. C. (1960), Pielou, E. C. (1960), Bardwell, G. E. (1961), Bartko, J. J. (1962), Cassie, R. M. (1962), Bardwell, G. E., and Crow, E. L. (1964), Crow, E. L., and Bardwell, G. E. (1965), Kemp, C. D., and Kemp, A. W. (1965).
Class.: B-P:mb:G-BM

" Student ". 1919. 2520
An explanation of deviations from Poisson's law in practice. *Biometrika*, **12**, 211–15.
Users: Thorndike, F. (1926), Bowen, M. F. (1947), Hoel, P. G. (1947), Bliss, C. I., and Fisher, R. A. (1953), David, F. N., and Moore, P. G. (1954), Turner, M. E., and Eadie, G. S. (1957), Jensen, P. (1959), Waters, W. E., and Henson, W. R. (1959), Griffiths, J. C. (1960).
Class.: P-B-NB:cm-gf:G

Studer, H. 1966. 2521
Prüfung der Annäherung der exakten χ^2 Verteilung durch die stetige χ^2 Verteilung. *Metrika*, **11**, 55–78.
Class.: B-M:gf:G

Subrahmaniam, K. 2522
On a general class of contagious distributions: the Pascal-Poisson distribution. Paper 356, Dept. Biostatist., Johns Hopkins Univ., pp. 21.
Class.: GP-GB-COB-CONB:mb-a-gf:G-BM

Subrahmaniam, K. 1963. 2523
A test for " intrinsic " correlation in the theory of accident proneness (Abstract). *Ann. Math. Statist.*, **34**, 1628.

Subrahmaniam, K. 1964. 2524
On a general class of contagious distributions and Pascal-Poisson distribution (Abstract). *Ann. Math. Statist.*, **35**, 462.

Subrahmaniam, K. 1965a. 2525
On a property of the binomial distribution (Abstract). *Ann. Math. Statist.*, **36**, 1086.

Subrahmaniam, K. 1965b. 2526
A note on estimation in the truncated Poisson. *Biometrika*, **52**, 279–82.
Review: STMA **6** (1965), 1148.
Class.: TCP:pe:G

Subrahmaniam, K. 1966. 2527
A test for " intrinsic correlation " in the theory of accident proneness. Paper 355, Dept. Biostatist. Johns Hopkins Univ. pp. 16.
Class.: MB:mb-tp-a:A

Subrahmaniam, K. 1966a. 2528
A test for intrinsic correlation in the theory of accident proneness. *J. Roy. Statist. Soc. Ser. B*, **28**, 180-89, Corrigenda: **28**, 585.
Class.: MP-MNB:mb-mi:G-A

Subrahmaniam, K. 1966b. 2529
On the general class of contagious distributions: The Pascal-Poisson distribution. *Trabajos Estadist.*, **17**, 109-28.
Class.: GB-GP-GNB:c-a-gf:G-BM

Subrahmaniam, K., and Saleh, A. K. MD. E. 1965. 2530
Decomposition of a mixture of two Poisson distributions (Abstract). *Ann. Math. Statist.*, **36**, 1087.

Sugimori, Makato. 1961. 2531
Binomial probabilistic sequential circuits (Japanese). *Elec. Comm. Lab. Tech. J.*, **10**, 657–81.
Review: MR **24B** (1962), 237.

Sugino, T., and Kono, T.
See Kono, T., and Sugino, T. (1958).

Sugiyama, H. 1952. 2532
On the asymptotic behaviour of $\Sigma p_m{}^2$ in case of certain probability distributions. I. *Math. Japon.*, **2**, 187–92.
Review: ZBL **41** (1953), 360.
User: Sugiyama, H. (1955).
Class.: B-P-MI:mi:G

Sugiyama, H. 1955. 2533
On the asymptotic behaviour of $\Sigma p_m{}^2$ in case of certain probability distributions. II. *Math. Japon.*, **3**, 121–26.
Review: ZBL **66** (1957), 375.
Class.: B-P-M:mi:G

Sukhatme, P. V. 1937. 2534
The problem of *K* samples for Poisson population. *Proc. National Inst. Sci. India*, **3**, 297–305.
User: Bennett, B. M. (1959).

Sukhatme, P. V. 1938. 2535
On the distribution of χ^2 in samples of the Poisson series. *J. Roy. Statist. Soc. Supple.* **5**, 75–79.

Users: Cochran, W. G. (1940), Hoel, Paul G. (1943), Seal, H. L. (1949), Lancaster, H. O. (1950), Lancaster, H. O. (1952), Cochran, W. G. (1954), Okamoto, M. (1955), Rao, C. R., and Chakravarti, I. (1956), Bennett, B. M. (1959), Chakravarti, I. M., and Rao, C. R. (1959), Bennett, B. M. (1962a), Nicholson, W. L. (1963).

Sukhatme, P. V., and Panse, V. G. 1943. 2536
Size of experiments for testing sera or vaccines. *Indian J. Vet. Sci. and Animal Husbandry*, **13**, 75–82.

Sukhatme, S. B. 1960. 2537
Nonparametric tests for location and scale parameters in a mixed model with discrete and continuous variables (Abstract). *Ann. Math. Statist.*, **31**, 529.

Sukhatme, S. 1962. 2538
Some non-parametric tests for location and scale parameters in a mixed model of discrete and continuous variables. *J. Indian Soc. Agric. Statist.*, **14**, 121–37.
Review: MR **29** (1965), 1006.

Sukhatme, S. 1964. 2539
A c-sample non-parametric test for location in a mixed model of continuous and discrete variables. *J. Indian Soc. Agri. Statist.*, **16**, 202–11.
Review: STMA **7** (1966), 527.

Suranyi, J. 1956. 2540
On a problem of old Chinese mathematics. *Publ. Math. Debrecen*, **4**, 195–97.
Review: MR **18** (1957), 4.
Class.: MI:mi:G

Sutcliffe, M. I., Mainland, D., and Herrera, L.
See Mainland, D., Herrera, L., and Sutcliffe, M. I. (1956).

Suter, Glenn W. 1951. 2541
Theory of regression for a binomial distributed variate. Master's Thesis, Virginia Polytechnic Inst.

Suzuki, Yukio, and Govindarajulu, Z.
See Govindarajulu, Z., and Suzuki, Yukio (1963).

Svedberg, The. 1922. 2542
Ett Bidrag till de statistiska metodernas användning inom växtbiologien. *Svensk. Botanisk Tidskrift*, **16**, 1–8.

Sverdrup, E. 1951. 2543
Om punktestimering av sannsynligheter på grunnlag av et tilfeldig utvalg (On point-estimation of probabilities on the basis of random sample). Memo., Univ. Oslo, Inst. Econ., pp. 18.

Swineford, Frances. 1948. 2544
A table for estimating the significance of the difference between correlated percentages. *Psychometrika*, **13**, 23–25.
Review: BA **22** (1948), 19898.
Class.: MB:tp-tc:G

Sylwester, D. L., Gaffey, W. R., Manheimer, D. I., and Mellinger, G. D.
See Mellinger, G. D., Sylwester, D. L., Gaffey, W. R., and Manheimer, D. I. (1965).

Sylwester, David, and Boen, J. R.
See Boen, J. R., and Sylwester, David (1966).

T

Taga, Yasushi. 1964. 2545
On high order moments of the number of renewals. *Ann. Inst. Statist. Math. Tokyo*, **15**, 187–96.

Takács, L. 1951. 2546
Occurrence and coincidence phenomena in case of happenings with arbitrary distribution law of duration. *Acta. Math. Acad. Sci. Hungar.*, **2**, 275–98.

Takács, L. 1954. 2547
On secondary processes generated by a Poisson process and their applications in physics. *Acta. Math. Acad. Sci. Hungar.*, **5**, 203–36.
User: Takács, L. (1956b).

Takács, L. 1955. 2548
On processes of happenings generated by means of a Poisson process. *Acta. Math. Acad. Sci. Hungar.*, **6**, 81–99.
User: Takács, L. (1958b).

Takács, L. 1955a. 2549
On stochastic processes connected with certain physical recording apparatuses. *Acta. Math. Acad. Sci. Hungar.*, **6**, 363–380.

Takács, L. 1956a. 2550
On secondary stochastic processes generated by recurrent processes. *Acta. Math. Acad. Sci. Hungar.*, **7**, 17–29.
User: Takács, L. (1956b).

Takács, L. 1956b. 2551
On the generalization of Erlang's formula. *Acta. Math. Acad. Sci. Hungar.*, **7**, 419–33.
User: Takács, L. (1957c), Takács, L. (1958c).
Class.: MI:pc:G

Takács, L. 1956c. 2552
On a probability problem arising in the theory of counters. *Proc. Cambridge Philos. Soc.*, **52**, 488–98.
User: Takács, L. (1958b).

Takács, L. 1956d. 2553
On the sequence of events, selected by a counter from recurrent process events. *Theor. Probability Appl.*, **1**, 90–102.
User: Anselone, Philip M. (1960).

Takács, L. 1957a. 2554
Uber die Wahrscheinlichkeitstheoretische Behandlung der Anodenstromschwankungen von Elektronenrokren. *Acta. Phys. Acad. Sci. Hungar.*, **7**, 25–50.

Takács, L. 1957b. 2555
On some probability problems concerning the theory of counters. *Acta. Math. Acad. Sci. Hungar.*, **8**, 127–38.

Takács, L. 1957c. 2556
On a probability problem concerning telephone traffic. *Acta. Math. Acad. Sci. Hungar.*, **8**, 319–24.

Takács, L. 1957d. 2557
On a queueing problem concerning telephone traffic. *Acta. Math. Acad. Sci. Hungar.*, **8**, 325–35.
Users: Finch, P. D. (1959a), Finch, P. D. (1959b).

Takács, L. 1957e. 2558
On secondary stochastic processes generated by a multi-dimensional Poisson process (Hungarian). *Magyar Tud. Akad. Mat. Kutató Int. Közl.*, **2**, 71–80.
Review: ZBL **87** (1961), 333.
User: Prékopa, A. (1958).

Takács, L. 1957f. 2559
On a stochastic process concerning some waiting time problems (Russian summary). *Teor. Verojatnost. i Primenen*, **2**, 92–105.
Review: ZBL **81** (1959), 133.
User: Takács, L. (1957c).

Takács, L. 1958a. 2560
On a coincidence problem concerning telephone traffic. *Acaa. Math. Acad. Sci. Hungar.*, **9**, 45–81.
Review: ZBL **85** (1961), 126.

Takács, L. 1958b. 2561
On a probability problem in the theory of counters. *Ann. Math. Statist.*, **29**, 1257–63.

Takács, L. 1958c. 2562
On a combined waiting time and loss problem concerning telephone traffic (Hungarian). *Ann. Univ. Sci. Budapest Sec. Math.*, **1**, 73–82.
Review: IJA **1** (1959), 144.

Takács, L. 1958d. 2563
On a general probability theorem and its application in the theory of the stochastic processes. *Proc. Cambridge Philos. Soc.*, **54**, 219–24.
Review: ZBL **80** (1959), 352.

Takács, L. 1961. 2564
Stochastic processes with balking in the theory of telephone traffic. *Bell System Tech. J.*, **40**, 795–820.

Takács, L. 1961a. 2565
Charles Jordan, 1871-1959. *Ann. Math. Statist.*, **32**, 1–11.
Class.: MI:mi:G

Takács, L. 1961b. 2566
On a coincidence problem concerning particle counters. *Ann. Math. Statist.*, **32**, 739–56.

Takács, L. 1961c. 2567
The transient behaviour of a single server queueing process with recurrent input and Gamma service time. *Ann. Math. Statist.*, **32**, 1286–98.
Review: STMA **5** (1964), 984; ZBL **111** (1964-65), 331.
User: Takács, L. (1963a).
Class.: P-MI:pc:G

Takács, L. 1961d. 2568
Transient behaviour of single-server queueing processes with Erlang input. *Trans. Amer. Math. Soc.*, **100**, 1–28.
Review: MR **31** (1966), 957.
User: Takács, L. (1963a).

Takács, L. 1962. 2569
A generalization of the ballot problem and its application in the theory of queues. *J. Amer. Statist. Assoc.*, **57**, 327–37.
User: Takács, L. (1962).
Class.: P-MI:pc-mi:G

Takács, L. 1962a. 2570
A combinatorial method in the theory of queues. *J. Soc. Indust. Appl. Math.*, **10**, 691–94.
Review: ZBL **118** (1965–66), 135.

Takács, L. 1963a. 2571
The stochastic law of the busy period for a single-server queue with Poisson input. *J. Math. Analysis Appl.*, **6**, 33–42.
Review: ZBL **117** (1965), 360.

Takács, L. 1964. 2572
Combinatorial methods in the theory of queues (French summary). *Rev. Inst. Internat. Statist.*, **32** 207–19.
Review: MR **31** (1966), 499.

Takács, L. 1965. 2573
A moment problem. *J. Austral. Math. Soc.*, **5**, 487–90.
Review: STMA **7** (1966), 470.
Class.: MI:m:G

Takács, L., and Renyi, Alfred.
See Rényi, Alfred, and Takács, L. (1952).

Takahasi, K. 1961. 2574
Model for the estimation of the size of a population by using capture-recapture method. *Ann. Inst. Statist. Math.*, **12**, 237–48.
User: Holgate, P. (1966a).

Tallis, G. M. 1962. 2575
The use of a generalized multinomial distribution in the estimation of correlation in discrete data. *J. Roy. Statist. Soc. Ser. B*, **24**, 530–34.
Review: MR **26** (1963), 1344; ZBL **114** (1965), 116; STMA **5** (1964), 913.
Users: Tallis, G. M. (1964), Patil, G. P., and Bildikar, S. (1966a).
Class.: OMR:mb-pe:G

Tallis, G. M. 1964. 2576
Further models for estimating correlation in discrete data. *J. Roy. Statist. Soc. Ser. B*, **26**, 82–85.
Review: MR **29** (1965), 1233; STMA **7** (1966), 139.
Class.: OMR-GP:osp:G

Tanaka, Masao. 1961–62. 2577
On a confidence interval of given length for the parameter of the binomial and the Poisson distributions. *Ann. Inst. Statist. Math.*, **13**, 201–15.
Review: MR **26** (1963), 1343; STMA **5** (1964), 83.

Tanner, J. C. 1951. 2578
The delay to pedestrians crossing a road. *Biometrika*, **38**, 383–92.
Users: Adams, W. F. (1936), Garwood, F. (1940), Raff, M. S. (1951), Miller, A. J. (1961), Weiss, G. H. (1963), Haight, F. (1965b).

Tanner, J. C. 1953. 2579
A problem of interference between two queues. *Biometrika*, **40**, 58–69.
Users: Haight, F. (1959), Haight, F., and Breuer, M. (1960), Haight, F. (1961a), Miller, A. J. (1961), Takács, L. (1962), Takács, L. (1963a).

Tanner, J. C. 1958. 2580
A problem in the combination of accident frequencies. *Biometrika*, **45**, 331–42.
Review: IJA **1** (1959), 59; ZBL **88** (1961), 123.
Class.: MI:tp:A

Tanner, J. C. 1961. 2581
A derivation of Borel-distribution. *Biometrika*, **48**, 222–24.
Users: Takács, L. (1962), Takács, L. (1963a).
Class.: MI:pc:G

Tanner, J. C., and Garwood, F.
See Garwood, F., and Tanner, J. C. (1956).

Tanton, M. T. 1965. 2582
Problems of live trapping and population estimation for the wood mouse, *Apodemus sylvaticus* (L). *J. Animal Ecol.*, **34**, 1–22.
User: Holgate, P. (1966a).

Tate, Merle W. 1951. 2583
A note on common mistakes in testing significance of a proportion. *J. Educ. Res.*, **44**, 551–53.
Review: PA **26** (1952), Ab.No. 41.

Tate, Robert F. 1954. 2584
Correlation between a discrete and a continuous variable. Point-Biserial correlation. *Ann. Math. Statist.*, **25**, 603–07.
Review: ZBL **56** (1955), 367.
Users: Olkin, I., and Tate, R. F. (1961).

Tate, R. F., and Goen, R. L. 1958. 2585
Minimum variance unbiased estimation for the truncated Poisson distribution. *Ann. Math. Statist.*, **29**, 755–65.
Review: MR **20** (1959), 63; ZBL **86** (1961), 354.
Users: Cohen, A. C. (1959e), Cohen, A. C. (1960b), Cohen, A. C. (1960e), Cacoullos, T. (1961), Cohen, A. C. (1961a), Patil, G. P. (1961b), Cacoullos, T. (1962), Hughes, E. J. (1962), Patil, G. P. (1963a), Bardwell, G. E., and Crow, E. L. (1964), Crow, E. L., and Bardwell, G. E. (1965), Tweedie, M. C. K. (1965).
Class.: TCP:pe:G

Tate, R. F., and Olkin, Ingram. 1960. 2586
Multivariate correlation models with mixed discrete and continuous variables (Summary). *J. Amer. Statist. Assoc.*, **55**, 373.
Users: Patil, G. P., and Bildikar, S. (1966a).

Tate, R. F., and Olkin, Ingram.
See Olkin, Ingram, and Tate, R. L. (1961).

Taylor, C. J. 1961. 2587
The application of the negative binomial distribution to stock control problems. *Operational Research Quarterly*, **12**, 81–88.
Class.: NB:mi:E

Taylor, L. R. 1961. 2588
Aggregation, variance and the mean. *Nature*, **189**, 732–35.
User: Harcourt, D. G. (1965).

Taylor, R. J. 1955. 2589
On the use of an auxiliary variable in the transformation of discrete data. Master's Thesis, Virginia Polytechnic Inst.

Taylor, R. J., and David, H. A. 1962. 2590
A multi-stage procedure for the selection of the best of several populations. *J. Amer. Statist. Assoc.*, **57**, 785–96.
Class.: B:srp:BM

Taylor, W. F. 1956. 2591
Problems in contagion. *Proc. 3rd Berkeley Symp. Math. Statist. Prob.*, **4**, 167–79.
Review: ZBL **70** (1957), 151.
Class.: MI:mb:A

Teicher, H. 1952. 2592
On the multivariate Poisson distribution (Abstract). *Ann. Math. Statist.*, **23**, 144.

Teicher, H. 1954a. 2593
On the factorization of distributions. *Ann. Math. Statist.*, **25**, 769–74.
Review: MR **16** (1955), 377; ZBL **56** (1955), 359.
Users: Dwass, M., and Teicher, H. (1957), Teicher, H. (1958a), Ramachandran, B. (1961a), Edwards, C. B. (1962).
Class.: OBR-M:mi:G

Teicher, H. 1954b. 2594
On the convolution of distributions. *Ann. Math. Statist.*, **25**, 775–78.
Review: MR **16** (1955), 377.
Users: Crow, E. L. (1958a), Teicher, H. (1960a), Teicher, H. (1961), Jacobsen, Robert Leland (1966).
Class.: MI-P:mi:G

Teicher, H. 1954c. 2595
On the multivariate Poisson distribution. *Skand. Aktuarietidskr.*, **37**, 1–9.
Review: MR **17** (1956), 983; ZBL **56** (1955), 358.
Users: Dwass, M., and Teicher, H. (1957), Edwards, C. B., and Gurland, J. (1960), Ahmed, M. S. (1961), Edwards, C. B., and Gurland, J. (1961), Edwards, C. B. (1962), Fuchs, C. E., and David, H. (1965a), Patil, G. P., and Bildikar, S. (1966a).

Teicher, H. 1955. 2596
An inequality on Poisson probabilities. *Ann. Math. Statist.*, **26**, 147–49.
Review: MR **17** (1955), 722; ZBL **64** (1956), 382.
Users: Samuels, S. M. (1964), Samuels, S. M. (1965).
Class.: P:mi:G

Teicher, H. 1958. 2597
On the mixture of distributions. Tech. Rep. 1, Purdue Univ.
Class.: MI-COP:mi:G

Teicher, H. 1958a. 2598
Sur les puissances de fonctions caractéristiques. *C. R. Acad. Sci. Paris*, **246**, 694–96.
Review: ZBL **80** (1959), 120.

Teicher, H. 1960a. 2599
On the mixture of distributions. *Ann. Math. Statist*, **31**, 55–73.
Review: ZBL **107** (1964), 135; IJA **2** (1961), 267.
Users: Daniels, H. E. (1961), Teicher, H. (1961), Blischke, W. R. (1962), Teicher, H. (1963), Blischke, W. R. (1964), Patil, G. P. (1964), Blischke, W. R. (1965), Jacobson, Robert Leland (1966), Patil, G. P. (1966b).
Class.: MI-COP:mi:G

Teicher, H. 1960b. 2600
Identifiability of mixtures (Abstract). *Ann. Math. Statist.*, **31**, 243.

Teicher, H. 1961. 2601
Identifiability of mixtures. *Ann. Math. Statist.*, **32**, 244–48.
Review: IJA **4** (1963), 415.
Users: Blischke, W. R. (1962), Tucker, H. G. (1962a), Teicher, H. (1963), Tucker, H. (1963), Blischke, W. R. (1964), Patil, G. P. (1964), Robbins, H. (1964), Blischke, W. R. (1965), Jacobsen, Robert Leland (1966).

Teicher, H. 1963. 2602
Identifiability of finite mixtures. *Ann. Math. Statist.*, **34**, 1265–69.
Users: Blischke, W. R. (1964), Barndorff-Nielson, O. (1965), Blischke, W. R. (1965), Maritz, J. S. (1966).
Class.: MI-B:mi:G

Teicher, H. 1965. 2603
On random sums of random vectors. *Ann Math. Statist.*, **36**, 1450–58.
Review: MR **31** (1966), 738.

Teicher, H., and Dwass, Meyer.
See Dwass, Meyer, and Teicher, H. (1957).

Teichroew, G. 1955. 2604
Numerical analysis research unpublished statistical tables. *J. Amer. Statist. Assoc.*, **50**, 550–56.
Class.: M-P:tc:G

Terao, S. 1949. 2605
On the distribution of combined pedestrians (Japanese). *J. Apply. Phys.*, **18**,

Tester, John R., and Sinif, Donald B. 1965. 2606
Aspects of animal movement and home range data obtained by telemetry. *Trans. Thirtieth N. Amer. Wildlife and Natural Resources Conference*, pp. 379–92.
Class.: P-G:gf:BM

Teugels, J. 1963-64. 2607
Selective sampling in k-dimensional distributions. *Simon Stevin*, **37**, 55–70.
Review: MR **31** (1966), 501; STMA **6** (1965), 96.
Class.: MI:mi:G

Thatcher, A. R. 1964. 2608
Relationship between Bayesian and confidence limits for predictions. *J. Roy. Statist. Soc. Ser. B*, **26**, 176–210.
Review: MR **30** (1965), 1008; STMA **7** (1966), 91.
Users: Bartholomew, D. J. (1965), Plackett, R. L. (1966).
Class.: B:ie:G

Thedeen, T. 1964. 2609
A note on the Poisson tendency in traffic distribution. *Ann. Math. Statist.*, **35**, 1823–24.

Thiele, T. N. 1931. 2610
The theory of observations. *Ann. Math. Statist.*, **2**, 165–308.
User: Curtiss, J. H. (1941).
Class.: B:mi:G

Thionet, P. 1963. 2611
Sur le moment d'ordre (—1) de la distribution binomiale tronquée application à l'echantillonnage de Hajek.
Review: MR **31** (1966), 149; ZBL **117** (1965), 148.
Class.: B:m-pe:G

Thoday, J. M., Catcheside, D. G., and Lea, D. E.
See Catcheside, D. G., Lea, D. E., and Thoday, J. M. (1945-6).

Thomas, H. A., Jr. 1952. 2612
On averaging results of coliform tests. *J. Boston Soc. Civil Engrs.*, **39**, 253–70.
User: Harris, E. K. (1958).

Thomas, J. B., Ghent, A. W., and Fraser, D. A.
See Ghent, A. W., Fraser, D. A., and Thomas, J. B. (1957).

Thomas, M. 1949. 2613
A generalization of Poisson's binomial limit for use in ecology. *Biometrika*, **36**, 18–25.
Users: Anscombe, F. J. (1950a), Barnes, H., and Marshall, S. M. (1951), Barnes, H., and Stanbury, F. (1951), Johnson, N. L. (1951), Thomas, M. (1951) Cain, S. A., and Evans, F. C. (1952), Skellam, J. G. (1952), Thomson, G. W. (1952), Bliss, C. I., and Fisher, R. A. (1953), MacFadyen, A. (1953), David, F. N., and Moore, P. G. (1954), Moore, P. G. (1954b), Thompson, H. R. (1954), Barton, D. E. (1957), Pielou, E. C. (1957), Bliss, C. I. (1958), Hairston, N. G. (1959), Morisita, M. (1959), Pruess, K. P., and Weaver, C. R. (1959), Waters, W. E., and Henson, W. R. (1959), Pielou, D. P. (1960a), Katti, S. K. (1960d), Pielou, E. C. (1960), Cassie, R. M. (1962), Ito, Nakamura, Kondo, Miyashita, and Nakamura (1962), Tsao, C. M. (1962), Shiyomi, M., and Nakamura, K. (1964), Holgate, P. (1965).
Class.: T:mb-gf:BM

Thomas, M. 1951. 2614
Some tests for randomness in plant populations. *Biometrika*, **38**, 102–11.
Review: ZBL **42** (1952), 380.
Users: Bennett, B. M. (1956), Hetz, and Klinger (1958), Nicholson, W. L. (1961).
Class.: T:tp-h:BM

Thomas, R. E., Roberts, H. R., and McCall, C. H., Jr.
See Roberts, H. R., McCall, C. H., Jr., and Thomas, R. E. (1958).

Thomissen, F. X. 1956. 2615
The frequency of industrial accidents (Dutch summary). *Statistica Neerlandica*, **10**, 163–76.

Thompson, Catherine M. 1941. 2616
Tables of percentage points of the incomplete beta-function. *Biometrika*, **32**, 151–81.
Review: BA **17** (1943), Ab. No. 19512.
Users: Scheffe, H. (1943–46), Wise, E. (1950), Sandelius, M. (1951c), Blom, G. (1954), Goodman, L. A. (1954), Patnaik, P. B. (1954), Crow, E. L. (1956), Bol'shev, L. N. (1960), Leon, F. C., Hayman, G. E., Chu, J. T., and Topp, C. W. (1960).
Class.: B:tc:G

Thompson, H. R. 1954. 2617
A note on contagious distributions. *Biometrika*, **41**, 268–71.
Review: MR **16** (1955), 54; ZBL **55** (1955), 121.
Users: Skellam, J. G. (1958b), Thompson, H. R. (1958).
Class.: NB-N-MI:mb:G

Thompson, H. R. 1956. 2618
Distribution of distance to nth neighbor in a population of randomly distributed individuals. *Ecology*, **37**, 391–94.
Users: Hairston, N. G. (1959), Waters, W. E., and Henson, W. R. (1959), Holgate, P. (1965).
Class.: P:h:BM

Thompson, H. R. 1958. 2619
The statistical study of plant distribution patterns using a grid of quadrats. *Austral, J. Bot.*, **6**, 322–43.
User: Greig-Smith, P. (1961).

Thompson, J. R., and Thomson, G. H.
See Thomson, G. H., and Thompson, J. R. (1915).

Thompson, K. R. 1955. 2620
Spatial point processes, with applications to ecology. *Biometrika*, **42**, 102–15.

Thompson, Keith H. 1962. 2621
Estimation of the proportion of vectors in a natural population of insects. *Biometrics*, **18**, 568–78.
Review: IJA **4** (1963), 721.

Thompson, William E., and Springer, Melvin D.
See Springer, Melvin D., and Thompson, William E. (1965).

Thompson, W. E., and Springer, M. D.
See Springer, M. D., and Thompson, W. E. (1966).

Thomson, G. H., and Thompson, J. R. 1915. 2622
Outlines of a method for the quantatitive analysis of writing vocabularies. *Brit. J. Psychol.*, **8**, 52–69.
Class.: MI:pe:L

Thomson, G. W. 1952. 2623
Measures of plant aggregation based on contagious distributions. *Contri. Lab. Vert. Biol. Univ. Mich.*, **53**, 1–17.
Users: Bliss, C. I., and Fisher, R. A. (1953), Clark, P. J., and Evans, F. C. (1954a), Comita, G. W., and Comita, J. J. (1957), Hairston, N. G. (1959).
Class.: N-T:gf:BM

Thorndike, Frances. 1926. 2624
Applications of Poisson's probability summation. *Bell System Tech. J.*, **5**, 604–24.
Users: Dandekar, V. M. (1955), Duker, S. (1955), Martin, L. (1961).
Class.: P:mi-gf:P-BM-E

Thornton, H. G., Mackenzie, W. A., and Fisher, R. A.
See Fisher, R. A., Thornton, H. G., and MacKenzie, W. A. (1922).

Thyregod, P., and Hald, A.
See Hald, A., and Thyregod, P. (1965).

Tiago de Oliveira, J. 1952a. 2625
Sur le calcul des moments de la réciproque d'une variable aléatoire positive de Bernouilli et Poisson. *An. Fac. Ci. Porto*, **36**, 165–68.
Review: MR **15** (1954), 969; ZBL **49** (1959), 215.

Tiago de Oliveira, J. 1952b. 2626
A note on a special case of inverse binomial sampling. *Separata da revista*, **2** (2), 111–14.
Review: MR **14** (1953), 995.
Class.: G:pe-tp:G

Tiago de Oliveira, J. 1952c. 2627
A note on a special case of inverse binomial sampling. *Univ. Lisboa. Revista Fac. Ci. A. Ci. Mat.*, **2**, 111–14.
Review: MR **14** (1953), 995; ZBL **49** (1959), 101.
Class.: G:pe:tp:G

Tiago de Oliveira, J. 1952d. 2628
Tests for the equality of proportions in a multinomial population. *Univ. Lisboa. Revista Fac. Ci. A. Ci. Mat.*, **2**, 197–200.
Review: MR **14** (1953), 995; ZBL **49** (1959), 100.
Users: Patil, G. P., and Bildikar, S. (1966a).
Class.: M:tp:G

Tiago de Oliveira, J. 1954a. 2629
Composite distributions. Their application to ecology (Portuguese). *Ciência (Lisboa)*, **4** (9–10), 81–87.

Tiago de Oliveira, J. 1954b. 2630
Composite distributions and its application to some ecological problems. *Univ. Lisboa. Revista Fac. Ci. A. Ci. Mat.*, **3** (2), 171–75.
Review: MR **16** (1955), 153; ZBL **56** (1955), 382.
Class.: MI:mb:BM

Tiago de Oliveira, J. 1965. 2631
Some elementary tests for mixtures of discrete distributions. *Classical and Contagious Discrete Distributions*, Ed. G. P. Patil, Statistical Publishing Society, Calcutta and Pergamon Press, pp. 379–84.
Class.: MI-P:tp:G

Tikkha, R. N. 1962. 2632
Persistence probability of the spells of hot and cold days at Gorakhpur. *Agra Univ. J. Res.*, **10**, 69–79.
Class.: MI:mi:P

Tiku, M. L. 1964. 2633
A note on the negative moments of a truncated Poisson variate. *J. Amer. Statist. Assoc.*, **59**, 1220–24.
Review: MR **30** (1965), 1003; ZBL **124** (1966), 113.
Class.: P:m-a:G

Tiner, J. D. 1954. 2634
The fraction of *Peromyscus leucopus* fatalities caused by racoon ascarid larvae. *J. Mammalogy*, **35**, 589–92.

Tippett, L. H. C. 1932. 2635
A modified method of counting particles. *Proc. Roy. Soc. Ser. A*, **137** (832), 434–46.
Review: BA **9** (1935), Ab. No. 4474.
Users: Thomas, M. (1949), Moore, P. G. (1952), Rider, P. R. (1953), Cohen, A. C. (1954), Moore, P. G. (1954a), Murakami, M., Asai, A., and Kawamura, M. (1954), Wadley, F. M. (1954), Moore, P. G. (1956b), Tate, R. F., and Goen, R. L. (1958), Patil, G. P. (1959), Patil, G. P. (1961b), Hughes, E. J. (1962).

Tippett, L. H. C. 1955. 2636
Statistical methods in textile research, uses of the binomial and Poisson distributions. *J. Textile Inst.*, **26**, T13–T50.
Class.: B-P:mi:E

Tippett, L. H. C. 1958. 2637
A guide to acceptance sampling. *Appl. Statist.*, **7**, 133–48.
User: Hald, A. (1960).
Class.: B:sqc:E

Ţiţeica, Şerban. 1939. 2638
Sur un problème de probabilités. *Bull. Math. Phys. École Polytechn. Bucarest*, **10**, 57–64.
Review: ZBL **23** (1941), 56.

Tocher, K. D. 1950. 2639
Extension of Neyman-Pearson theory of tests to discontinuous variates. *Biometrika*, **37**, 139–44.
Review: MR **12** (1951), 193; ZBL **40** (1951), 75.
Users: Pearson, E. S. (1950), Lancaster, H. O. (1952), Birnbaum, A. (1953), Birnbaum, A. (1954b), David, F. N. (1955a), Stevens, W. L. (1957), Blyth, C. R., and Hutchinson, D. W. (1960), Burkholder, D. L. (1960), Ellner, H. (1963), Thatcher, A. R. (1964).
Class.: MI-B-P:tp:ctp:G

Tocher, K. D. 1954. 2640
Symposium on Monte Carlo methods: The application of automatic computers to sampling experiments. *J. Roy. Statist. Soc. Ser. B*, **16**, 39–75.
Review: ZBL **55** (1955), 369.

Toledo, M. 1963. 2641
Markov chains and birth and death processes (Spanish). *Estadíst. Española No.*, **18**, 29–45.
Review: STMA **5** (1964), 757.
Class.: P-MI:pc:G

De Toledo Piza, Alfonso P. 1951. 2642
Betrachtungen über das geometrische Verteilungsgesetz (Portuguese; Spanish summary). *Trabajos Estadíst.*, **2**, 79–104.
Review: ZBL **43** (1952), 131.

Topp, C. W., Leone, F. C., Hayman, G. E., and Chu, J. T.
See Leone, F. C., Hayman, G. E., Chu, J. T., and Topp, C. W. (1960).

Tortorici, P. 1963. 2643
An elementary demonstration of the tendency, towards the normal law, of the deviation distribution law in the problem of repeated proofs (Italian). *Giorn. Ist. Ital. Attuari*, **26**, 52–63.
Review: STMA **5** (1964), 580.
Class.: B:a:G

Toulmin, G. H., and Good, I. J.
See Good, I. J., and Toulmin, G. H. (1956).

Tranquilli, G. B. 1961. 2644
The cograduation table and the dissimilarity between independent Bernoullian samples (Italian). *Statistica*, **21**, 713–66.
Review: STMA **6** (1965), 954.
User: Herzel, A. (1963c).
Class.: B:mi-ctp:G

Trawinski, B. J. 1965. 2645
General form of the probability function associated with paired-comparison experiments. *Classical and Contagious Discrete Distributions*, Ed. G. P. Patil, Statistical Publishing Society, Calcutta and Pergamon Press, pp. 459–64.
Class.: MI:mi:G

Tripp, Clarence A., Weinberg, George H., and Fluckiger, Fritz A.
See Weinberg, George H., Fluckiger, Fritz A., and Tripp, Clarence A. (1960).

Tritter, A. L., Freimer, M., and Gold, B.
See Freimer, M., Gold, B., and Tritter, A. L. (1959).

Trucco, Ernesto. 1957. 2646
Note on a combinatorial problem. *Bull. Math. Biophys.*, **19** (4), 309–36.
Review: BA **32** (1958), Ab. No. 18471.
Class.: MI:mi:G

Truesdell, C. 1947. 2647
A note on the Poisson-Charlier functions. *Ann. Math. Statist.*, **18**, 450–54.
Class.: P:mi:G

Trybula, S. 1957. 2648
On a problem of prognosis (Russian summary). *Bull. Acad. Polon. Sci. III*, **5**, 859–62.
Review: MR **19** (1958), 991.

Trybula, S. 1958a. 2649
Some problems of simultaneous minimax estimation. *Ann. Math. Statist.*, **29**, 245–53.
Review: MR **20** (1959), 64; ZBL **87** (1961), 142.
Class.: MH-M:pe:G

Trybula, S. 1958b. 2650
O minimaksowej estymacji parametrów w rozkladzie wielomianowym. *Zastos. Mat.*, **3**, 307–22.
Review: MR **21** (1960), 75.
Notes: English translation; On the minimax estimation of the parameters in a multinomial distribution. *Selected Translations Math. Statist. and Prob.*, **3**,(1962).

Trybula, A. 1959a. 2651
The estimation of frequency in a population of elements belonging to classes not represented in the sample (Polish; Russian and English summaries). *Zastos. Mat.*, **4**, 244–48.
Review: IJA **1** (1960), 407; ZBL **94** (1962), 134.

Trybula, S. 1959b. 2652
Estimation taking into account the error of the controller (Polish). *Zastos. Mat.*, **4**, 249–54.
Review: IJA **1** (1960), 466.
Class.: B:sqc-pe:E

Trybula, S. 1962. 2653
On the minimax estimation of the parameters in a multinomial distribution. *Selected Translations Math. Statist. and Prob.*, **3**, 225–38.
Review: MR **27** (1964), 829.

Trybula, S. 1964. 2654
Properties of the hypergeometric distribution connected with Bayes' rule. *Bull. Acad. Polon. Sci. Ser. Sci. Math. Astronom. Phys.*, **12**, 753–56.
Review: MR **30** (1965), 665.

Tsao, C. K. 1956. 2655
Distribution of the sum in random samples from a discrete population. *Ann. Math. Statist.*, **27**, 703–12.
Review: ZBL **73** (1960), 354.
User: Govindarajulu, Z. (1965).
Class.: MI:tc:G

Tsao, C. K. 1965a. 2656
A moment generating function of the hypergeometric distributions. *Classical and Contagious Discrete Distributions*, Ed. G. P. Patil, Statistical Publishing Society, Calcutta and Pergamon Press, pp. 75–78.
Class.: MH-IH:m:G

Tsao, C. K. 1965b. 2657
Distribution of the product in random samples from a finite population. *Classical and Contagious Discrete Distributions*, Ed. G. P. Patil, Statistical Publishing Society, Calcutta and Pergamon Press, pp. 427–36.
Class.: MI:os-tc:G

Tsao, C. M. 1962. 2658
A general class of discrete distributions and mixtures of distributions. Ph.D. Thesis, Univ. Oregon.

Tsaregradskii, I. P.
See Caregradskii, I. P. (1958).

Tucker, H. 1955. 2659
Tests of contagion and time effect in accident proneness (Abstract). *Ann. Math. Statist.*, **26**, 162.

Tucker, H. G. 1963a. 2660
An estimate of the compounding distribution of a compound Poisson distribution. *Theor. Probability Appl.*, **8**, 195–200.
User: Haight, F. (1965a).
Class.: COP:pe:G

Tucker, H. G. 1963b. 2661
An estimate of the compounding distribution of a compound Poisson distribution (Russian summary). *Teor. Verojatnost i Primenen*, **8**, 211–16.
Review: MR **27** (1964), 828; STMA **6** (1965), 466.
Class.: COP:pe:G

Tucker, H. G., and Stratton, H. H., Jr.
See Stratton, H. H., Jr., and Tucker, H. G. (1964).

Tufo, Thomas. 1963. 2662
Approximations to the hypergeometric by the binomial and to the binomial by Poisson. Master's Thesis, Florida State Univ.
User: Govindarajulu, Z. (1965).

Tukey, J. W. 1948. 2663
Non-parametric estimation, III. Statistically equivalent blocks and multivariate tolerance regions—the discontinuous case. *Ann. Math. Statist.*, **19**, 30–39.

Tukey, John W. 1949a. 2664
Moments of random group size distributions. *Ann. Math. Statist.*, **20**, 523–39.
Review: ZBL **35** (1950), 93.
Users: Barton, D. E., and David, F. N. (1959a).
Class.: B-P:mb-m:P-BM

Tukey, J. W., and Freeman, M. F.
See Freeman, M. F., and Tukey, J. W. (1950).

Tukey, John W., and Mosteller, F.
See Mosteller, F., and Tukey, John W. (1949).

Tumanyan, S. H. 1955. 2665
Asymptotic investigation of the multinomial probability distribution (Russian; Armenian summary). *Akad. Nauk. Armjan. SSR. Dokl.*, **20**, 65–74.
Review: MR **17** (1956), 47; ZBL **65** (1956), 111.
Users: Patil, G. P., and Bilkidar, S. (1966a).

Tumanyan, S. H. 1956. 2666
Asymptotic distribution of the χ^2 criterion when the number of observations and number of groups increase simultaneously. *Theor. Probability Appl.*, **1**, 117–31.
Class.: M:a:G

Turner, F. M., and Chamberlain, A. C.
See Chamberlain, A. C., and Turner, F. M. (1952).

Turner, M. E., and Eadie, G. S. 1957. 2667
The distribution of red blood cells in the hemacytometer. *Biometrics*, **13**, 485–95.
User: Hamaker, H. C. (1958).
Class.: B-P-MH:mb:BM

Turner, M. E. and Ham, W. T., Jr. 1960. 2668
Target theory. Pre-print 3, Dept. of Biophys. and Biometry, Medical College of Virginia, Richmond, Va., pp. 9.
Class.: P-NB:mb-pe:BM

Tweedie, M. C. K. 1945. 2669
Inverse statistical variates. *Nature*, **155**, 453.
Users: Barnard, G. A. (1946), Tweedie, M. C. K. (1946), Tweedie, M. C. K. (1947), Anscombe, F. J. (1949b), Sandelius, M. (1951a), Tweedie, M. C. K. (1952), Tweedie, M. C. K. (1956), deGroot, M. H. (1959), Nadler, J. (1960), Wason, M. T. (1965).
Class.: B-NB:osp:G

Tweedie, M. C. K. 1946. 2670
The regression of the sample variance on the sample mean. *J. London Math. Soc.*, **21**, 22–28.
User: Tweedie, M. C. K. (1965).
Class.: MI:osp:G

Tweedie, M. C. K. 1947. 2671
Functions of a statistical variate with given means, with special reference to Laplacian distributions. *Proc. Cambridge Philos. Soc.*, **43**, 41–49.
Users: Lehmann, E. L., and Stein, C. (1949), Tweedie, M. C. K. (1952), Tweedie, M. C. K. (1956), Gart, J. J. (1959), Bardwell, G. E. and Crow, E. L. (1964), Govindarajuly, Z. (1965), Tweedie, M. C. K. (1965).
Class.: PS:osp:G

Tweedie, M. C. K. 1952. 2672
The estimation of parameters from sequentially sampled data on a discrete distribution. *J. Roy. Statist. Soc. Ser. B*, **14**, 238–45.
Review: ZBL **48** (1953), 121.
Users: Mosimann, J. E. (1963), Tweedie, M. C. K. (1965), Bhat, B. R., and Kulkarni, N. V. (1966b).
Class.: MI:se:G

Tweedie, M. C. K. 1953. 2673
The covariances of frequencies from a multinomial distribution under a sequential sampling rule (Abstract). *Ann. Math. Statist.*, **24**, 142.

Tweedie, M. C. K. 1956. 2674
Some statistical properties of inverse Gaussian distributions. *Virginia J. Sci. N.S.*, **7**, 160–65.
User: Tweedie, M. C. K. (1965).

Tweedie, M. C. K. 1965. 2675
Further results concerning expectation-inversion technique. *Classical and Contagious Discrete Distributions*, Ed. G. P. Patil, Statistical Publishing Society, Calcutta and Pergamon Press, pp. 195–218.
Class.: PS-B-P-NB-TCP:pe-m-osp:G

U

Ugolini, Giovanni, B. 1938. 2676
I procedimenti statistici e la idrologia. *Atti l. Congr. Un. Mat. Ital.* 693–98.
Review: ZBL **20** (1939), 41.

Uhlmann, W. 1966. 2677
Vergleich der hypergeometrischen mit der Binomial-Verteilung. *Metrika*, **10**, 145–58.
Class.: H-B-P:tp-a:G

Ungar, Peter. 1960. 2678
The cutoffpoint for group testing. *Comm. Pure Appl. Math.*, **13**, 49–54.
Review: MR **22** (1961), 352.
Class.: MI:mi:G

United States Army Ordnance Corps. 1952. 2679
Tables of the cumulative binomial probabilities. Pamphlet ORDP 20-1, Office Tech. Services, Dept. Commerce, Order No. PB111389, Washington, D. C.
User: Crow, E. L. (1956).

Upholt, W. M. 1942. 2680
The use of the square root transformation and analysis of variance with contagious distributions. *J. Econ. Entomol.*, **35**, 536–43.
Review: BA **17** (1943), Ab. No. 8054.
Users: Kono, Utida, Yosida, and Watanabe (1952), Utida, Kōno, Watanabe, and Yosida (1952).

Upholt, W. M. 1944. 2681
The power of the analysis of variance with the Poisson distribution. *J. Econ. Entomol.*, **37**, 717.

Upholt, W. M., and Craig, R. 1940. 2682
A note on the frequency distribution of black scale insects. *J. Econ. Entomol.*, **33**, 113–14.
User: Subrahmaniam, K. (1964).

Uppuluri, V. R. Rao, and Bowman, K. O. 1966. 2683
Likelihood ratio test criterion for small samples from multinomial distributions (Abstract). *Biometrics*, **22**, 650–51.

Urbanik, K. 1956. 2684
Remarks on the maximum number of bacteria in a population (Polish; Russian and English summaries). *Zastos. Mat.*, **2**, 341–48.

Urbanik, K. 1958. 2685
Poisson distributions on compact topological groups. *Colloq. Math.*, **6**, 13–24.
Review: MR **21** (1960), 553; ZBL **97** (1962), 133.
Class.: P-COP:osp:G

Urbanik, K., and Fisz, M.
See Fisz, M., and Urbanik, K. (1955).

Urbanik, K., and Fisz, M.
See Fisz, M., and Urbanik, K. (1956).

Urbanik, K., and Steinhaus, H.
See Steinhaus, H., and Urbanik, K. (1959).

Ury, Hans K. 1966. 2686
A table of sums of discrete right triangular random variables (or, alternatively, of a measure of rank differences between two particular objects) (Abstract). *Ann. Math. Statist.*, **37**, 1415.

Usai, G. 1950. 2687
Valor medio della potenza di una variable casuale nel problema delle prove ripetute. *Atti. Accad. Gioenia Catania*, **6** (8), 1–7.
Review: MR **12** (1951), 190; ZBL **38** (1951), 287.

Upensky, J. V. 1931. 2688
On Ch. Jordan's series for probability. *Ann. of Math.*, **32**, 306–12.
User: Boas, R. P. Jr. (1949).

Utida, S. 1943a. 2689
Studies on experimental population of the Azuki bean weevil, *Callosobruchus chinensis* (L.). VIII. Statistical analysis of the frequency distribution of the emerging weevils on beans. *Mem. Coll. Agri. Kyoto Imperial Univ. No.* 54 (*Ent. Series.* 10), 1–22.
Class.: P:id-gf:BM

Utida, S. 1943b. 2690
Studies on experimental population of the Azuki bean weevil, *Callosobruchus chinensis* (L.). IX. General consideration and summary of the serial reports from I to VIII. *Mem. Coll. Agri. Kyoto Imperial Univ. No.* 54 (*Ent. Series.* 10), 23–40.

Utida, S. 1950. 2691
On the equilibrium state of the interacting population of an insect and its parasite. *Ecology*, **31**, 165–75.
Class.: MI:mi:BM

Utida, S., Kōno, T., Watanabe, S., and Yosida, T. 2692
1952.
Pattern of spatial distribution of the common cabbage butterfly, Pieris rapae in a cabbage farm. Pattern of the spatial distribution of insects. 1st report (Japanese; English summary). *Res. Population Ecol.*, **1**, 49–64.
Users: Kobayashi, S. (1957), Kobayashi, S. (1965), Kobayashi, S. (1966).
Class.: P-GP:gf:BM

Utida, Synnro, Yosida, Tosiharu, Watanabe, Syozi, and Kōno, Tatsuro.
See Kōno, Tatsuro, Utida, Synnro, Yosida, Tosiharu, and Watanabe, Syozi (1952).

V

Vagholkar, M. K. 1959. 2693
The process curve and the equivalent mixed binomial with two components. *J. Roy. Statist. Soc. Ser. B*, **21**, 63–66.
Review: IJA **1** (1960), 467.
Users: Vagholkar, M., and Wetherill, G. (1960), Hald, A. (1963b), Blischke, W. R. (1965).
Class.: COB:sqc:E

Vagholkar, M. K., and Wetherill, G. B. 1960. 2694
The most economical binomial sequential probability ratio test. *Biometrika*, **47**, 103–09.
Review: ZBL **91** (1962), 308; STMA **5** (1964), 147.
User: Wetherill, B. G. (1961).
Class.: B:tp:G

Vajani, L. 1954. 2695
L'interpretazione della distribuzione delle malattie nei caseggiati mediante gli schemi di Yule e Polya. *Bull. Inst. Internat. Statist.*, **34** (3), 406–412.
Review: ZBL **57** (1956), 360.

Vajani, L. 1960. 2696
A relation on the moments of a geometrical distribution. *Riv. Int. Sci. Econ. Comm.*, **7**, 758–64.
Review: IJA **3** (1962), 16.
Class.: G:m:G

Vajani, L. 1961. 2697
Probability of an even or odd number of events in some distributions (Italian). *Giorn. Economisti*, **20**, 245–51.
Review: IJA **4** (1963), 23.
Class.: B-P-G:osp:G

Vajani, L. 1962a. 2698
Probability distribution and convergent series (Italian). *Giorn. Economisti*, **21**, 646–55.
Review: STMA **6** (1965), 39.
Class.: MI:mi:G

Vajani, L. 1962b. 2699
Statistical methods for the measurement of the mean variability ratio (Italian). *Riv. Int. Sci. Econ. Comm.*, **9**, 428–38.
Review: IJA **4** (1963), 417.

Vajda, Stefan. 1939. 2700
Die Wahrscheinlichkeit einer bestimmten Auszahlungssumme. *Skand. Aktuarietidskr.*, 10–21.
Review: ZBL **21** (1940), 147.

Van Der Gon Denier, J. J., and Strackee, J.
See Strackee, J., and van der Gon Denier, J. J. (1962).

Van der Waerden, B. L. 1939. 2701
Vertrauensgrenzen für unbekannte Wahrscheinlichkeiten. *Ber Verh. Sächs. Akad. Wiss. Leipzig*, **91**, 213–28.
Review: MR **1** (1940), 249.

Van der Waerden, B. L. 1960. 2702
Sampling inspection as a minimum loss problem. *Ann. Math. Statist.*, **31**, 369–84.
Review: IJA **2** (1961), 139.
Class.: P:sqc:E

Van der Waerden, B. L. 1965. 2703
Sequential sampling inspection as a minimum problem (German). *Zeit. Wahrscheinlichkeitsth.*, **4**, 187–202.
Review: STMA **7** (1966), 378.
Class.: B:tp:G

Van der Waerden, B. L., Egg, K., and Rust, H.
See Egg, K., Rust, H., and Van der Waerden, B. L. (1965).

Van Eeden, C. 1955. 2704
A sequential test with three possible decisions for comparing two unknown probabilities, based on groups of observations. *Rev. Inst. Internat. Statist.*, **23**, 20–28.
Review: MR **18** (1957), 243.

Van Eeden, Constance. 1965. 2705
Conditional limit distributions for the entries in a $2 \times k$ contingency table. *Classical and Contagious Discrete Distributions*, Ed. G. P. Patil, Statistical Publishing Society, Calcutta and Pergamon Press, pp. 123–26.
Class.: MH-M-B-P:a:G

Van Eeden, C., and Runnenburg, J. T. 1960. 2706
Conditional limit distributions for the entries in a 2×2 table (Dutch). *Statistica Neerlandica*, **14**, 111-26.
Review: IJA **2** (1961), 23.
Class.: M-B-P:a:G

Van Eeden, C., and Runnenburg, J. T. 1960a. 2707
Conditional limit distributions for the entries in a 2×2 table (Dutch). *Report S254 Statist. Dept., Math. Centre, Amesterdam.*
Review: IJA **2** (1961), 23.
Class.: M-B-P:a:G

Van Eeden, C., and Bloemena, A. R. 1960b. 2708
On probability distributions arising from points on a lattice. *Report S257 Statist. Dept., Math. Centre, Amesterdam.*
Review: IJA **1** (1960), 542.
Class.: MI-P:mb-a:G

Van Elteren, Ph., and Gerrits, H. J. M. 1961. 2709
A queueing problem occurring at measurements of the threshold value of the human eye (Dutch). *Statistica Neerlandica*, **15**, 385–401.
Review: IJA **3** (1962), 456.
Class.: P:pc:BM

Van Heerden, D. F. I. 1961. 2710
Studie van 'n groep diskrete waarskynlikheidsfunksies geassosieer met die terme van positiewe magreekse. Ph.D. Thesis, Univ. S. Africa.
Users: Van Heerden, D. F. I., and Gonin, H. T. (1966).

Van Heerden, D. F. I., and Gonin, H. T. 1966. 2711
The orthogonal polynomials of power series probability distributions and their uses. *Biometrika*, **53**, 121–28.
Class.: PS-LS-NB:mi-pe:G

Van Klinken, J. 1957. 2712
Statistical methods to inquire if the risk of accidents has changed. *Het Verzek.-Arch., Actuar. Bij.*, **34**, 17–30.
Review: ZBL **81** (1959), 366.

Van Klinken, J. 1958. 2713
Method for interval-estimation of numbers of future invalidity-pensionholders and the corresponding present value. *Het Verzek.-Arch., Actuar. Bij.*, **35**, 57–61.
Review: ZBL **84** (1960), 368.

Van Klinken, J. 1959. 2714
On some estimation problems with regard to the Poisson-distribution and the χ^2-minimum method. *Mitt. Verein. Schweiz. Versich.-Math.*, **59**, 297–306.
Review: MR **22** (1961), 183; ZBL **95** (1962), 335.

Van Klinken, J. 1959a. 2715
Applications of tests based on the Poisson model. *Internat. Z. Versicherungsmath. Statist. Probl. Soz. Sicherheit*, **3**, 17–29.
Review: ZBL **89** (1961), 164.

Van Klinken, J. 1959b. 2716
The χ^2-minimum method in the case of Poisson-variables. *Het Verzek.-Arch., Actuar. Bij.*, **36**, 69–75.
Review: ZBL **96** (1962), 127.

Van Klinken, J. 1960. 2717
Note on a test for the multiple Poisson-distribution. *Actuar. Studiën*, **1**, 16–27.
Review: ZBL **92** (1962), 366.

Van Klinken, J. 1961. 2718
A method of inquiring whether the Gamme-distribution represents the frequency-distribution of industrial accident costs. *Actuar. Studiën*, **2**, 83–92.
Review: IJA **3** (1962), 530; ZBL **102** (1963), 153.
Class.: P:a:G-S

Van Klinken, J., and Prins, H. J. 1954. 2719
Survey of testing and estimation methods with respect to the Poisson distribution (Dutch). *Math. Centrum Amsterdam. Statist. Afdeling Rap. S*133, 1–77.
Review: MR **16** (1955), 383.

Van Ryzin, J. 1966. 2720
Repetitive play in finite statistical games with unknown distributions. *Ann. Math. Statist.*, **37**, 976–94.
Class.: MI:pe:G

Van Veen, S. C., and Bottema, O.
See Bottema, O., and Van Veen, S. C. (1943).

Van Wijngaarden, A. 1950. 2721
Table of the cumulative symmetric binomial distribution. *Nederl. Akad. Wetensch. Proc. Ser. A*, **53**, 857–68.

Van Wormer, T. A., and Simon, H. A.
See Simon, H. A., and Van Wormer, T. A. (1963).

Varangot, V. 1947. 2722
Een eenvoudige Afleiding van de Verdeeling van Poisson. *Statistica Neerlandica*, **1**, 161.

Varley, G. C., and Finney, D. J.
See Finney, D. J., and Varley, G. C. (1955).

Vasilas, J. N., Peterson, R. O., and Fitzpatrick, R.
See Fitzpatrick, R., Vasilas, J. N., and Peterson, R. O. (1953).

Vasudevan, R., and Ramakrishnan, A.
See Ramakrishna, A., and Vasudevan, R. (1957).

Vaulot, A. E. 1931. 2723
Application du calcul des probabilités à l'exploitation téléphonique: Formule de Poisson et applications. *Rev. Gen. Electr.*, **30**, 173–75.

Verhagen, A. M. W. 1965. 2724
Recurrent events in sequences of independent binomial trials with alternating probability of success. Tech. Paper No. 20 CSIRO Div. Math. Statist. pp. 12.
Review: STMA **6** (1965), 1237.
Class.: B:pe:G

Vessereau, A. 1959. 2725
Sur les conditions d'application du criterium χ^2 de Pearson. *Bull. Inst. Internat. Statist.*, **36** (3), 87–101.
Review: ZBL **94** (1962), 141.

Vidwans, S. M. 1964. 2726
A note on the negative binomial distribution. *Biometrika*, **51**, 264–65.
Review: MR **30** (1965), 809; STMA **6** (1965), 76.
User: Haight, F. (1965b).

Vietoris, L. 1961. 2727
An estimate concerning a sampling distribution (German). *Monatsh. Math.*, **65**, 287–90.
Review: IJA **3** (1962), 323.
Class.: B:pe:G

Viktorova, I. I., and Cistyakov, V. P. 1965. 2728
Asymptotic normality in a problem on balls when the probabilities of falling into different boxes are different (Russian; English summary). *Teor. Verojatnost. i Primenen.*, **10**, 162–67.

Viktorova, I. I., and Cistyakov, V. P. 1966. 2729
Some generalizations of the empty boxes test (Russian; English summary). *Teor. Verojatnost. i Primenen.*, **11**, 306–43.
Class.: OMR:tp-a:G

Viktorova, I. I., and Chistyakov, V. P.
See Chistyakov, V. P., and Viktorova, I. I. (1965).

Vincze, I., and Csáki, E.
See Csáki, E., and Vincze, I. (1963).

Vistelius, A. B., and Sarmanov, O. V. 1947. 2730
Stochastische Begründung einer geologisch wichtigen Wahrscheinlichkeitsverteilung (Russian). *Doklady Akad. Nauk SSSR*, **58**, 631–34.
Review: ZBL **28** (1948), 153.

Vogel, Walter. 1960. 2731
Ein Irrfahrten-Problem und seine Anwendung auf die Theorie der sequentiellen Versuchs-Pläne. *Arch. Math.*, **11**, 310–20.
Review: MR **22** (1961), 695.

Vogel, W. 1961. 2732
Sequentielle Versuchs-Pläne. *Metrika*, **4**, 140–57.
Review: ZBL **102** (1963), 144.

Vogel, W. 1962. 2733
Models of probability (German). *Math. Phys. Semesterber.*, **9**, 169–89.
Review: IJA **4** (1963), 394.

Volodin, I. N. 1965a. 2734
On the distinction between the Pólya and the Poisson distributions when a large number of small samples is available. *Theor. Probability Appl.*, **10**, 335-38.
Class.: PO-P:cm-a:G

Volodin, I. N. 1965b. 2735
On the distinction between the Poisson and Pólya distributions when a large number of small samples is available (Russian; English summary). *Teor. Verojotnost. i Primenen*, **10**, 364–67.
Review: MR **31** (1966), 1136.

Von Mises, R. 1939. 2736
An inequality for the moments of a discontinuous distribution. *Skand. Aktuarietidskr.*, **22**, 32–36.
Review: MR **1** (1940), 22; ZBL **21** (1940), 147.

Von Mises, R. 1939a. 2737
Über Aufeilungs- und Besetzungs- Wahrscheinlichkeiten. *Rev. Fac. Sci. Univ. Istanbul* N.S. 4 (1–2), 145–63.
Review: ZBL **21** (1940), 145.
User: Békéssy, A. (1963).

Von Schelling, H. 1937. 2738
Fehlerrechnung bei biologischen Messungen. *Naturwissenschaften*, **25**, 699–700.
User: von Schelling H. (1941a).

Von Schelling, H. 1964a. 2739
Bermerkungen zur Verteilung von Pascal. *Naturwissenschaften*, **29**, 517–18.
Review: ZBL **25** (1942), 416.

Von Schelling, H. 1941b. 2740
Bemerkungen zur Geschwistermethode. *Z. Menschl. Verbungs-u. Kostitutionslehre*, **25**, 391–97.
Review: ZBL **25** (1942), 418.
Class.: B:pe:G

Von Schelling, H. 1942. 2741
Eine Formel für die Teilsummen gewisser hypergeometrischer Reihen und deren Bedeutung für die Wahrscheinlichkeitstheorie. Naturwissenschaften, **30**, 757–58.

Review: MR **7** (1946), 128; ZBL **27** (1943), 338.
User: von Schelling, H. (1949a).

Von Schelling, H. 1944. 2742
Zur Deutung auslesefreier Zwillingserhebungen. *Z. Menschl. Verebungs-u. Konstitutionslehre*, **27**, 778–81.
Review: ZBL **60** (1957), 317.

Von Schelling, H. 1949a. 2743
A formula for the partial sums of some hypergeometric series. *Ann. Math. Statist.*, **20**, 120–22.
Review: ZBL **41** (1952), 250.
User: von Schelling, H. (1950).
Class.: MI:mb-osp:G

Von Schelling, H. 1949b. 2744
Coupon collecting for unequal probabilities. *Amer. Math. Monthly*, **56**, 306–11.
Review: ZBL **55** (1955), 370.

Von Schelling, H. 1950. 2745
A second formula for the partial sum of hypergeometric series having unity as the fourth argument. *Ann. Math. Statist.*, **21**, 458–60.
Class.: MI:c:G

Von Schelling, H. 1951. 2746
Distribution of the ordinal number of simultaneous events which last during a finite time. *Ann. Math. Statist.*, **22**, 452–55.
Review: ZBL **43** (1952), 133.
Class.: MI:mb-m-a:G

Von Schelling, H. 1955. 2747
Statistische Modelle als Hilfsmittel der Naturbeschriebung. *Mitteilungsbl. Math. Statist.*, **7**, 173–92.
Review: ZBL **66** (1957), 118.

Von Schelling, H., and Gumbel, E. L.
See Gumbel, E. J., and von Schelling, H. (1950).

Vora, R. B. 1966. 2748
A two-sided slippage test for Poisson variates (Abstract). *Ann. Math. Statist.*, **37**, 1067.

W

Wadley, F. M. 1943. 2749
Statistical treatment of percentage counts. *Science* **98**, 536–38.
Review: SA **47A** (1944), Ab. No. 1005.
Class.: P:pe:BM

Wadley, F. M. 1945. 2750
An application of the Poisson series to some problems of enumerations. *J. Amer. Statist. Assoc.*, **40**, 85–92.
User: Wadley, F. M. (1954).
Class.: P:gf-h:BM

Wadley, F. M. 1950. 2751
Notes on the form of distribution of insect and plant populations. *Ann. Ent. Soc. America*, **43**, 581–86.
Users: Kono, Utida, Yosida, and Watanabe (1952), Bliss, C. I., and Fisher, R. A. (1953), Wadley, F. M. (1954), Waters, W. (1959), Harcourt, D. G. (1960), Kuno, E. (1963).
Class.: P-N-NB:gf:BM

Wadley, F. M. 1954. 2752
Limitations of the "zero method" of population counts. *Science*, **119**, 689–90.
Users: Wood, J., Davis, D., and Komarek, E. (1958).
Class.: P:pe:BM

Wadley, F. M., and Davis, E. G.
See Davis, E. G., and Wadley, F. M. (1949).

Walford, L. A., and Winsor, C. P.
See Winsor, C. P., and Walford, L. A. (1936).

Walker, A. M. 1950. 2753
Sequential sampling formulae for a binomial population. *J. Roy. Statist. Soc. Ser. B.*, **12**, 301–07.
Review: MR **14** (1953), 569; ZBL **41** (1952), 459.

Walker, M. G. 1942. 2754
A Mathematical analysis of the distribution in maize of Heliothis armigera Hbn. *Canad. J. Res. D*, **20**, 235–61.
Users: Waters, W. E., and Henson, W. R. (1959).

Walker, T. J., Jr. 1957. 2755
Ecological studies of the arthropods associated with certain decaying materials in four habitats. *Ecology*, **38**, 262–76.
Class.: B:tp:BM

Wallace, David L. 1959. 2756
Conditional confidence level properties. *Ann. Math. Statist.*, **30**, 864–76.
Review: IJA **2** (1961), 303.
User: Pratt, J. W. (1961).
Class.: B-P:ie:G

Wallace, David L., and Mosteller, F.
See Mosteller, F., and Wallace, David L. (19).

Wallace, David L., and Mosteller, F.
See Mosteller, F., and Wallace, David L. (1963).

Wallace, David L., and Mosteller, Frederick.
See Mosteller, Frederick, and Wallace, David L. (1964).

Wallington, P. A., and Shenton, L. R.
See Shenton, L. R., and Wallington, P. A. (1962).

Wallis, W. Allen. 1936. 2757
The Poisson distribution and the supreme court. *J. Amer. Statist. Assoc.*, **31**, 376–80.
Class.: P:gf:O

Wallwork, J. A., Rodriguez, J. G., and Ibarra, E. L.
See Ibarra, E. L., Wallwork, J. A., and Rodriguez, J. G. (1965).

Walsh, J. E. 1952. 2758
Large-sample validity of the binomial distribution for lives with unequal mortality rates. *Skand. Aktuarietidskr.*, **35**, 11–15.
Review: ZBL **47** (1953), 136.
Users: Walsh, J, E. (1955c), Walsh, J. E. (1956), Eisenberg, H. B., Geoghagen, R. M., and Walsh, J. R. (1962), Eisenberg, H. B., Geoghagen, R. M., and Walsh, J. E. (1963).
Class.: B:mi:BM

Walsh, J. E. 1953a. 2759
Actuarial validity of the binomial distribution for large numbers of lives with small mortality probabilities (Abstract). *Ann. Math. Statist.*, **24**, 681.

Walsh, J. E. 1953b. 2760
The Poisson distribution as a limit of dependent binomial distributions with unequal probabilities (Abstract). *Ann. Math. Statist.*, **24,** 689.
User: Govindarajulu, Z. (1965).

Walsh, J. E. 1954. 2761
Analytic tests and confidence intervals for the mean value, probabilities, and percentage points of a Poisson distribution. *Sankhyā*, **14,** 25–38.
Review: MR **16** (1955), 383; ZBL **57** (1956), 115.
Users: Walsh, J. E. (1955b), Crow, E. L. (1959).
Class.: P:tp-ie:G

Walsh, J. E. 1955a. 2762
Approximate probability values for observed number of successes from statistically independent binomial events with unequal probabilities (Abstract). *Ann. Math. Statist.*, **26,** 162.

Walsh, J. E. 1955b. 2763
The Poisson distribution as a limit for dependent binomial events with unequal probabilities. *J. Operations Res. Soc. America*, **3,** 198–209.
Review: MR **16** (1955), 938.
Users: Fitzpatrick, R. (1958), Walsh, J. E. (1964), Eisenberg, H. B. *et al.* (1966).
Class.: B-P:a:G-S-E

Walsh, J. E. 1955c. 2764
Approximate probability values for observed number of " successes " from statistically independent binomial events with unequal probabilities. *Sankhyā*, **15,** 281–90.
Review: ZBL **67** (1958), 117.
Users: Patil, G. P. (1959, David, H. A. (1960), Eisenberg, H. B., Geoghagen, R. M., and Walsh, J. E. (1962), Eisenberg, H. B., Geoghagen, R. M., and Walsh, J. E. (1963).
Class.: B:tp-ie:G

Walsh, J. E. 1956. 2765
Actuarial validity of the binomial distribution for large numbers of lives with small mortality probabilities. *Skand. Aktuarietidskr.*, **39,** 39–46.
Review: ZBL **74** (1962), 148.
Class.: B:mi:BM

Walsh, J. E. 1960. 2766
Nonparametric linear estimation of common median of symmetrical populations from symmetrically censored observations. *Sankhyā*, **22,** 295–300.
Review: IJA **2** (1961), 670.
Class.: MI:pe:G

Walsh, J. E. 1962. 2767
Bounded probability properties of Kolmogorov-Smirnov and similar statistics for discrete data. SP-848, Systems Development Corp., California, 7 p.
Class.: MI:ie-ctp:G

Walsh J. E. 1963a. 2768
Bounded probability properties of Kolmogorov-Smirnov and similar statistics for discrete data. *Ann. Inst. Statist. Math.*, **15,** 153–58.
Review: MR **30** (1965), 665; STMA **6** (1965), 568.

Walsh, J. E. 1963b. 2769
Simultaneous confidence intervals of differences of classification probabilities. *Biom. Zeit.*, **5,** 231–34.
Review: STMA **5** (1964), 619.
Class.: M:ctp-ie:G

Walsh, J. E. 1964. 2770
Approximate distribution of extremes for nonsample cases. *J. Amer. Statist. Assoc.*, **59,** 429–36.
Review: ZBL **123** (1966), 371.

Walsh, J. E., Eisenberg, H. B., and Geoghagen, R. R. M.
See Eisenberg, H. B., Geoghagen, R. R. M., and Walsh, J. E. (1962).

Walsh, J. E., Eisenberg, H. B., and Geoghagen, R. R. M.
See Eisenberg, H. B., Geoghagen, R. R. M., and Walsh, J. E. (1963).

Walsh, John E., Eisenberg, Herbert B., and Geoghagen, Randolph R. M.
See Eisenberg, Herbert B., Geoghagen, Randolph, and Walsh, John E. (1966).

Wani, J. K. 1966. 2771
Moment relations for some discrete distributions (Abstract). *Ann. Math. Statist.*, **37,** 768.

Wani, J. K., and Patil, G. P.
See Patil G. P., and Wani, J. K. (1965).

Wani, J. K., and Patil, G. P.
See Patil, G. P., and Wani, J. K. (1966).

Wani, J. K., Patil, G. P., and Kamat, A. R.
See Patil, G. P., Kamat, A. R., and Wani, J. K. (1964).

Warner, Stanley L. 1965. 2772
Randomized response: a survey technique for eliminating evasive answer bias. *J. Amer. Statist. Assoc.*, **60,** 63–69.
Class.: B:pe:S

Warren, William G. 1962. 2773
Contributions to the study of spatial point processes. PhD. Thesis, Univ. North Carolina.

Wasan, M. T. 1962. 2774
Minimax estimate of an inverse-binomial parameter (Abstract). *Ann. Math. Statist.*, **33,** 1501.

Wasan, M. T.1963. 2775
Sequential optimum procedures for unbiased estimation of a binomial parameter (Abstract). *Ann. Math. Statist.*, **34,** 1129–30.

Wasan, M. T. 1964. 2776
Sequential optimum procedures for unbiased estimation of a binomial parameter. *Technometrics*, **6,** 259–71.
Review: MR **32** (1966), 542; STMA **6** (1965), 890.
Class.: B:se:G

Wasan, M. T. 1965a. 2777
Asymptotic normality of binomial sequential stopping rules (Abstract). *Ann. Math. Statist.*, **36,** 1609.

Wasan, M. T. 1965b. 2778
Sequential estimation of a binomial parameter. *Classical and Contagious Discrete Distributions*, Ed. G. P. Patil, Statistical Publishing Society, Calcutta and Pergamon Press, pp. 263–72.
Class.: B:se:G

Watanabe, H. 1956. 2779
On the Poisson distribution. *J. Math. Soc. Japan.*, **8**, 127–34.
Review: MR **19** (1958), 70; ZBL **73** (1960), 352.

Watanabe, Syozi, Kono, Tatsuro, Utida, Synnro, and Yosida, Tosiharu.
See Kono, Tatsuro, Utida, Synnro, Yosida, Tosiharu, and Watanabe, Syozi (1952).

Watanabe, Y. 1954. 2780
Bimodal distributions. *J. Gakugei Tokushima Univ.*, **5**, 29–38.

Waters, W. E. 1955. 2781
Sequential sampling in forest insect surveys. *Forest Sci.*, **1**, 68–79.
Users: Bliss, C. I. (1958), Bliss, C. I., and Owen, A. R. G. (1958), Harcourt, D. G. (1960), Kuno, E. (1963), Kobayashi, S. (1965).
Class.: B-NB-P:tp:BM

Waters, W. E. 1958. 2782
The ecological significance of aggregated distributions with special reference to forest insects. Ph.D. Thesis, Yale Univ.

Waters, W. E. 1959. 2783
A quantitative measure of aggregation in insects. *J. Econ. Entomol.*, **52**, 1180–84.
User: Kobayashi, S. (1965).

Waters, W. E. 1964. 2784
The ecological significance of aggregation in forest insects. *Proc. 11th Internat. Congr. Ent.*, Vienna, 1960.

Waters, W. E., and Henson, W. R. 1959. 2785
Some sampling attributes of the negative binomial distribution with special reference to forest insects. *Forest Sci.*, **5**, 397–412.
User: Henson, W. R. (1959).
Class.: NB:gf-mi:BM

Waters, W. E., and Henson, W. R. 1959a. 2786
The ecological significance of aggregated distributions with special reference to forest insects. *Forest Sci.*, **5**, 397-412.
User: Henson, W. R. (1959).

Watson, G. S. 1959. 2787
Some recent results in chi-square goodness-of-fit tests. *Biometrics*, **15**, 440–68.
Review: IJA **1** (1960), 391.

Watson, G. S. 1965. 2788
The distribution of organisms. *Biometrics*, **21**, 543–50.

Watson, G. S., and Bloch, D. A.
See Bloch, D. A., and Watson, G. S. (1966).

Watson, G. S., and Leadbetter, M. R. 1963. 2789
On the estimation of the probability density, I. *Ann. Math. Statist.*, **34**, 480–491.
Class.: MI-G-P:mi:G

Watt, K. E. F. 1956. 2790
The choice and solution of mathematical models for predicting and maximizing the yield of a fishery. *J. Fisheries Res. Board Canada*, **13**, 613–45.

Weaver, C. R., and Pruess, K. P.
See Pruess, K. P., and Weaver, C. R. (1959).

Webb, W. B., and Jones, E. R. 1953. 2791
Some relations between two statistical approaches to accident proneness. *Psychol. Bull.*, **50**, 133–36.
User: Fitzpatrick, R. (1958).
Class.: P-MP:cm:A

Weber, Erna. 1958–59. 2792
Das Ergebnis-Folge-Verfahren (Sequenzanalyse). Grundlagen und Anwendungen (1. Fortsetzung) (Russian, English and French summaries). *Wiss. Z. Humboldt-Univ. Berlin Math.-Nat. Reihe*, **8**, 519–34.
Review: MR **22** (1961), 1964.

Weesakul, B., and Yeo, G. F. 1963. 2793
Some problems in finite dams with an application to insurance risk. *Zeit. Wahrscheinlichkeitsth.*, **2**, 135–46.
Review: STMA **5** (1964), 759.
Class.: G:pc:S

Weibull, C. 1958. 2794
The distribution of reciprocal choices in sociometric tests. Stat. Inst. Univ. Gothenburg, Publ. 1958, 4, pp. 16.
Review: MR **20** (1959), 1105.

Weida, Frank M. 1935. 2795
On certain distribution functions when the law of the universe is Poisson's first law of error. *Ann. Math. Statist.*, **6**, 102–10.

Weiler, H. 1959. 2796
Sex-ratio and birth control. *Amer. J. Sociology*, **65**, 298–99.

Weiler, H. 1964. 2797
A significance test for simultaneous quantal and quantitative responses. *Technometrics*, **2**, 273–85.
Review: STMA **6** (1965), 921.
Class.: MI:tp:G

Weiler, H. 1965. 2798
The use of incomplete beta functions for prior distributions in binomial sampling. *Technometrics*, **7**, 335–47.
Review: STMA **7** (1966), 571.
Class.: NH:pe-tp-tc:G

Weinberg, George H., Fluckiger, Fritz A., and Tripp, Clarence A. 1960. 2799
A proposed variation of the matching technique. *Psychometrika*, **25**, 291–95.

Weiner, S., Anderson, R. L., Binet, F. E., and Leslie, R. T.
See Binet, F. E., Leslie, R. T., Weiner, S., and Anderson, R. L. (1956).

Weintraub, Sol. 1962. 2800
Cumulative binomial probabilities. *J. Assoc. Comput. Mach.*, **9**, 405–07.
Review: MR **27** (1964), 819; ZBL **107** (1964), 135.
Class.: B:c:G

Weintraub, Sol. 1963. 2801
Tables of the cumulative binomial probability distribution for small values of p. Free Press of Glencoe, New York; Collier-MacMillan, London.

Weiss, G. 1962. 2802
On the pedestrian queueing problem. *Bull. Inst. Internat. Statist.*, **39**, (4) 163–67.
Review: STMA **5** (1964), 512; ZBL **114** (1965), 94.
Class.: MI:pc-mb:G

Weiss, G. H. 1963. 2803
An analysis of pedestrian queueing. *J. Res. Nat. Bur. Standards Sect., B* **67**, 229–43.
Review: STMA **6** (1965), 756.
Class.: MI:pc:G

Weiss, Irving. 1958. 2804
Limiting distributions in some occupancy problems. *Ann. Math. Statist.*, **29**, 878–84.
Users: Jones, N. L. (1959), Békéssy, A. (1963), Sevast'yanov, B. A., and Chistyakov, V. P. (1964), Chistyakov, V. P. (1964b).

Weiss, Lionel. 1962. 2805
A sequential test of the equality of probabilities in a multinomial distribution. *J. Amer. Statist. Assoc.*, **57**, 769–74.
Review: ZBL **114** (1965), 104; STMA **7** (1966), 115.
Class.: M:tp:G

Weiss, L., and Freeman, D.
See Freeman, D., and Weiss, L. (1964).

Welch, B. L., and Jennett, W. J.
See Jennett, W. J., and Welch, B. L. (1939).

Welch, B. L., and Morant, G. M.
See Morant, G. M., and Welch, B. L. (1939).

Welch, P. D. 1965. 2806
On the busy period of a facility which serves customers of several types. *J. Roy. Statist. Soc. Ser. B*, **27**, 361–70.
Review: STMA **7** (1966), 420.
Class.: MI:pc:G

Welch, R. W. 1959. 2807
Some properties of the negative multinomial distribution. Master's Thesis, Pennsylvania State Univ.

Welsh, D. J. A., and Morgan, R. W.
See Morgan, R. W., and Welsh, D. J. A. (1965).

Wembersie, A., and Lellouch, J.
See Lellouch, J., and Wembersie, A. (1966).

Wesler, Oscar. 1959. 2808
A classification problem involving multinomials. *Ann. Math. Statist.*, **30**, 128–33.
Review: MR **21** (1960), 837; IJA **1** (1959), 185; ZBL **89** (1961), 145.
Class.: M:tp:G

Westenberg, J. 1947. 2809
Mathematics of pollen diagrams. *Nederl. Akad. Wetensch. Proc.* **50**, 509–20.
Review: ZBL **30** (1949), 310.

Westoff, Charles F., Potter, Robert G., Jr., and Sagi, Philip C.
See Potter, Robert G., Jr., Sagi, Philip C., and Westoff, Charles F. (1962).

Westwood, J. C. N., Phipps, P. H., and Boulter, E. A. 1957. 2810
The titration of vaccinia virus on the chorioallantoic membrane of the developing chick embryo. *J. Hyg. Camb.*, **55**, 123–39.

Wetherill, G. B. 1960a. 2811
Some remarks on the Bayesian solution of the single sample inspection scheme. *Technometrics*, **2**, 341–52.
Review: MR **22** (1961), 1032; IJA **2** (1961), 718; ZBL **97** (1962), 137.
Users: Wetherill, G. B. (1961), Samuel, E. (1963).
Class.: COB:sqc:E

Wetherill, G. B. 1961. 2812
Bayesian sequential analysis. *Biometrika*, **48**, 281–92.
Review: STMA **6** (1965), 256.
Users: Freeman, D., and Weiss, L. (1964), Pratt, John W. (1966).
Class.: B:tp:G

Wetherill, G. B., and Vagholkar, M. K.
See Vagholkar, M. K., and Wetherill, G. B. (1960).

Wette, R. 1959. 2813
Zur biomathematischen Begründung der Verteilung der Elemente taxonomischer Einheiten des natürlichen Systems in einer logarithmischen Reihe. *Biom. Zeit.*, **1**, 44–50.
Review: ZBL **108** (1964), 326; IJA **1** (1959), 149.
Class.: LS:mb-pc:BM

Wheeler, R. E. 1956. 2814
A variable probability distribution function. *Ann. Math. Statist.*, **27**, 196–99.
Review: MR **17** (1956), 863.
Class.: OBR:mb:G

Whisler, B. F., Mosher, W. W., Jr., and Haight, Frank A.
See Haight, Frank A., Whisler, B. F., and Mosher, W. W., Jr. (1961).

Whitaker, Lucy. 1914–15. 2815
On the Poisson law of small numbers . *Biometrika*, **10**, 36–71.
Users: Cole, L. C. (1946), Binet, F. E. (1953), Bliss, C. I., and Fisher R. A. (1953), Dandekar, V. M. (1955), Turner, M. E., and Eadie, G. S. (1957), Jensen, P. (1959), Waters, W. E., and Henson, W. R. (1959), Bardwell, G. E. (1961).
Class.: P:gf:G

White, G. M. 1959. 2816
Electronic probability generator. *Rev. Sci. Instrum.*, **30**, 825–29.
Review: SA **62**A (1959), Ab. No. 11712.

White, Robert F., and Graca, Joseph G. 1958. 2817
Multinomially grouped response times for the quantal response bioassay. *Biometrics*, **14**, 462–88.
Review: IJA **1** (1959), 94.

White, Robert P., and Greville, T. N. E. 1959. 2818
On computing the probability that exactly k of n independent events will occur . *Soc. Actuar. Trans.*, **11**, 88–99.
Review: MR **21** (1960), 1233.
Class.: MI:c:G

Whiteford, P. B. 1949. 2819
Distribution of woodland plants in relation to succession and clonal growth. *Ecology*, **30**, 199–208.
Users: Curtis, J., and McIntosh, R. (1950), Cain, S. A., and Evans, F. C. (1952), Thomson, G. W. (1952), Bray, J. R. (1962).
Class.: P:h:BM

Whitin, T. M., and Hadley, G.
See Hadley, G., and Whitin, T. M. (1961).

Whitlock, S. C., and Eberhardt, L. 1956. 2820
Large-scale dead deer surveys: methods, results and management implications. *Trans. N. Amer. Wildlife Conf.*, **21**, 555–66.

Whittaker, J. M. 1937. 2821
The shot effect for showers. *Proc. Cambridge Philos. Soc.*, **33**, 451–58.

Whittaker, R. H. 1952. 2822
A study of summer foliage insect communities in the Great Smoky Mountains. *Ecol. Monogr.*, **22**, 1–44.
Class.: LS-MI:mi:BM

Whittaker, R. H. 1956. 2823
Vegetation of the Great Smoky Mountains. *Ecol. Monogr.*, **26**, 1–80.
Class.: B:mi:BM

Whittaker, R. H., and Fairbanks, C. W. 1958. 2824
A study of plankton copepod communities in the Columbia Basin, southeastern Washington. *Ecology*, **39**, 46–65.
User: Hairston, N. G. (1959).
Class.: MI:mi:BM

Whittinghill, Maurice, and Potthoff, Richard F.
See Potthoff, Richard F., and Whittinghill, Maurice (1965).

Whittinghill, Maurice, and Potthoff, Richard F.
See Potthoff, Richard F., and Whittinghill, Maurice (1966).

Whittle, P. 1959. 2825
Quadratic forms in Poisson and multinomial variables. *J. Austral. Math. Soc.*, **1**, 233–40.
Review: MR **30** (1965), 809; IJA **1** (1960), 592.
Class.: P-M:mi:G

Whittlesey, John R. B. 1963. 2826
Incomplete gamma functions for evaluating Erlang process probabilities. *Math. Comp.*, **17**, 11–17.

Whittlesey, John R. B., and Haight, Frank A. 2827
1961–62.
Counting distributions for an Erlang process. *Ann. Inst. Statist. Math.*, **13**, 91–103.
Review: MR **25** (1963), 698; STMA **6** (1965), 757.
Users: Whittlesey, John R. B. (1963), Gaight, F. (1965b).

Widrig, T. M., and Holmes, R. W.
See Homes, R. W., and Widrig, T. M. (1956).

Wierzbowska, T., and Andrezejewski, R.
See Andrezejewski, R., and Wierzbowska, T. (1961).

Wiezorke, Bernard. 1955. 2828
Fehlerabschätzung zum Binomialpapier von Mosteller-Tuckey. *Mitteilungsbl. Math. Statist.*, **7**, 66–70.
Review: ZBL **64** (1956), 140.

Wiid, A. J. B. 1957–58. 2829
On the moments and regression equations of the fourfold negative and fourfold negative factorial binomial distributions. *Proc. Roy. Soc. Edinburgh Sec. A*, **65**, 29–34
Review: ZBL **86** (1961), 133.
User: Bennett, B. M. (1962).
Class.: MNB-MIH:m-osp:G

Wiid, A. J. B., and Steyn, H. S. 1956. 2830
Uitbreidings van die Binomial en Faktoriaal-Binomiale stellings. *Tydskr. Wet. Kuns.*, **16**, 210–17.

Wild, A. J. B., and Steyn, H. S.
See Steyn, H. S., and Wiid, A. J. B. (1958).

Wijngaarden, A., and Kaarsemaker, L.
See Kaarsemaker, L., and Wijngaarden, A. (1952).

Wilenski, H., and Przyborowski, J.
See Przyborowski, J., and Wilenski, H. (1935).

Wilenski, H., and Przyborowski, J.
See Przyborowski, J., and Wilenski, H. (1939).

Wilkins, C. A. 1960. 2831
On two queues in parallel. *Biometrika*, **47**, 198–99.
Review: IJA **2** (1961), 590.
Class.: MI:pc:G

Wilkinson, G. N. 1961. 2832
Note: Estimation of proportion from zero-truncated binomial data. *Biometrics*, **17**, 153–59.
Review: ZBL **97** (1962), 144.
Class.: TCB:pe:G-BM

Wilkinson, R. I. 1942. 2833
The combination of probability curves in engineering. *Trans. Amer. Inst. Elect. Engrs.*, **61**, 953–63.
Review: SA **47**A (1944), Ab. No. 324.

Wilks, S. S. 1938. 2834
Shortest average confidence intervals from large samples. *Ann. Math. Statist.*, **9**, 166–75.

Wilks, S. S. 1959–60. 2835
Recurrence of extreme observations. *J. Austral. Math. Soc.*, **1**, 106–12.
Review: IJA **1** (1960), 381.
Class.: MI:NB:mi:G

Wilks, S. S. 1960. 2836
A two-stage scheme for sampling without replacement. *Bull. Inst. Internat. Statist.*, **37** (2), 241–48.
Review: IJA **2** (1961), 140; ZBL **90** (1961), 111.
Class.: MH:pe:G

Wilks, S. S. 1962. 2837
Mathematical Statistics. John Wiley and Sons, New York, 644 pp.

Wilks, S. S., and Daly, J. F. 1939. 2838
An optimum property of confidence regions associated with the likelihood function. *Ann. Math. Statist.*, **10**, 225–35.
Review: ZBL **23** (1941), 59.

Williams, C. B. 1939–40. 2839
A note on the statistical analysis of sentence length as a criterion of literary style. *Biometrika*, **31**, 365–61.

Williams, C. B. 1944a. 2840
The number of publications written by biologists. *Ann. Eugenics*, **12**, 143–46.
Users: Riddell, W. J. B. (1944), Williams, C. B. (1947b), Kendall, D. G. (1948), Williams, C. B. (1954), Patil, G. P. (1959), Patil, G. P. (1962b), Nelson, W. C., and David, H. A. (1964).
Class.: G-LS:gf:O

Williams, C. B. 1944b. 2841
Some applications of the logarithmic series and the index of diversity to ecological problems. *J. Ecol.*, **32**, 1–44.
Users: Harrison, J. L. (1943), Williams, C. B. (1947), Williams, C. B. (1947b), Williams, C. B. (1947c), Ashby, E. (1948), Kendall, D. G. (1948), Quenouille,

M. H. (1949), Williams, C. B. (1949), Anscombe, F. J. (1950a), Bliss, C. I., and Fisher, R. A. (1953), Brian, M. V. (1953), Myers, E., and Chapman, V. (1953), Williams, C. B. (1953), Hairston, N. G. (1959), Patil, G. P. (1962b), Siromoney, G. (1962), Tsao, C. M. (1962), Clark, P. J., Eckstrom, P. T., and Linden, L. C. (1964), Nelson, W. C., and David, H. A. (1964), Patil, G. P., and Wani, J. K. (1965).
Class.: LS:gf:BM

Williams, C. B. 1946. 2842
Yule's 'Characteristic' and the 'Index of Diversity.' *Nature*, **157**, 482.

Williams, C. B. 1947a. 2843
The generic relations of species in small ecological communities. *J. Animal Ecol.*, **16**, 11–18.
Users: Williams, C. B. (1947b), Williams, C. B. (1949), Bagenal, T. B. (1951), Williams, C. B. (1954), Hairston, N. G. (1959), Nelson, W. C., and David, H. A. (1964).

Williams, C. B. 1947b. 2844
The logarithmic series and its application to biological problems. *J. Ecol.*, **34**, 253–72.
Review: BA **22** (1948), 2807.
Users: Williams, C. B. (1947), Kendall, D. G. (1948), Archibald, E. E. A. (1949a), Quenouille, M. H. (1949), Williams, C. B. (1949), Williams, C. B. (1950), Barnes, H., and Stanbury, F. (1951), Hunter, G. C., and Quenouille, M. H. (1952), Williams, C. B. (1952), Brian, M. V. (1953), Myers, E., and Chapman, V. (1953), Williams, C. B. (1953), Ramabhadran, V. K. (1954), Williams, C. B. (1954), Herdan, G. (1957), Hairston, N. G. (1959), Darwin, J. H. (1960), Siromoney, G. (1962), Nelson, W. C., and David, H. A. (1964), Williamson, E., and Bretherton, M. (1964).
Class.: LS:gf-mi:BM

Williams, C. B. 1947c. 2845
The logarithmic series and the comparison of island floras. *Proc. Linn. Soc. Lond.*, **158**, 104–08.
Users: Williams, C. B. (1949), Williams, C. B. (1950), Williams, C. B. (1954), Nelson, W. C., and David, H. A. (1964).
Class.: LS:ctp:BM

Williams, C. B. 1949. 2846
Jaccard's generic coefficient and coefficient of floral community, in relation to the logarithmic series and the index of diversity. *Ann. Bot. Lond.*, *N.S.*, **13**, 53–58.
Users: Williams, C. B. (1950), Williams, C. B. (1954).
Class.: LS:mi:BM

Williams, C. B. 1950. 2847
The application of the logarithmic series to the frequency of occurrence of plant species in quadrats. *J. Ecol.*, **38**, 107–38.
Users: Williams, C. B. (1954), Dahl, E. (1960), Darwin, J. H. (1960), Nelson, W. C., and David, H. A. (1964).
Class.: LS:gf-mi:BM

Williams, C. B. 1952. 2848
Sequences of wet and of dry days considered in relation to the logarithmic series. *Quart. J. Roy. Meteorol. Soc.*, **78**, 91–96.
Users: Ramabhadran, V. K. (1954), Williams, C. B. (1954), Srinivasan, T. R. (1956), Siromoney, G. (1962), Patil, G. P., and Bildikar, S. (1966a).
Class.: LS:gf:P

Williams, C. B. 1953. 2849
The relative abundance of different species in a wild animal population. *J. Animal Ecol.*, **22**, 14–31.
Users: Williams, C. B. (1954), Hairston, N. G. (1959), Clark, P. J., Eckstrom, P. T., and Linden, L. C. (1964), Nelson, W. C., and David, H. A. (1964), Bliss, C. I. (1965).

Williams, C. B. 1954. 2850
The statistical outlook in relation to ecology. *J. Ecol.*, **42**, 1–13.

Williams, C. B. 1956. 2851
Studies in the history of probability and statistics. IV. A note on an early statistical study of literary style. *Biometrika*, **43**, 248–56.
User: Herdan, G. (1958).
Class.: MI:mi:L

Williams, C. B. 1964. 2852
Patterns in the balance of nature and related problems in quantitative ecology. Academic Press, London and New York.

Williams, C. B., Fisher, R. A., and Corbet, A. S.
See Fisher, R. A., Corbet, A. S., and Williams, C. B. (1943).

Williams, Clyde M., and Overall, John E.
See Overall, John E., and Williams, Clyde M. (1961).

Williams, E. J. 1961a. 2853
The distribution of larvae of randomly moving insects. *Austral. J. Biol. Sci.*, **14**, 598–604.
User: Watson, G. (1956).

Williams, E. J. 1961b. 2854
Fitting a geometric progression to frequencies. *Biometrics*, **17**, 584–606.
Review: IJA **3** (1962), 580.
User: Patil, G. P. (1962c).
Class.: P:pe-gf:G

Williams, J. S., Fletcher, N. T., and Nelson, A. C., Jr.
See Nelson, A. C., Jr., Williams, J. S., and Fletcher, N. T. (1963).

Williamson, Eric, and Bretherton, Michael H. 1963. 2855
Tables of the negative binomial probability distribution. Wiley, London.

Williamson, Eric, and Bretherton, Michael H. 1964. 2856
Tables of the logarithmic series distribution. *Ann. Math. Statist.*, **35**, 284–97.
Review: MR **35** (1964), 336.
Users: Nelson, W. C., and David H. A. (1964), Chatfield, C., Ehrenberg, A. S. C., and Goodhardt, G. J. (1965), Chatfield, C., Ehrenberg, A. S. C., and Goodhardt, C. (1966), Patil, G. P., and Bildikar, S. (1966a).
Class.: LS:tc:G

Willis, D. M. 1964. 2857
The statistics of a particular non-homogeneous Poisson process. *Biometrika*, **51**, 399–404.
Review: STMA **6** (1965), 1064.
Class.: P:pc:G-P

Wilson, Edwin B. 1927. 2858
Probable inference, the law of succession, and statistical inference. *J. Amer. Statist. Assoc.*, **22**, 209–12.
Class.: B:ie:G

Wilson, T. R., and Katz, Leo.
See Katz, Leo, and Wilson, T. R. (1956).

Winokur, Herbert S., Jr., and Margolin, Barry H.
See Margolin, Barry H., and Winokur, Herbert S., Jr. (1966).

Winsor, C. P., and Walford, L. A. 1936. 2859
Sampling variations in the use of plankton nets. *J. Conseil*, **11**, 190–204.
Users: Sette, O. E., and Ahlstrom, E. H. (1948), Barnes, H., and Marshall, S. M. (1951).

Winsten, C. B. 1959. 2860
Geometric distributions in the theory of queues. *J. Roy. Statist. Soc. Ser. B*, **21**, 1–22.
Review: MR **21** (1960), 1238; IJA **1** (1960), 519.
Class.: G:pc:G
Notes: Distribution; *J. Roy. Statist. Soc. Ser. B*, **21**, 22–35.

Winter, A. 1935. 2861
On convergent Poisson convolutions. *Amer. J. Math.*, **57**, 827.

Wintner, Aurel. 1945. 2862
The moment problem of enumerating distributions. *Duke Math. J.*, **12**, 23–25.
Class.: MI:m:G

Wintner, Aurel, and van Kampen, E. R.
See van Kampen, E. R., and Wintner, Aurel (1939).

Wise, M. E. 1946. 2863
The use of the negative binomial distribution in an industrial sampling problem. *J. Roy. Statist. Soc. Suppl.*, **8**, 202–11.
Review: MR **9** (1948), 49.
Users: Wise, E. (1950), Patil, G. P. (1960c).
Class.: NB:sqc:E

Wise, M. E. 1950. 2864
The incomplete beta function as a contour integral and a quickly converging series for its inverse. *Biometrika*, **37**, 208–18.

Wise, M. E. 1954. 2865
A quickly convergent expansion for cumulative hypergeometric probabilities, direct and inverse. *Biometrika*, **41**, 317–29.
Review: MR **16** (1955), 600; ZBL **58** (1958), 343.
Users: Wise, M. E. (1954a), Sandiford, P. J. (1960), Gart, J. J. (1962), Bol'shev, L. N. (1964a), Govindarajulu, Z. (1965), Johnson, N. L. (1965).
Class.: H:mi:G
Notes: Correction; *Biometrika*, **42**, 277.

Wise, M. E. 1954a. 2866
The ratio of two factorials and some fundamental probabilities. *Nederl. Akad. Wetensch. Proc. Ser. A*, **57**, 513–21.
Review: ZBL **58** (1958), 343.
User: Wise, M. E. (1963).

Wise, M. E. 1963. 2867
Multinomial probabilities and the χ^2 and X^2 distributions. *Biometrika*, **50**, 145-54.
Review: STMA **5** (1964), 41.
Users: Patil, G. P., and Bildikar, S. (1966a).
Class.: M:a-gf:G
Note: Correction; *Biometrika*, **50**, (1963) 546.

Wise, M. E. 1964. 2868
A complete multinomial distribution compared with the χ^2 approximation and an improvement of it. *Biometrika*, **51**, 277–81.
Review: MR **30** (1965), 1003; STMA **6** (1965), 468.

Wishart, J. 1949. 2869
Cumulants of multivariate multinomial distributions. *Biometrika*, **36**, 47–58.
Review: ZBL **32** (1950), 73.
Users: Wiid, A. J. B. (1957–8), Tsao, C. M. (1962), Mosimann, J. E. (1963), Capobianco, M. F. (1964), Patil, G. P. (1965), Patil, G. P. (1965), Patil, G. P., and Bildikar, S. (1966a).
Class.: M-MB-MI:osp:G

Wishart, John. 1952. 2870
Moment coefficients of the k-statistics in samples from a finite population. *Biometrika*, **39**, 1–13.
Review: ZBL **46** (1953), 357.

Wishart, J. 1956. 2871
An approximation to the binomial distribution. *Inc. Statist.*, **6**, 124–31.
Class.: B:a:G

Wisniewski, T. K. M. 1966. 2872
Another statistical solution of a combinatorial problem. *Amer. Statistician*, **20**, 25.
Class.: NB-MI:osp:G

Wohlschlag, D. E. 1952. 2873
Estimation of fish populations in a fluctuating reservoir. *California Fish and Game*, **38**, 63–72.
Class.: P:ie:BM

Wohlschlag, D. E. 1954. 2874
Mortality rates of whitefish in an arctic lake. *Ecology*, **35**, 388–96.
Class.: M-NB:pe:BM

Wold, Herman. 1965. 2875
Bibliography on time series and stochastic processes. Oliver and Boyd, Edinburgh and London, 516 pp.

Wolfowitz, J. 1944. 2876
Asymptotic distribution of runs up and down. *Ann. Math. Statist.*, **15**, 163–72.
Users: Olmstead, P. S. (1946), Govindarajulu, Z. (1965).

Wolfowitz, J. 1946. 2877
On sequential binomial estimation. *Ann. Math. Statist.*, **17**, 489–93.
Users: Savage, L. J. (1947), Wolfowitz, J. (1947), Plackett, R. L. (1948), Wasan, M. T. (1965), deGroot, M. H. (1959).
Class.: B:se:G

Wolfowitz, J. 1947. 2878
Consistency of sequential binomial estimates. *Ann. Math. Statist*, **18**, 131–35.
Review: ZBL **32** (1950), 420.
Class.: B:se:G

Wolfowitz, J., and Kiefer, J.
See Kiefer, J., and Wolfowitz, J. (1956).

Wolfowitz, J., and Kiefer, J.
See Kiefer, J., and Wolfowitz, J. (1959).

Woll, J. W., Jr. 1959. 2879
Homogeneous stochastic processes. *Pacific J. Math.*, **9**, 293–325.
Class.:

Wood, J. E., Davis, D. E., and Komarek, E. V. 2880
1958.
The distribution of fox populations in relation to vegetation in southern Georgia. *Ecology*, **39**, 160–62.
Class.: P:gf:BM

Woodbury, Max A. 1949. 2881
On a probability distribution. *Ann. Math. Statist.*, **20**, 311–13.
Review: MR **10** (1949), 720; ZBL **41** (1952), 250.
Users: Rutherford, R. S. G. (1954), Tsao, C. M. (1962).
Class.: OBR:mb:G

Woods, H. M., and Greenwood, M.
See Greenwood, M., and Woods, H. M. (1919).

Worcester, J. 1954. 2882
How many organisms. *Biometrics*, **10**, 227–34.

Wynn, A. H. A., Maguire, B. A., and Pearson, E. S.
See Maguire, B. A., Pearson, E. S., and Wynn, A. H. A. (1952).

Wynn, A. H. A., Maguire, B. A., and Pearson, E. S.
See Maguire, B. A., Pearson, E. S., and Wynn, A. H. A. (1953).

Y

Yamamoto, S., Satomi, H., Outi, Y., Okada, T., and Kuno, E.
See Kuno, E., Yamamoto, S., Satomi, H., Outi, Y., and Okada, T. (1963).

Yamasaki, Mitsuru, and Ishii, Goro.
See Ishii, Goro, and Yamasaki, Mitsuru (1960–61).

Yamato, H., Yanagawa, T., Kudô, A., and Okuma, A.
See Kudô, A., Okuma, A., Yamato, H., and Yanagawa, T. (1965).

Yassky, D. 1962. 2883
A model for the kinetics of phage attachment to bacteria in suspension. *Biometrics*, **18**, 185–91.
Class.: MI-P-B-G:mb:BM

Yazima, T., Kitagawa, T., and Huruya, S.
See Kitagawa, T., Huruya, S., and Yazima, T. (1942).

Yeh, Andrew T., Fu, Yumin, and Srinath, Mandyam D.
See Fu, Yumin, Srinath, Mandyam, D., and Yeh, Andrew T. (1965).

Yeo, G. F., and Weesakul, B.
See Weesakul, B., and Yeo, G. F. (1963).

Yntema, L. 1954. 2884
Einiges zur Wahrscheinlichkeitsansteckung. *Ned. Verzek. Actuar. Bij.*, **31**, 86–91.

Yoneda, K. 1962. 2885
Estimations in some modified Poisson distributions. *Yokohama Math. J.*, **10**, 63–95.
Review: ZBL **124** (1966), 100.
Class.: OPR:pe-gf:G

Yoshimura, I. 1963a. 2886
A moment recurrence relation and its application to multinomial distributions. *Rep. Statist. Appl. Res. Un. Japan. Sci. Engrs.*, **10**, 1–14.

Yoshimura, I. 1963b. 2887
A moment recurrence relation and its application to multinomial distributions and others. *Rep. Statist. Appl. Res. Un. Japan. Sci. Engrs.*, **10**, 137–50.
Review: MR **27** (1964), 595.

Yoshimura, I. 1964a. 2888
Unified system of cumulant recurrence relations. *Rep. Statist. Appl. Res. Un. Japan. Sci. Engrs.*, **11**, 1–8.
Review: MR **30** (1965), 1002.
Users: Patil, G. P., and Bildikar, S. (1966a).

Yoshimura, I. 1964b. 2889
A complementary note on the multivariate moment recurrence relation. *Rep. Statist. Appl. Res. Un. Japan. Sci. Engrs.*, **11**, 9–12.
Review: MR **30** (1965), 1002.
Users: Patil, G. P., and Bildikar, S. (1966a).

Yoshimura, I., Shimizu, Ryoichi, and Sibuyo, Masaaki.
See Sibuyo, Masaaki, Yoshimura, I., and Shimizu, Ryoichi (1964).

Yosida, T. 1954. 2890
The relation between the population density and the pattern of distribution of the rice-plant skipper, *parara guttata Bremer et Grey* (Japanese; English summary). *Ova-Kontyu*, **9**, 129–34.

Yosida, Tosiharu, Watanabe, Syozi, Kono, Tatsuro, and Utida, Synnro.
See Kono, Tatsuro, Utida, Synnro, Yosida, Tosiharu, and Watanabe, Syozi (1952).

Young, D. A., Owen, D. B., Gilbert, E. J., and Steck, G. P.
See Owen, D. B., Gilbert, E. J., Steck, G. P., and Young, D. A. (1959).

Young, D. H. 1962. 2891
Two alternatives to the standard χ^2-test of the hypothesis of equal cell frequencies. *Biometrika*, **49**, 107–16.
Review: STMA **6** (1965), 174.
Class.: M:tp:G

Young, D. H., and Johnson, N. L.
See Johnson, N. L., and Young, D. H. (1960).

Young, H., Neess, J., and Emlen, J. T., Jr. 1952. 2892
Heterogeneity of trap response in a population of house mice. *J. Wildlife Managem.*, **16**, 169–80.
User: Geiss, A. D. (1955).

Youngs, J. W. T., Geisler, M. A., and Mirkovich, 2893
A. R. 1954.
Confidence intervals for Poisson parameters in logistics research. Res. Memo. RM-1357, RAND Corp.

Youtz, C., and Mosteller, F.
See Mosteller, F., and Youtz, C. (1961).

Yueh, M. I. 1959. 2894
On the problem of M/M/n in the theory of queues (Chinese). *Acta Math. Sinica*, **9**, 494–502.
Review: IJA **2** (1961), 215.
Class.: MI:pc:G

Yule, G. U. 1912–14. 2895
Fluctuations of sampling in Mendelian ratios. *Proc. Cambridge Philos. Soc.*, **17**, 425–32.
Class.: B:gf:BM

Yule, G. U. 1924. 2896
A mathematical theory of evolution based on the conclusions of Dr J. C. Willis, F.R.S. *Phil. Trans. Roy. Statist. Soc. B*, **213**, 21–97.
Users: Feller (1949), Simon, H. A. (1955), Bulmer, M. G. (1958b), Simon, H. (1960), Bartko, J. J. (1961b), Irwin, J. O. (1963), Nelson, W. C., and David, H. A. (1964).

Yule, G. U. 1938. 2897
On sentence-length as a statistical characteristic of style in prose: with application to two cases of disputed authorship. *Biometrika*, **30**, 363–90.
User: Williams, C. B. (1956).
Class.: MI:mi:L

Yule, G. U. 1944. 2898
The statistical study of literary vocabulary. Uni. Press, Cambridge.

Yule, G. U., and Chambers, E. G.
See Chambers, E. G., and Yule, G. U. (1941).

Yule, G. U., and Greenwood, M.
See Greenwood, M., and Yule, G. U. (1920).

Z

Zacks, S., and Kander, Z.
See Kander, Z., and Zacks, S. (1966).

Zeller, K., and Meyer-König, W.
See Meyer-König, W., and Zeller, K. (1959).

Zellner, A., and Lee, T. H. 1963. 2895-2904
Joint estimation of relationships involving discrete random variables. Paper 6307 Systems Formulation and Methodology Workship, Univ. Wisconsin.
User: Huang (1965).

Zellner, Arnold, and Lee, Tong Hun. 1965. 2900
Joint estimation of relationships involving discrete random variables. *Econometrica*, **33**, 382–94.
Review: MR **31** (1966), 502.

Zipf, G. K. 1932. 2901
Selected studies of the principle of relative frequency in language. Harvard Univ. Press.

Zipf, G. K. 1949. 2902
Human behaviour and the principle of least effort; an introduction to human ecology. Addision-Wesley Press, Cambridge Mass.

Zippin, C. 1958. 2903
The removal method of population estimation. *J. Wildlife Managem.*, **22**, 82–90.

Zolotarev, V. M. 1964. 2904
Distribution of the length of a queue and of the number of operating servers in a system of Erlang type with random breakdowns and renewal of servers (Russian). *Trudy Mat. Inst. Steklov.*, **71**, 51–61.
Review: MR **30** (1965), 1002.
Class.: MI:pc:G

Zubrzycki, S., and Aczel, J.
See Aczel, J., and Zubrzycki, S. (1956).

Zubrzycki, S., and Steinhaus, H.
See Steinhaus, H., and Zubrzycki, S. (1957).

Zubrzycki, S., and Steinhaus, H.
See Steinhaus, H., and Zubrzycki, S. (1958).

Chapter 3

CLASSIFICATION ACCORDING TO DISTRIBUTIONS AND INFERENCE

3.1 Introduction

In this chapter, the entries of the bibliography of Chapter 2 are classified according to the 44 classes of distributions and 22 classes of statistical inference as listed in Section 2.2. The classification here is complete only to the extent of the individual entries for which classification is available in Chapter 2. It is hoped that one would come to know of most of the entries not classified here through the users listings available for the individual entries in Chapter 2.

3.2 Classification for Binomial Distribution

(1) *Anova and Transformations:* 62, 150, 194, 222, 224, 302, 309, 314, 512, 560, 576, 577, 671, 696, 895, 918, 947, 948, 1010, 1380, 1625, 1853, 2335.

(2) *Approximations, Asymptotics:* 22, 102, 159, 161, 213, 228, 314, 347, 394, 423, 460, 467, 470, 482, 647, 759, 830, 855, 862, 864, 1088, 1097, 1195, 1227, 1364, 1394, 1535, 1567, 1593, 1701, 1703, 1783, 1784, 1914, 2043, 2044, 2312, 2357, 2390, 2643, 2677, 2705, 2706, 2707, 2763, 2871.

(3) *Comparison of Models:* 493, 1219, 2053, 2520.

(4) *Comparison of Two Populations:* 72, 266, 275, 517, 518, 578, 586, 604, 627, 639, 687, 696, 1062, 1285, 1364, 1423, 1453, 1626, 1689, 1922, 2033, 2152, 2500, 2639, 2644.

(5) *Computations:* 467, 470, 1205, 1505, 1709, 1780, 1811, 2252, 2344, 2456, 2800.

(6) *Goodness of Fit:* 194, 269, 419, 475, 508, 511, 512, 543, 581, 599, 781, 849, 990, 1004, 1034, 1092, 1120, 1319, 1448, 1566, 1592, 1657, 1845, 1859, 1963, 1964, 2009, 2043, 2080, 2179, 2368, 2520, 2521, 2895.

(7) *Homogeneity:* 143, 161, 224, 225, 236, 403, 475, 511, 575, 637, 2088, 2111, 2179, 2441, 2458.

(8) *Index of Dispersion:* 224.

(9) *Interval Estimation:* 4, 149, 278, 314, 321, 446, 450, 462, 475, 572, 573, 658, 799, 847, 855, 940, 981, 1218, 1246, 1248, 1698, 1806, 1899, 1922, 1974, 2063, 2080, 2095, 2121, 2236, 2309, 2331, 2343, 2479, 2503, 2505, 2608, 2756, 2764, 2858.

(10) *Miscellaneous:* 6, 10, 50, 120, 171, 242, 280, 431, 432, 433, 476, 493, 511, 543, 580, 719, 755, 878, 988, 997, 1040, 1042, 1043, 1062, 1079, 1089, 1103, 1158, 1275, 1276, 1304, 1320, 1412, 1413, 1585, 1616, 1617, 1636, 1788, 1849, 1875, 1885, 1892, 1924, 1955, 2047, 2113, 2133, 2163, 2164, 2165, 2170, 2171, 2190, 2238, 2282, 2322, 2360, 2408, 2532, 2533, 2602, 2610, 2636, 2644, 2758, 2765, 2823.

(11) *Model Building:* 117, 161, 493, 849, 967, 1023, 1394, 1510, 1859, 2052, 2374, 2519, 2664, 2667, 2883.

(12) *Moments:* 95, 337, 476, 646, 660, 854, 927, 960, 1204, 1273, 1361, 1394, 1410, 1423, 1448, 1511, 1514, 1900, 2237, 2265, 2315, 2489, 2491, 2611, 2664, 2675.

(13) *Order Statistics:* 2, 646, 1144, 1145, 2382.

(14) *Other Structural Properties:* 129, 561, 770, 781, 1157, 1329, 1401, 1669, 1670, 1782, 1841, 1957, 2023, 2024, 2026, 2173, 2475, 2482, 2637, 2669, 2697, 2740.

(15) *Point Estimation:* 43, 97, 106, 157, 228, 280, 435, 438, 462, 529, 538, 543, 578, 595, 599, 601, 644, 689, 726, 798, 822, 831, 847, 862, 891, 916, 935, 980, 1015, 1021, 1023, 1026, 1205, 1212, 1285, 1368, 1448, 1450, 1451, 1459, 1631, 1657, 1698, 1711, 1718, 1855, 1856, 1879, 1891, 1902, 1947, 2003, 2009, 2095, 2109, 2256, 2287, 2298, 2308, 2315, 2350, 2416, 2442, 2456, 2458, 2475, 2484, 2518, 2652, 2724, 2727, 2772.

(16) *Processes and Chains:* 25, 69, 543, 1164, 1482, 1496, 1693, 1710, 1799, 1877, 1879, 1881, 2455.

(17) *Regression and Prediction:* 29, 1327.

(18) *Selection and Ranking Problems:* 1145, 1147, 2034, 2590.

(19) *Sequential Estimation:* 54, 84, 462, 631, 725, 1526, 1706, 2093, 2297, 2776, 2778, 2877, 2878.

(20) *SQC and AS:* 397, 404, 540, 715, 761, 763, 799, 874, 916, 1112, 1128, 1194, 1195, 1197, 1198, 1199, 1200, 1201, 1202, 1701, 1794, 1795, 2005, 2330, 2405, 2517, 2652.

(21) *Tabulation and Charts:* 72, 84, 321, 347, 460, 473, 511, 569, 572, 627, 658, 1120, 1128, 1194, 1201, 1250, 1364, 1612, 1626, 1685, 1698, 1853, 1974, 2133, 2143, 2179, 2261, 2616.

(22) *Test on Parameters:* 25, 42, 54, 68, 85, 117, 119, 250, 256, 278, 295, 406, 572, 579, 601, 603, 604, 630, 696, 706, 718, 770, 830, 847, 855, 1019, 1020, 1213, 1286, 1298, 1372, 1414, 1459, 1486, 1505, 1559, 1560, 1586, 1593, 1637, 1684, 1685, 1689, 1698, 1705, 1706, 1776, 1777, 1778, 1779, 1780, 1781, 2063, 2093, 2121, 2122, 2248, 2261, 2296, 2362, 2484, 2485, 2486, 2494, 2639, 2677, 2694, 2703, 2755, 2764, 2781, 2812.

3.3. Classification for Borel-Tanner Distribution

(10) *Miscellaneous:* 1832.

(12) *Moments:* 1177.

(21) *Tabulation and Charts:* 1177.

3.4. Classification for Compound Binomial Distribution

(2) *Approximations, Asymptotics:* 1151, 1570, 2522.

(3) *Comparison of Models:* 2372.

(5) *Computations:* 1727, 1728, 2411, 2412, 2413.

(6) *Goodness of Fit:* 508, 1092, 1151, 1241, 1436, 1438, 1490, 1658, 1659, 1729, 1730, 1763, 1860, 2481, 2522.

(7) *Homogeneity:* 2249.

(8) *Index of Dispersion:* 2372.

(10) *Miscellaneous:* 749, 1658, 2007.

(11) *Model Building:* 1151, 1490, 1658, 1659, 1660, 1860, 2522.

(12) *Moments:* 1965, 1966.

(14) *Other Structural Properties:* 1590, 1965, 2178, 2481.

(15) *Point Estimation:* 112, 113, 298, 299, 1241, 1436, 1438, 1490, 1660, 1727, 2178, 2249, 2412, 2413.

(16) *Processes and Chains:* 1799.

(17) *Regression and Prediction:* 1966.

(20) *SQC and AS:* 134, 2693, 2811.
(21) *Tabulation and Charts:* 1436.
(22) *Test on Parameters:* 114.

3.5. Classification for Compound Geometric Distribution

(6) *Goodness of Fit:* 2083.
(11) *Model Building:* 2083.

3.6. Classification for Compound Multinomial Distribution

(10) *Miscellaneous:* 2021.
(14) *Other Structural Properties:* 1846.
(15) *Point Estimation:* 1846.

3.7. Classification for Compound Negative Binomial Distribution

(2) *Approximations, Asymptotics:* 782, 1437, 2522.
(5) *Computations:* 2412, 2413.
(6) *Goodness of Fit:* 1437, 2522.
(11) *Model Building:* 2086, 2522.
(12) *Moments:* 782, 1437.
(14) *Other Structural Properties:* 1847.
(15) *Point Estimation:* 1437, 1847, 2412, 2413.

3.8. Classification for Compound Negative Multinomial Distribution

(10) *Miscellaneous:* 2021.

3.9. Classification for Compound Poisson Distribution

(2) *Approximations, Asymptotics:* 993.
(6) *Goodness of Fit:* 249, 306, 1530, 1860, 2328, 2451.
(10) *Miscellaneous:* 513, 863, 941, 2007, 2124, 2599, 2597.
(11) *Model Building:* 263, 863, 1268, 1860, 2451.
(12) *Moments:* 1402, 1966, 2328.
(14) *Other Structural Properties:* 263, 1150, 1679, 2123, 2128, 2325, 2685.
(15) *Point Estimation:* 941, 2661, 2662.
(16) *Processes and Chains:* 8, 1175, 1402, 2065, 2123, 2125.
(17) *Regression and Prediction:* 1966.

3.10. Classification for Discrete Lognormal Distribution

(3) *Comparison of Models:* 493.
(10) *Miscellaneous:* 65, 493.
(11) *Model Building:* 493.

3.11. Classification for Generalized Binomial Distribution

(2) *Approximations, Asymptotics:* 2522, 2529.
(5) *Computations:* 2529.
(6) *Goodness of Fit:* 588, 598, 1492, 2084, 2522, 2529.
(11) *Model Building:* 2522.
(15) *Point Estimation:* 588, 598, 1492, 2084.
(21) *Tabulation and Charts:* 588.

3.12. Classification for Generalized Binomial Distribution of Poisson

(2) *Approximations, Asymptotics:* 1088, 1284, 1700,
(9) *Interval Estimation:* 2378.
(10) *Miscellaneous:* 1953.
(12) *Moments:* 551, 1965.
(14) *Other Structural Properties:* 685, 1290, 1965, 2300.
(16) *Processes and Chains:* 1041.
(20) *SQC and AS:* 1700, 1701.

3.13. Classification for Generalized Geometric Distribution

(6) *Goodness of Fit:* 2083.
(11) *Model Building:* 2083.

3.14. Classification for Generalized Multinomial Distribution

3.15. Classification for Generalized Negative Binomial Distribution

(2) *Approximations, Asymptotics:* 1088, 2529.
(5) *Computations:* 2529.
(6) *Goodness of Fit:* 2529.

3.16. Classification for Generalized Poisson Distribution

(2) *Approximations, Asymptotics:* 1151, 1437, 1467, 2529.
(3) *Comparison of Models:* 1443.

(5) *Computations:* 2411, 2412, 2413, 2529.
(6) *Goodness of Fit:* 249, 1151, 1436, 1437, 1438, 1443, 1467, 1529, 1530, 1531, 1533, 1543, 1692, 1763, 1890, 2080, 2328, 2529, 2692.
(9) *Interval Estimation:* 2080.
(10) *Miscellaneous:* 863, 1681, 1983.
(11) *Model Building:* 863, 1151, 1443.
(12) *Moments:* 1437, 1467, 2328.
(14) *Other Structural Properties:* 1150, 1463, 1467, 1679, 2328, 2576.
(15) *Point Estimation:* 1153, 1437, 1438, 1436, 1467, 1890, 2412, 2413, 2431.
(21) *Tabulation and Charts:* 1436.

3.17. Classification for Geometric Distribution

(2) *Approximations, Asymptotics:* 49, 188.
(3) *Comparison of Models:* 638.
(6) *Goodness of Fit:* 427, 508, 587, 1025, 1305, 2083, 2085, 2087, 2487, 2606, 2840.
(9) *Interval Estimation:* 571.
(10) *Miscellaneous:* 2170, 2172, 2789.
(11) *Model Building:* 49, 327, 2083, 2085, 2086, 2883.
(12) *Moments:* 587, 2696.
(14) *Other Structural Properties:* 103, 654, 659, 870, 1670, 2024, 2415, 2697.
(15) *Point Estimation:* 529, 587, 1718, 2626, 2627.
(16) *Process and Chains:* 458, 682, 1165, 2065, 2793, 2860.
(19) *Sequential Estimation:* 2297.
(21) *Tabulation and Charts:* 571, 587.
(22) *Test on Parameters:* 2296, 2626, 2627.

3.18. Classification for Hypergeometric Distribution

(2) *Approximations, Asymptotics:* 468, 709, 710, 830, 1088, 1593, 1914, 2043, 2044, 2312, 2337, 2677.
(3) *Comparison of Models:* 1219.
(4) *Comparison of Two Populations:* 1285, 1423.
(5) *Computations:* 468.
(6) *Goodness of Fit:* 1964, 2043.
(9) *Interval Estimation:* 1447, 1973.
(10) *Miscellaneous:* 280, 1082, 1086, 1383, 2048, 2113, 2170, 2322, 2865.
(11) *Model Building:* 1466.
(12) *Moments:* 96, 1410, 1423, 1466, 2045, 2237, 2266, 2489.
(14) *Other Structural Properties:* 1973.
(15) *Point Estimation:* 280, 523, 528, 1916, 1015, 1285, 1466, 2307, 2416.
(19) *Sequential Estimation:* 1526.
(20) *SQC and AS:* 562, 715, 916, 1198, 1971, 1973.
(21) *Tabulation and Charts:* 562.
(22) *Test on Parameters:* 523, 830, 1593, 2677.

3.19. Classification for Inverse Factorial Series Distribution

(6) *Goodness of Fit:* 1341.
(22) *Test on Parameters:* 1341.

3.20. Classification for Inverse Hypergeometric Distribution

(10) *Miscellaneous:* 1753, 2151.
(11) *Model Building:* 1466.
(12) *Moments:* 1410, 1466, 2656.
(15) *Point Estimation:* 106, 1466, 2307.

3.21. Classification for Logarithmic Series Distribution

(1) *Anova and Transformations:* 1395.
(4) *Comparison of Two Populations:* 2845.
(5) *Computations:* 1090.
(6) *Goodness of Fit:* 306, 418, 534, 535, 567, 792, 897, 1305, 1395, 2128, 2161, 2489, 2840, 2841, 2844, 2847, 2848.
(10) *Miscellaneous:* 65, 77, 78, 675, 1265, 1431, 1868, 2021, 2170, 2711, 2822, 2844, 2846, 2847.
(11) *Model Building:* 151, 534, 535, 897, 1469, 2813.
(12) *Moments:* 1410, 2121, 2031.
(14) *Other Structural Properties:* 129, 2031, 2146, 2178, 2444, 2482.
(15) *Point Estimation:* 308, 897, 1999, 2008, 2003, 2020, 2028, 2031, 2178, 2711.
(16) *Processes and Chains:* 151, 1469, 2813.
(21) *Tabulation and Charts:* 2028, 2031, 2856.

3.22. Classification for Miscellaneous Distributions

(1) *Anova and Transformations:* 150, 193, 194, 886, 1119, 1255, 1394.
(2) *Approximations, Asymptotics:* 49, 142, 156, 164, 188, 204, 205, 206, 209, 210, 290, 353, 355, 357, 358, 429, 439, 440, 500, 506, 550, 555, 681, 768, 775, 837, 838, 899, 945, 973, 974, 975, 992, 1059, 1131, 1140, 1352, 1400, 1418, 1437, 1446, 1455, 1467, 1535, 1574, 1653, 1703, 1724, 1772, 1831, 1952, 2185, 2204, 2219, 2323, 2347, 2349, 2380, 2708, 2746.
(3) *Comparison of Models:* 183, 493, 667, 1443, 2082.
(4) *Comparison of Two Populations:* 254, 517, 933, 1062, 1283, 1359, 1423, 2639, 2767.
(5) *Computations:* 1059, 1505, 2745, 2818.
(6) *Goodness of Fit:* 154, 194, 195, 196, 469, 542, 581, 602, 605, 688, 781, 1120, 1249, 1305, 1375, 1425, 1437, 1443, 1448, 1467, 1495, 1614, 1654, 1724, 1901, 2082, 2128, 2219, 2367, 2369, 2370, 2420, 2424, 2425, 2451.
(7) *Homogeneity:* 439, 440, 703, 717, 1521, 2389.
(9) *Interval Estimation:* 527, 1069, 1070, 1863, 1899, 2095, 2767.
(10) *Miscellaneous:* 20, 65, 73, 136, 173, 176, 177, 178, 179, 180, 204, 207, 208, 227, 270, 291, 300, 313, 392, 457, 465, 476, 483, 493, 497, 499, 545, 554, 574, 605, 607, 622, 657, 663, 665, 666, 669, 677, 690, 692, 697, 699, 707, 739, 754, 756, 777, 813, 844, 845, 863, 922, 930, 939, 946, 988,

1006, 1007, 1011, 1012, 1062, 1064, 1065, 1071, 1074, 1078, 1094, 1239, 1263, 1266, 1272, 1274, 1280, 1336, 1340, 1420, 1431, 1468, 1475, 1557, 1591, 1616, 1633, 1634, 1632, 1682, 1683, 1719, 1752, 1762, 1775, 1813, 1855, 1882, 1884, 1885, 1909, 1970, 2037, 2109, 2157, 2190, 2199, 2205, 2206, 2207, 2214, 2223, 2234, 2243, 2244, 2245, 2248, 2304, 2311, 2321, 2322, 2354, 2385, 2426, 2445, 2471, 2491, 2532, 2540, 2565, 2569, 2594, 2597, 2599, 2602, 2607, 2632, 2645, 2646, 2678, 2691, 2698, 2789, 2822, 2824, 2835, 2851, 2897.

(11) *Model Building:* 49, 151, 162, 165, 183, 240, 493, 500, 650, 683, 684, 688, 690, 691, 730, 737, 863, 872, 1007, 1039, 1054, 1072, 1140, 1174, 1268, 1331, 1338, 1356, 1358, 1394, 1425, 1443, 1446, 1449, 1495, 1677, 1720, 1724, 1901, 1904, 1908, 1942, 1948, 1976, 1985, 2082, 2169, 2183, 2185, 2365, 2367, 2417, 2420, 2421, 2422, 2423, 2424, 2451, 2591, 2617, 2630, 2708, 2743, 2746, 2802, 2883.

(12) *Moments:* 156, 164, 284, 330, 476, 550, 705, 743, 908, 1140, 1356, 1357, 1359, 1394, 1406, 1415, 1423, 1428, 1433, 1437, 1441, 1442, 1448, 1467, 1495, 1989, 2089, 2349, 2417, 2492, 2573, 2746, 2862.

(13) *Order Statistics:* 2, 1274, 1493, 2323, 2332, 2333, 2657.

(14) *Other Structural Properties:* 103, 164, 240, 317, 659, 770, 781, 999, 1140, 1150, 1411, 1428, 1433, 1449, 1467, 1668, 1670, 1904, 2023, 2024, 2026, 2120, 2319, 2370, 2402, 2417, 2670, 2743, 2869, 2872.

(15) *Point Estimation:* 100, 109, 289, 300, 319, 320, 435, 516, 525, 527, 529, 545, 574, 601, 602, 683, 684, 688, 730, 737, 828, 891, 934, 945, 954, 980, 1047, 1052, 1067, 1070, 1072, 1249, 1278, 1328, 1400, 1437, 1439, 1448, 1450, 1467, 1495, 1520, 1718, 1891, 1902, 1975, 1990, 2095, 2177, 2256, 2361, 2387, 2399, 2417, 2622, 2720, 2766.

(16) *Processes and Chains:* 16, 18, 70, 116, 151, 209, 210, 211, 217, 218, 240, 254, 281, 283, 290, 312, 391, 412, 501, 506, 556, 650, 652, 682, 742, 768, 778, 876, 880, 881, 882, 883, 899, 922, 924, 978, 1033, 1035, 1036, 1045, 1049, 1058, 1060, 1140, 1164, 1167, 1168, 1169, 1175, 1254, 1259, 1455, 1506, 1507, 1513, 1515, 1516, 1518, 1614, 1693, 1799, 1802, 1824, 1825, 1826, 1904, 1920, 1942, 2098, 2118, 2212, 2446, 2455, 2514, 2551, 2569, 2567, 2581, 2641, 2802, 2803, 2806, 2831, 2894, 2904.

(17) *Regression and Prediction:* 516, 2510.

(18) *Selection and Ranking Problems:* 1014, 1829.

(19) *Sequential Estimation:* 64, 2297, 2672.

(20) *SQC and AS:* 133, 1550, 1551, 1548, 1863.

(21) *Tabulation and Charts:* 835, 876, 1120, 1169, 1314, 1359, 1428, 1571, 1948, 2060, 2657.

(22) *Test on Parameters:* 26, 181, 300, 451, 582, 601, 737, 769, 770, 1069, 1286, 1372, 1374, 1415, 1505, 1857, 2406, 2580, 2631, 2639, 2655, 2797.

3.23. Classification for Multinomial Distribution

(1) *Anova and Transformations:* 745, 2275.

(2) *Approximations, Asymptotics:* 901, 905, 1053, 1059, 1293, 1386, 1382, 1561, 1593, 1783, 1784, 2215, 2666, 2705, 2706, 2707, 2867.

(3) *Comparison of Models:* 638.

(4) *Comparison of Two Populations:* 260, 261, 1940, 2769.

(5) *Computations:* 1053, 1059.

(6) *Goodness of Fit:* 273, 548, 549, 1203, 1256, 1295, 1323, 1692, 1854, 1939, 2130, 2176, 2521, 2867.
(7) *Homogeneity:* 2111.
(9) *Interval Estimation:* 1038, 1075, 2769.
(10) *Miscellaneous:* 199, 428, 544, 623, 628, 663, 1103, 1280, 1785, 1878, 1945, 2021, 2106, 2130, 2149, 2426, 2533, 2593, 2825.
(11) *Model Building:* 1929, 1946, 2052, 2376, 2480.
(12) *Moments:* 640, 1382, 2012, 2326, 2480.
(13) *Order Statistics:* 163, 1561.
(14) *Other Structural Properties:* 1329, 1946, 2012, 2016, 2130, 2326, 2480, 2482, 2483, 2869.
(15) *Point Estimation:* 88, 97, 187, 255, 544, 1382, 1504, 1821, 1946, 2012, 2174, 2175, 2177, 2483, 2649, 2874.
(16) *Processes and Chains:* 494, 1075, 1260.
(17) *Regression anE Prediction:* 2510.
(19) *Sequential Estimation:* 631.
(20) *SQC and AS:* 636.
(21) *Tabulation and Charts:* 1915, 1916, 2153, 2604.
(22) *Test on Parameters:* 28, 253, 256, 257, 521, 745, 951, 1213, 1292, 1293, 1382, 1386, 1593, 1607, 1916, 2130, 2176, 2247, 2628, 2708, 2805, 2891.

3.24. Classification for Multivariate Binomial Distribution

(2) *Approximations, Asymptotics:* 2513.
(10) *Miscellaneous:* 656, 1044, 1688.
(11) *Model Building:* 2527.
(12) *Moments:* 2513.
(14) *Other Structural Properties:* 1664, 2869.
(17) *Regression and Prediction:* 2513.
(21) *Tabulation and Charts:* 2544.
(22) *Test on Parameters:* 2527, 2544.

3.25. Classification for Multivariate Hypergeometric Distribution

(2) *Approximations, Asymptotics:* 2046, 2508, 2509, 2512, 2513, 2705.
(10) *Miscellaneous:* 663.
(11) *Model Building:* 2667.
(12) *Moments:* 2158, 2159, 2160, 2507, 2508, 2509, 2513, 2656.
(14) *Other Structural Properties:* 2508.
(15) *Point Estimation:* 2649, 2836.
(17) *Regression and Prediction:* 2507, 2510, 2513.

3.26. Classification for Multivariate Hypergeometric Related Distributions

(2) *Approximations, Asymptotics:* 1217.
(9) *Interval Estimation:* 1458.
(11) *Model Building:* 1458, 2326.
(14) *Other Structural Properties:* 2326.
(20) *SQC and AS:* 1217.

3.27. Classification for Multivariate Inverse Hypergeometric Distribution

(2) *Approximations, Asymptotics:* 2508, 2509, 2512.
(12) *Moments:* 2507, 2508, 2509, 2529.
(14) *Other Structural Properties:* 2508, 2829.
(17) *Regression and Prediction:* 2507, 2510.

3.28. Classification for Multivariate Negative Binomial Distribution

(2) *Approximations, Asymptotics:* 2513.
(6) *Goodness of Fit:* 810, 811.
(10) *Miscellaneous:* 2528.
(11) *Model Building:* 2528.
(12) *Moments:* 1756, 2513, 2829.
(14) *Other Structural Properties:* 2829.
(17) *Regression and Prediction:* 2513.

3.29. Classification for Multivariate Poisson Distribution

(2) *Approximation, Asymptotics:* 965.
(3) *Comparison of Models:* 911, 1717, 2791.
(6) *Goodness of Fit:* 474, 1716.
(10) *Miscellaneous:* 474, 794, 1044, 1958, 2528.
(12) *Moments:* 474, 2528.
(14) *Other Structural Properties:* 1564.
(15) *Point Estimation:* 794.
(17) *Regression and Prediction:* 474.
(22) *Test on Parameters:* 26.

3.30. Classification for Multivariate Power Series Distribution

(12) *Moments:* 2012.
(14) *Other Structural Properties:* 2012, 2016.
(15) *Point Estimation:* 2012.

3.31. Classification for Negative Binomial Distribution

(1) *Anova and Transformations:* 62, 63, 194, 305, 310, 314, 696, 1230, 1255, 1321, 1581.
(2) *Approximations, Asymptotics:* 148, 161, 314, 506, 1088, 1395, 1574, 1701, 2056, 2137.
(3) *Comparison of Models:* 318, 493, 911, 2519.
(4) *Comparison of Two Populations:* 232, 696, 817, 1423.

(5) *Computations:* 1205, 1727, 1728, 1923.
(6) *Goodness of Fit:* 17, 75, 194, 249, 305, 400, 418, 534, 535, 602, 781, 817, 818, 893, 1092, 1120, 1230, 1262, 1321, 1349, 1375, 1395, 1448, 1532, 1533, 1543, 1580, 1727, 1845, 1889, 1890, 1963, 2000, 2080, 2137, 2141, 2193, 2263, 2407, 2519, 2751, 2755.
(7) *Homogeneity:* 161, 231, 1581.
(8) *Index of Dispersion:* 700, 1580.
(9) *Interval Estimation:* 314, 541, 2063, 2080.
(10) *Miscellaneous:* 65, 73, 146, 147, 476, 493, 513, 580, 749, 863, 922, 1873, 1892, 2133, 2170, 2231, 2426, 2460, 2587, 2711, 2785, 2835.
(11) *Model Building:* 151, 161, 400, 493, 534, 817, 818, 863, 1023, 1100, 1534, 2326, 2617, 2668.
(12) *Moments:* 221, 337, 476, 804, 1410, 1423, 1448, 1756, 1921, 1963, 2675.
(14) *Other Structural Properties:* 129, 770, 781, 1381, 1669, 1670, 1713, 1841, 1995, 2024, 2146, 2178, 2326, 2402, 2482, 2669, 2675, 2872.
(15) *Point Estimation:* 63, 305, 308, 310, 387, 388, 389, 600, 602, 700, 804, 817, 818, 887, 893, 980, 1023, 1153, 1156, 1205, 1206, 1207, 1321, 1328, 1439, 1448, 1542, 1727, 1867, 1890, 1891, 2000, 2003, 2178, 2274, 2308, 2397, 2398, 2401, 2403, 2675, 2711, 2874.
(16) *Processes and Chains:* 151, 506, 817, 922, 969, 1164, 1376, 1693, 2167.
(17) *Regression and Prediction:* 74, 818, 819, 1321, 1379.
(19) *Sequential Estimation:* 725.
(20) *SQC and AS:* 645, 874, 1701, 2005, 2863.
(21) *Tabulation and Charts:* 232, 388, 389, 1120, 1121, 2133, 2137.
(22) *Test on Parameters:* 226, 696, 770, 1124, 1234, 1574, 2063, 2296, 2781.

3.32. Classification for Negative Hypergeometric Distribution

(2) *Approximations, Asymptotics:* 344, 782.
(4) *Comparison of Two Populations:* 983.
(5) *Computations:* 1656.
(6) *Goodness of Fit:* 1657, 1859, 2448.
(9) *Interval Estimation:* 1458.
(10) *Miscellaneous:* 344.
(11) *Model Building:* 1337, 1458, 1466, 1656, 1859, 2448.
(12) *Moments:* 782, 1466, 1656.
(14) *Other Structural Properties:* 1656, 2024.
(15) *Point Estimation:* 1466, 1657, 2396, 2448, 2798.
(20) *SQC and AS:* 1316.
(21) *Tabulation and Charts:* 2798.
(22) *Test on Parameters:* 2798.

3.33. Classification for Negative Multinomial Distribution

(2) *Approximations, Asymptotics:* 1574, 2512.
(3) *Comparison of Models:* 318, 911.

(6) *Goodness of Fit:* 75, 534, 535, 810, 811, 1716, 1771.
(10) *Miscellaneous:* 1945, 2021.
(11) *Model Building:* 534, 535, 1771, 1904.
(12) *Moments:* 2012.
(14) *Other Structural Properties:* 1904, 2012, 2016.
(15) *Point Estimation:* 255, 1206, 2012.
(16) *Processes and Chains:* 1904.
(19) *Sequential Estimation:* 2510.
(22) *Test on Parameters:* 1574, 2511.

3.34. Classification for Neyman's Type A Distribution

(1) *Anova and Transformations:* 194, 2137.
(2) *Approximations, Asymptotics:* 775, 1724, 1726, 2512.
(3) *Comparison of Models:* 141, 493,
(5) *Computations:* 1727, 1728.
(6) *Goodness of Fit:* 140, 141, 194, 196, 605, 606, 773, 927, 1120, 1375, 1724, 1727, 1763, 1890, 1901, 2081, 2137, 2193, 2263, 2623, 2751.
(7) *Homogeneity:* 927.
(8) *Index of Dispersion:* 174.
(10) *Miscellaneous:* 65, 73, 493, 605, 863, 988, 1306.
(11) *Model Building:* 493, 863, 1724, 1901, 2454, 2617.
(12) *Moments:* 1965.
(14) *Other Structural Properties:* 155, 1965.
(15) *Point Estimation:* 308, 773, 1306, 1727, 1890, 2081, 2395.
(21) *Tabulation and Charts:* 773, 1120, 1122, 2137.

3.35. Classification for other Binomial Related Distributions

(2) *Approximations, Asymptotics:* 672, 2281.
(3) *Comparison of Models:* 2053.
(6) *Goodness of Fit:* 680, 803, 849, 1730, 2281.
(9) *Interval Estimation:* 2107.
(10) *Miscellaneous:* 953, 1073, 1832, 2151, 2172, 2593.
(11) *Model Building:* 510, 680, 803, 805, 849, 953, 2281, 2814.
(12) *Moments:* 701, 702, 714, 890, 970, 1361.
(14) *Other Structural Properties:* 970, 2281.
(15) *Point Estimation:* 510, 680.
(16) *Processes and Chains:* 172, 507, 890, 970.
(21) *Tabulation and Charts:* 510.

3.36. Classification for other Hypergeometric Related Distributions

(1) *Anova and Transformations:* 1465.
(2) *Approximations, Asymptotics:* 1236.
(6) *Goodness of Fit:* 1456, 1457, 1465.
(7) *Homogeneity:* 234.

(9) *Interval Estimation:* 1236.
(10) *Miscellaneous:* 2151.
(11) *Model Building:* 1456, 1466, 1467.
(12) *Moments:* 1236, 1466, 2502.
(15) *Point Estimation:* 1236, 1466, 1465.
(22) *Test on Parameters:* 2038.

3.37. Classification for other Multinomial Related Distributions

(2) *Approximations, Asymptotics:* 1053, 1350, 2729.
(5) *Computations:* 1053.
(6) *Goodness of Fit:* 694, 1578, 2130, 2367.
(10) *Miscellaneous:* 1043, 2130.
(11) *Model Building:* 1578, 2367, 2575.
(14) *Other Structural Properties:* 1325, 2130, 2576.
(15) *Point Estimation:* 643, 2575.
(21) *Tabulation and Charts:* 167, 1916, 2015.
(22) *Test on Parameters:* 1916, 2130, 2729.

3.38. Classification for other Poisson Related Distributions

(2) *Approximations, Asymptotics:* 899, 901, 905.
(3) *Comparison of Models:* 1443.
(6) *Goodness of Fit:* 131, 680, 1166, 1443, 2490, 2885.
(10) *Miscellaneous:* 2172, 2447.
(11) *Model Building:* 131, 680, 861, 1443, 1840, 2490.
(12) *Moments:* 131, 701, 702, 711, 1166, 2384, 2490.
(14) *Other Structural Properties:* 294, 336, 338, 2490.
(15) *Point Estimation:* 131, 593, 661, 680, 711, 899, 1378, 2384, 2490, 2885.
(16) *Processes and Chains:* 1545, 1546, 2167, 2490.

3.39. Classification for Poisson Distribution

(1) *Anova and Transformations:* 62, 150, 193, 194, 224, 305, 314, 512, 530, 576, 632, 671, 696, 947, 1321, 1333, 1335, 1820, 1853.
(2) *Approximations, Asymptotics:* 148, 156, 159, 161, 188, 204, 209, 210, 314, 394, 506, 537, 550, 632, 672, 830, 838, 839, 903, 932, 964, 965, 1088, 1284, 1350, 1394, 1395, 1418, 1527, 1593, 1653, 1700, 1701, 1703, 1724, 1772, 1977, 2380, 2515, 2633, 2677, 2705, 2706, 2707, 2708, 2718, 2734, 2763.
(3) *Comparison of Models:* 39, 141, 318, 638, 911, 1219, 2082, 2372, 2520, 2734, 2791.
(4) *Comparison of Two Populations:* 275, 277, 341, 342, 517, 632, 933, 1062, 1297, 1423, 2033, 2064, 2140, 2150, 2500, 2639.

(5) *Computations:* 1728, 1727, 1809.

(6) *Goodness of Fit:* 17, 93, 140, 141, 194, 243, 244, 269, 304, 305, 469, 511, 512, 543, 552, 581, 599, 605, 606, 626, 712, 713, 781, 894, 927, 1004, 1092, 1120, 1229, 1312, 1321, 1349, 1385, 1388, 1389, 1448, 1529, 1530, 1531, 1532, 1533, 1543, 1566, 1580, 1692, 1724, 1727, 1730, 1763, 1845, 1890, 1895, 1963, 1964, 2051, 2080, 2082, 2127, 2128, 2138, 2165, 2179, 2193, 2216, 2263, 2345, 2367, 2407, 2520, 2606, 2624, 2689, 2692, 2750, 2751, 2757, 2815, 2854, 2880.

(7) *Homogeneity:* 161, 191, 224, 511, 575, 626, 746, 927, 1108, 1307, 1488, 1649, 1680, 1817, 2111, 2179, 2618, 2819.

(8) *Index of Dispersion:* 174, 220, 223, 224, 686, 894, 1422, 1580, 1917, 1936, 2372, 2689.

(9) *Interval Estimation:* 4, 314, 322, 341, 342, 446, 450, 632, 662, 712, 847, 986, 1022, 2063, 2064, 2080, 2138, 2220, 2305, 2306, 2310, 2331, 2464, 2506, 2756, 2761, 2873.

(10) *Miscellaneous:* 73, 144, 202, 204, 242, 280, 329, 472, 476, 484, 511, 543, 565, 580, 605, 669, 699, 719, 749, 794, 922, 988, 1062, 1272, 1304, 1387, 1573, 1576, 1619, 1642, 1817, 1885, 1892, 2113, 2124, 2129, 2133, 2163, 2164, 2166, 2170, 2171, 2190, 2191, 2322, 2460, 2515, 2532, 2533, 2569, 2594, 2596, 2624, 2636, 2647, 2789, 2825.

(11) *Model Building:* 151, 161, 967, 1023, 1394, 1534, 1724, 1731, 1929, 2082, 2367, 2519, 2664, 2667, 2668, 2708, 2883.

(12) *Moments:* 156, 337, 476, 550, 660, 774, 1160, 1273, 1333, 1394, 1410, 1423, 1448, 1921, 1963, 1965, 2237, 2326, 2489, 2492, 2633, 2664, 2675.

(13) *Order Statistics:* 2, 1142.

(14) *Other Structural Properties:* 103, 129, 332, 338, 346, 498, 536, 659, 770, 781, 786, 1160, 1329, 1381, 1401, 1544, 1665, 1668, 1670, 1822, 1965, 2004, 2016, 2023, 2026, 2024, 2124, 2146, 2173, 2178, 2326, 2392, 2415, 2482, 2675, 2685, 2697.

(15) *Point Estimation:* 97, 135, 157, 171, 280, 304, 305, 435, 529, 543, 552, 585, 599, 601, 794, 847, 891, 916, 980, 1023, 1028, 1156, 1321, 1377, 1378, 1448, 1450, 1488, 1542, 1611, 1718, 1727, 1890, 1902, 1984, 2003, 2064, 2091, 2178, 2246, 2250, 2298, 2305, 2306, 2310, 2345, 2350, 2352, 2455, 2465, 2668, 2675, 2749, 2752, 2854.

(16) *Processes and Chains:* 151, 209, 210, 277, 409, 471, 501, 506, 543, 765, 766, 876, 922, 1164, 1165, 1327, 1346, 1376, 1499, 1508, 1704, 1731, 1792, 1830, 2118, 2152, 2334, 2386, 2392, 2467, 2499, 2569, 2641, 2709, 2857.

(17) *Regression and Prediction:* 29, 827, 985, 1321.

(18) *Selection and Ranking Problems:* 1014.

(19) *Sequential Estimation:* 1022, 1526, 1706, 2297, 2432.

(20) *SQC and AS:* 135, 916, 1128, 1198, 1700, 1701, 1794, 1795, 1941, 2380, 2702.

(21) *Tabulation and Charts:* 322, 473, 511, 662, 771, 774, 827, 1120, 1128, 1251, 1335, 1853, 2133, 2153, 2179, 2476, 2515, 2604.

(22) *Test on Parameters:* 28, 277, 304, 601, 770, 771, 830, 847, 1019, 1264, 1298, 1593, 1706, 2063, 2122, 2138, 2296, 2362, 2631, 2639, 2677, 2761, 2781.

3.40. Classification for Polya Distribution

(2) *Approximations, Asymptotics:* 2239, 2734.

(3) *Comparison of Models:* 183, 2734.

(10) *Miscellaneous:* 946, 2113.
(11) *Model Building:* 183.
(12) *Moments:* 2489.

3.41. Classification for Power Series Distribution

(2) *Approximation, Asymptotics:* 1088.
(6) *Goodness of Fit:* 2000.
(10) *Miscellaneous:* 2445, 2711.
(12) *Moments:* 1410, 1489, 1556, 1921, 1998, 2675.
(14) *Other Structural Properties:* 1556, 1998, 2178, 2671, 2675, 2711.
(15) *Point Estimation:* 1556, 1997, 2000, 2001, 2002, 2003, 2029, 2178, 2274, 2675.
(21) *Tabulation and Charts:* 2001.

3.42. Classification for Thomas Distribution

(1) *Anova and Transformations:* 2137.
(3) *Comparison of Models:* 141, 493.
(6) *Goodness of Fit:* 140, 141, 2081, 2137, 2263, 2613, 2623.
(7) *Homogeneity:* 2614.
(10) *Miscellaneous:* 65, 493.
(11) *Model Building:* 493, 2613.
(15) *Point Estimation:* 308, 2081.
(21) *Tabulation and Charts:* 2037.
(22) *Test on Parameters:* 2614.

3.43. Classification for Truncated or Censored Binomial Distribution

(2) *Approximations, Asymptotics:* 1774.
(5) *Computations:* 885.
(6) *Goodness of Fit:* 508, 2000.
(7) *Homogeneity:* 1208.
(10) *Miscellaneous:* 563.
(12) *Moments:* 1083, 1085, 1774, 2502.
(15) *Point Estimation:* 885, 1997, 2000, 2001, 2002, 2228, 2381, 2383, 2832.
(21) *Tabulation and Charts:* 885, 1091, 2001.

3.44. Classification for Truncated or Censored Negative Binomial Distribution

(2) *Approximations, Asymptotics:* 1081, 1084.
(6) *Goodness of Fit:* 1249, 1305, 1494, 2295.

(10) *Miscellaneous:* 1696.
(12) *Moments:* 1081, 1084, 2232, 2295.
(15) *Point Estimation:* 398, 600, 1249, 1494, 2228, 2295, 2381.
(21) *Tabulation and Charts:* 2232.

3.45. Classification for Truncated or Censored Poisson Distribution

(1) *Anova and Transformations:* 1819.
(6) *Goodness of Fit:* 511, 592, 594, 888, 1249, 1305, 1349, 2179.
(7) *Homogeneity:* 511, 2179.
(8) *Index of Dispersion:* 700.
(9) *Interval Estimation:* 462.
(10) *Miscellaneous:* 463, 511, 563.
(11) *Model Building:* 239.
(12) *Moments:* 2675.
(14) *Other Structural Properties:* 2675.
(15) *Point Estimation:* 27, 462, 584, 589, 591, 592, 594, 596, 643, 700, 772, 888, 1178, 1249, 1339, 1454, 1814, 1816, 1997, 2002, 2094, 2226, 2274, 2381, 2526, 2585, 2675.
(19) *Sequential Estimation:* 462.
(21) *Tabulation and Charts:* 167, 511, 589, 591, 594, 596, 1091, 1454, 2179.

Chapter 4

ADDENDA

4.1. Dictionary

In this section, we propose to cover explicitly a few important zero-truncated discrete distributions.

A-1. The Zero-truncated Distribution

Suppose that a non-negative integer-valued random variable X^* has the range $T^* = TU\{0\}$ where $0 \notin T$, and has the pf given by $P^*(x)$, $x \in T^*$. Then, the corresponding zero-truncated distribution has T for its range and has the pf

$$P(x) = \frac{P^*(x)}{(1-\phi)}, \qquad x \in T$$

where $\phi = P^*(0)$

$$G(z) = \frac{G^*(z) - G^*(0)}{(1-\phi)}$$

$$\mu = \frac{\mu^*}{(1-\phi)}, \qquad \sigma^2 = \frac{\sigma^{2*} - \mu\phi\mu^*}{(1-\phi)}$$

where $G^*(z)$, μ^* and σ^{2*} are respectively the pgf, the mean and the variance of the original distribution defined by P^* on T^*.

A-2. The Zero-truncated Binomial Distribution

(The positive binomial distribution)

The zero-truncated binomial distribution with parameters n and p has the pf

$$P(x) = \frac{\binom{n}{x} p^x q^{n-x}}{(1-q^n)} \qquad \begin{array}{l} x = 1, 2, \ldots, n. \\ n = 1, 2, \ldots \\ 0<p<1,\ q = 1-p. \end{array}$$

$$G(z) = \frac{(q+pz)^n - q^n}{(1-q^n)}.$$

$$\mu = \frac{np}{(1-q^n)}, \quad \sigma^2 = \mu q(1-\mu q^n).$$

Some inter-relations: (i) The zero-truncated binomial distribution with parameters n and p is a power series distribution with the series function $(1+\theta)^n - 1$, where $\theta = \dfrac{p}{(1-p)}$, which in itself is a zero-truncated distribution obtained from the power series distribution with the series function $(1+\theta)^n$.

(ii) As $n \to \infty$ and $p \to 0$ such that $np \to \lambda$, $0 < \lambda < \infty$, the zero-truncated binomial distribution with parameters n and p tends to the zero-truncated Poisson distribution with parameter λ.

Some references: Finney (1947-49); Patil (1962d).

A-3. The Zero-truncated Poisson Distribution

The zero-truncated Poisson distribution with parameter λ has the pf

$$P(x) = \frac{1}{(e^{\lambda}-1)} \cdot \frac{\lambda^x}{x!} \qquad \begin{array}{l} x = 1, 2, \ldots \\ 0 < \lambda < \infty \end{array}$$

$$G(z) = \frac{e^{\lambda z}-1}{e^{\lambda}-1}.$$

$$\mu = \frac{\lambda}{1-e^{-\lambda}}, \quad \sigma^2 = \mu(1-\mu e^{-\lambda}).$$

Some inter-relations: (i) The zero-truncated Poisson distribution with parameter λ is a power series distribution with the series function $e^{\theta}-1$, where $\theta = \lambda$, which in itself is a zero-truncated distribution obtained from the power series distribution with the series function e^{θ}.

(ii) The zero-truncated Poisson distribution with parameter λ is the limit of the zero-truncated binomial distribution with parameters n and p as $n \to \infty$ and $p \to 0$ such that $np \to \lambda$, $0 < \lambda < \infty$.

(iii) The zero-truncated Poisson distribution with parameter λ is the limit of the zero-truncated negative binomial distribution with parameters k and p as $k \to \infty$ and $p \to 1$ such that $k(1-p) \to \lambda$, $0 < \lambda < \infty$.

(iv) The zero-truncated negative binomial distribution with parameters k and p results if the zero-truncated Poisson distribution with parameter $-k \log p$ is generalized by the logarithmic distribution with parameter $1-p$.

Some references: Cohen (1954); Finney and Varley (1955); Rider (1953).

A-4. The Zero-truncated Negative Binomial Distribution

(The decapitated negative binomial distribution)

The zero-truncated negative binomial distribution with parameters k and p is defined by the pf

$$P(x) = \frac{\dfrac{\Gamma(k+x)}{x!\Gamma(k)} p^x q^k}{(1-p^k)} \qquad \begin{array}{l} x = 1, 2, \ldots \\ 0 < k < \infty \\ 0 < p < 1,\ q = 1-p. \end{array}$$

$$G(z) = \frac{p^k[(1-qz)^{-k}-1]}{(1-p^k)}.$$

$$\mu = \frac{kq}{p(1-p^k)} \qquad \sigma^2 = \frac{\mu}{p}(1-\mu p^{k+1}).$$

Some inter-relations: (i) The zero-truncated negative binomial distribution with parameters k and p is a power series distribution with the series function $(1-\theta)^{-k}-1$, where $\theta = 1-p$, which in itself is a zero-truncated distribution obtained from the power series distribution with the series function $(1-\theta)^{-k}$.

(ii) As $k \to 0$, the zero-truncated negative binomial distribution with parameters k and p tends to the logarithmic distribution with parameter $1-p$.

(iii) As $k \to \infty$ and $p \to 1$ such that $k(1-p) \to \lambda$, $0 < \lambda < \infty$, the zero-truncated negative binomial distribution with parameters k and p tends to the zero-truncated Poisson distribution with parameter λ.

(iv) The zero-truncated negative binomial distribution with parameters k and p results if the zero-truncated Poisson distribution with parameter $-k \log p$ is generalized by the logarithmic distribution with parameter $1-p$.

Some references: Rider (1962b); Sampford (1955).

4.2. Bibliography

A

Ahmad, Munir. 1964. A-1
Estimation of some parameters when a few observations are missing I. *Proc. Pakistan Statist. Assoc.*, **12**, 53-60.

Ahmad, Munir, and Kudô, Akio. 1967. A-2
Modified and partially truncated Poisson distribution. *Bull. Inst. Statist. Res. Training, Univ. Dacca*, **1**, 82-90.

Anderson, T. W., and Burstein, Herman. 1967. A-3
Approximating the upper binomial confidence limit. *J. Amer. Statist. Assoc.*, **62**, 857-61.

Armitage, P. 1967. A-4
Addendum to Armitage, P. (1966). The Chi-square test for heterogeneity, after adjustment for stratification. *J. Roy. Statist. Soc. Ser. B*, **29**, 197.

B

Bancroft, T. A., and Kale, B. K.
See Kale, B. K., and Bancroft, T. A. (1967).

Bartholomew, D. J. 1967. A-5
Hvpothesis testing when the sample size is treated as a random variable. *J. Roy. Statist. Soc. Ser. B*, **29**, 53-82.

Bartko, J. J. 1967. A-6
Errata to Bartko, J. J. (1966). *Technometrics*, **8**, 345-50. *Technometrics*, **9**, 498.

Bartko, J. J., and Watterson, G. A. 1965. A-7
Some problems of statistical inference in absorbing Markov chains. *Biometrika*, **52**, 127-38.

Bartlett, N. S., and Govindarajulu, Z. 1967a. A-8
Selecting a subset containing the best hypergeometric population (Abstract). *Ann. Math. Statist.*, **38**, 953.

Bartlett, N. S., and Govindarajulu, Z. 1967b. A-9
Selection procedures for negative binomial populations (Abstract). *Technometrics*, **9**, 185.

Barton, D. E. 1967a. A-10
Comparison of sample sizes in inverse binomial sampling. *Technometrics*, **9**, 337-39. Correction: p. 695.

Barton, D. E. 1967b. A-11
Completed runs of length k above and below median. *Technometrics*, **9**, 683-94.

Barton, D. E., and David, F. N. 1967. A-12
Four-letter words: the distribution of pattern frequencies in ring permutations. *J. Roy. Statist. Soc. Ser. B*, **29**.

Benishay, Haskel. 1967. A-13
Testing the effect of treatment or time on a binomial population (Abstract). *J. Amer. Statist. Assoc.*, **62**, 724.

Bennett, J. H. 1967. A-14
A general class of enumerations arising in genetics. *Biometrics*, **23**, 517-37.

Berger, Agnes, and Gold, Ruth Z. 1967. A-15
On estimating recessive frequencies from truncated samples. *Biometrics*, **23**, 356-60.

Berger, Agnes, Gold, Ruth Z., and Berman, Simeon M.
See Gold, Ruth Z., Berman, Simeon M., and Berger, Agnes (1967).

Berman, Simeon M., Berger, Agnes, and Gold, Ruth Z.
See Gold, Ruth Z., Berman, Simeon M., and Berger, Agnes (1967).

Berthet, P., and Gérard, G.
See Gérard, G., and Berthet, P. (1966).

Bhatt, N. M., and Phatak, A. G.
See Phatak, A. G., and Bhatt, N. M. (1967).

Bhattacharya, S. K. 1967. A-16
A result on accident proneness. *Biometrika*, **54**, 324-25.

Bildikar, Sheela, and Patil, Ganapati P. 1967.
See Patil, Ganapati P., and Bildikar, Sheela (1967).

Binet, F. E., and Hudson, B.
See Hudson, B., and Binet, F. E. (1965).

Birnbaum, Allan, and Eisenhart, Churchill.
See Eisenhart, Churchill, and Birnbaum, Allan (1967).

Bloch, Daniel A., and Watson, Geoffrey S. 1967. A-17
A Bayesian study of the multinomial distribution. *Ann. Math. Statist.*, **38**, 1423-35.

Bol'shev, L. N., and Loginov, E. A. 1966. A-18
Interval estimates in the presence of nuisance parameters. *Theor. Probability Appl.*, **11**, 82-94.

Bowman, K. O. and Shenton, L. R. 1967. A-19
Remarks on estimation problems for the parameters of the Neyman type A distribution. Tech. Rep. ORNL-4102, Oak Ridge National Lab., Oak Ridge, Tennessee, p. 58.

Bowman, K. O., and Shenton, L. R.
See Shenton, L. R., and Bowman, K. O. (1967).

Brown, J. L. Jr., and Patil, G. P. 1967. A-20
On the statistical independence of joint Gaussian variables after non linear transformation. *IEEE Trans. on Information Theory*, **IT-13**, 123-24.

Burstein, Herman, and Anderson, T. W.
See Anderson, T. W., and Burstein, Herman (1967).

Bush, R. R., and Mosteller, F. 1953. A-21
A stochastic model with applications to learning. *Ann. Math. Statist.*, **24**, 559-86.

C

Campbell, Robert W. 1967. A-22
The analysis of numerical change in Gypsy Moth populations. Forest Science Monograph 15, Soc. Amer. Foresters, p. 33.

Chistyakov, V. P., and Viktorova, I. I.
See Viktorova, I. I., and Chistyakov, V. P. (1966).

Cogburn, Robert. 1967. A-23
Stringent solutions to statistical decision problems. *Ann. Math. Statist.*, **38**, 447-63.

D

Darling, D. A., and Robbins, H. 1967. A-24
Finding the size of a finite population. *Ann. Math. Statist.*, **38**, 1392-98.

David, F. N., and Barton, D. E.
See Barton, D. E., and David, F. N. (1967).

David, H. A., and Nelson, W. C.
See Nelson, W. C., and David, H. A. (1967).

Downton, F. 1967. A-25
A note on the ultimate size of a general stochastic epidemic. *Biometrika*, **54**, 314-16.

Dwass, Meyer. 1967. A-26
A theorem about infinitely divisible distributions (Abstract). *Ann. Math. Statist.*, **38**, 970.

E

Eberhardt, L. L. 1967. A-27
Some developments in ' Distance sampling ' (French summary on p. 581). *Biometrics*, **23**, 207-16.

Eisenhart, Churchill, and Birnbaum, Allan. 1967. A-28
Anniversaries in 1966-67 of interest to statisticians. Part II: Tercentennials of Arbuthnot and DeMoivre. *Amer. Statist.*, **21**, (3), 22-29.

Ericson, W. A. 1967a. A-29
An example of discrepancies in inferences under non-informative stopping rules. *Biometrika*, **54**, 329-30.

Ericson, W. A. 1967b. A-30
Optimal sample design with nonresponse. *J. Amer. Statist. Assoc.*, **62**, 63-78.

F

Feibes, Walter, and Jen, Frank C.
See Jen, Frank C., and Feibes, Walter (1967).

Flatto, Leopold. 1967. A-31
Infinitely divisible distributions on cyclic groups. *Amer. Math. Monthly*, **74**, 255-61.

G

Gamkrelidze, N. G. 1966. A-32
On the speed of convergence in the local limit theorem for lattice distributions. *Theor. Probability Appl.*, **11**, 114-25.

Gart, John J., and Zweifel, James R. 1967. A-33
On the bias of various estimators of the logit and its variance with application to quantal bioassay. *Biometrika*, **54**, 181-87.

Gedanken, Irving L. 1967. A-34
A simple approximation to the hypergeometric probability: Case (N, n, k, 0) (Abstract). *J. Amer. Statist. Assoc.*, **62**, 729.

Gérard, G., and Berthet, P. 1966. A-35
A statistical study of microdistribution of Oribatei (Acari). Part II: The transformation of the data (Russian abstract). *Oikos*, **17**, 142-49.

Gold, Ruth Z., and Berger, Agnes.
See Berger, Agnes, and Gold, Ruth Z. (1967).

Gold, Ruth Z., Berman, Simeon M., and Berger, Agnes. 1967. A-36
On the question of whether a disease is familial. *J. Amer. Statist. Assoc.*, **62**, 409-20.

Goldman, Jay R. 1967. A-37
Stochastic point processes: limit theorems. *Ann. Math. Statist.*, **38**, 771-79.

Good, I. J. 1967. A-38
A Bayesian significance test for multinomial distributions. *J. Roy. Statist. Soc. Ser. B*, **29**, 199.

Govindarajulu, Z., and Bartlett, N. S.
See Bartlett, N. S., and Govindarajulu, Z. (1967).

Govindarajulu, Zakkula, and Makabe, Hajime.
See Makabe, Hajime, and Govindarajulu, Zakkula (1967).

Greenwood, J. A. 1940. A-39
The first four moments of a general matching problem. *Ann. Eugenics*, **10**, 290-92.

Greville, T. N. E. 1938. A-40
Exact probabilities for the matching hypothesis. *J. Parapsychology*, **2**, 55-59.

Greville, T. N. E. 1941. A-41
The frequency distribution of a general matching problem. *Ann. Math. Statist.*, **12**, 350-54.

Greville, T. N. E. 1943. A-42
Frequency distributions of ESP scores for certain selected call-patterns. *J. Parapsychology*, **7**, 272-76.

Gupta, Shanti S., and Nagel, Klaus. 1966. A-43
On selection and ranking procedures and order statistics from multinomial distribution. Mimeograph Ser. No. 77, Dept. Statist. Purdue Univ., pp. 53.

Gupta, Shanti S., and Studden, William J. 1967. A-44
Some selection and ranking procedures with applications to preliability roblems (Abstract). *Technometrics*, **9**, 187.

Gurland, John. 1967. A-45
An inequality satisfied by the expectation of the reciprocal of a random variable. *Amer. Statist.*, **21** (2), 24-25.

Gurland, John, and Hinz, Paul.
See Hinz, Paul, and Gurland, John (1967).

Guttman, Irwin. 1967. A-46
The use of the concept of a future observation in goodness-of-fit problems. *J. Roy. Statist. Soc. Ser. B*, **29**, 83-100.

H

Hader, R. J. 1967. A-47
Random roommate pairing of negro and white students. *Amer. Statist.*, **21** (5), 24-26.

Hald, A. 1967. A-48
The determination of single sampling attribute plans with given producer's risk and consumer's risk. *Technometrics*, **9**, 401-15.

Haller, B. 1945. A-49
Verteilungsfunctionen und ihre Auszeichnung durch Functionalgleichungen. *Mitt. Verein. Schweiz. Versich-Math.*, **45**, 97-163.
Note: Part III of the work translated into English by Kalaba, R. E.
See Kalaba, R. E. (1953).

Hinz, Paul, and Gurland, John. 1967. A-50
Simplified techniques for estimating parameters of some generalized Poisson distributions. *Biometrika*, **54**, 555-66.

Holla, M. S. 1967. A-51
On a Poisson-inverse Gaussian distribution. *Metrika*, **11**, 115-21.

Hudson, B., and Binet, F. E. 1956. A-52
The accuracy of eosinophil counts. *Austral. J. Expt. Biol. Med. Sci.*, **34**, 479-84.

Hudson, D. J. 1967. A-53
Likelihood intervals, 1: introduction, and the binomial case (Abstract). *Ann. Math. Statist.*, **38**, 1934.

I

Ibragimov, I. A. 1966. A-54
On the accuracy of Gaussian approximation to the distribution functions of sums of independent variables. *Theor. Probability Appl.*, **11**, 559-79.

Irwin, J. O. 1967. A-55
William Allen Whiteworth and a hundred years of probability. *J. Roy. Statist. Soc. Ser. A*, **130**, 147-76.

Ivchenko, G. I., and Medvedev, Yu. I. 1966. A-56
Asymptotic behavior of the number of complexes in a classical allocation problem. *Theor. Probability Appl.*, **11**, 619-26.

J

Jen, Frank C., and Feibes, Walter. 1967. A-57
A permutation theorem. *Amer. Statist.*, **21** (3), 31.

Jogdeo, Kumar. 1967. A-58
Monotone convergence of binomial probabilities with an application to maximum likelihood estimation. *Ann. Math. Statist.*, **38**, 1583-86.

Johnson, N. L. 1967. A-59
Note on a uniqueness relation in certain accident proneness models. *J. Amer. Statist. Assoc.*, **62**, 288-89.

Joshi, S. W., and Patil, G. P.
See Patil, G. P., and Joshi, S. W. (1966).

K

Kalaba, R. E. 1953. A-60
A summary of known distribution functions. Tech. Rep. T-27, RAND Corporation, Santa Monica, California. p. 29.
Note: See Haller, B. (1945).

Kale, B. K. and Bancroft, T. A. 1967. A-61
Inference for some incompletely specified models involving normal approximations to discrete data (French summary on pp. 585-86). *Biometrics*, **23**, 335-48.

Kalinin, V. M. 1967. A-62
Convergent and asymptotic expansions of probability distributions (Russian; English summary). *Teor. Verojatnost. i Primenen.*, **12**, 24-38.

Katti, S. K. 1967. A-63
Infinite divisibility of integer valued random variables. *Ann. Math. Statist.*, **38**, 1306-08.

Kemp, C. D. 1967. A-64
On a contagious distribution suggested for accident data (French summary on p. 582). *Biometrics*, **23**, 241-55.

Kemp, Kenneth W. 1967. A-65
Formal expressions which can be used for determination of the operating characteristic and average sample number of a simple sequential test. *J. Roy. Statist. Soc. Ser. B*, **29**, 248-62.

Kolchin, V. F. 1966. A-66
The speed of convergence to the limit distributions in the classical ball problem. *Theor. Probability Appl.*, **11**, 128-40.

Kolčin, V. F. 1967. A-67
A case of uniform local limit theorems with changing lattice in a classical problem with balls (Russian; English summary). *Teor. Verojatnost. i Primenen.*, **12**, 62-72.

Koopmans, L. H., and Lowry, Dorothy C. 1957. A-68
Estimation of mean and variance of a distribution of binomial probabilities (Abstract). *Biometrics*, **13**, 551.

Krengel, Ulrich. 1967. A-69
A problem on random points in a triangle. *Amer. Math. Monthly*, **74**, 8-14.

Krutchkoff, R. G., and Rutherford, J. R.
See Rutherford, J. R., and Krutchkoff, R. G. (1967).

Kudô, Akio, and Ahmad, Munir.
See Ahmad, Munir, and Kudô, Akio (1967).

Kullback, S. 1967. A-70
The two concepts of information. *J. Amer. Statist. Assoc.*, **62**, 685-86.

L

Lane, R. O. D., and Whittle, P.
See Whittle, P., and Lane, R. O. D. (1967).

Lewis, Peter A. W. 1967. A-71
Non-homogeneous branching Poisson process. *J. Roy. Statist. Soc. Ser. B*, **29**, 343-54.

Loginov, E. A., and Bol'shev, L. N.
See Bol'shev, L. N., and Loginov, E. A. (1966).

Lowry, Dorothy C., and Koopmans, L. H.
See Koopmans, L. H., and Lowry, Dorothy C. (1957).

Lunghi, G. 1967-68. A-72
A theoretical approach to the study of genetic parameters of histo-compatibility in man. *Ann. Human Genetics*, **31**, 85.

M

Mahamunulu, D. M. 1967. A-73
A note on regression in the multivariate Poisson distribution. *J. Amer. Statist. Assoc.*, **62**, 251-58.

Makabe, Hajime, and Govindarajulu, Zakkula. 1967. A-74
Approximations to the generalized negative binomial distribution (Abstract). *Ann. Math. Statist.*, **38**, 957.

Mantel, Nathan. 1967. A-75
Assumption-free estimators using U statistics and a relationship to the Jackknife method (French summary). *Biometrics*, **23**, 567-71.

Marcus, Allan H. 1967. A-76
A multivariate immigration with multiple death process and applications to lunar crators. *Biometrika*, **54**, 251-61.

Margolin, Barry H., and Winokur, Herbert S. 1967. A-77
Exact moments of the order statistics of the geometric distribution and their relation to inverse sampling and reliability of redundant systems. *J. Amer. Statist. Assoc.*, **62**, 915-25.

Matthes, T. K., and Truax, D. R. 1967. A-78
Tests of composite hvpotheses for the multivariate exponential family. *Ann. Math. Statist.*, **38**, 681-97.

Medvedev, Yu. I., and Ivchenko, G. I.
See Ivchenko, G. I., and Medvedev, Yu. I. (1966).

Meyer, Richard M. 1967. A-79
Some limit theorems for dependent 'rare' events (Abstract). *Ann. Math. Statist.*, **38**, 964.

Milch, Paul R. 1967. A-80
The extended negative hypergeometric distribution (Preliminary report). *Ann. Math. Statist.*, **38**, 965.

Moore, R. H., and Zeigler, R. K. 1967. A-81
The use of non-linear regression methods for analyzing sensitivity and quantal response data. *Biometrics*, **23**, 563-66.

Morris, Carl. 1967a. A-82
Admissible Bayes procedures and classes of epsilon Bayes procedures for testing hypotheses in a multinomial distribution (Abstract). *Ann. Math. Statist.*, **38**, 1597.

Morris, Carl. 1967b. A-83
A class of epsilon Bayes tests of a simple null hypotheses on many parameters in exponential families (Abstract). *Ann. Math. Statist.*, **38**, 1597.

Mosteller, F., and Bush, R. R.
See Bush, R. R., and Mosteller, F. (1953).

Moy, Shu-Teh C. 1967. A-84
Extensions of a limit theorem of Everett, Ulam and Harris on multitype branching process with countably many types. *Ann. Math. Statist.*, **38**, 992-99.

N

Nagel, Klaus, and Gupta, Shanti S.
See Gupta, Shanti S., and Nagel, Klaus (1966).

Nathan, Gad. 1967. A-85
Outcome probabilities for a record matching process with complete invariant information. *J. Amer. Statist. Assoc.*, **62**, 454-69.

Nelson, W. C., and David, H. A. 1967. A-86
The logarithmic distribution: a review. *Virginia J. Sci. (N.S.)*, **18**.

Noether, Gottfried E. 1967. A-87
Wiloxon confidence intervals for location parameters in the discrete case. *J. Amer. Statist. Assoc.*, **62**, 184-88.

O

Ord, J. K. 1967a. A-88
On a system of discrete distributions. *Biometrika*, **54**, 649-56.

Ord, J. K. 1967b. A-89
Graphical methods for a class of discrete distributions. *J. Roy. Statist. Soc. Ser. A*, **130**, 232-38.

Ord, J. K. 1967c. A-90
On families of discrete distributions. Ph.D. Thesis, Univ. London.

P

Pathak, K. B., and Singh, S. N.
See Singh, S. N., and Pathak, K. B. (1967).

Patil, G. P., and Bildikar, Sheela. 1967. A-91
Multivariate logarithmic series distribution as a probability model in population and community ecology and some of its properties. *J. Amer. Statist. Assoc.*, **62**, 655-74.

Patil, G. P., and Brown, J. L. Jr.
See Brown, J. L., Jr., and Patil, G. P. (1967).

Patil, G. P., and Joshi, S. W. 1966. A-92
Bibliography of classical and contagious discrete distributions. Tech. Rep., Aerospace Res. Lab., Wright-Patterson Air Force Base, pp. 270.

Paulson, Edward. 1967. A-93
Sequential procedures for selection of the best one of several binomial populations. *Ann. Math. Statist.*, **38**, 117-23.

Penkov, B., and Theodorescu, R. 1967. A-94
Über die spieltheoretische Behaldung von Prüfplänen. *Metrika*, **11**, 122-26.

Peritz, Eric. 1966-67. A-95
A statistical study of intrauterine selection factors related to the ABO system. *Ann. Human Genetics*, **30**, 259-71.

Phatak, A. G., and Bhatt, N. M. 1967. A-96
Estimation of the fraction defective in curtailed sampling plans by attributes. *Technometrics*, **9**, 219-28.

Pollak, Edward. 1967. A-97
The limiting behaviour of a branching process (Abstract). *Biometrics*, **23**, 384.

Prairie, R. R. 1967. A-98
Probit analysis as a technique for estimating the reliability of a simple system. *Technometrics*, **9**, 197-203.

R

Robbins, H., and Darling, D. A.
See Darling, D. A., and Robbins, H. (1967).

Roberts, Harry V. 1967. A-99
Informative stopping rules and inferences about population size. *J. Amer. Statist. Assoc.*, **62**, 763-75.

Rothman, E. 1967. A-100
Testing for clusters in a Poisson process (Abstract). *Ann. Math. Statist.*, **38**, 965.

Rutemiller, Herbert C. 1967. A-101
Estimation of the probability of zero failures in m binomial trials. *J. Amer. Statist. Assoc.*, **62**, 272-77.

Rutherford, J. R., and Krutchkoff, R. G. 1967. A-102
The empirical Bayes approach: estimating the prior distribution. *Biometrika*, **54**, 326-28.

S

Sandelius, Martin. 1967. A-103
A note on the variance of a discrete uniform distribution. *Amer. Statist.*, **21** (5), 21.

Schreider, Yu. A. 1967. A-104
On a possibility of theoretical deduction of statistical laws of a (printed) test (Russian). *Probl. Peredachi Informatzii*, **3**, 57-63.

Sevast'yanov, B. A. 1966. A-105
Limit theorems in a scheme for allocation of particles in cells. *Theor. Probability Appl.*, **11**, 614-19.

Sevast'yanov, B. A. 1967. A-106
Convergence of the distribution of the number of empty boxes to Gaussian and Poisson processes in a classical problem with balls (Russian; English summary). *Teor. Verojatnost. i Primenen.*, **12**, 144-54.

Shanbhag, D. N. 1966. A-107
On congestion systems with negative exponential desired service time distributions. *Ann. Inst. Statist. Math. (Tokyo)*, **18**, 223-28.

Shenton, L. R., and Bowman, K. O. 1967. A-108
Remarks on large sample estimators for some discrete distributions. *Technometrics*, **9**, 587-98.

Shenton, L. R., and Bowman, K. O.
See Bowman, K. O., and Shenton, L. R. (1967).

Singh, S. N., and Pathak, K. B. 1967. A-109
On the distribution of the number of conceptions (Abstract). *Ann. Math. Statist.*, **38**, 1315.

Smith, Cedric A. B. 1967-68. A-110
Notes on gene frequency estimation with multiple alleles. *Ann. Human Genetics*, **31**, 99-107.

Solov'ev, A. D. 1966. A-111
A combinatorial identity and its application to the problem concerning the first occurrence of a rare event. *Theor. Probability Appl.*, **11**, 276-82.

Staff, P. J. 1967. A-112
The displaced Poisson distribution—Region B. *J. Amer. Statist. Assoc.*, **62**, 643-54.

Studden, William J., and Gupta, Shanti S.
See Gupta, Shanti S., and Studden, William J. (1967).

T

Takács, Lajos. 1967. A-113
On the method of inclusion and exclusion. *J. Amer. Statist. Assoc.*, **62**, 102-13.

Theodorescu, R., and Penkov, B.
See Penkov, B., and Theodorescu, R. (1967).

Truax, D. R., and Matthes, T. K.
See Matthes, T. K., and Truax, D. R. (1967).

Turner, Malcolm E. 1967. A-114
Toward an algebra of convolution and compounding. Biometry Preprint Series, Preprint No. 8, Dept. Biometry, Emory Univ., Atlanta, Georgia, pp. 27.

Tweedie, M. C. K. 1967. A-115
A mean-square-error characterization of binomial-type distributions. *Ann. Math. Statist.*, **38**, 620-23.

V

Viktorova, I. I., and Chistyakov, V. P. 1966. A-116
Some generalizations of the test of empty boxes. *Theor. Probability Appl.*, **11**, 270-76.

W

Watson, Geoffrey S., and Bloch, Daniel A.
See Bloch, Daniel A., and Watson, Geoffrey S. (1967).

Watterson, G. A., and Bartko, J. J.
See Bartko, J. J., and Watterson, G. A. (1965).

Whittle, P., and Lane, R. O. D. 1967. A-117
A class of situations in which a sequential estimation procedure is non-sequential. *Biometrika*, **54**, 229-34.

Winokur, Herbert S., and Margolin, Barry H.
See Margolin, Barry H., and Winokur, Herbert S. (1967).

Y

Young, D. H. 1967. A-118
A note on the first two moments of the mean deviation of the symmetrical multinomial distribution. *Biometrika*, **54**, 312-14.

Z

Zeigler, R. K., and Moore, R. H.
See Moore, R. H., and Zeigler, R. K. (1967).

Zweifel, James R., and Gart, John J.
See Gart, John J., and Zweifel, James R. (1967).